MILITARY DIASPORAS

Military Diasporas proposes a new research approach to analyze the role of foreign military personnel as composite and partly imagined para-ethnic groups.

These groups not only buttressed a state or empire's military might but crucially connected, policed, and administered (parts of) realms as a transcultural and transimperial class while representing the polity's universal or at least cosmopolitan aspirations at court or on diplomatic and military missions. Case studies of foreign militaries with a focus on their diasporic elements include the Achaemenid Empire, Ptolemaic Egypt, and the Roman Empire in the ancient world. These are followed by chapters on the Sassanid and Islamic occupation of Egypt, Byzantium, and the Latin Aegean (Catalan Company) to Iberian Christian noblemen serving North African Islamic rulers, Mamluks, and Italian Stradiots, followed by chapters on military diasporas in Hungary, the Teutonic Order including the Sword Brethren, and the Swiss military. The volume thus covers a broad band of military diasporic experiences and highlights aspects of their role in the building of state and empire from Antiquity to the late Middle Ages and from Persia via Egypt to the Baltic.

With a broad chronological and geographic range, this volume is the ideal resource for upper-level undergraduates, postgraduates, and scholars interested in the history of war and warfare from Antiquity to the sixteenth century.

Georg Christ is a senior lecturer in medieval and early modern history at the University of Manchester, UK, and a general staff officer in the Swiss Army. His work focuses on relations between Venice and the Mamluk Empire including the role of diasporas in transmediterranean connectivity.

Patrick Sänger is professor of ancient history at the Westfälische Wilhelms-Universität Münster, Germany. His research focuses on the administrative, legal, and social history of the Hellenistic and Roman world with a specific interest in Egypt and in migration and integration issues.

Mike Carr is a lecturer in late medieval history at the University of Edinburgh, UK, and a leading specialist on the history of papal trade policies and crusading in the late medieval Mediterranean.

MILITARY DIASPORAS

Building of Empire in the Middle East and Europe (550 BCE–1500 CE)

Edited by Georg Christ, Patrick Sänger, and Mike Carr

Routledge
Taylor & Francis Group
LONDON AND NEW YORK

Cover image: agefotostock/Alamy Stock Photo

First published 2023
by Routledge
4 Park Square, Milton Park, Abingdon, Oxon OX14 4RN

and by Routledge
605 Third Avenue, New York, NY 10158

Routledge is an imprint of the Taylor & Francis Group, an informa business

British Library Cataloguing-in-Publication Data
A catalogue record for this book is available from the British Library

Library of Congress Cataloging-in-Publication Data
Names: Christ, Georg, editor. | Sänger, Patrick, editor. | Carr, Mike, 1984- editor.
Title: Military diasporas : building of empire in the Middle East and Europe (550 BCE-1500 CE) / [edited by] Georg Christ, Patrick Sänger, and Mike Carr.
Other titles: Building of empire in the Middle East and Europe (550 BCE-1500 CE)
Description: Abingdon, Oxon ; New York, NY : Routledge, 2022. | Includes bibliographical references and index.
Identifiers: LCCN 2022021158 (print) | LCCN 2022021159 (ebook) | ISBN 9781032157566 (hardback) | ISBN 9781032157573 (paperback) | ISBN 9781003245568 (ebook)
Subjects: LCSH: Military history, Ancient. | Foreign enlistment--Europe--History--To 1500. | Foreign enlistment--Middle East--History--To 1500. | Military art and science--History--Medieval, 500-1500.
Classification: LCC U29 .M55 2022 (print) | LCC U29 (ebook) | DDC 355.02/0940902--dc23/eng/20220505
LC record available at https://lccn.loc.gov/2022021158
LC ebook record available at https://lccn.loc.gov/2022021159

ISBN: 978-1-032-15756-6 (hbk)
ISBN: 978-1-032-15757-3 (pbk)
ISBN: 978-1-003-24556-8 (ebk)

DOI: 10.4324/9781003245568

Typeset in Bembo
by KnowledgeWorks Global Ltd.

CONTENTS

PREFACE

This book is the late fruit of an international workshop on military diasporas held in Heidelberg in autumn 2013. It concentrated on military groups and organizations and their political, economic, and cultural impact on the building, defending, shaping, toppling, and connecting of empires.[1] The workshop, the fourth in a series of workshops on trans-European diasporas in the pre-modern period,[2] showed that military groups were particularly interesting examples of diasporic communities because of their considerable potential to mobilize, transfer, and project personnel and capital between regions and empires.

The workshop, however, shaped its own questions as well and made us think more about the role of military diasporas in imperial politics, including courts, and state formation more generally. The book approaches the theme in a transcultural perspective bringing together studies of military diasporic groups in Europe and the Middle East produced by scholars from different disciplinary and national backgrounds. We thank the DAAD/MÖB for the generous support of this project, the Central European University and in particular Professor Katalin Szende for their excellent cooperation, the Universität Heidelberg, and especially Professors Julia Burkhardt-Dücker and Andrea Jördens for hosting and organizing the workshop. We also thank the anonymous reviewers of the manuscript, who provided critical but extremely rigorous and helpful feedback that, so we hope, significantly improved this volume. Finally, we would like to thank Mareike Stanke for her editorial support. All remaining errors and shortcomings are ours.

We are sad that Jacob Klingner, who generously supported and encouraged this book project when first offered to De Gruyter, is no longer with us to see its completion, and we are grateful to Routledge and, in particular, Izzy Voice and Laura Pilsworth to step in and see through the publication of this volume in a spirit of ever collegial efficiency and friendly support even in the face of longer and longer delays.

The editors

Notes

1 "Military Diasporas and Diasporic Regimes in East Central Europe and the Eastern Mediterranean 500–1800" (Heidelberg, October 2013), https://www.ceu.edu/sites/default/files/attachment/event/8647/abstracts-military-diasporas-and-diasporic-regimes.pdf accessed 22.08.2022.
2 Katalin Szende and Georg Christ, "Trans-European Diasporas: Migration, Minorities, and the Diasporic Experience in East Central Europe and the Eastern Mediterranean from the Late Antiquity to the Early Modern Era—Project Report," *Annual of Medieval Studies at CEU* 20 (2014): 296–305.

INTRODUCTION

Military Diasporas and Diasporic Groups as Engines of Empire: An Introductory Research Agenda

Georg Christ and Patrick Sänger[1]

Pre-modern Military diasporic groups were crucial to boost the military potential of a ruler or polity, thus enabling state formation and empire building. This might seem paradoxical as we associate the modern state with the nation state (whereby state and empire cannot be distinguished sharply) and national militaries.[2] The soldiers engineering empire, however, were often and to varying degrees foreign and diasporic.

Military diasporas are here defined as ethnically defined groups of soldiers serving far away from their place of origin but typically retaining some links to it. These groups not only fought wars but also policed, represented, and administered states/empires and contributed crucially to a trans-local cultural framework of rule. Their foreign status helped to keep them apart from local particular interests and made them a suitable projector of the ruler's or central governing agency's will. Serving in secondary roles as administrators or policemen, either after returning home or retired in their land of service, diasporic and transnational soldiers continued to shape polities by controlling, securing, and challenging both the state/empire and local polities. Thus they significantly intervened in the formation of atomized and submissive populations in which the modern state is rooted.[3]

"Foreign" in the context of pre-modern empires should not be misunderstood as necessarily meaning "from outside the empire", which—in the case of a universal empire—would be strictly speaking impossible. If the empire has more limited aspirations (e.g. to rule over an ecumene), it might make sense to distinguish between militaries from within versus those from beyond a "limes", but the constitutive element of a diasporic group in our understanding can only be relative foreignness, that is a palpable feeling of being foreign to the area of operation and separated from the place of origin regardless of its formal political

DOI: 10.4324/9781003245568-1

affiliation. Historically, the term diaspora indeed means trans-imperial rather than inter-imperial dispersion of an ethnicity, e.g. the Jewish diaspora within the Roman Empire.

Before giving an overview of the contributions in this volume, we will illustrate how the continuous importance of military diasporas has been overshadowed by the modern equation of nation state and national military service, discuss the state of research regarding foreign military groups, define military diasporas/ as well as diasporic groups, and detail the proposed explanatory framework.

Problem: The Anomaly of National Service

While hardly denied or unknown, diasporic militaries' contribution to state and empire formation has not been considered explicitly. The reason might be that a diasporic focus somewhat goes against the narrative of the nation state as a national project carried out by soldiers of this very nation. Private military firms of the likes of Blackwater or Executive Outcome are seen as much as an aberration as the French Foreign Legion is seen as an exception. The connection between nation state and military, however, is a relatively recent and hardly ever fully completed development. It began tentatively in the seventeenth century with a move towards bringing the foreign troops "home" followed by eighteenth-century militia systems, the Prussian recruitment cantons and a first attempt at conscription in Russia in 1705.[4] It became more radical with the French Revolution's military expansion. The *levée en masse* of 1793, culminating in the general conscription of 1799, seemed to produce a congruence of nation-state-military, although only the 18–25 year-old men were affected by it as French-citizen-soldiers.[5] Most European states followed suit and militaries of the nineteenth and twentieth centuries thus were characterized by similar phenomena of mass mobilization/conscription and linking of civic rights and military service.[6] Yet the full use of the population fit for military service was neither always practical nor desirable. There was strong parliamentary resistance against the Prussian military reforms aiming at a better use of conscription 1858–1862. Also, the subsequent 2nd German Reich (Empire) was reluctant to use the full potential of its manpower due to concerns that it would not be able to recruit enough sufficiently qualified officer trainees to deal with the armed masses.[7] The solution of beefing an army's ranks with officers from abroad, however, was increasingly barred in national armies.

Because of the clear link between German nation building and military build-up in the 2nd Empire, the more diasporic Prussian military-cultural legacy is sometimes—paradoxically—linked to nationalist and national social militarism. This is telling: The Prussian concept of serving a non-national dynastic state and hence a diasporic military not tied to a nation but a somewhat abstract state and the king had become unfamiliar to the twentieth-century observer. More importantly, the victors of World War Two drew on the concept of the nation state themselves.[8] Focusing on the obvious connection between militarism and nationalism more generally was undesirable. By linking Nazi Germany to a specific "Prussian militarism", however, the more troubling discussion about the nation state's inherent militarism could be postponed.[9] The militaries remained national and the idea of a transnational European Defence Community was rejected in 1950 under fierce resistance against a supranational military command and the idea of joining ranks with the former foe.[10]

The steadfast adherence to the doctrine of the nation state and its national military thus prevented and continues to hinder alternative military structures for Europe. European defence cooperation around the European Defence Agency, the Common European Foreign and Security Policy (CFSP), and, in particular, the Common Security and Defence Policy (CSDP) is very limited and European military structures are weak (EU battle groups, EGF, EBCGA or FRONTEX). Defence remains essentially the remit and responsibility of the member states, while NATO, somewhat contradictorily, is accepted as a transnational institution ultimately responsible for the defence of Europe.[11] NATO's clear commitment to national armies as building blocks of cooperation might go some way in explaining this seeming contradiction.

Increasingly, however, these Western national armies are less national in their composition. Conscription or national service has been abandoned with a nationalized social contract moving from serving the nation to being served by the national welfare state in exchange for permanent war-style taxation.[12] Therefore, many allegedly national militaries, due to recruitment problems, are becoming increasingly multi-national, although this is not properly acknowledged. Not only do the US armed forces increasingly include foreigners but so do many European militaries.[13] Regardless of the percentage of non-national service personnel, armies are becoming more diasporic. Even the Swiss Army, which is still based on the national service model excluding non-citizens from its ranks, is in many ways a multi-ethnic force: A great number of soldiers are not Swiss-born and have a mother tongue that is not a national language.[14] Also, the opponents faced by many Western militaries e.g. in Afghanistan, Iraq, North Africa, or domestically are increasingly diasporic and not serving a nation state but a religion, a leader, an idea.[15]

It seems that politicians find it difficult to acknowledge that, if not the nation state, certainly the national military has widely ceased to exist. Considering European demographic developments combined with the reluctance of many "old" Europeans to do military service, it might be timely to consider whether, perhaps, a European armed force open to non-Europeans, including trans-Mediterranean immigrants, lies ahead of us—distantly mirroring the late Roman army. A look at the late Roman army or the multi-ethnic, -lingual, and -religious militaries of Frederick the Great's Prussia or of the Austro-Hungarian Empire or more remote examples of non-national, composite and multi-ethnic armies might thus be instructive to overcome the impasse of national-military thinking: We should analyze former armed forces in their transnational compositions.

Yet the prevalent national-military thinking tends to essentialize soldiers both to their supposed and official ethnicity and to their role as soldiers, i.e. fighters. Soldiers, however, were more than that as they became part of a diasporic network: Grouped primarily but not exclusively with their co-locals they joined composite armies of various kinds and at various distances from their place of origin. In these composite forces, they formed multiple links with local populations and other military diasporic groups. They thus not only administered power *manu militari* but contributed to state/empire and nation building projects in other ways too. As foreign, allochthonous groups they protected rulers as bodyguards, they policed local populations, they represented and symbolized the universal reach of the regime at court and on embassies. They projected state/imperial power from the centre to the various parts of the realm as administrators in state/imperial bureaucracies, which were increasingly detached from traditional tribal or local powerbases. Meanwhile, they could also influence their homelands by tying them in multiple ways to emerging states/empires. Furthermore, military diasporas could boost their homelands' economic and fiscal base as well as their military potential while stabilizing social and economic structures: It drained the land of excess young men and thus allowed for a controlled social mobility while generating income for the respective polity and middlemen élite that facilitated this military service, which in turn stabilized oligarchic governmental structures.[16]

State of Research

The state of research on military diasporas is quickly narrated. Diaspora studies are a wide and fairly established field that produces not only a continuous stream of studies on the classic three "big" diasporas: The Jewish, the Greek, and the Armenian but also on Chinese, Tibetan, Italian, Pakistani, Indian and many more diasporas. Indeed, the term has proliferated to such an extent that Rogers Brubaker rightly questioned its analytical value in his seminal essay "The 'diaspora' diaspora".[17] Military diasporas, however, have hardly been studied, at least not under this heading. One of the very few references found is in Chang's study on Yunnanese in Thailand, which emphasizes the need to study this phenomenon.[18] Even if the term appears, it is rarely with reference to a diaspora studies' explanatory framework.[19] The only military diaspora explicitly called so, seems to be the Irish military diaspora.[20]

This is not to say that military diasporas and diasporic groups (for this distinction, see below) have not been considered from the perspective of military and political history. Erik-Jan Zürcher's edited volume on military labour, for instance, is path-breaking for our purpose, although limited to the period post-1500.[21] Typically, military diasporic groups are addressed as mercenaries, a focus which already becomes evident when looking at the Ancient World. Nubanda Mardune entered the service of Sargon of Akkad as captain of the Amories in the twenty-fourth century BCE.[22] The famous Nubian archers served the Egyptian Old, Middle and New Kingdom armies (third–second millennium BCE). Cretan mercenaries might or might not be identified as one of King David's two bodyguards (the Cherethites and Pelethites, c. 1000 BCE).[23]

Greece and the neighbouring Aegean region were homelands par excellence of diasporic mercenaries. As early as in the Archaic and Classical periods (eight–fourth century BCE), we encounter groups of Hellenic soldiers who left the Greek cultural zone and sought their fortune in the service of Middle Eastern rulers.[24] In the course of the campaigns of Alexander the Great (334–324 BCE), a Macedonian-Greek army (combining the Greek heavy infantry, i.e. the phalanx formation, with cavalry units) conquered the Middle Eastern kingdoms. In the settlements and cities that were established in Alexander's huge empire and—after his death in 323 BCE—in the Hellenistic kingdoms of the Diadochi (Alexander's generals and successors) from the end of the fourth to the first century BCE, allochthonous mercenaries or professional soldiers played a dominant role and (former) military men formed a major part of the new Greco-Macedonian ruling élite.[25] Indeed, the recruitment of diasporic mercenaries undoubtedly reached its peak in this "axial" period,[26] marking an essential feature of Hellenism that distinguishes this era from the preceding classical period and its formative warrior protagonists, the citizen-soldiers of the Greek cities or poleis. The mass deployment of mercenaries of different ethnicities was a decisive component in transforming the Greek world of individual poleis and their power struggles into a more "global" structure of imperial kingdoms. Military diasporic groups hailed from foreign realms or specific regions within the kingdom and some of them were strategically resettled within the realm to stabilize Diadochi rule.[27]

Mobility and migration also characterized the Roman army, which, however, never operated on a large scale with foreign troops (which were usually only temporarily recruited for specific operations). In republican times, its permanent (infantry) core was formed as a citizen militia. It transformed into a standing army consisting of imperial subjects as professional soldiers at the beginning of the imperial period, thus abandoning the militia system.[28] With its élite troops, the heavily armed infantry legions of Roman citizens, Rome conquered a vast territory, which in imperial times (starting at the end of the first century BCE) stretched from Britain to the Middle East (defeating the afore-mentioned Hellenistic kingdoms), and, thus, united the shores of the Mediterranean into a single empire. Stationed in camps of varying sizes, Roman soldiers were spread across this huge empire in order to guarantee its internal security and protect its external borders. These patterns of dissemination and mobility—including the settlements of retired soldiers first from Italy than with the expansion of the recruitment base from the entire empire—created a network of Roman diasporic groups across the empire. As Roman "cultural agents" in the provinces and on the peripheries of the empire they—through processes of transculturation (see below)—contributed significantly to the formation of a common Roman imperial culture (whose emergence is not to be seen as contradictory or antithetical but rather complementary to an ongoing ethnic identity).

To support the legions, the Roman army increasingly made use of foreign soldiers. During republican times, namely, from the end of the third/beginning of the second century BCE, its deficiency in cavalry and light-armed special forces was compensated by ethnic contingents of Cretan archers and Balearic slingers or Gallic and Iberian horsemen. Such auxiliary troops (*auxilia*) were recruited on a regular basis, starting with the reign of Augustus, the first Roman emperor, among the *peregrini*, the freeborn inhabitants of the Roman provinces without Roman citizenship. The importance of the *auxilia* for imperial politics cannot be overestimated. After the troop reductions at the beginning of Augustan rule, they became an integral part of the standing army and represented a welcome and necessary addition to the Roman formations both quantitatively and qualitatively. They allowed a relatively unproblematic doubling and, at the same time, an enhancement of the Roman military potential, especially regarding cavalry (*ala*) and ballistic support, whereby auxiliary troops retained (at least initially) their specific weaponry and fighting style (e.g. the Syrian archers) and thus widened the range of armament and tactics.[29] At the same time, the deployment of the *auxilia* relieved the state treasury since less funds were required for their upkeep compared to the more expensive legions. As far as security policy was concerned, the enlistment of the *auxilia*, which was mainly carried out by force under Augustus, was an effective means of weakening regions of the empire that had just been pacified or conquered by withdrawing the segment of the population most capable of armed resistance and stationing them in other parts of the empire. This policy was initially reflected in the ethnic homogeneity of the individual *auxilia* units, whose soldiers originated from a certain tribe or region and thus can be considered diasporic groups. The homogeneity of *auxilia*, however, changed fundamentally during the imperial period. Persistent geographical separation from their former homelands and a shift to local recruitment to supplement the units increasingly diversified their ethnic composition. The social objective behind the creation of the *auxilia* and their actual effect on the imperial population can be subsumed under the catchword "Romanization" (in the sense of transculturation). Through the Roman *auxilia*, the *peregrini* were offered an opportunity to participate in the imperial project and to become familiar with the Roman way of life and the Latin language. At the end of their service, the auxiliary soldiers were granted Roman citizenship so that their male descendants could serve in the legions.[30]

Until the end of the imperial period, the Roman army thus included diasporic groups but without including contingents from beyond the imperial borders, i.e. so-called barbarians.[31] This would only change in Late Antiquity (where the distinction between legions and *auxilia* had become obsolete after emperor Caracalla's *Constitutio Antoniniana,* which granted Roman citizenship to the *peregrini* in 212 CE). In the fourth and fifth centuries, the military leadership of the Western Roman Empire had to ensure that a Roman field army remained operational under internal and external pressure[32]; Aetius was the latest in a line of Roman generals who mustered barbarian contingents (most prominently Goths and Huns) into the army to gain the upper-hand in the power struggles of the late Roman Empire.[33] Especially the western part of the Roman Empire, whose borders were under stronger pressure than those in the east, had to rely to a large extent on recruits from beyond its borders.[34] For it became increasingly difficult to find recruits among the Roman population—probably not because of a general unwillingness to fight but rather because large areas of the empire lay desolate due to devastation and population decline. The manpower problem was also tackled by the integration of German *foederati*. This raised the "symbiosis" between the Western Roman Empire and the Germanic tribes to a new, albeit dangerous, level. The *foederati* were given a closed settlement area on Roman soil to secure the imperial border under their own leaders. Thus virtually independent diasporic groups or foreign allies had been created on Roman territory. This development contributed massively to the disintegration of the Western Roman Empire, which culminated in its fall commonly associated with the year 476. The Eastern Roman Empire by contrast was much more stable in its administrative and military structures. Even so, the Sasanians—the last Persian empire before the rise of Islam—conquered the provinces of the Orient in the second and third quarters of the seventh century. Although the lost territories could be reconquered by the Byzantine emperor Heraclius (610–641), they were then definitely lost in the Arab invasion that began in the 30s of the seventh century.[35]

Diasporic militaries have been studied with regard to mercenaries in medieval Western Europe, notably in the edited volume *Mercenaries and Paid Men* by John France, which provides a good overview of the state of research on, often ethnically defined, mercenary groups.[36] The general literature emphasizes an alleged shift from feudal retinues to mercenary armies.[37] Military obligations, be they owed to the fleet in coastal areas or to the prince's feudal cavalry inland, could be fulfilled by sending a representative or by direct monetary payment. This, then, led to the rise of mercenary armies financed by this very tax revenue. The classical argument was that money replaced feudal allegiance as the driver for military service: Werner Sombart argued military leaders were among the first capitalistic entrepreneurs in pre-modern Europe.[38] It should not be overlooked, however, that monetary payment also played an important role in financing feudal retinues. Moreover, mercenaries continued to be imbedded in webs of allegiances. Even the Renaissance Italian *condottieri* did not serve merely for money. Rather, family allegiance, feudal ties, and regional affiliation played a crucial role in their decision whom to serve.[39]

Various pre-modern ethnic mercenary groups have been extensively researched: Vikings,[40] Brabançons and other Flemish mercenaries in England in the twelfth century,[41] Normans hiring Welsh and other foreign mercenaries,[42] the Varangian Guard in Byzantium,[43] or the Swiss.[44] Occasionally mentioned are Genoese, Pisan, and Muslim sailors serving Christian rulers such as the Angevins but also the Ottomans.[45]

The military orders have no doubt received the most attention, although they have not been systematically studied as military diasporic communities (for instance, with regard to the different *langues* ("tongues" i.e. nations) of the Hospitallers). Military orders indeed were characterized by a multi-national dynamic somewhat akin to current European institutions, marked by power struggles between ethnic groups stifling the organizations' effectiveness while providing a platform for transnational communication. Military orders also mobilized and transferred considerable amounts of capital and personnel and maintained a constant flow of money, goods, personnel, and knowledge between their eastern possessions and Europe. At the same time, they ruled over lands, which often intersected with trade routes from East to West and South to North and became involved in the respective trade.[46]

The Egyptian Mamluk and the Turkish-Ottoman military diasporas were entangled with mercantile networks and the respective diasporic groups controlling the supply of slave soldiers and providing the necessary resources to fund them.[47] Yet how these links along with shared ideals and codes of conduct might have provided a common denominator among officially opposed military diasporas (such as Mamluks and Hospitallers) remains to be studied. It might help to explain the co-habitation of Cyprus or Rhodes and Mamluk Egypt in the fifteenth century,[48] but also why diasporic military élites remained separate from their subjects, be it in Egypt, Syria, or Rhodes.[49]

More attention has been given to early modern ethnic regiments, for instance, the Irish in the service of Catholic rulers (France, Spain, the Holy Roman Empire) but also of the Stuart king Charles II although not from a deliberately diasporic perspective. Studies emphasize the importance of their Catholic confession for the Irish and how it could top monetary incentives or personal-dynastic allegiances. Nevertheless, they were sometimes suspected to plot with fellow British, i.e. Scottish soldiers.[50] Similarly, the well-studied Swiss units serving various European rulers were recruited from Protestant and Catholic cantons depending on the ruler's confession. In France, however, Swiss of both confessions served, and there was a meta-confessional Swiss allegiance. The service contracts thus stipulated that Swiss units could not be sent into battle against other Swiss.[51] The Scottish military diaspora of the early modern period has been studied, e.g. by Steve Murdoch and Alexia Grosjean.[52] A biography of Marshal Schomberg by Michael Glozier, as well as his collection of essays edited with David Onnekink focus on Huguenot soldiering.[53] Huguenots were important in spreading the latest military knowledge and modernizing armies. They thus contributed significantly to state formation by drawing on their experience in the more advanced French army before being dispersed from France and taking service abroad.[54]

For the modern and contemporary periods (which are not the focus of this book), we find a few studies on Sudanese, Fijian, and Ghurka soldiers.[55] A plethora of literature of varying quality deals with the French Foreign Legion, for which diasporic groups are constitutive to the point of jeopardizing overall cohesion, cf. the so-called Spaniards' indiscipline of 1840 or the high-jacking of a train by German legionnaires in 1908. British legionnaires' preference to serve in the para regiment, by contrast, seems to be seen as a boost to morale and combat power.[56] "Transnational soldiers"[57] have also been noted in Russian and Austro-Hungarian armed forces. International volunteers have been studied for the Italian Risorgimento,[58] the Balkan Wars, and the Russian and the Spanish Civil Wars.[59] National socialist Germany attempted to recruit a multi-ethnic volunteer force against "Communism", and Jewish and non-Jewish foreign volunteers fought in the Israeli-Arab wars, while transnational mercenaries were also employed in various (post-)colonial conflicts.[60] Yet the closest to an application of a conceptual framework lent from diaspora studies is an application of the concept of diasporic intersection to the US military serving the nation abroad and the problems related to their coming back "home".[61] So we see, once again, that while there are many studies on ethnic or migrating militaries, these have rarely been studied as diasporic groups.[62]

Questions and Explanatory Framework

This book seeks to address this lacuna by analyzing various ethnic and para-ethnic military groups as military diasporas. We define military diasporas as groups of combatants with a shared ethnic (or "imagined", para-ethnic)[63] background serving in foreign lands that were more or less far removed from their place of origin—regardless of whether they formally belonged to the same state/empire or not. Typically these diasporas thereby served rulers or governments of another ethnicity—if they were not ruling themselves, such as, for instance, the Mamluks in Egypt and Syria or serving higher authorities such as God, the idea of the Islamic *umma* or the Church/the Pope. In military diasporas, we suggest, ethnic categories will typically matter for the identity politics of the group and the land of origin will remain important in the imagination or as a place of recruitment and return. We distinguish between military diasporas and single military diasporic groups, although the former term—where appropriate (e.g. in Andrade's piece)—can stand in for both, i.e., for the single diasporic groups of the same ethnic moniker in different places as well as for the respective diaspora of x or y they are collectively forming consisting of these diasporic groups z1, z2, and z3.

We thus suggest a broader but also more nuanced definition of diaspora than William Safran: Groups dispersed abroad from a homeland centre, to which they yearn to return while remaining separate from the host society.[64] This latter definition suits ethno-religious diasporas such as the Jewish or Armenian that had no, or very limited, control over their "homeland". For military diasporas, this was not always the case. Their military strength could preserve a status of relative political independence of their homeland e.g. in the case of the Swiss, and thus significantly impact state formation and empire building not only in the countries of service but also at home. Still, the homeland is part also of the military diasporas' identities while they are serving abroad, but it is different from the spiritual yearning for Jerusalem, and in many cases, return is not only a dream but also a reality (for those surviving however traumatized, wounded, and maimed they may have been).

Why should one use the term diaspora at all? Has it not become completely meaningless in its wider and wider circles of applications and with its programmatic undertones?[65] Would "mercenary" not be a suitable and already familiar alternative for the topic at hand?[66] While, traditionally, mercenaries are defined as troops foreign and paid, Stephen Morillo shows that this equation is misleading. There are not only paid non-diasporic troops (such as many contemporary Western armies) but also diasporic troops which, although typically paid, were not in it for that pay so much as for other reasons, such as political power in the case of the Mamluks in Egypt, or an ideologically motivated enterprise/career such as Crusade (military orders) or jihad (ghazis).[67] Migrant soldier, however, is too unspecific and de-emphasizes both the prospects of more permanent resettlement in the country of service and the ties between the latter and the homeland. Military transhumance (used in Christ/Weltert's chapter) would again focus on the migratory element and especially seasonal and recurring patterns, which (although crucial) are only a marginal part of the diasporic experience. Finally, ethnonyms used as autonyms, i.e. "Swiss" or "Scots", can reify national identity and overshadow the fact that groups, in fact, were of more transnational composition than their name might suggest. The diasporic framework shall thus help to critically interrogate such categories with regard to processes of group formation and identity construction within a context of transimperial connectivity and transculturation affecting and being facilitated by such military groups.

It is thus for lack of a suitable alternative that we recur to "diaspora" unless the reader prefers "allochthonous military group operating under a (para-)ethnic group identity". Regardless of the terminology, we propose to explore to which extent the comparative focus on such groups and the methodological combination of diaspora studies and military history can help exploring social cohesion, allegiance, identity, and motivations in pre-modern militaries, the tension fields within which they operated and their contribution to processes of empire building/state formation. Furthermore, we want to explore possible allegiances and connections that could be negotiated by foreign military personnel between homeland and land of service and the respective liege lords. It helps to foreground foreign soldiers' role beyond their core military function in policing, representing, and administrating state/empire. Thereby we ought to pay attention to processes of transculturation, i.e. how deculturation and acculturation intermingled, were embraced, endured but also resisted. This resulted in transcultural modes of adaptation between the place of origin and land of service enabling transcultural and thus thick and flexible transregional and -imperial connectivities. We argue that such a diasporic focus will shed light on and foster a more holistic understanding of pre-modern militaries. It highlights how military diasporic groups were crucial elements of state formation while creating and maintaining—among other things—links between polities horizontally and vertically (or, anachronistically speaking, internally and diplomatically).

State/empire building involves violent processes to extract resources and to control/monopolize violence and power. It requires a reliable force to project military might from the governing centre to the polity's parts and to safeguard its centre of gravity (ruler, capital). It needs to be both loyal to the centre and apart from its opponents; the recipients of violence. There are different options of how to generate such forces capable of projecting military or administrative power while remaining loyal to the dispatching ruler or administration. One is to form a transimperial élite from scratch, e.g. by recruiting children into closed institutions where they can be educated (de-cultured and then ac-cultured) into this role, such as the Janissaries, Mamluks, or, ideally, members of (military) orders[68]; another option is to rely on a domestic powerbase (*Hausmacht*: Family, clan, tribe, or town). While the latter option, according to the fourteenth-century North African historian Ibn Khaldūn, prevails in the funding phases of empires/states, the former dominates subsequently.[69] A third option is a variant of the first: The recruitment of diasporic military groups either as slaves or mercenaries/professional soldiers, whereby their foreignness keeps them separate from the targeted population while they gain access to the ruling (military) élite. Thinking of an alternative voluntary, bottom-up process of state formation/empire building, even such spontaneous political upscaling, as for instance, observable in medieval Italian communes, can lead to locked contests for power between different factions as carriers of bottom-up military power. The cities thus often resorted to recruiting a *podestà* or *signore* from outside, who first relied on his *Hausmacht* but then increasingly on diasporic groups to recruit an allochthonous mercenary police force to maintain law and order and break factional stalemates.[70]

In either case, these enforcers needed to be uprooted from their original context. In compensation for lost traditional (family, tribal) support, employing rulers tied these relatively isolated military diasporic groups to their person and thus enhanced and protected personal and dynastic power. The soldiers' new homeland was the ruler's household. Responding only and directly to the ruler, they could serve as a force tipping the power balance amidst factional fights in the latter's favour. This impacted on their armament and military organization. Often they were used in a Praetorian (close personal protection and thus infantry role) and/or policing, power projecting, and civil/tax enforcement role (light/heavy cavalry). Their new point of reference being courtly politics, there was a strong incentive to take an active part in it. Hence, military diasporas, while summoned for the protection of rulers and their empires, could become the most dangerous threat to them. A further challenge in the employment of diasporic groups was that their ethnic or para-ethnic bonding fostered solidarity across the wider diaspora, i.e. groups of the same ethnic identity serving different rulers, which on the up-side provided links across lines of conflict.

Military diasporas also enabled rulers to upscale their armed forces by extending and differentiating their recruitment base. Departing from a model where a prince drew on his own land's military potential (feudal retinue for heavy cavalry and rural militias for infantry), the size of the armed forces would be limited by the size of the population. The trend towards exempting populations from military service in exchange for financial contributions restricted the recruitment base furthermore. Yet rich princes could compensate by tapping into the military resources of other lands, typically those characterized by surplus population and conditions of scarcity. Such "marginal" lands could export mercenaries, e.g. densely populated Flanders (twelfth century) or the Swiss Alps (esp. from the fourteenth century onwards).[71] Diasporic soldiers were thus an important element of population relocation in the context of state formation and empire building. They not only projected power but also unleashed a broader dynamic of drawing diverse populations (both geographically and socially) into imperial/state projects of acculturation, transculturation, and thus cultural homogenization.

In order to appreciate the importance of military diasporic groups, we propose the following questions:

1. **How diasporic** was a military diasporic group? I.e. was the ethnic eponym or autonym merely a **moniker of corporate identity**[72] or did it reflect a deeper affiliation with a homeland and its diaspora? How mono-ethnic were ethnically identified diasporic groups? What, by contrast, was the role of ethnicity in officially non-ethnic military diasporic groups (such as Mamluks, Janissaries, *ghilmān*, members of military orders)?
2. What was the **military edge**, the speciality of this group in terms of tactics, armament and equipment? The Swiss infantry, for instance, developed a particular style of dealing with heavy cavalry attacks, which made them an attractive asset on European battlefields of the late fifteenth century. Other examples would include the German *Landsknechte*, the Gascon infantry, the *stradioti* (Albanian) light cavalry, etc.—some of which are analyzed in the following chapters. Out of these regional contingents evolved different military specialities that soon were not the prerogative of ethnic specialists anymore. This was, for instance, the case with the, originally Polish, lancers or the, originally Hungarian, Hussars that, while retaining distinctive "ethnic" dress, were recruited domestically. The originally Swiss speciality of the long pike became commonplace in many infantry units including the Spanish *Tercios*, which increasingly recruited internationally, e.g. in Italy.[73] The specific equipment, armament, and tactics of diasporic units were an asset for the employers as their tactics, at first, came as a nasty surprise to enemies not accustomed to them. Yet even after this surprise element had worn off and their ethnic character had become increasingly blurred, they continued to exist as part of the force-mix of increasingly diversified and bigger armed forces. Thus, they complemented the hard-core military capabilities such as heavy cavalry in the Middle Ages, heavy infantry in antiquity or line

infantry in the early modern period. To what extent did such ethnic and tactical diversification require a stronger investment in training of military élites to ensure effective joint-strike capability and interoperability between the different corps?

3. How did a priori territorial militias,[74] typically associated with a local context and being particularly effective in it, become military diasporic contingents removed from their context and thus suitable as enforcers of central power? To which extent does this also apply to noble contingents (including non-nobles in auxiliary roles), although they might acculturate more easily based on shared élite status?[75] In other words how did military diasporas transform through but also facilitate processes of acculturation, deculturation and transculturation?[76]
4. How, and to what extent, was a diasporic group **integrated** into the wider military/police apparatus and culture of the host society?[77] To what extent did armed bodies such as standing armies but also navies, police corps, and bureaucracies tend to be diasporic? What were the implications with regard to **professionalization/diversification**, **centralization**, **build-up/upscaling** of such bodies, and, hence, **state formation and empire building** both in the places of service and at "home"? How did it affect demographics and social structures both of the host and home societies? How did diasporic solidarities shape military communities in contrast or parallel to professional collegiality within and across different armed forces? Were soldiers primarily part of the ruler's "family", a multi-ethnic military, or part of their overarching respective (e.g. Cretan, Flemish, Genoese, Swiss, Irish) diasporas? What was the role of meta-ethnic, e.g. confessional, allegiances? The other trait of their *esprit de corps* arguably was military-technical professionalism that could transcend diasporic allegiance or even constitute a new diasporic, in this case multi-ethnic, military allegiance. Professionals on both sides of conflict lines developed a shared identity, which was arguably strongest in technical branches such as the navy, engineering, and artillery (or among pilots in the modern period). These relatively small élites would be characterized by strong transnational and -imperial connectivities, in the early modern period also by shared readings and to an extent languages, and high mobility between enemy lines, e.g. Orban, the renegade master of Ottoman artillery at the siege of Constantinople (1453).[78]

Overview

The sketches presented here reflect the research foci of the participants in the project. While we succeeded in extending the breath of the volume by adding further contributions (Klinkott, Andrade, Jaspert, Loiseau, Whelan, Carr/Grant), we unfortunately did not succeed mustering contributions on the Hospitallers, Templars, Janissaries, and the early modern Scottish and Irish military diasporas. It also would be desirable to consider diasporic elements in the

Viking expansion, Spanish Tercios or French and northern European diasporic militaries, among others.

Nevertheless, combining studies of Persian, Mediterranean, and European military diasporas from antiquity to the late Middle Ages, we provide a first glance at military diasporas in the *longue durée*. The essays will explore military diasporas with a particular focus on state formation/empire-building on the macro and micro level, i.e. in the securing and defending of empires as well as in courtly politics, while also considering—where appropriate—the "transimperial" (or rather inter-imperial) mobility of diasporic groups and their contribution to the linking of states/empires.

The volume proceeds in a roughly chronological order. Hilmar Klinkott explores ethnic militaries in the Achaemenid Empire. He examines two groups of potential military diasporas—ethnically defined (non-Persian) troops on the one hand and Persian contingents on the other—and confronts the findings with the world view of the kings.

Patrick Sänger examines the military diaspora in Hellenistic Egypt ruled by the Ptolemaic dynasty between 323 and 30 BC (when Egypt was incorporated into the Roman Empire) through papyrological evidence. After Alexander the Great had occupied Egypt in 332 BC, civilians and soldiers from the Greek-speaking world came to Egypt and were grouped according to ethnic categories. Of these, the *cleruchs*, the members of the regular army of the Ptolemies, and the mercenaries or professional soldiers, are investigated: While the *cleruchs* present themselves as preservers of a (common) Greek cultural identity, mercenaries/professional soldiers seem to express an ethnic identity (e.g. Boeotian, Cretan, Jewish). Most prominent among the latter are the *politeumata,* administrative units based on (semi-autonomous) ethnic communities of mercenaries/professional soldiers, with which the Ptolemies may have intended to create an integrative (urban) counterpart to the cleruchic settlements based on individual land grants.

Nathanael Andrade discusses Syrian military diasporas in the Roman Imperial period (27 BC–AD 284). He establishes the types of Syrian practices that can be considered diasporic and argues that there were military diasporas of Syrians but not an integrated Syrian military diaspora. Andrade scrutinizes how Syrian diasporic traits were reconfigured as unit traditions, appropriated and redefined by military and civilian networks of non-Syrians, or integrated into local municipal cultures.

The next two articles investigate military diasporas in Late Antiquity. Mariana Bodnaruk's case study focuses on the political and social role of the Roman generals (*magistri militum*) Stilicho, Constantius III, and Aetius, who made their careers in the second half of the fourth and the first half of the fifth centuries. Based on the epigraphic evidence of Late Antique Rome, she illuminates a decisive turning point in Roman history: The fate of the empire relied more and more on the generals' ability to raise or command troops that were not genuinely "Roman" but diasporic; *foederati* recruited from outside and settled within the Roman Empire.

Lajos Berkes studies the Persian/Sasanian and Islamic occupation of Egypt in the first half of the seventh century. Drawing on papyri, Berkes analyzes how the new masters conquered and ruled Egypt addressing administrative continuities and discontinuities. He points out that Persians and Arabs relied heavily on the local structures of the former Byzantine administration. Furthermore, it can be noticed that the Persian and Arabic occupying forces tended to separate themselves from the indigenous population as, for instance, evidenced by the foundation of Fustāt, a separate new city in the vicinity of the Roman military camp of Babylon.

The second part of the book is devoted to the Medieval Mediterranean. It opens with a study of Byzantine military diasporas during the Komnenoi revolt of 1081 by Roman Shliakhtin based on the Alexiad by Anna Komnena. Against the backdrop of economic growth and military defeat (Manzikert 1071), diasporic mercenary units grew in importance. Military groups hailed from Scandinavia and Rus (serving in the Varangian guard), but included also Pechenegs (Hungarian light cavalry), "Franks" (garrison troops), and "Iberians" (guarding the eastern provinces of the Empire), thus mirroring the routes of trade connecting Byzantium with the wider world. Although these diasporas raised concerns, they were a necessity both in internal contests for power and for the defence of the empire.

Mike Carr and Alasdair Grant explore another military diaspora in the Byzantine realm: The Catalan company as a diasporic group based on a common homeland and patron saint: Saint George. The integration of the Catalans into the officially universal Byzantine realm did not pose problems per se (cf. Klinkott for an earlier period) and was achieved by investing the leaders with titles and thus including them into the courtly taxonomy. Yet ethnic differentiation did seem to play a pivotal role in the Catalan company's rule of their territorial base, the Duchy of Athens, where the Catalan and wider Latin ruling élite treated Greeks as second-class citizens if not slaves.

Nikolas Jaspert explores Christian *alcayts* mercenaries serving Muslim rulers in Medieval North Africa and al-Andalus. Focusing on the question of affiliation across the confessional divide, Jaspert analyzes to which extent they remained loyal to their former feudal overlords without jeopardizing their new allegiance to their Muslim employers. Their diasporic status and Christian faith seemed not to have negatively affected their "embeddedness" into the Muslim host society but enabled them to serve the ruler in matters delicate in terms of Islamic law such as collection of (non-canonical) taxes and thus might have given them a competitive edge.

Julien Loiseau analyzes diasporic military groups in the Mamluk sultanate. The Mamluks, as "slave" soldiers imported from beyond the Black Sea, manned not only the core of the heavy cavalry and thus the backbone of the armed forces of Egypt and Syria but also its military-administrative élite including the ruling dynasty. The first generations of Turkic Mamluks were part of the much wider Turkic military diaspora serving in armed forces across the Islamic world. As a first diasporic transformation, Mongol horsemen were cautiously integrated into the Mamluk force, then the Circassians rose to power. The Ottoman conquest of Syria and Egypt cut short the third transformation: Europeans (and Africans), who came to serve in the Mamluk army.

Nicholas C. J. Pappas describes a much-underexplored military diaspora: The *stradioti* as a double-diasporic Albano-Greek military community. He describes how the expansion of Ottoman power in south-eastern Europe in the fifteenth and sixteenth centuries compelled this former Byzantine soldiery to find refuge and new employment. They became a light cavalry force in the Greek holdings of the Venetian realm but also in Naples, Sicily, and all over central Europe and thus part of most European armies' force-mix as lancers/uhlans. Their competitive edge was their light armament, their hit-and-run tactics, and the low cost of their pay and equipment compared to heavy cavalry. Pappas highlights how their military prowess gave them particular prestige, which allowed them to successfully lobby for often-discriminated Greek diasporic communities.

The third part of the book will look at military diasporas in central Europe. László Veszprémy studies military diasporas in the kingdom of Hungary in the twelfth and thirteenth centuries. He explores the light-cavalry auxiliaries such as the Pechenegs, Székelys, and Cumans through contemporary chronicles, in which they were described with contempt—probably because they were seen as representatives of old tactics by the Hungarian military élite, which by this time had adopted a mainstream European heavy cavalry role. Teutonic and Hospitaller knights also operated in Hungary but in a heavy cavalry role.

Christopher Mielke analyzes the role of medieval queens in shaping military diasporas. After appreciating the diplomatic importance of their' retinues, he focuses on the political role of dynastic marriages and finally the relationship between royal women, crusaders, and military orders. Mielke outlines the declining importance of knights in the queens' retinues and of troops supporting the policy of a "foreign" king outside their homeland as a result of a dynastic marriage and partly explains this development by the growing importance of dowries given in cash or supplies.

Verena Schenk zu Schweinsberg is dealing with the *Schwertbrüderorden* (Order of the Brethren of the Sword) in thirteenth century Livonia. Based on the "Livonian Rhymed Chronicle" (*Livländische Reimchronik*), an extensive vernacular source written within the Order, Schenk analyzes how the isolated military order, in competition with missionaries and bishops and in need of constant crusade to justify its existence, developed an almost symbiotic relationship with its enemies, who are described in a remarkably respectful way thus blurring the lines between friend and foe; Christian and heathen.

Mark Whelan explores how the diasporic character of the Teutonic Order (closely linked to the above-mentioned *Schwertbrüderorden*) shaped its rule in Prussia. Whelan focuses on the order's ability to transplant the technologies and techniques of their native German lands to Prussia. He shows how enduring links with the knights' homelands were crucial for the running of the order, not least by replenishing the order's ranks from the principal recruiting grounds in Franconia and Swabia.

Anna Katharina Weltert and Georg Christ explore Swiss militaries abroad focusing on the Cold Winter Expedition of 1511. The campaign was rather exceptional: It was not an infantry-only mercenary contingent but a Swiss expeditionary force. The expedition is analyzed using a methodological framework of mercenary service ranging from cantonal contingents within the framework of a military alliance to unregulated individual service. The authors argue that the campaign showed the limits of independent Swiss military power projection hampered by both the cantons' unwillingness to shoulder the necessary costs but also distrust and conflicts of interest. Thus, the Swiss tended to renounce on independent military power projection, to accept the auxiliary military diasporic role carved out for them by the big powers, and to enter into respective alliances, e.g. with France.

Outlook: Towards a Diasporic Deconstruction of the National Military

The present volume provides a range of studies both temporally and geographically tied together by the overarching focus on the diasporic element in military groups. Many promising lines of enquiry emerged and were pursued, while many more remain to be explored including Vikings, Flemish mercenaries, other military orders, Croats, "Cossacks", Genoese and other Italian military including naval specialists, diasporic groups in Spanish imperial forces, Greek sailors (some of them "colonial" subjects) in the Venetian navy, or other Islamic diasporic militaries such as janissaries and Abbasid *ghilmān* slave soldiers.

A military diasporic focus can also foster a conversation between diverged disciplines, namely diaspora studies and military history. This enables us to rethink foreign militaries' contribution to state/empire formation. For state formation is not only the result of autochthonous military-fiscal build-up but very much of the movement of people and their skills first needed to enable a polity to secure and control regions, to administer them, and to extract taxes. While mainstream military history tends to overlook that some military forces, such as heavy cavalry in the late Middle Ages, served frequently in the tax enforcement role,[79] economic and political history focus less on the military build-up that was not only the result but also the pre-condition for forming centralized fiscal-military states. This volume seeks to shed light on some of these blind spots but can only be a modest beginning.

While the term military diaspora focuses on connectedness between soldiers of similar origin and ethnic affiliation in different armies, the concept of a military diasporic group targets the respective groups in their local contexts. Thus we can explore the importance and handling of ethnic eponyms in the management of group identity. Diasporic militaries often were sought out not only because they were foreign or apart but also because of their specific military skills and their strong esprit de corps rooted in common origin and ethnicity. They thus contributed to a division of labour and specialization in the military, transfer of military technology and respective professionalization but also to numerically increased armies, which were crucial in the context of medieval and especially early modern wars of expansion.

It becomes clear that military diasporas crucially contributed to the rise and maintenance of standing armies that are hence marked by what we might call the diasporic-nationalist paradox of modern armies: An increasingly transcultural, homogenized military culture, characterized by similar mind-sets, codes of conduct, ranks, customs, language, structures, armament, tactics, and even uniforms, coexists with the powerful deployment of national narratives and imaginaries that emphasize the respective army's supposed uniqueness setting it apart from its (essentially identical) foe. Even this emphasis on national ideology is, of course, a shared feature of this overarching military "transculture" that serves also to suppress, to "de-culture" the force's very own soldiers' natural group feelings based on common origin or (tribal) ethnicity and to replace them with a homogenized, overarching "national" identity shared across the armed forces binding the thus atomized soldiers directly to the military apparatus.

Notes

1 We are grateful to the anonymous reviewers and Adrian Wettstein for critical, meticulous, and most helpful comments on this introduction. All remaining errors are ours.

2 To distinguish between state and empire would make sense within a model of layered statehood whereby an empire would be on a higher hierarchical level than a state (which could be roughly equated with a kingdom) and typically encompassing a bigger territory and population composed of several states/kingdoms (for an application of this model to the Mamluk Empire see Georg Christ, "Settling Accounts with the Sultan: Cortesia, Zemechia and Venetian Fiscality in Fifteenth-Century Alexandria," in *Trajectories of State Formation across Fifteenth-Century Islamic West-Asia: Eurasian Parallels, Connections and Divergences*, ed. Jo van Steenbergen [Leiden: Brill, 2020], 319–352, accessed 3 July 2020, https://doi-org.manchester.idm.oclc.org/10.1163/9789004431317_012). We would rather reject such a distinction if we critically evaluated Pomeranz's definition, which seems to postulate that the only difference between state and empire is the latter differentiating modes of rule in its different parts. Pomeranz does not fail to see that such differentiation marks also states but sees it as "temporary failing" rather than constitutive (Kenneth Pomeranz, "Social History and World History: From Daily Life to Patterns of Change," *Journal of World History* 18 [2007]: 69–98, here 87–88, accessed 10 August 2021, https://www.jstor.org/stable/20079411), cf., however, Kumar's take on modern states as "empires in miniature" (Krishan Kumar, "Nation-States as Empires, Empires as Nation-States: Two Principles, One Practice?," *Theory and Society* 39, no. 2 [2010]: 119–143, here 119) or, indeed, as Kumar reminds us, Hannah Arendt, *The Origins of Modern Totalitarianism*, 2nd ed. (New York: Meridian Books, 1958), 153. Hence, for our purpose it seems appropriate to use the two almost interchangeably or as a double expression with oblique (state/empire). In this sense, as a fuzzy term for a higher-level polity marked by a conspicuous projection of military/policing power towards its constituent parts or other entities it may also include cities (city states) such as Bologna or Venice that deployed diasporic troops internally as police forces, Trevor Dean, "Police Forces in Late Medieval Italy: Bologna, 1340–1480," *Social History* 44, no. 22 (2019): 151–172, accessed 24 March 2022, https://doi.org/10.1080/03071022.2019.1579974.

3 Robert Ian Moore, *The Formation of a Persecuting Society: Power and Deviance in Western Europe, 950–1250* (Oxford: Basil Blackwell, 2006).

4 Nir Arielli, and Bruce Collins, "Introduction: Transnational Military Service since the Eighteenth Century," in *Transnational Soldiers: Foreign Military Enlistment in the Modern Era*, ed. iid. (Basingstoke: Palgrave Macmillan, 2013), 1–12, here 1; on the "bringing home" of foreign regiments: Martin Kitchen, "Elites in Military History," in *Elite Military Formations in War and Peace*, ed. A. Hamish Ion and Keith Neilson (Westport, CT: Praeger, 1996), 7–30, here 17.

5 The *levée en masse* was not that massive with 300k men, Alan Forrest, "L'armée de l'an II: La levée en masse et la création d'un mythe républicain," *Annales historiques de la Révolution française* 335 (2004): 111–130, here 115, accessed 16 November 2017, http://www.jstor.org/stable/41889474.

6 James J. Sheehan, *Where Have All the Soldiers Gone? The Transformation of Modern Europe* (Boston, MA: Mariner Books, 2009).

7 Herbert Rosinski, *Die Deutsche Armee: Eine Analyse* (Düsseldorf: Econ Verlag, 1970), 88–92; Detlef Bald, *Der deutsche Offizier: Sozial- und Bildungsgeschichte des deutschen Offizierkorps im 20. Jahrhundert* (München: Bernard & Graefe, 1982).

8 Even increasingly, the Soviet armed forces mobilized the idea of the Great Patriotic War: *Welikaja Otetschestwennaja woina.*

9 Sebastian Haffner, *Preußen ohne Legende* (Herrsching: Manfred Pawlak, 1985), 334–335.

10 Michel Dumoulin, *La communauté européenne de défense, leçons pour demain? = The European Defence Community, Lessons for the Future?* Euroclio 15, Études et documents (Bruxelles: PIE Peter Lang, 2000).

11 The EU battle groups were instituted in 2005 (fully operational 1 January 2017) but have not yet been deployed. The deployment was hampered (despite opportunities) by the slow decision-making process. Enthusiasm for the battle groups also suffered because the providing states had to cover all the costs. Most recently, this seems to change with a common funding mechanism envisaged as part of the Global EU Strategy on Security and Foreign Policy, which, however, is still far off from establishing proper supranational structure and European armed forces. Staff capabilities of the EU Military Staff remain limited. EUFOR operations (peace keeping, e.g. in Bosnia, Chad, etc.) have been typically led from national operations centres and only since 2007 has there also been a dedicated EU Operations Centre (not permanently staffed though). The Euro Corps is, despite its name, a binational Franco-German cooperation subsequently expanded to include also Belgium, Spain, and Luxembourg and is not subordinate to the EU; the European Gendarmerie Force (EGF or EUROGENDFOR), likewise, is not a supranational EU force but a multi-national cooperation between gendarmeries (militarized police forces) of Italy, Spain, Portugal, France, Romania, and Poland that operated as such only in training missions in Afghanistan, except for one relief support effort in Haiti in 2010. Frontex (officially European Agency for the Management of Operational Cooperation at the External Borders) was founded as a relatively weak coordinator agency in 2005 but is planned to become more robust after its relaunch as the European Border and Coast Guard Agency (EBCGA) in 2016 with a proper budget and its own personnel. Responsibility for the management of the EU's borders, however, remains a shared responsibility of EU and member states. While traits of a European military identity emerge especially among officers with language skills who believe in international cooperation, overall, the military identity remains strongly national, Frédéric Mérand, "Dying for the Union? Military Officers and the Creation of a European Defence Force," *European Societies* 5, no. 3 (2012): 253–282, accessed 24 March 2022, https://doi.org/10.1080/1461669032000111306.

12 Sheehan, *Where Have All the Soldiers Gone?*

13 For the US: Beth Bailey, "Soldiering as Work: The All-Volunteer Force in the United States," in *Fighting for a Living: A Comparative History of Military Labour, 1500–2000*, ed. Erik-Jan Zürcher (Amsterdam: Amsterdam University Press, 2014), 581–612, accessed 19 July 2021, http://www.jstor.org/stable/j.ctt6wp6pg.22; for Germany:

Thorsten Jungholt, "Bundeswehr wirbt um Schulabbrecher und Ausländer," *Welt*, 1 December 2016, accessed 19 April 2018, https://www.welt.de/politik/deutschland/article159898506/Bundeswehr-wirbt-um-Schulabbrecher-und-Auslaender.html.

14 The field chaplaincy service, although moving towards a meta-(or, rather, post-)confessional understanding of its role, still draws exclusively on ministers from Christian confessions.

15 Nir Arielli and Bruce Collins, "Conclusions: Jihadists, Diaspora and Professional Contractors: The Resurgence of Non-State Recruitment since the 1980s," in *Transnational Soldiers: Foreign Military Enlistment in the Modern Era*, ed. iid. (Basingstoke: Palgrave Macmillan, 2013), 250–256; see also the special issue of the *Journal of Medieval Military History* 14 (2016), ed. Jochen Schenk, on proxy actors/irregular warfare, esp. Schenk's article "'New Wars' and Medieval Warfare: Some Terminological Considerations," 149–162.

16 Cf. chapter Weltert/Christ in this volume.

17 Rogers Brubaker, "The 'Diaspora' Diaspora," *Ethnic and Racial Studies* 28, no. 1 (2005): 1–19, accessed 11 July 2017, http://www.sscnet.ucla.edu/soc/faculty/brubaker/Publications/29_Diaspora_diaspora_ERS.pdf.

18 Wen-Chin Chang, "Home Away from Home: Migrant Yunnanese Chinese in Northern Thailand," *International Journal of Asian Studies* 3, no. 1 (2006): 49–76, here 70, 73.

19 E.g. Mark Villegas, "Currents of Militarization, Flows of Hip-Hop: Expanding the Geographies of Filipino American Culture," *Journal of Asian American Studies* 19, no. 1 (2016): 25–46. Kostas Buraselis, "Ambivalent Roles of Centre and Periphery. Remarks on the Relation of the Cities of Greece with the Ptolemies until the End of Philometor's Age," in *Centre and Periphery in the Hellenistic World*, ed. Per Bilde et al. (Aarhus: Aarhus University Press, 1993), 251–270, at 258 use of term "military diaspora" to describe the masses of immigrant soldiers in Hellenistic Egypt; for a more nuanced approach, see Patrick Sänger's article in this volume.

20 Wikipedia https://en.wikipedia.org/wiki/Irish_military_diaspora, and similar online products; for other rare instances of the term being applied to the premodern context, see Grozier on the Huguenot military diaspora n. 53 below or Leo Schelbert, "Swiss Diaspora," in *Encyclopedia of Diasporas: Immigrant and Refugee Cultures around the World*, ed. Melvin Ember, Carol R. Ember, and Ian Skoggard (New York: Springer, 2005).

21 Erik-Jan Zürcher, ed., *Fighting for a Living: A Comparative History of Military Labour, 1500–2000* (Amsterdam: Amsterdam University Press, 2014).

22 Stephen Morillo, "Mercenaries, Mamluks and Militia: Towards a Cross-Cultural Typology of Military Service," in *Mercenaries and Paid Men: The Mercenary Identity in the Middle Ages*, ed. John France (Leiden: Brill, 2008), 243–260, here 243.

23 Cf. the German expression of *Krethi und Plethi* for these two Levantine ethnic groups meaning "everybody and their dog," a motley crowd, see Itamar Singer, "Egyptians, Canaanites, and Philistines in the Period of the Emergence of Israel," in *From Nomadism to Monarchy: Archaeological and Historical Aspects of Early Israel*, ed. Israel Finkelstein and Nadav Na'aman (Jerusalem: Israel Exploration Society, 1994), 282–338; Hermann Schult, "Ein inschriftlicher Beleg für 'Plethi'?," *Zeitschrift des Deutschen Palästina-Vereins* 81, no. 1 (1965): 74–79.

24 Cf. e.g. Marco Bettalli, *I mercenari nel mondo greco I: Dalle origini alla fine del V sec. a.C.* (Pisa: Ed. ETS, 1995); id., *Mercenari: Il mestiere della armi nel mondo greco antico; Età arcaica e classica* (Rome: Carocci, 2013); Ludmila P. Marinovič, *Le mercenariat grec au IVe siècle av n.è. et la crise de la polis*, (Paris 1988); Matthew Trundle, *Greek Mercenaries: From the Late Archaic Period to Alexander* (London: Routledge, 2004).

25 Cf. e.g. Guy T. Griffith, *The Mercenaries of the Hellenistic World* (Cambridge: Cambridge University Press, 1935); Marcel Launey, *Recherches sur les armées hellénistiques* (Paris: De Boccard, 1949/1950; 2nd ed. 1987). Fresh perspectives open in Patrick Sänger and Sandra Scheuble-Reiter, eds., *Söldner und Berufssoldaten in der griechischen Welt: Politische und soziale Gestaltungsräume*, Historia Einzelschriften (Stuttgart: Franz Steiner, 2022).

26 Karl Jaspers, *Vom Ursprung und Ziel der Geschichte* (Basel: Schwabe, 2017), engl.: *Origin and Goal of History* (London: Routledge Revivals, 2011).
27 A case study on Ptolemaic Egypt is the subject of Patrick Sänger's article in this volume.
28 On the Roman army see in general, e.g. Paul Erdkamp, ed., *A Companion to the Roman Army* (Malden, MA: Blackwell, 2007); Yann Le Bohec, *L'armée romaine sous le Haut-Empire*, 3rd ed. (Paris: Picard, 2002); id., *L'armée romaine sous le Bas-Empire*, Antiquité—Synthèses 11 (Paris: Picard, 2006); see further, e.g. James N. Adams, "The Language of the Vindolanda Writing Tablets: An Interim Report," *Journal of Roman Studies* 85 (1995): 86–134; Geza Alföldy, Brian Dobson, and Werner Eck, eds. *Kaiser, Heer und Gesellschaft in der Römischen Kaiserzeit: Gedenkschrift für Eric Birley*, Heidelberger althistorische Beiträge und epigraphische Studien 31 (Stuttgart: Steiner, 2000); Anthony R. Birley, "A New Tombstone from Vindolanda," *Britannia* 29 (1998): 299–306; id., *Garrison Life at Vindolanda: A Band of Brothers* (Stroud: Tempus 2002); Alan K. Bowman, *Life and Letters from the Roman Frontier* (New York: Routledge 1998); Ian Haynes, *Blood of the Provinces: The Roman* Auxilia *and the Making of Provincial Society from Augustus to the Severans* (Oxford: Oxford University Press, 2013); John C. Mann, *Legionary Recruitment and Veteran Settlement during the Principate*, Institute of Archaeology, London; Occasional Publication 7 (London: Routledge 1983); Ramsay MacMullen, "The Legion as a Society," *Historia* 33, no. 4 (1984): 440–456; Grace Simpson, *Britons and the Roman Army: A Study of Wales and the Southern Pennines in the 1st–3rd Centuries* (London: Gregg, 1964).
29 On the military diasporas of Syrians, see Nathanael Andrade's article in this volume.
30 For this characterization of the Roman *auxilia*, Patrick Sänger, "Augustus—Herr über 28 Legionen: Das militärische Erbe der Republik und die kaiserzeitliche Armee," in *Krieg in der antiken Welt*, ed. Gerfried Mandl and Ilia Steffelbauer (Essen: Magnus, 2007), 64–84, here 74–75.
31 For a comparable situation in the Persian Achaemenid Empire, see Hilmar Klinkott's article in this volume.
32 This is the historical setting of Mariana Bodnaruk's article in this volume.
33 Kelly DeVries, "Medieval Mercenaries: Methodology, Definitions, and Problems," in *Mercenaries and Paid Men: The Mercenary Identity in the Middle Ages*, ed. John France (Leiden: Brill, 2008), 43–60, here 47–49.
34 On the Roman army in Late Antiquity see also the excellent overview by Bernhard Palme, "Feldarmee und Grenzheer: Das römische Militär in der Spätantike," in *Krieg in der antiken Welt*, ed. Gerfried Mandl and Ilia Steffelbauer (Essen: Magnus, 2007), 85–113.
35 In Egypt, papyrus documents provide a good understanding of how the Sasanian and Arabic conquerors—who were confronted with a diasporic situation on the territory of the former Byzantine Empire—acted to establish their new rule; see Lajos Berkes' article in this volume.
36 John France, ed., *Mercenaries and Paid Men: The Mercenary Identity in the Middle Ages* (Leiden: Brill, 2008).
37 Stephen D. B. Brown, "Military Service and Monetary Reward in the Eleventh and Twelfth Centuries," *History* 74 (1989): 20–38.
38 Werner Sombart, *Der Bourgeois: Zur Geistesgeschichte des modernen Wirtschaftsmenschen* (München: Duncker & Humblot, 1913), 79–80, cf. id., *Krieg und Kapitalismus*, Studien zur Entwicklungsgeschichte des modernen Kapitalismus 2 (München: Verlag von Duncker & Humblot, 1913).
39 Brown, "Military Service;" Michael E. Mallett, "Mercenaries," in *Medieval Warfare: A History*, ed. Maurice Hugh Keen (Oxford: Oxford University Press, 1999), 209–229, here 215, 219 and passim; Georg Christ, "Introduction: The Rise of the Business Sphere in the Middle Ages," in *A Cultural History of Business*, vol. 2, *Middle Ages 800–1450*, ed. Georg Christ and Catherine Casson (London: Bloomsbury, forthcoming).

40 Judith Jesch, *The Viking Diaspora* (London: Routledge, Taylor & Francis Group, 2015).

41 John D. Hosler, "Revisiting Mercenaries under Henry Fitz Empress, 1167–1188," in *Mercenaries and Paid Men: The Mercenary Identity in the Middle Ages*, ed. John France (Leiden: Brill, 2008), 33–42; Steven Isaac, "The Problem with Mercenaries," in *The Circle of War in the Middle Ages: Essays on Medieval Military and Naval History*, ed. Donald J. Kagay and L. J. Andrew Villalon (Woodbridge [UK]: Boydell Press, 1999), 101–110; cf. also Jean-Denis G. G. Lepage, *Medieval Armies and Weapons in Western Europe* (Jefferson: McFarland, 2005); Kelly De Vries, "Medieval Mercenaries: Methodology, Definitions, and Problems," in *Mercenaries and Paid Men: The Mercenary Identity in the Middle Ages*, ed. John France (Leiden: Brill, 2008), 43–60 here 51–52.

42 Richard Philip Abels and Bernhard S. Bachrach, *The Normans and their Adversaries at War: Essays in Memory of C. Warren Hollister* (Woodbridge [UK]: Boydell & Brewer, 2001).

43 H. R. Ellis Davidson, *The Viking Road to Byzantium* (Crows Nest: Allen & Unwin, 1976); Nicholas Charles J. Pappas, "English Refugees in the Byzantine Armed Forces: The Varangian Guard and Anglo-Saxon Ethnic Consciousness," 23 June 2014, accessed 8 June 2018, http://deremilitari.org/2014/06/english-refugees-in-the-byzantine-armed-forces-the-varangian-guard-and-anglo-saxon-ethnic-consciousness/; Mark C. Bartusis, *The Late Byzantine Army: Arms and Society 1204–1453* (Philadelphia: University of Pennsylvania Press, 1992), 273; see also the contribution by Roman Shliakhtin in this volume.

44 Walter Schaufelberger, *Marignano: Strukturelle Grenzen eidgenössischer Militärmacht zwischen Mittelalter und Neuzeit,* Schriftenreihe der Schweizerischen Gesellschaft für militärhistorische Studienreisen 11 (Frauenfeld: Ed. ASMZ im Huber Verlag, 1993), 60–83; Arnold Esch, "Mit Schweizer Söldnern auf dem Marsch nach Italien: Das Erlebnis der Mailänderkriege 1510–1515 nach bernischen Akten," *Quellen und Forschungen aus italienischen Archiven und Bibliotheken* 70 (1990): 348–440; Wolfgang Friedrich von Mülinen, *Geschichte der Schweizer-Söldner* (Bern: Huber, 1887); Werner Hirzel, *Die Schweizer in fremden Diensten*, reprint from *Zeitschrift für Heereskunde* (Coppet: Fondation pour l'histoire des Suisses à l'étranger, 1978).

45 Lawrence V. Mott, "The Battle of Malta 1283: Prelude to Disaster," in *The Circle of War in the Middle Ages: Essays on Medieval Military and Naval History*, ed. Donald J. Kagay and L. J. Andrew Villalon (Woodbridge [UK]: Boydell Press, 1999), 145–172, here 147; id., "Serving in the Fleet: Crews and Recruitment Issues in the Catalan-Aragonese Fleets during the War of the Sicilian Vespers (1282–1302)," *Journal of Medieval Encounters* 13 (2006): 56–77; Georg Christ, "News from the Aegean: Antonio Morosini Reporting on Ottoman-Venetian Relations in the Wake of the Battle of Gallipoli (Early 15th Century)," in *The Transitions from the Byzantine to the Ottoman Era in the Romania in the Mirror of Venetian Chronicles*, ed. Sebastian Kolditz (Rome: Viella, 2018), 87–110, here 93, 94 n. 29, 96.

46 Malcolm Barber, "Supplying the Crusader States: The Role of the Templars," in *The Horns of Hattin*, ed. Benjamin Z. Kedar (Jerusalem: Yad Izhak Ben-Zvi, 1992), 315–326; Teresa Sartore Senigaglia, "Latin Rhodes and European Malta: Two Fortress Islands at the Frontier of the Western World" (PhD diss., University of Heidelberg, 2013); Theresa M. Vann, "The Exchange of Information and Money between the Hospitallers of Rhodes and Their European Priories in the Fourteenth and Fifteenth Centuries," in *International Mobility in the Military Orders (Twelfth to Fifteenth Centuries): Travelling on Christ's Business*, ed. Jochen Burgtorf and Helen J. Nicholson (Cardiff: Univ. of Wales Press, 2006), 35–47.

47 On the Mamluk military diasporas, see Loiseau in this volume as well as various studies by David Ayalon; for Ottomans, see, for instance, Gilles Veinstein, "On the Ottoman Janissaries (Fourteenth-Nineteenth Centuries)," in *Fighting for a Living: A Comparative History of Military Labour, 1500–2000*, ed. Erik-Jan Zürcher (Amsterdam: Amsterdam University Press, 2014), 115–134, accessed 19 July 2021, http://www.jstor.org/stable/j.ctt6wp6pg.7.

48 Benjamin Arbel, "Venetian Cyprus and the Muslim Levant, 1473–1570," in *Kypros kai oi Staurophories* [Cyprus and the Crusades], ed. Nicholas S. H. Coureas (Nicosia: Cyprus Research Centre and SSCLE, 1995), 159–185; Georg Christ, ed., *Commerce and Crusade: The Mamluk Empire and Cyprus in a Euro-Mediterranean Perspective* (Leuven: Peeters, forthcoming).

49 David Ayalon, "The Muslim City and the Mamluk Military Aristocracy," *Proceedings of the Israel Academy of Sciences and Humanities* 2 (1968): 311–329.

50 Ciarán Óg O'Reilly, "The Irish Mercenary Tradition in the 1600s," in *Mercenaries and Paid Men: The Mercenary Identity in the Middle Ages*, ed. John France (Leiden: Brill, 2008), 383–394; cf. Tim Newark, *The Fighting Irish: The Story of the Extraordinary Irish Soldier* (London: Constable & Robinson, 2012) and note 20 above; cf. also Muríosa Prendergast, "Scots Mercenary Forces in Sixteenth Century Ireland," in *Mercenaries and Paid Men: The Mercenary Identity in the Middle Ages*, ed. John France (Leiden: Brill, 2008), 363–381.

51 See also above, note 44; Richard Feller, "Bündnisse und Söldnerdienst 1515–1798," in *Schweizer Kriegsgeschichte* ed. M. Feldmann and H. G. Wirz, part 2, no. 6, chap. 3 (Bern: Oberkriegskommissariat [Druckschriftenverwaltung], 1916), 5–60; Rudolph Jaun, Pierre Streit, and Hervé de Weck, eds., *Schweizer Solddienst: Neue Arbeiten, neue Aspekte = Service étranger suisse: Nouvelles études, nouveaux aspects; Colloque de l'Association suisse d'histoire et de science militaires (ASHSM), Centre d'histoire et de prospective militaires (CHPM), Zurich, 9–10 octobre 2009* (Birmensdorf: Schweizerische Vereinigung für Militärgeschichte und Militärwissenschaft, 2010); Hans Rudolf Fuhrer and Robert-Peter Eyer, ed., *Schweizer in "Fremden Diensten": Verherrlicht und verurteilt* (Zürich: Verlag Neue Zürcher Zeitung, 2006).; Philippe Henry, "Service étranger," in *Historisches Lexikon der Schweiz (HLS)*, Schweizerische Akademie der Geistes- und Sozialwissenschaften, 2017, accessed 11 May 2020, https://hls-dhs-dss.ch/fr/articles/008608/2017-12-08/ and Alain-Jacques Czouz-Tornare, "Reisläufer," also in *Historisches Lexikon der Schweiz (HLS)*, s.v.; see also various contributions in Erik-Jan Zürcher, ed., *Fighting for a Living*, André Holenstein's current project on Swiss military entrepreneurs "Militärunternehmertum & Verflechtung: Strukturen, Interessenlagen und Handlungsräume in den transnationalen Beziehungen des Corpus Helveticum in der frühen Neuzeit," accessed 11 August 2021, https://www.hist.unibe.ch/forschung/forschungsprojekte/militaerunternehmertum__verflechtung/index_ger.html.

52 Steve Murdoch and Alexia Grosjean, *Alexander Leslie and the Scottish Generals of the Thirty Years' War, 1618–1648*, Warfare, Society and Culture 9 (London: Pickering & Chatto, 2014); Alexia Grosjean, *An Unofficial Alliance: Scotland and Sweden 1569–1654* (Leiden: Brill, 2003); David Worthington, *Scots in Habsburg Service, 1618–1648*, History of Warfare 21 (Leiden: Brill, 2004); Matthew Glozier, "Scots in the French and Dutch Armies during the Thirty Years' War," in *Scotland and the Thirty Years' War 1618–1648*, ed. Steve Murdoch (Leiden: Brill, 2001), 117–141; James Miller, "The Scottish Mercenary as a Migrant Labourer in Europe, 1550–1650," in *Fighting for a Living: A Comparative History of Military Labour, 1500–2000*, ed. Erik-Jan Zürcher (Amsterdam: Amsterdam University Press, 2014), 169–200, accessed 19 July 2021, http://www.jstor.org.manchester.idm.oclc.org/stable/j.ctt6wp6pg.9, as well as Leonhard and Way in the same volume.

53 Matthew Glozier, *Marshal Schomberg, 1615–1690: "The Ablest Soldier of His Age": International Soldiering and the Formation of State Armies in Seventeenth-Century Europe* (Brighton: Sussex Academic Press, 2005); this work has been harshly criticized by K. A. J. McLay in *French History*, 20 (2006): 114–117 for what seem rather small mistakes. His take on military diasporas and state formation is not treated in the review.

54 See above and esp.: Matthew Glozier and David Onnekink, "Introduction," in *War, Religion and Service: Huguenot Soldiering, 1685–1713*, ed. iid., Politics and Culture in North-Western Europe 1650–1720 (Aldershot: Ashgate, 2007), 1–8, here 6.

55 Timothy Parsons, "'Kibra Is Our Blood': The Sudanese Military Legacy in Nairobi's Kibera Location, 1902–1968," *The International Journal of African Historical Studies* 30, no. 1 (1997): 87–122; Christine Liava'a, *Qaravi Na'i Tavi = They Did Their Duty: Soldiers from Fiji in the Great War* (Auckland: Polygraphia, 2009) n.v.; Teresia K. Teaiwa, "Same Sex, Different Armies: Sexual Minority Invisibility among Fijians in the Fiji Military Forces and British Army," in *Gender on the Edge: Transgender, Gay, and Other Pacific Islanders*, ed. Niko Besnier and Kalissa Alexeyeff (Honolulu: University of Hawai'i Press, 2014), 266–292.

56 Douglas Porch, "The French Foreign Legion: The Mystique of Elitism," in *Elite Military Formations in War and Peace*, ed. A. Hamish Ion and Keith Neilson (Westport, CT: Praeger, 1996), 117–133; for Britons in the Legion, more specifically in the 2e régiment étranger de parachutistes (which seems to be the preferred unit), see Simon Murray, *Legionnaire: The Real Life Story of an Englishman in the French Foreign Legion* (London: Sidgwick & Jackson, 2001 [1st ed. 1978]); Alex Lochrie, *Fighting for the French Foreign Legion: Memoirs of a Scottish Legionnaire* (Barnsley: Pen & Sword, 2009).

57 Arielli and Collins, eds., *Transnational Soldiers.*

58 Ferdinand Nicolas Göhde, "A New Military History of the Italian Risorgimento and Anti-Risorgimento: The Case of 'Transnational Soldiers'," *Modern Italy* 19, no. 1 (January 2014), 21–39; see also Marcella Pellegrino Sutcliffe, "British Red Shirts: A History of the Garibaldi Volunteers (1860)," in *Transnational Soldiers: Foreign Military Enlistment in the Modern Era*, ed. Nir Arielli and Bruce Collins (Basingstoke: Palgrave Macmillan, 2013), 202–218.

59 Cf. Arielli and Collins, eds., *Transnational Soldiers*; see also id., "Introduction," 1–12, here 1–8.

60 Lennart Westberg, Petter Kjellander and Geir Brenden, eds., *III. Germanic SS Panzer-Korps: The History of Himmler's Favourite SS-Panzer-Korps, 1943–1945* (Warwick: Helion, 2019); Nir Arielli, "When Are Foreign Volunteers Useful? Israel's Transnational Soldiers in the War of 1948 Re-Examined," *The Journal of Military History* 78, no. 2 (2014): 703–724; Miles Larmer, "Of Local Identities and Transnational Conflict: The Katangese Gendarmes and Central-Southern Africa's Forty-Years War, 1960–1999," in *Transnational Soldiers: Foreign Military Enlistment in the Modern Era*, ed. Nir Arielli and Bruce Collins (Basingstoke: Palgrave Macmillan, 2013), 160–178 and Nir Arielli and Bruce Collins, "Conclusions: Jihadists, Diaspora and Professional Contractors: The Resurgence of Non-State Recruitment since the 1980s," in *Transnational Soldiers: Foreign Military Enlistment in the Modern Era*, ed. iid. (Basingstoke: Palgrave Macmillan, 2013), 250–256.

61 Joshua Ewalt and Jessy Ohl, "'We Are Still in the Desert': Diaspora and the (De) Territorialisation of Identity in Discursive Representations of the US soldier," *Culture and Organization* 19, no. 3 (2013): 209–226.

62 See also studies in Erik-Jan Zürcher, ed., *Fighting for a Living.*

63 Benedict Richard O'Gorman Anderson, *Imagined Communities: Reflections on the Origin and Spread of Nationalism*, rev. ed. (London: Verso, 1991).

64 We thus suggest a broader but also more nuanced definition than the one William Safran proposed (groups dispersed abroad from a homeland-centre, to which they yearn to return while remaining separate from the host society), William Safran, "Diasporas in Modern Societies: Myths of Homeland and Return," *Diaspora: A Journal of Transnational Studies* 1 (1991): 83–99, summary on https://muse.jhu.edu/article/443574/summary, accessed 28 September 2017: "1) they, or their ancestors, have been dispersed from a specific original 'center' to two or more 'peripheral,' or foreign, regions; 2) they retain a collective memory, vision, or myth about their original homeland—its physical location, history, and achievements; 3) they believe that they are not—and perhaps cannot be—fully accepted by their host society and therefore feel partly alienated and insulated from it; 4) they regard their ancestral homeland as their true, ideal home and as the place to which they or their descendants would (or should) eventually return—when conditions are appropriate; 5) they believe that they

should, collectively, be committed to the maintenance or restoration of their original homeland and to its safety and prosperity; and 6) they continue to relate, personally or vicariously, to that homeland in one way or another, and their ethno-communal consciousness and solidarity are importantly defined by the existence of such a relationship. In terms of that definition, we may legitimately speak of the Armenian, Maghrebi, Turkish, Palestinian, Cuban, Greek, and perhaps Chinese diasporas at present and of the Polish diaspora of the past, although none of them fully conforms to the 'ideal type' of the Jewish Diaspora".

65 Brubaker, "The 'Diaspora' Diaspora," note 17 above.

66 Michael E. Mallett, "Mercenaries," in *Medieval Warfare: A History*, ed. Maurice Hugh Keen (Oxford: Oxford University Press, 1999), 209–229, here 210.

67 Stephen Morillo, "Mercenaries, Mamluks and Militia: Towards a Cross-Cultural Typology of Military Service," in *Mercenaries and Paid Men: The Mercenary Identity in the Middle Ages*, ed. John France (Leiden: Brill, 2008), 243–260, here 243.

68 Cf. the founding of military/chivalric orders in the service of monarchs to stabilize dynastic rule. Their crusading purpose increasingly became mere symbolism. An early example is Peter I of Lusignan's Order of the Sword (1347).

69 ᶜAbd ar-Raḥmān Ibn Khaldûn, *The Muqaddimah: An Introduction to History*, ed. Franz Rosenthal, 3 vols. (New York: Pantheon Books, 1958).

70 Dean, "Police Forces in Late Medieval Italy," 151–172.

71 Walter Schaufelberger, "*Montales et bestiales homines sine domino*: Der alpine Beitrag zum Kriegswesen in der spätmittelalterlichen Eidgenossenschaft," in *Krieg und Gebirge: Der Einfluss der Alpen und des Juras auf die Strategie im Laufe der Jahrhunderte = La guerre et la montagne: L'influence des Alpes et du Jura sur la stratégie à travers les siècles = La guerra e la montagna: L'influsso delle Alpi e del Jura sulla strategia nel corso dei secoli*, Revue internationale d'histoire militaire 65, Sonderausg. (Hauterive: Ed. Gilles Attinger, 1988), 105.

72 The ethnic eponym "Circassian" stood for a power group in the Mamluk system that also included many non-Circassians, Julien Loiseau, "Soldiers Diaspora or Cairene Nobility? The Circassians in the Mamluk Sultanate," in *Union in Separation: Diasporic Groups and Identities in the Eastern Mediterranean (1100–1800)*, ed. Georg Christ et al. (Rome: Viella, 2015), 207–217; the same use of ethnic eponyms for corporate identity was proposed for Huns, Saxons, Varangians, Brabaçons, Flemings, the Catalan company, etc. De Vries, "Medieval Mercenaries: Methodology, Definitions, and Problems."

73 Charles Oman, *A History of the Art of War in the Sixteenth Century* (London: Methuen & Co, 1937), 59, 61, 62.

74 'Telluric', i.e. earth-bound, see Carl Schmitt, *Theorie des Partisanen: Zwischenbemerkung zum Begriff des Politischen* (Berlin: Duncker & Humblot, 1963).

75 Cf. for the composition of the lance as the four men base unit: Mark Whelan, "Taxes, Wagenburgs and a Nightingale: The Imperial Abbey of Ellwangen and the Hussite Wars, 1427–1435," *The Journal of Ecclesiastical History* 72 (2021): 751–777, accessed 26 February 2022, https://doi.org/10.1017/S0022046920002602.

76 Fernando Ortiz, *Contrapunteo cubano del tabaco y el azúcar: (advertencia de sus contrastes agrarios, económicos, históricos y sociales, su etnografía y su transculturación)*, edited by Enrico Mario Santí, Letras hispánicas, 5th ed. (Madrid: CÁTEDRA/ Música Mundana Maqueda, 2002).

77 Cf. Georg Christ, "Diasporas and Diasporic Communities in the Eastern Mediterranean: An Analytical Framework," in *Union in Separation: Diasporic Groups and Identities in the Eastern Mediterranean (1100–1800)*, ed. id. et al. (Rome: Viella, 2015), 19–36.

78 Marios Philippides, "Urban's Bombard(s), Gunpowder, and the Fall of Constantinople (1453)," *Byzantine Studies/Études byzantines*, n.s., 4 (1999): 1–67.

79 Cf. Jaspert's contribution on Christian militias in North African service; for the Mamluk Empire, see the cursory remark in Stuart James Borsch, *The Black Death in Egypt and England: A Comparative Study* (Austin: University of Texas Press, 2005), 32.

References

Abels, Richard Philip, and Bernhard S. Bachrach. *The Normans and their Adversaries at War: Essays in Memory of C. Warren Hollister.* Woodbridge: Boydell & Brewer, 2001.

Adams, James N. "The Language of the Vindolanda Writing Tablets: An Interim Report." *Journal of Roman Studies* 85 (1995): 86–134.

Alföldy, Geza, Brian Dobson, and Werner Eck, eds. *Kaiser, Heer und Gesellschaft in der Römischen Kaiserzeit: Gedenkschrift für Eric Birley.* Heidelberger althistorische Beiträge und epigraphische Studien 31. Stuttgart: Steiner, 2000.

Anderson, Benedict Richard O'Gorman. *Imagined Communities: Reflections on the Origin and Spread of Nationalism.* Rev. ed. London: Verso, 1991.

Arbel, Benjamin. "Venetian Cyprus and the Muslim Levant, 1473–1570." In *Kypros kai oi Staurophories* [Cyprus and the Crusades], edited by Nicholas S. H. Coureas, 159–185. Nicosia: Cyprus Research Centre and SSCLE, 1995.

Arendt, Hannah. *The Origins of Modern Totalitarianism.* 2nd ed. New York: Meridian Books, 1958.

Arielli, Nir, and Bruce Collins. "Conclusions: Jihadists, Diaspora and Professional Contractors: The Resurgence of Non-State Recruitment since the 1980s." In *Transnational Soldiers: Foreign Military Enlistment in the Modern Era*, edited by iid., 250–256. Basingstoke: Palgrave Macmillan, 2013.

Arielli, Nir, and Bruce Collins. "Introduction: Transnational Military Service since the Eighteenth Century." In *Transnational Soldiers: Foreign Military Enlistment in the Modern Era*, edited by iid., 1–12. Basingstoke: Palgrave Macmillan, 2013.

Arielli, Nir. "When Are Foreign Volunteers Useful? Israel's Transnational Soldiers in the War of 1948 Re-Examined." *The Journal of Military History* 78, no. 2 (2014): 703–724.

Ayalon, David. "The Muslim City and the Mamluk Military Aristocracy." *Proceedings of the Israel Academy of Sciences and Humanities* 2 (1968): 311–329.

Bailey, Beth. "Soldiering as Work: The All-Volunteer Force in the United States." In *Fighting for a Living: A Comparative History of Military Labour, 1500–2000*, edited by Erik-Jan Zürcher, 581–612. Amsterdam: Amsterdam University Press, 2014. Accessed 19 July 2021. http://www.jstor.org/stable/j.ctt6wp6pg.22.

Bald, Detlef. *Der deutsche Offizier: Sozial- und Bildungsgeschichte des deutschen Offizierkorps im 20. Jahrhundert.* München: Bernard & Graefe, 1982.

Barber, Malcolm. "Supplying the Crusader States: The Role of the Templars." In *The Horns of Hattin*, edited by Benjamin Z. Kedar, 315–326. Jerusalem: Yad Izhak Ben-Zvi, 1992.

Bartusis, Mark C. *The Late Byzantine Army: Arms and Society 1204–1453.* Philadelphia: University of Pennsylvania Press, 1992.

Bettalli, Marco. *I mercenari nel mondo greco I: Dalle origini alla fine del V sec. a. C.* Pisa: Ed. ETS, 1995.

Bettalli, Marco. *Mercenari: Il mestiere della armi nel mondo greco antico; Età arcaica e classica.* Rome: Carocci, 2013.

Birley, Anthony R. "A New Tombstone from Vindolanda." *Britannia* 29 (1998): 299–306.

Birley, Anthony R. *Garrison Life at Vindolanda: A Band of Brothers.* Stroud: Tempus, 2002.

Borsch, Stuart James. *The Black Death in Egypt and England: A Comparative Study.* Austin: University of Texas Press, 2005.

Bowman, Alan K. *Life and Letters from the Roman Frontier.* New York: Routledge, 1998.

Brown, Stephen D. B. "Military Service and Monetary Reward in the Eleventh and Twelfth Centuries." *History* 74 (1989): 20–38.

Brubaker, Rogers. "The 'Diaspora' Diaspora." *Ethnic and Racial Studies* 28, no. 1 (2005): 1–19. Accessed 11 July 2017. http://www.sscnet.ucla.edu/soc/faculty/brubaker/Publications/29_Diaspora_diaspora_ERS.pdf.

Buraselis, Kostas. "Ambivalent Roles of Centre and Periphery: Remarks on the Relation of the Cities of Greece with the Ptolemies until the End of Philometor's Age." In *Centre and Periphery in the Hellenistic World*, edited by Per Bilde, Troels Endberg-Pedersen, Lise Hannestad, Jan Zahle, and Klavs Randsborg, 251–270. Aarhus: Aarhus University Press, 1993.

Chang, Wen-Chin. "Home Away from Home: Migrant Yunnanese Chinese in Northern Thailand." *International Journal of Asian Studies* 3, no. 1 (2006): 49–76.

Christ, Georg. "Diasporas and Diasporic Communities in the Eastern Mediterranean: An Analytical Framework." In *Union in Separation: Diasporic Groups and Identities in the Eastern Mediterranean (1100–1800)*, edited by Georg Christ, Franz-Julius Morche, Roberto Zaugg, Wolfgang Kaiser, Stefan Burkhardt, and Alexander Daniel Beihammer, 19–36. Rome: Viella, 2015.

Christ, Georg. "Introduction: The Rise of the Business Sphere in the Middle Ages." In *A Cultural History of Business*. Vol. 2, *Middle Ages 800–1450*, edited by id. and Catherine Casson. London: Bloomsbury, forthcoming.

Christ, Georg. "News from the Aegean: Antonio Morosini Reporting on Ottoman-Venetian Relations in the Wake of the Battle of Gallipoli (Early 15th Century)." In *The Transitions from the Byzantine to the Ottoman Era in the Romania in the Mirror of Venetian Chronicles*, edited by Sebastian Kolditz, 87–110. Rome: Viella, 2018.

Christ, Georg. "Settling Accounts with the Sultan: Cortesia, Zemechia and Venetian Fiscality in Fifteenth-Century Alexandria." In *Trajectories of State Formation across Fifteenth-Century Islamic West-Asia: Eurasian Parallels, Connections and Divergences*, edited by Jo van Steenbergen, 319–352. Leiden: Brill, 2020. Accessed 3 July 2020. https://doi-org.manchester.idm.oclc.org/10.1163/9789004431317_012.

Christ, Georg, ed. *Commerce and Crusade: The Mamluk Empire and Cyprus in a Euro-Mediterranean Perspective*. Leuven: Peeters, forthcoming.

Czouz-Tornare, Alain-Jacques. "Reisläufer." In *Historisches Lexikon der Schweiz (HLS)*. Schweizerische Akademie der Geistes- und Sozialwissenschaften, 2011. https://hls-dhs-dss.ch/ accessed 24 March 2022. s.v.

Davidson, H. R. Ellis. *The Viking Road to Byzantium*. Crows Nest: Allen & Unwin, 1976.

Dean, Trevor. "Police Forces in Late Medieval Italy: Bologna, 1340–1480." *Social History* 44, no. 22 (2019): 151–172. Accessed 24 March 2020. https://doi.org/10.1080/03071022.2019.1579974.

DeVries, Kelly. "Medieval Mercenaries: Methodology, Definitions, and Problems." In *Mercenaries and Paid Men: The Mercenary Identity in the Middle Ages*, edited by John France, 43–60. Leiden: Brill, 2008.

Dumoulin, Michel. *La communauté européenne de défense, leçons pour demain?* [The European Defence Community, Lessons for the Future?] Euroclio 15, Études et documents. Bruxelles: PIE Peter Lang, 2000.

Erdkamp, Paul, ed. *A Companion to the Roman Army*. Malden: Blackwell, 2007.

Esch, Arnold. "Mit Schweizer Söldnern auf dem Marsch nach Italien: Das Erlebnis der Mailänderkriege 1510–1515 nach bernischen Akten." *Quellen und Forschungen aus italienischen Archiven und Bibliotheken* 70 (1990): 348–440.

Ewalt, Joshua, and Jessy Ohl. "'We Are Still in the Desert': Diaspora and the (De) Territorialisation of Identity in Discursive Representations of the US Soldier." *Culture and Organization* 19, no. 3 (2013): 209–226.

Feller, Richard. "Bündnisse und Söldnerdienst 1515–1798." In *Schweizer Kriegsgeschichte* edited by M. Feldmann, and H. G. Wirz, part. 2, no. 6, chap. 3, 5–60. Bern: Oberkriegskommissariat (Druckschriftenverwaltung), 1916.

Forrest, Alan. "L'armée de l'an II: La levée en masse et la création d'un mythe républicain." *Annales historiques de la Révolution française* 335 (2004): 111–130. Accessed 16 November 2017. http://www.jstor.org/stable/41889474.

France, John, ed. *Mercenaries and Paid Men: The Mercenary Identity in the Middle Ages.* Leiden: Brill, 2008.

Fuhrer, Hans Rudolf, and Robert-Peter Eyer, eds. *Schweizer in "Fremden Diensten": Verherrlicht und verurteilt.* Zürich: Verlag Neue Zürcher Zeitung, 2006.

Glozier, Matthew. *Marshal Schomberg, 1615–1690: "The Ablest Soldier of His Age": International Soldiering and the Formation of State Armies in Seventeenth-Century Europe.* Brighton: Sussex Academic Press, 2005.

Glozier, Matthew. "Scots in the French and Dutch Armies During the Thirty Years' War." In *Scotland and the Thirty Years' War 1618–1648*, edited by Steve Murdoch, 117–141. Leiden: Brill, 2001.

Glozier, Matthew, and David Onnekink. "Introduction." In *War, Religion and Service: Huguenot Soldiering, 1685–1713*, edited by iid., 1–8. Politics and Culture in North-Western Europe 1650–1720. Aldershot: Ashgate, 2007.

Glozier, Matthew, and David Onnekink, eds. *War, Religion and Service: Huguenot Soldiering, 1685–1713.* Politics and Culture in North-Western Europe 1650–1720. Aldershot: Ashgate, 2007.

Göhde, Ferdinand Nicolas. "A New Military History of the Italian Risorgimento and Anti-Risorgimento: The Case of 'Transnational Soldiers'." *Modern Italy* 19, no. 1 (January 2014): 21–39.

Griffith, Guy T. *The Mercenaries of the Hellenistic World.* Cambridge: Cambridge University Press, 1935.

Grosjean, Alexia. *An Unofficial Alliance: Scotland and Sweden 1569–1654.* Leiden: Brill, 2003.

Haffner, Sebastian. *Preußen ohne Legende.* Herrsching: Manfred Pawlak, 1985.

Haynes, Ian. *Blood of the Provinces: The Roman Auxilia and the Making of Provincial Society from Augustus to the Severans.* Oxford: Oxford University Press. 2013.

Henry, Philippe. "Service étranger." In *Historisches Lexikon der Schweiz (HLS).* Schweizerische Akademie der Geistes- und Sozialwissenschaften, 2017. Accessed 11 May 2020. https://hls-dhs-dss.ch/fr/articles/008608/2017-12-08/. s.v.

Hirzel, Werner. *Die Schweizer in fremden Diensten.* Coppet: Fondation pour l'histoire des Suisses à l'étranger, 1978.

Holenstein, André. "Militärunternehmertum & Verflechtung: Strukturen, Interessenlagen und Handlungsräume in den transnationalen Beziehungen des Corpus Helveticum in der frühen Neuzeit (Forschungsprojekt Ordinariat Holenstein)." Accessed 11 August 2021. https://www.hist.unibe.ch/forschung/forschungsprojekte/militaerunternehmertum__verflechtung/index_ger.html.

Hosler, John D. "Revisiting Mercenaries under Henry Fitz Empress, 1167–1188." In *Mercenaries and Paid Men: The Mercenary Identity in the Middle Ages*, edited by John France, 33–42. Leiden: Brill, 2008.

Ibn Khaldūn, ʿAbd ar-Raḥmān. *The Muqaddimah: An Introduction to History*, edited by Franz Rosenthal. 3 vols. New York: Pantheon Books, 1958.

Isaac, Steven. "The Problem with Mercenaries." In *The Circle of War in the Middle Ages: Essays on Medieval Military and Naval History*, edited by Donald J. Kagay and L. J. Andrew Villalon, 101–110. Woodbridge: Boydell Press, 1999.

Jaspers, Karl. *Vom Ursprung und Ziel der Geschichte.* Basel: Schwabe, 2017; engl.: *Origin and Goal of History.* Routledge Revivals, 2011.

Jaun, Rudolph, Pierre Streit, and Hervé de Weck, eds. *Schweizer Solddienst: Neue Arbeiten, neue Aspekte = Service étranger suisse: Nouvelles études, nouveaux aspects; Colloque de l'Association suisse d'histoire et de science militaires (ASHSM), Centre d'histoire et de prospective militaires (CHPM), Zurich, 9–10 octobre 2009*. Birmensdorf: Schweizerische Vereinigung für Militärgeschichte und Militärwissenschaft, 2010.

Jungholt, Thorsten. "Bundeswehr wirbt um Schulabbrecher und Ausländer." *Welt*, December 1, 2016. Accessed 19 April 2018. https://www.welt.de/politik/deutschland/article159898506/Bundeswehr-wirbt-um-Schulabbrecher-und-Auslaender.html.

Kitchen, Martin. "Elites in Military History." In *Elite Military Formations in War and Peace*, edited by A. Hamish Ion, and Keith Neilson, 7–30. Westport: Praeger, 1996.

Kumar, Krishan. "Nation-States as Empires, Empires as Nation-States: Two Principles, One Practice?" *Theory and Society* 39, no. 2 (2010): 119–143.

Larmer, Miles. "Of Local Identities and Transnational Conflict: The Katangese Gendarmes and Central-Southern Africa's Forty-Years War, 1960–1999." In *Transnational Soldiers: Foreign Military Enlistment in the Modern Era*, edited by Nir Arielli and Bruce Collins, 160–178. Basingstoke: Palgrave Macmillan, 2013.

Launey, Marcel, *Recherches sur les armées hellénistiques*. Bibliothèque des écoles françaises d'Athènes et de Rome 169. Paris: De Boccard, 1949/1950; 2nd ed. 1987: Réimpression avec addenda et mise à jour, en postface par Yvon Garlan, Philippe Gauthier, Claude Orrieux.

Le Bohec, Yann. *L'armée romaine sous le Haut-Empire*. 3rd ed. Paris: Picard, 2002.

Le Bohec, Yann. *L'armée romaine sous le Bas-Empire*. Antiquité—Synthèses 11. Paris: Picard, 2006.

Lepage, Jean-Denis G. G. *Medieval Armies and Weapons in Western Europe*. Jefferson: McFarland, 2005.

Liava'a, Christine. *Qaravi Na'i Tavi = They Did Their Duty: Soldiers from Fiji in the Great War*. Auckland: Polygraphia, 2009.

Lochrie, Alex. *Fighting for the French Foreign Legion: Memoirs of a Scottish Legionnaire*. Barnsley: Pen & Sword, 2009.

Loiseau, Julien. "Soldiers Diaspora or Cairene Nobility? The Circassians in the Mamluk Sultanate." In *Union in Separation: Diasporic Groups and Identities in the Eastern Mediterranean (1100–1800)*, edited by Georg Christ, Franz-Julius Morche, Roberto Zaugg, Wolfgang Kaiser, Stefan Burkhardt, and Alexander Daniel Beihammer, 207–217. Rome: Viella, 2015.

MacMullen, Ramsay. "The Legion as a Society." *Historia* 33, no. 4 (1984): 440–456.

Mallett, Michael E. "Mercenaries." In *Medieval Warfare: A History*, edited by Maurice Hugh Keen, 209–229. Oxford: Oxford University Press, 1999.

Mann, John C. *Legionary Recruitment and Veteran Settlement during the Principate*. Institute of Archaeology, London; Occasional Publication 7. London: Routledge, 1983.

Marinovič, Ludmila P. *Le mercenariat grec au IVe siècle av n.è. et la crise de la polis*. Paris: Les Belles Lettres, 1988.

McLay, K. A. J. "Review of M. Glozier, Marshal Schomberg…" *French History* 20, no. 1 (2006): 114–117.

Mérand, Frédéric. "Dying for the Union? Military Officers and the Creation of a European Defence Force." *European Societies* 5, no. 3 (2012): 253–282. Accessed 24 March 2022. https://doi.org/10.1080/1461669032000111306.

Miller, James. "The Scottish Mercenary as a Migrant Labourer in Europe, 1550–1650." In *Fighting for a Living: A Comparative History of Military Labour, 1500–2000*, edited by Erik-Jan Zürcher, 169–200. Amsterdam: Amsterdam University Press, 2014. Accessed 19 July 2021, http://www.jstor.org.manchester.idm.oclc.org/stable/j.ctt6wp6pg.9.

Moore, Robert Ian. *The Formation of a Persecuting Society: Power and Deviance in Western Europe, 950–1250.* Oxford: Basil Blackwell, 2006.

Morillo, Stephen. "Mercenaries, Mamluks and Militia: Towards a Cross-Cultural Typology of Military Service." In *Mercenaries and Paid Men: The Mercenary Identity in the Middle Ages*, edited by John France, 243–260. Leiden: Brill, 2008.

Mott, Lawrence V. "The Battle of Malta 1283: Prelude to Disaster." In *The Circle of War in the Middle Ages: Essays on Medieval Military and Naval History*, edited by Donald J. Kagay and L. J. Andrew Villalon, 145–172. Woodbridge: Boydell Press, 1999.

Mott, Lawrence V. "Serving in the Fleet: Crews and Recruitment Issues in the Catalan-Aragonese Fleets during the War of the Sicilian Vespers (1282–1302)." *Journal of Medieval Encounters* 13 (2006): 56–77.

Mülinen, Wolfgang Friedrich von. *Geschichte der Schweizer-Söldner.* Bern: Huber, 1887.

Murdoch, Steve, and Alexia Grosjean. *Alexander Leslie and the Scottish Generals of the Thirty Years' War, 1618–1648.* Warfare, Society and Culture 9. London: Pickering & Chatto, 2014.

Murray, Simon. *Legionnaire: The Real Life Story of an Englishman in the French Foreign Legion.* London: Sidgwick & Jackson, 2001; 1st ed. 1978.

Newark, Tim. *The Fighting Irish: The Story of the Extraordinary Irish Soldier.* London: Constable & Robinson, 2012.

O'Reilly, Ciarán Óg. "The Irish Mercenary Tradition in the 1600s." In *Mercenaries and Paid Men: The Mercenary Identity in the Middle Ages*, edited by John France, 383–394. Leiden: Brill, 2008.

Oman, Charles. *A History of the Art of War in the Sixteenth Century.* London: Methuen & Co, 1937.

Ortiz, Fernando. *Contrapunteo cubano del tabaco y el azúcar (advertencia de sus contrastes agrarios, económicos, históricos y sociales, su etnografía y su transculturación).* Edited by Enrico Mario Santí. Letras hispánicas, 5th ed. Madrid: CÁTEDRA/Música Mundana Maqueda, 2002.

Palme, Bernhard. "Feldarmee und Grenzheer: Das römische Militär in der Spätantike." In *Krieg in der antiken Welt*, edited by Gerfried Mandl and Ilia Steffelbauer, 85–113. Essen: Magnus, 2007.

Pappas, Nicholas Charles J. "English Refugees in the Byzantine Armed Forces: The Varangian Guard and Anglo-Saxon Ethnic Consciousness." June 23, 2014. Accessed 8 June 2018. http://deremilitari.org/2014/06/english-refugees-in-the-byzantine-armed-forces-the-varangian-guard-and-anglo-saxon-ethnic-consciousness/.

Parsons, Timothy. "'Kibra Is Our Blood': The Sudanese Military Legacy in Nairobi's Kibera Location, 1902–1968." *The International Journal of African Historical Studies* 30, no. 1 (1997): 87–122.

Pellegrino Sutcliffe, Marcella. "British Red Shirts: A History of the Garibaldi Volunteers (1860)." In *Transnational Soldiers: Foreign Military Enlistment in the Modern Era*, edited by Nir Arielli and Bruce Collins, 202–218. Basingstoke: Palgrave Macmillan, 2013.

Philippides, Marios. "Urban's Bombard(s), Gunpowder, and the Fall of Constantinople (1453)." *Byzantine Studies/Études byzantines*, n.s., 4 (1999): 1–67.

Pomeranz, Kenneth. "Social History and World History: From Daily Life to Patterns of Change." *Journal of World History* 18 (2007): 69–98. Accessed 10 August 2021. https://www.jstor.org/stable/20079411.

Porch, Douglas. "The French Foreign Legion: The Mystique of Elitism." In *Elite Military Formations in War and Peace*, edited by A. Hamish Ion and Keith Neilson, 117–133. Westport: Praeger, 1996.

Prendergast, Muríosa. "Scots Mercenary Forces in Sixteenth Century Ireland." In *Mercenaries and Paid Men: The Mercenary Identity in the Middle Ages*, edited by John France, 363–381. Leiden: Brill, 2008.

Rosinski, Herbert. *Die Deutsche Armee: Eine Analyse*. Düsseldorf: Econ Verlag, 1970.

Safran, William. "Diasporas in Modern Societies: Myths of Homeland and Return." *Diaspora: A Journal of Transnational Studies* 1 (1991): 83–99. Accessed 28 September 2017. https://muse.jhu.edu/article/443574/summary.

Sänger, Patrick. "Augustus—Herr über 28 Legionen: Das militärische Erbe der Republik und die kaiserzeitliche Armee." In *Krieg in der antiken Welt*, edited by Gerfried Mandl and Ilia Steffelbauer, 64–84. Essen: Magnus, 2007.

Sänger, Patrick, and Sandra Scheuble-Reiter, eds. *Söldner und Berufssoldaten in der griechischen Welt: Politische und soziale Gestaltungsräume*. Historia Einzelschriften 269. Stuttgart: Franz Steiner, 2022.

Schaufelberger, Walter. "*Montales et bestiales homines sine domino*: Der alpine Beitrag zum Kriegswesen in der spätmittelalterlichen Eidgenossenschaft." In *Krieg und Gebirge: Der Einfluss der Alpen und des Juras auf die Strategie im Laufe der Jahrhunderte = La guerre et la montagne: L'influence des Alpes et du Jura sur la stratégie à travers les siècles = La guerra e la montagna: L'influsso delle Alpi e del Jura sulla strategia nel corso dei secoli*. Revue internationale d'histoire militaire 65, Sonderausg. Hauterive: Ed. Gilles Attinger, 1988.

Schaufelberger, Walter. *Marignano: Strukturelle Grenzen eidgenössischer Militärmacht zwischen Mittelalter und Neuzeit*. Schriftenreihe der Schweizerischen Gesellschaft für militärhistorische Studienreisen 11. Frauenfeld: Ed. ASMZ im Huber Verlag, 1993.

Schelbert, Leo. "Swiss Diaspora." In: *Encyclopedia of Diasporas: Immigrant and Refugee Cultures Around the World*, edited by Melvin Ember, Carol R. Ember, and Ian Skoggard. S. v. New York: Springer, 2005.

Schenk, Jochen. "'New Wars' and Medieval Warfare: Some Terminological Considerations." *In Journal of Medieval Military History* 14 (2016): 149–162.

Schmitt, Carl. *Theorie des Partisanen. Zwischenbemerkung zum Begriff des Politischen*. Berlin: Duncker & Humblot, 1963.

Schult, Hermann. "Ein inschriftlicher Beleg für 'Plethi'?" *Zeitschrift des Deutschen Palästina-Vereins* 81, no. 1 (1965): 74–79.

Sartore Senigaglia, Teresa. "Latin Rhodes and European Malta: Two Fortress Islands at the Frontier of the Western World." PhD diss., University of Heidelberg, 2013.

Sheehan, James J. *Where Have All the Soldiers Gone? The Transformation of Modern Europe*. Boston: Mariner Books, 2009.

Simpson, Grace. *Britons and the Roman Army: A Study of Wales and the Southern Pennines in the 1st–3rd Centuries*. London: Gregg, 1964.

Singer, Itamar. "Egyptians, Canaanites, and Philistines in the Period of the Emergence of Israel." In *From Nomadism to Monarchy: Archaeological and Historical Aspects of Early Israel*, edited by Israel Finkelstein and Nadav Na'aman, 282–338. Jerusalem: Israel Exploration Society, 1994.

Sombart, Werner. *Der Bourgeois: Zur Geistesgeschichte des modernen Wirtschaftsmenschen*. München: Duncker & Humblot, 1913.

Sombart, Werner. *Krieg und Kapitalismus. Studien zur Entwicklungsgeschichte des modernen Kapitalismus 2*. München: Verlag von Duncker & Humblot, 1913.

Szende, Katalin, and Georg Christ. "Trans-European Diasporas: Migration, Minorities, and the Diasporic Experience in East Central Europe and the Eastern Mediterranean from the Late Antiquity to the Early Modern Era—Project Report." *Annual of Medieval Studies at CEU* 20 (2014): 296–305.

Teaiwa, Teresia K. "Same Sex, Different Armies: Sexual Minority Invisibility among Fijians in the Fiji Military Forces and British Army." In: *Gender on the Edge: Transgender, Gay, and Other Pacific Islanders*, edited by Niko Besnier and Kalissa Alexeyeff, 266–292. Honolulu: University of Hawai'i Press, 2014.

Trundle, Matthew. *Greek Mercenaries: From the Late Archaic Period to Alexander.* London: Routledge, 2004.

Vann, Theresa M. "The Exchange of Information and Money between the Hospitallers of Rhodes and Their European Priories in the Fourteenth and Fifteenth Centuries." In *International Mobility in the Military Orders (Twelfth to Fifteenth Centuries): Travelling on Christ's Business*, edited by Jochen Burgtorf and Helen J. Nicholson, 35–47. Cardiff: University of Wales Press, 2006.

Veinstein, Gilles. "On the Ottoman Janissaries (Fourteenth-Nineteenth Centuries)." In *Fighting for a Living: A Comparative History of Military Labour, 1500–2000*, edited by Erik-Jan Zürcher, 115–134. Amsterdam: Amsterdam University Press, 2014. Accessed 19 July 2021. http://www.jstor.org/stable/j.ctt6wp6pg.7.

Villegas, Mark. "Currents of Militarization, Flows of Hip-Hop: Expanding the Geographies of Filipino American Culture." *Journal of Asian American Studies* 19, no. 1 (2016): 25–46.

Westberg, Lennart, Petter Kjellander and Geir Brenden, eds. *III. Germanic SS Panzer-Korps: The History of Himmler's Favourite SS-Panzer-Korps, 1943–1945.* Warwick: Helion, 2019.

Whelan, Mark. "Taxes, Wagenburgs and a Nightingale: The Imperial Abbey of Ellwangen and the Hussite Wars, 1427–1435." *The Journal of Ecclesiastical History* 72 (2021): 751–777. Accessed 26 February 2022. https://doi.org/10.1017/S0022046920002602.

Worthington, David. *Scots in Habsburg Service, 1618–1648.* History of Warfare 21. Leiden: Brill, 2004.

Zürcher, Erik-Jan, ed. *Fighting for a Living: A Comparative History of Military Labour, 1500–2000.* Amsterdam: Amsterdam University Press, 2014.

1

MILITARY DIASPORAS IN AN ACHAEMENID PERSPECTIVE

Hilmar Klinkott

The concept of a military diaspora is rooted in the idea that military forces served outside the "homeland". These military forces were garrisoned at "outposts" overseas, which can be exemplified by Spartan garrisons on the Aegean Islands or on the southern and western coast of Asia Minor under the command of harmosts (military commanders). Consequently, this concept assumes that a perception and individual identification with a kind of "homeland" really existed. In general, the Achaemenid empire was a multicultural and multiethnic entity that recruited soldiers from all over its territory. All these multiethnic forces, in both army and navy, served in different ways: On the one hand, there were royal forces, directly under the command of the Great King and the *strategoi* he appointed.[1] These units were hired in the whole empire and were rarely stationed in their regions of origin. Instead, they were highly flexible in terms of their deployment and location and independent from regular administration in the satrapies. Therefore, Xenophon remarks that "as Cyrus then effected his organization, even so unto this day all garrisons under the king, are kept up" as a precautious measurement against satrapal revolts.[2]

On the other hand, in the satrapies, the governors (the satraps) possessed military forces of their own, usually recruited from the local population and used to guarantee the security of the province. These troops also seemed to be part of the military array the satrap had to convene and bring to a central, trans-satrapal meeting point. As known for the array of the "people at the sea", in this case, the satraps and their troops were placed under the command of the so-called *karanos*.[3] These military collecting points seemed to be important instruments in the case of royal empire-wide arrays; the last one happened under Xerxes I for the western expedition to Greece.

Finally, the Great King and his satraps possessed forces, occasionally hired and partially paid by "private" income, usually called mercenaries.[4] It is known

DOI: 10.4324/9781003245568-2

that above all, the satraps in western Asia Minor possessed their own, difficult to control military power in addition to the official royal and satrapal contingents.[5]

With this structure in mind, military diasporas could take place on various levels and in different conditions. It is clear that the related diaspora feeling is closely linked to the individual perspective and articulation.[6] It does not have to be linked to foreign regions but can also exist within the borders of the empire due to different ethnic contexts and identities. But does the concept of a "military diaspora" make sense in the perspective of an empire like that of the Achaemenids? This question shall be investigated by identifying two groups of potential military diasporas associated with the Achaemenid empire—ethnically defined (non-Persian) troops on the one hand and Persian contingents on the other hand—and by comparing the findings with the world view of the Achaemenid kings.

Ethnically Defined (Non-Persian) Military Diasporas within the Empire?

Such an approach lends itself well to understanding the so-called Greek mercenaries, serving far away from Greece and distributed all over the empire. In particular, the invasion of Alexander provides detailed information on the number and stationing of Greek soldiers in the Achaemenid Empire.[7] The fact that most Greek "mercenaries" tried to return to Greece after Alexander refused to incorporate them into his own army proves the perception of an original Greek "homeland".[8] On the other hand, Greek mercenaries were provided with private land by the Great King. Memnon, Mentor, Chabrias, or Themistocles, for example, received huge estates in Asia Minor.[9] Certainly, from a Greek point of view, these soldiers lived in a diaspora;[10] but from an Achaemenid perspective, they were integrated into the empire. Xen., *An.* 7.8.9–18 shows that these settled Greeks, such as Gongylos from Eretria,[11] or Procles, a descendent of Demaratos,[12] seem to be estate-holders with local military power in the same—legal and technical, but certainly not social—way as the settled Persians Asidates and Itamenes (see below).

In view of the Greek mercenaries, it could be suggested that the Great King was interested in a substantial social and administrative incorporation of armed forces through permanent settlement. In Babylonia, the so-called (elam.) *kurtaš*/ *(bab.) gardu*[13] received some land for their maintenance and were structured and organized in military categories—the "bowland", "horse land", and "chariot land" (a model that could partially be compared with the Hellenistic *kleruchoi*). These Babylonian military settlers were included in larger territorial and fiscal units, so-called *ḫaṭru*, theoretically with the duty to keep on standing in military preparedness.[14] Even in the case of the Babylonian *kurtaš*, it is, however, hard to define military diasporas: Greek, Carian, Lydian (from Sardis), Phrygian, or Tyrian settlements in Babylonia can be seen as diasporic,[15] mostly from a western perspective, but the basic military character of these settlements cannot be proven.[16] Thus, the Babylonian and Persian *kurtaš* settlements certainly represent a suitable system for military incorporation, in general, but not a specific and exclusive model for the settling of diasporic military units, in particular. This is especially the case at the edges of the empire, for example, in western Asia Minor, where land-grants like the *kurtaš* model are attested,[17] and it certainly might have been connected with the local defence system of fortresses and garrisons.[18]

Royal troops regularly served in a kind of military diaspora, far away from local origins, but still inside the Achaemenid empire. Late Babylonian cuneiform tablets inform us about the movement of obviously royal detachments in the territory of the empire.[19] In Egypt, the papyri from Elephantine probably attest to royal detachments in the garrisons of Syene and Elephantine; they were comprised of Caspian, Chwarezmian, Egyptian, Aramean/Judean, and Babylonian detachments under Persian and Median commanders.[20] Sales contracts, site boundaries, and marriage contracts illustrate that these foreign soldiers obviously stayed for a longer time and settled in that location.[21] The military conditions as well as the status of their employment unfortunately remain unclear. Nevertheless, it is evident that these troops stayed far away from their original homeland. Individual foreign soldiers—Greek, Judean, and Nubians (Medjai)—are also attested (besides local Egyptian troops) for the north-eastern and southern frontiers of Egypt.[22] In general, all these foreign soldiers were recruited from a broader regional context, which can hardly be described as a system of military service in a diaspora-like character. Pétigny pointed out that the sites of the Egyptian garrisons and fortresses in the Achaemenid period continued on much older installations.[23] Therefore, these garrisons were not newly installed "Persian" outposts in the diaspora of the empire, but in an architectural sense, traditional fortifications of the Egyptian territory. Since it is unclear when the occupying forces were recruited, these garrisons cannot be used as a reliable example of diasporas, which would have been a result of Persian recruitment policies.

At least in the case of the Jewish detachments at Syene, it might be possible to trace their roots. The papyri from Elephantine prove that they were comprised of settled members of the local Jewish community, which used the ethnic designation Jew/Jewish in its proper sense for official legal contracts.[24] Unfortunately, it is not clear from the texts whether we are dealing with a diaspora for military reasons or a longer established religious and/or ethnic diaspora group that also provided the local garrison. An Aramaic papyrus from Elephantine under Darius II emphasizes that the Jahû-temple was built "in the days of the kings of Egypt" and was protected by Cambyses, "when Cambyses entered Egypt".[25] If this statement is historically accurate, the roots of the Jewish community in Elephantine would exist before the Persian conquest. In this case, the Persian garrison would have taken over older structures of military border protection and recruited the garrison's crew from the local population, namely from the long-established Jewish community in Elephantine. Consequently, although the example of Elephantine can prove the existence of a Jewish (military) diaspora in southern Egypt, it cannot prove a specific military diaspora founded by the Achaemenids.

With regard to the garrison system in Asia Minor,[26] the so-called Hyrcanean horse-riders seem to provide another example for soldiers detached from their homeland (in the south of the Caspian Sea). Garrisoned in Lydia in western Asia Minor, they did not serve under satrapal command: According to Xen., *An.* 7.8.12–19, it was not the satrap, but the Persian landholders Asidates and Itamenes who brought military support with the additional detachment of the Hyrcanian riders.[27] When Xenophon attacked the land estates of Asidates centred about the so-called *pyrgoi/tyrseis* ("Turmgehöft"), he had to besiege it because of the strength of its fortification and the many fighters on the towers (ἄνδρας πολλοὺς καὶ μαχίμους ἔχουσα).[28] Unfortunately, Xenophon does not say if these fighters on the tower were locals or Persians like Asidates. By light and sound signals,[29] they informed their Persian neighbour, Itamenes, who assisted Asidates with his own military forces, including heavily armed Assyrians, Hyrcanian riders, and approximately 80 royal mercenaries from Komania, as well as 800 lightly armed soldiers, and others from Parthenion, Apollonias and adjacent sites, and some more riders.[30] The Assyrian hoplites and the Hyrcanian riders from Lydian Komania are difficult to explain in detail: Xenophon's report remains unclear about the exact status, circumstances, and conditions of their service. On the one hand, they are mentioned in the context of "private" troops of the Persian landholder (τὴν ἑαυτοῦ δύναμιν) and, on the other hand, of "mercenaries" under royal command (βασιλέως μιθοφόροι). Therefore, these Assyrian and Hyrcanian soldiers theoretically could have been part of Itamenes' "own" troops, special units of royal mercenaries, or regular members of satrapal or royal detachments.

Apparently, Cyrus II settled some of his Median and Hyrcanean allies in Mesopotamia, but probably also in Asia Minor.[31] Thus, Hyrcanean riders must have been garrisoned for a long time in Lydia, if not constantly.[32] However, whether we can identify them as a settled military diaspora is open to debate because it is unclear if the members of the troop stayed permanently at this place or rotated at regular intervals. If there are patterns behind these Hyrcaneans similar to those considered in connection with the Jewish soldiers of the garrison at Elephantine, we should expect permanent and enduring settlements that could explain the concept of the so-called Hyrcanean plain, which is well documented in literary sources.[33] Unfortunately, an epigraphic proof of the ethnonym is still missing. It is also questionable if the term "Hyrcaneans" was still used in an ethnic sense. Comparable with the designation "Perses tes epigones" that occurred in Ptolemaic Egypt,[34] we have to clarify if "Hyrcanean rider" and "Assyrian hoplite" are ethnic identifications or terms of military specification. In naming a special military unit without any particular ethnic connotation, the conceptual background for a diaspora-like understanding, the long distance from the individual homeland, would be missing. However, if the fortified estates of Asidates and Itamenes were composed of ethnic contingents, they could in principle be classified as military diaspora(s). But does the concept of "diaspora" make any sense at all in political terms?

In the context of a multicultural empire like the Achaemenid one, it is difficult to define "homeland" and "diaspora". Both terms are connected with the perception of inside and outside. Chwarezmians in Egypt, Carians in Babylonia, or Hyrcaneans in Lydia lived outside their original "homelands" (Chwarezmia,[35] Caria, Hyrcania). Even if these ethnic labels were used in their proper sense, these garrison communities nevertheless resided within the political entity of the Achaemenid empire. Thus, the political "homeland" differs from the cultural, ethnic, or linguistic one, and vice versa from the perception of a diaspora. Finally, due to the lack of sources, it is impossible to define the transition process when foreignness is transformed into the integrative understanding of a new home.

Beyond that, the particular "military" background intensifies the legal and political aspects of a diaspora understanding: In a narrow sense, a military diaspora is intrinsically linked to a political entity. Therefore, Achaemenid troops in the territory of the empire are not legally in a diaspora, regardless of the large distances between the soldiers and their country of origin.[36] Of course, the individual self-perception of the soldiers may be independent of their legal and political status, but due to the lack of sources, this is difficult to assess.

A Late-Babylonian text from the Ebbabar archive of Sippar mentions some cavalrymen retuning from Egypt to Babylonia. The text from the fourth year of Darius (I.) speaks of "Tattanu and his horsemen, who returned from Egypt".[37] Another part of the same text tells of "38 shekel of silver for Šamaš-iddin and his horsemen who have come back from Egypt".[38] However, the context remains unclear: The text gives no information on how long the Babylonian horsemen stayed abroad, if they served in a garrison, as regulars or mercenaries, or if they participated in a military campaign.[39] Furthermore, the administrative text contains no reference to the perceptions and feelings of the soldiers, neither for Egypt nor, after their return, for Babylonia.

Persian Military Diasporas within the Empire and Beyond?

As we have seen, it is a logical consequence of an empire as vast as the Achaemenid one that long distances created a feeling of diaspora, and this circumstance should also be considered for Persian troops stationed outside their homeland. Before we turn to groups of Persian soldiers stationed within the borders of the Achaemenid empire, it should be stressed that there is actually no information on Persian troops stationed outside the empire or overseas, as known from the Lacedaimonian harmost system after the Peloponnesian war.[40] Of course, there were Persian soldiers who operated overseas and outside the empire; the best-known example is the expeditions ordered by Darius I and Xerxes I. Democedes of Croton, for example, launched a naval expedition from Sidon, which sailed to the Greek islands and mainland coasts and reaching as far as Tarentum in Italy and the Illyrian Japyges.[41] Democedes was joined by Phoenicians and Persians, apparently preparing the royal campaigns against Greece through the expedition report.[42] In the same way, Sataspes twice launched a naval expedition exploring the sea route westwards around Africa.[43] Although Sataspes as well as the Persian companions of Democedes served abroad on military missions, it would be going too far to define these expeditions, even under military protection, as service in a "diaspora"-like environment.

As to (hypothetical) diasporas beyond the empire's frontier, it has to be taken into consideration that according to the Achaemenid ideology of universal rulership,[44] there were no borders to the realm. By contrast, the Achaemenids began to create their own identity, of which Persia, as their homeland, was an important factor, but not the only one.[45] The royal representative inscriptions from Naqsh-i Rustam perfectly illustrate the Great King's understanding that the Ahura Mazda bestowed him as "king of the countries containing all races, king on this earth even far off".[46] In other words, in a royal understanding, there was no region outside the rule of the Great King.[47] Consequently, in principle, there were no external territories that were perceived as diasporic parts of the empire. Nevertheless, the old Persian royal inscriptions talk of some overseas regions. The so-called *dahyava*-list explicitly mentions "the Yauna beyond the sea" in DPe § 2, DSe § 4, and XPh § 3 and "the Saka beyond the sea" in DNa § 3, DSe § 4, and A^3Pb 24 as regions ruled by the empire.[48] Therefore, there was an understanding of overseas lands in the perception of the Great King—especially under the rule of Darius I and Xerxes I—although not beyond the empire but as an integral part of it.

Let us now take a look at the situation of the Persian troops within the empire and start with the military protection of the border. We know that a system of military outposts existed at the Aegean border of the empire: In the context of Persian long-distance communication, we are informed about so-called *naustathmoi*—naval posts, continuously garrisoned and in constant readiness for fire and light signals.[49] Unfortunately, there is no information about what kind of soldiers were stationed in these outposts—troops under the command of satraps or of special royal *strategoi*, local soldiers, or detachments from all over the empire—and how long a single garrison had been in service.

Xen., *An.* 7.8.12–15 informs us in detail that small military units, a network of local forts and fortified estates protected the western satrapies of Asia Minor.[50] This system of locally organized protection through fortified estates seems to be symptomatic of the Achaemenid Empire; we encounter it in Syene in Egypt as well as in Ionia, Lydia, and Lycia.[51] In the event of an attack, as Xenophon did with the siege of the Asidates estate, the local neighbours provided military assistance by quickly and efficiently bringing together troops from neighbouring garrisons. Only in a second step of defence was the next garrison in the hinterland informed and sent its available military support. Usually, these were royal forces, stationed in so-called *phrouria*, obviously under command of a Persian officer.

Despite Xenophon's episode, it is quite doubtful whether there really existed a system and dense network of military protection all along the empire's border. The defence systems rather focused either on local cities and settlements or on strategic positions controlling important routes in the empire.[52] Maybe, the garrisons in the border cities have to be connected with the title "those who go to the towns", known from the Aramaic papyri of Bactria.[53] Anyway, these *phrouria* were not oversea diasporas but were instead inside the Achaemenid empire and part of its administration in the satrapies.

A highly visible group of military personnel in diasporic positions, but within the empire, are the Persian officials in the satrapies. Besides the Persians in the administration of the satrapy,[54] of course, the Persian military commanders in the satrapies must also be considered. In Asia Minor, the *phrourarchoi*—the commanders of the garrisons—were generally Persians with Iranian names;[55] in Egypt, they appear under the title *rab ḥaylā'*,[56] in Arachosia, they are designated as *sgn byrt'*.[57] Elephantine provides a good local example for individual Persians, who were sent as governors and to command local soldiers.[58] This is confirmed by a demotic papyrus from Memphis, which contains the request of an Egyptian called Pediamūn regarding the delivery of lances to his "master", the Persian "chief of the army".[59] In Babylonia as well as in other satrapies, Persians were known as landholders.[60] Especially in Lydia, where there were Persian noblemen with large private estates.[61] The most famous one is certainly the former satrap Tissaphernes who held Ionia as a huge "private" possession under the satrapal government of the younger Cyrus.[62] Estate-holders like the already mentioned Asidates and Itamenes in Lydia represent a level of Persians on lower official positions, but with intensive local embedding. N. V. Sekunda assumed that these Persian noblemen made up the cavalry corps of the satraps.[63] Certainly, these Iranian landholders were part of a Persian community constituting the satrapal court,[64] as Xen., *Cyr.* 8.6.10–13 explains:

> (10) And he gave order to all the satraps he sent out to imitate him in everything that they saw him do: they were in the first place, to organize companies of cavalry and charioteers from the Persians who went with them and from the allies; to require as many as received lands and palaces to attend at the satrap's court (…) (11) "And whoever I find has the largest number of chariots to show and the largest number of the most efficient horsemen in proportion to his power", Cyrus added, "him will I honor as a valuable ally and as a valuable fellow-protector of the sovereignty of the Persians and myself. (…) (13) I try to do everything that I say you ought to do, And even as I bid you follow my example, so do you instruct those whom you appoint to office to follow yours".[65]

In this sense, the satrapal residences are imitations of the royal court in administrative, but also cultural and social terms, even literary, as illustrated by the poem of Symmachus for Xanthus in Lycia.[66] But even in these cases, the character and perception of a military diaspora cannot be proven. Persian officers and commanders within the empire were not in a diaspora, and their personal viewpoint was not supported by the sources. On the contrary, the Persian communities in the satrapies may have been—in Xenophon's sense—reflections of the residences of the Great King on a micro-political or provincial level. Indeed, they were not only pure imitations of the royal court (πάντα μιμεῖθαι)[67] but dynamic hubs of Persian domination on their own.[68] In this function, the residences with the satrapal court may have been centres which in principle redefined not a Persian, but an Achaemenid elite—with Persian participation—on a multicultural and multiethnic background.[69] Even Xenophon is very clear on this particular point saying: "They (i.e. the satraps in all parts of the empire) were in the first place, to organize companies of cavalry and charioteers from the Persians who went with them and from the allies".[70]

An Achaemenid palace found in the Caspian region confirms that even the Black Sea was fully exploited as a maritime hub to connect the adjacent territories.[71] Although most details remain unclear, the architecture clearly belongs to the so-called Achaemenid court style.[72] As a consequence, the owner of the building would have had close ties with the royal court, probably in some official capacity.[73] It is obvious that this official or Persian aristocratic owner and the building itself were under military protection.[74] It can be assumed that there was a garrison near the palace, but this cannot be proven. The fortress in Meydancikkalı in Cilicia, known for its reliefs in Persian court style, could be a parallel for the military character of such Achaemenid palace buildings.[75] However, neither in the case of Cilicia nor in that of the Caspian region does it seem appropriate to infer a diaspora character for the "outposts" in question.[76]

***

Ultimately one has to confront the question of where the diaspora and especially the "military diaspora" begins in the Achaemenid empire. The previous remarks have shown that this question—at least from the perspective of Achaemenid ideology—does not even arise: Because of the multi- and cross-cultural relations, but above all because of the political framework and self-understanding of the Great King's empire, there could be no form of military diaspora, neither inside nor outside the empire. As we have seen, any official perception of a diaspora was impossible in the royal worldview. As a result, a perception or feeling of diaspora, especially in a military context, could not be articulated or (officially) communicated. Even if there was an (individually felt) military diaspora, it could not be expected to be found in any official text; even non-Persian sources remain silent on this issue.

Abbreviations

Arr. *Anab.*	Arrian, *Anabasis.*
CII	*Corpus Inscriptionum Iranicarum.*

CT	*Cuneiform Texts from Babylonian Tablets in the British Museum.*
Curt.	Curtius Rufus, *Historiae Alexandri Magni Macedonis.*
Diod. Sic.	Diodorus, *Bibliotheca historica.*
*DN*a	Darius, *Naqš-e Rustam*, Inscription a.
*DS*e	Darius, *Susa*, Inscription e.
*DZ*c	Darius, *Suez*, Inscription c.
Hdt.	Herodotus, *Historiae.*
Plin. *HN*	Plinius, *Naturalis historia.*
Strab.	Strabo, *Geographica.*
Xen. *An.*	Xenophon, *Anabasis.*
Xen. *Cyr.*	Xenophon, *Cyropaedia.*
*XP*a	Xerxes, *Persepolis*, Inscription a.
*XP*b	Xerxes, *Persepolis*, Inscription b.
*XP*c	Xerxes, *Persepolis*, Inscription c.
*XP*d	Xerxes, *Persepolis*, Inscription d.
*XP*f	Xerxes, *Persepolis*, Inscription f.
*XP*h	Xerxes, *Persepolis*, Inscription h.

Notes

1 Concerning the royal forces, see Pierre Briant, *From Cyrus to Alexander: A History of the Persian Empire* (Winona Lake: Eisenbrauns, 2002), 783–792; Hilmar Klinkott, *Der Satrap: Ein achaimenidischer Amtsträger und seine Handlungsspielräume* (Frankfurt: Verlag Antike, 2005), 281–286.

2 Xen. *Cyr.* 8.6.14; see also Xen. *Cyr.* 6.1.1; Briant, *A History of the Persian Empire*, 66–67.

3 See Klinkott, *Der Satrap*, 321–330.

4 See for the problem of defining Achaemenid mercenaries: Hilmar Klinkott, "Söldner, Siedler und Soldaten? Zum Status von und Umgang mit multiethnischen Streitkräften im Achaimenidenreich," in *Shaping Politics and Society: Mercenaries in the Greek World*, ed. Patrick Sänger and Sandra Scheuble-Reiter (Stuttgart: Franz Steiner Verlag), forthcoming.

5 Concerning the problem especially under Artaxerxes III and his attempt of legal restriction: Gunter Seibt, *Griechische Söldner im Achaimenidenreich* (Bonn: Dr. Rudolf Habelt, 1977), 90–92.

6 For the practical expressions and facets of (i.e. Greek) military diasporas in Hellenistic Egypt, see the chapter of Patrick Sänger in this volume.

7 Seibt, *Griechische Söldner*; see also Christopher Tuplin, "*Mercenaries in the Achaemenid Empire*," in *Blackwell Companion to the Achaemenid Empire II*, ed. Bruno Jacobs and Robert Rollinger (Hoboken: Wiley Blackwell, 2021), 1183–1198; id., "Warlords and Mercenaries in the Achaemenid Empire," in *War, Warlord, and Interstate Relations in the Ancient Mediterranean*, ed. Toni Ñaco de Hoya and Fernando López-Sánchez (Leiden: Brill, 2017), 17–35.

8 For the perception of *patris* in context of Greek mercenaries, see Christopher Tuplin, "Xenophon, Isocrates and the Achaemenid Empire: History, Pedagogy and the Persian Solution to Greek Problems," *Trends in Classics* 10, no. 1 (2018): 45–46. See also Peter Funke, "Was ist der Griechen Vaterland? Einige Überlegungen zum Verhältnis von Raum und politischer Identität im antiken Griechenland," *Geographica Antiqua* 18 (2009): 123–131.

9 See Pierre Briant, "Dons de terre et de ville: L'Asie Mineure dans le context Achéménide," *Revue des Études Anciennes* 87 (1985): 58–64; on Memnon and the so-called Memnon's land, see Klinkott, "Söldner, Siedler und Soldaten?" For Greek mercenary generals, see also Matthew Trundle, *Greek Mercenaries from the Late Archaic Period to Alexander* (London: Routledge, 2004), 147–159.

10 However, from an imperial perspective the so-called mercenaries have been seen as military "specialists" and an integrated part of the Achaemenid empire: see Klinkott, "Söldner, Siedler und Soldaten?" For Greek soldiers at the service of the Great King during the conquest of Alexander, see Seibt, *Griechische Söldner*, 110–120; Trundle, *Greek Mercenaries*, 52, 70 concerning the loyalty of Greek mercenaries to Darius III (Arr., *Anab.* 3.24.5); compare also Arr., *Anab.* 3.21.4; Diod. Sic. 17.27.2; Curt. 5.8.4 with Trundle, *Greek Mercenaries*, 142; Jakob Seibert, *Die Eroberung des Perserreich durch Alexander d. Gr. Auf kartographischer Grundlage*, Tübinger Atlas des Vorderen Orients, Reihe B, 68 (Wiesbaden: Dr. Ludwig Reichert Verlag, 1985), 132–133.

11 Xen., *An.* 7.8.9.

12 Xen., *An.* 7.8.17.

13 See Matthew Stolper, *Entrepreneurs and Empire, The Murašu Archive, the Murašu Firm, and Persian Rule in Babylonia* (Istanbul: NINO, 1985), 56–59; Briant, *A History of the Persian Empire*, 429–431, 433–435.

14 In practice, the Kurtaš were released from active duty through payments: See Stolper, *Entrepreneurs and Empire*, 98–99; Christopher Tuplin, "Taxation and Death: Certainties in the Persepolis Fortification Archive?," in *L'archive des fortifications de Persépolis: État des questions et perspectives de recherches*, ed. Pierre Briant, Wouter Henkelman, and Matthew Stolper, Persika 12 (Paris: De Boccard, 2008), 317–386, at 317–318.

15 For ethnic determinations of Babylonian *ḫaṭru*, see Stolper, *Entrepreneurs and Empire*, 72–79. For Carian *kurtaš*, M. Jursa supposed the status of settled "mercenaries", probably from Egypt: Michael Jursa, "The Transition of Babylonia from the Neo-Babylonian Empire to Achaemenid Rule," in *Regime Change in the Ancient Near East and Egypt*, ed. Harriet Crawford (Oxford: Oxford University Press, 2007), 73–94, at 88 with linkage to Caroline Waerzeggers, "The Carians of Borsippa," *Iraq* 68 (2006): 1–22. To these Carian villages still in the time of Alexander: Diod. Sic. 17.110.3. In a similar way, settlements of Gezertians (people from the Palestine city of Gezer) and of Susaeans are attested in Babylonia: Michael Jursa, *Der Tempelzehnt in Babylonien vom siebenten bis zum dritten Jahrhundert v. Chr.*, Alter Orient und Altes Testament 254 (Münster: Ugarit Verlag, 1998), 25–26.

16 Otherwise, they are interpreted as settlements of deportees: see Chiara Matarese, *Deportationen im Perserreich in teispidisch-achaimenidischer Zeit*, Classica et Orientalia 27 (Wiesbaden: Harrassowitz, 2021), 39–44 (Carians), 65–74 (Milesians), 75–92 (Eretrians), 107–126 (Greeks and Lycians); Josef Wiesehöfer, "Deportations," in *Blackwell Companion to the Achaemenid Empire II*, ed. Bruno Jacobs and Robert Rollinger (Hoboken: Wiley Blackwell, 2021), 871–880; Robert Rollinger, "Between Deportation and Recruitment: Craftsmen and Specialists from the West in Ancient Near Eastern Empires (from Neo-Assyrian Times through Alexander III)," in *Infrastructure and Distribution in Ancient Economies*, ed. Bernhard Woytek (Vienna: Austrian Academy of Sciences Press, 2018), 425–444, at 430–438 calls these Greeks and Carians settled "specialists"; concerning the Carians in Babylonia, see Rollinger, "Between Deportation and Recruitment," 430; Klinkott, "Söldner, Siedler und Söldaten?".

17 See Stolper, *Entrepreneurs and Empire*, 71; Nicholas Victor Sekunda, "Achaemenid Colonization in Lydia," *Revue des Études Anciennes* 87 (1985): 7–29, at 27–28 (with the catalogue 18–25). For the Achaemenid land-tenure system in Lydia with structural continuation until early Hellenistic times, see Elspeth R. M. Dusinberre, *Aspects of Empire in Achaemenid Sardis* (Cambridge: Cambridge University Press, 2003), 123–127.

18 Xen., *An.* 7.8.9–18 reports on a network of fortified estates in possession of Gongylos, the Eretrian, in Pergamus, Asidates, Itamenes in Komania, Parthenion, and Apollonias, Procles in Halisarna and Teuthrania.

19 See for example *CT* 57.82; John MacGinnis, *The Arrow of the Sun: Armed Forces in Sippar in the First Millennium B.C.* (Dresden: ISLET-Verlag, 2012), 43, 495; Arminius Cornelius Valentinus Maria Bongenaar, *The Neo-Babylonian Ebbabar Temple at Sippar: Its Administration and Its Prosopography* (Istanbul: Nederlands Historisch-Archaeologisch Instituut, 1997), 133. Concerning the travel texts in the Persepolis archive, see Pierre Briant, "De Sardes à Suse," in *Asia Minor and Egypt: Old Cultures in a New Empire*, ed. Heleen Sancisi-Weerdenburg and Amélie Kuhrt, Achaemenid History 6 (Leiden: NINO, 1991), 67–82, at 69.

20 Bezalel Porten, *Archives from Elephantine: The Life of an Ancient Jewish Military Colony* (Berkeley: University of California Press, 1968), 29; Amaury Pétigny, "Des étrangers pour garder l'Égypte," in *L'armée en Égypte aux époches perse, ptolémaïque et romaine*, ed. Anne-Emmanuelle Veïsse and Stéphanie Wackenier (Geneva: Librairie Droz, 2014), 5–44, at 13. For the Chwarezmians/Chorazmians in Egypt, see also Michele Minardi, *Ancient Chorasmia: A Polity Between the Semi-Nomadic and Sedentary Cultural Areas of Central Asia* (Leuven: Peeters, 2015), 10.

21 Bezalel Porten, *The Elephantine Papyri in English: Three Millenia of Cross-Cultural Continuity and Change* (Leiden: Brill, 1996), 13 (on military duty), 23–25 (on marriages), 28 (chronological overview, 81–82 (on legal documents from the late sixth century B.C. onwards), on the Aramaic archives: 89–267. See also ibid., 262: B 50 from 401 B.C., which reports on "Malchiah son of Jashobia, an Aramean, hereditary-property-holder in Elephantine [the] fort[ress of the detach]ment of Nabukudu[rri]".

22 See Pétigny, "Des étrangers," 26–34.

23 Ibid., 7–24.

24 See Porten, *Elephantine Papyri*, 81, see for example 158–162: B 24; 220–222: B 39; 226–227: B 41; compare also 212–215: B 37 and much more.

25 Porten, *Elephantine Papyri*, 147: B 20; see also 148–149: B 21; Briant, *A History of the Persian Empire*, 55.

26 See Ismail Gezgin, "Defensive Systems in Aiolis and Ionia Regions in the Achaemenid Period," in *Achaemenid Anatolia: Proceedings of the First International Symposium on Anatolia in the Achaemenid Period, Bandirma 15–18 August 1997*, ed. Tomris Bakır et al. (Leiden: NINO, 2001), 182.

27 See Sekunda, "Achaemenid Colonization," 11–13.

28 Xen., *An.* 7.8.13.

29 For this regular system of communication inside the empire by a network of forts, see Josef Wiesehöfer, *Das antike Persien* (Zürich: Artemis & Winkler Verlag, 1994), 116–117; id., "Beobachtungen zum Handel des Achaimenidenreichs," *Münstersche Beiträge zur antiken Handelsgeschichte* 1 (1982): 5–16, at 8–9, 12; David Frank Graf, "The Persian Royal Road System," in *Continuity and Change*, ed. Heleen Sancisi-Weerdenburg, Amélie Kuhrt, and Margaret Cool Root, Achaemenid History 8 (Leiden: NINO, 1994), 167–189; Briant, "De Sardes à Suse," 67–75.

30 Xen., *An.* 7.8.15: … ἐκβοηθοῦσιν Ἰταμένης μὲν ἔχων τὴν ἑαυτοῦ δύναμιν, ἐκ Κομανίας δὲ ὁπλῖται Ἀσσύριοι καὶ Ὑρκάνιοι ἱππεῖς καὶ οὗτοι βασιλέως μισθοφόροι ὡς ὀγδήκκοντα, καὶ ἄλλοι πελτασταὶ εἰς ὀκτακοσίους, ἄλλοι δ' ἐκ Παρθενίου, ἄλλοι δ' ἐξ Ἀπολλωνίας καὶ ἐκ τῶν πλησίον χωρίων καὶ ἱππεῖς.

31 Xen., *Cyr.* 8.4.28. See Sekunda, "Achaemenid Colonization," 20.

32 See also Plin., *HN* 5.31; Tac., *Ann.* 2.47.4 who knows—obviously for Hellenistic Asia Minor—"Macedonians called Hyrcani". In detail, see Sekunda, "Achaemenid Colonization," 20.

33 See Sekunda, "Achaemenid Colonization," 20.

34 Katelijn Vandorpe, "Persian Soldiers and Persians of the Epigone: Social Mobility of Soldiers-Herdsmen in Upper Egypt," *Archiv für Papyrusforschung* 54 (2008): 87–108.

35 For Chwarezmia/Chorasmia and Chorasmians, see Minardi, *Ancient Chorasmia.*

36 This is even the case in the supposed citadel of Marakanda especially because it was a satrapal residence, see Vadim Mihailovic Masson, *Das Land der tausend Städte: Baktrien, Chorasmiern, Margiane, Parthien, Sogdien; Ausgrabungen in der südlichen Sowjetunion* (München: Pfriemer, 1981), 98–99; Klinkott, *Der Satrap*, 418.

37 *CT* 57.82; see MacGinnis, *Arrow of the Sun*, 43; Bongenaar, *The Neo-Babylonian Ebabbar Temple at Sippar*, 133.
38 *CT* 57.82, lines 6–8; see John MacGinnis, "The Role of Babylonian Temples in Contributing to the Army in the Early Achaemenid Empire," in *The World of Achaemenid Persia*, ed. John Curtis and St. John Simpson (London: Tauris, 2010), 495–502, at 495.
39 See Klinkott, "Söldner, Siedler und Soldaten?".
40 See for example Marie-Françoise Baslez, "Présence et traditions Iraniennes dans les cités de l'Égée," *Revue des Études Anciennes* 87 (1985): 137–155, without any source for regular Persian garrisons on the Greek Aegean islands. On the Lacedaimonian harmost system, see Lukas Thommen, *Lakedaimonion politeia: Die Entstehung der spartanischen Verfassung*, Historia Einzelschriften 103 (Stuttgart: Franz Steiner Verlag, 1996), 62–63; Georg Busolt, Adolf Bauer, and Iwan Müller, *Die griechischen Staats-, Kriegs- und Privataltertümer* (Nördlingen: C. H. Beck'sche Verlagsbuchhandlung, 1887), 206–207.
41 See Hdt, 4.42–44.
42 See Peter Kehne, "Zur Logistik des Xerxesfeldzuges 480 v. Chr.," in *Zu Wasser und zu Lande: Verkehrswege in der antiken Welt, Stuttgarter Kolloquium zur historischen Geographie des Altertums 7, 1999*, ed. Eckart Olshausen and Holger Sonnabend, Geographica Historica 17 (Stuttgart: Franz Steiner Verlag, 1999), 29–47, at 34–35; more critically Briant, *A History of the Persian Empire*, 143.
43 Hdt. 4.43.
44 See recently Robert Rollinger and Julian Degen, "Conceptualizing Universal Rulership: Considerations on the Persian Achaemenid Worldview and the Saka at 'the End of the World,'" in *Studien zu Geschichte und Kultur des alten Iran und seiner Nachbarn: Festschrift für R. Schmitt zum 80. Geburtstag*, ed. Hilmar Klinkott, Andreas Luther, and Josef Wiesehöfer, Oriens et Occidens 36 (Stuttgart: Franz Steiner Verlag, 2021), 187–224, at 189–190; Gregor Ahn, *Religiöse Herrscherlegitimation im Achämenidischen Iran*, Acta Iranica 31 (Leiden: E. J. Brill & Peeters Press, 1992), 258–272.
45 See Matthew Canepa, *The Iranian Expanse: Transforming Royal Identity Through Architecture, Landscape, and the Built Environment, 550 BCE–642 CE* (Oakland: University of California Press, 2018), 28–41, with other parts in Susa and Babylon.
46 *DNa* § 2; translation Rüdiger Schmitt, *The Old Persian Inscriptions of Naqsh-I Rustam and Persepolis (CII vol. I,2)* (London: School of Oriental and African Studies, 2000), 68; see also in: *DSe* 9–8; *DZc* 5. See also *XPa* § 2; *XPb* § 2; *XPc* § 2; *XPd* § 2; *XPf* § 2; *XPh* § 2: *"king of the countries containing many races, king on this great earth even far off"*; compare *XPj* § 1; according to the variation of the phrase see Ahn, *Religiöse Herrscherlegitimation*, 259–265; see also Robert Rollinger, "Dareios und Xerxes an den Rändern der Welt und die Inszenierung von Weltherrschaft: Altorientalisches bei Herodot," in *Herodots Quellen: Die Quellen Herodots*, ed. Boris Dunsch and Kai Ruffing, Classica et Orientalia 6 (Wiesbaden: Harrassowitz, 2013 [2014]), 95–116.
47 Compare Robert Rollinger and Kai Ruffing, "World View and Perception of Space," in *Aneignung und Abgrenzung: Wechselnde Perspektiven auf die Antithese von "Ost" und "West" in der griechischen Antike*, ed. Nicolas Zenzen, Tonio Hölscher, and Kai Trampedach (Heidelberg: Verlag Antike, 2013), 93–161, at 95–126, concerning other political entities outside the empire in Persian reflection. See also Robert Rollinger, "'Griechen' und 'Perser' im 5. und 4. Jahrhundert v. Chr. im Blickwinkel orientalischer Quellen, oder: Das Mittelmeer als Brücke zwischen Ost und West," in *Grenzen und Entgrenzungen: Historische und kulturwissenschaftliche Überlegungen am Beispiel des Mittelmeerraums*, ed. Beate Burtscher-Bechter et al., Saarbrücker Beiträge zur Vergleichenden Literatur- und Kulturwissenschaft 36 (Würzburg: Königshausen & Neumann, 2006), 125–153.
48 For this particular aspect, see Rollinger and Degen, "Conceptualizing Universal Rulership," 199–207; Hilmar Klinkott, "Der Großkönig und das Meer: Achaimenidisches Reichsverständnis in einem neuen Weltbild," in *Studien zu Geschichte und Kultur des alten Iran und seiner Nachbarn: Festschrift für R. Schmitt zum 80. Geburtstag*,

ed. Hilmar Klinkott, Andreas Luther, and Josef Wiesehöfer, Oriens et Occidens 36 (Stuttgart: Franz Steiner Verlag, 2021), 111–136, at 117–120, 123–124.

49 Hdt. 9.3.2; Klinkott, *Der Satrap*, 399, fn. 13.

50 Xen., *An.* 7.8.7–8 explains the way of Xenophon along the Mysian coastline.

51 For Ionian *pyrgoi*, see Jack Martin Balcer, "Fifth Century B.C. Ionia: A Frontier Redefined," *Revue des Études Anciennes* 87 (1985): 31–42, at 36; the estate of the Persian Asidates is an example for such a fortified *pyrgos* in Lydia: Sekunda, "Achaemenid Colonization," 11–13; for *pyrgoi*/"Turmgehöfte" in Lycia see Ulf Hailer, *Einzelgehöfte im Bergland von Yavu (Zentrallykien)* (Bonn: Dr. Rudolf Habelt Verlag, 2008).

52 See for example Xen., *An.* 1.1.6: Cyrus ordered the *phrourachoi* in the cities to recruit as many Peloponnesian soldiers as possible. See also Gezgin, "Defensive Systems," 181–183; Klinkott, *Der Satrap*, 414–418, 421–422; concerning Xenophon's terminology, see in detail ibid., 417.

53 Joseph Naveh and Shaul Shaked, *Aramaic Documents from Bactria (Fourth Century BCE) from the Khalili Collection* (London: The Khalili Family Trust, 2012), 29.

54 See for example the so-called **frataraka*, known from Egypt by only Iranian names: Josef Wiesehöfer, "PRTRK, RB ḤYL', SGN und MR': Zur Verwaltung Südägyptens in achaimenidischer Zeit," in *Asia Minor and Egypt: Old Cultures in a New Empire*, ed. Heleen Sancisi-Weerdenburg and Amélie Kuhrt, Achaemenid History 6 (Leiden: NINO, 1991), 305–309, at 305.

55 See for example Orontas in Sardis: Xen., *An.* 1.6.6; Mithrines in Sardis at the conquest of Alexander: Arr., *Anab.* 1.17.3; compare also Xen., *Cyr.* 8.6.1. In late-Achaemenid time also commanders with Greek names like Hegesistratus are known: Arr., *Anab.* 1.18.3. See the collection of Persian commanders in Briant, *A History of the Persian Empire*, 351; compare also Klinkott, *Der Satrap*, 417–418.

56 See Henry Sydney Smith and Amélie Kuhrt, "A Letter to a Foreign General," *Journal of Egyptian Archaeology* 68 (1982): 199–209 with pl. XX, at 208; Wiesehöfer, "Verwaltung Südägyptens," 305–309; for the equation of *rab ḥaylā'* and *phrourarchos*, see Pierre Grelot, *Documents Araméens d' Égypte* (Paris: Les Editions du Cerf, 1972), 45–46.

57 See Joseph Naveh and Shaul Shaked, "Ritual Texts or Traesury Documents?," *Orientalia* 42 (1973): 445–457; Wiesehöfer, "Verwaltung Südägyptens," 308. See also for Susa: Strab. 15.3.21. For the military terminology see Christopher Tuplin, "Persian Garrisons in Xenophon and Other Sources," in *Method and Theory*, ed. Amélie Kuhrt and Heleen Sancisi-Weerdenburg, Achaemenid History 3 (Leiden: NINO, 1988), 67–70; id., "The Persian Military Establishment in Western Asia Minor," in *Kelainai-Apameia Kibotos: Une métropole achéménide, hellénistique et romaine*, ed. Latife Summerer, Askold Ivantchik, and Alexander von Kienlin (Bordeaux: Ausonius éditions, 2016), 15–27.

58 For the case of the Persian governor of Syene, Widranga ordered Egyptian and "other troops" to destroy the Jewish temple on Elephantine: see Briant, *A History of the Persian Empire*, 603–607; see in detail, Pierre Briant, "Une curieuse affaire à Éléphantine en 419 av. n.é.: Widranga, le sanctuaire de Khnum et le temple de Yahweh," *Méditerranées* 6/7 (1996): 115–131.

59 For the text, see Smith and Kuhrt, "Letter to a Foreign General," 199–209 with pl. XX.

60 See also for example Nicholas Victor Sekunda, "Achaemenid Settlement in Caria, Lycia, and Greater Phrygia," in *Asia Minor and Egypt: Old Cultures in a New Empire*, ed. Heleen Sancisi-Weerdenburg and Amélie Kuhrt, Achaemenid History 6 (Leiden: NINO, 1991), 83–143.

61 Sekunda, "Achaemenid Colonization," 9–13. See also Dusinberre, *Aspects of Empire*, 119–120, 123–125.

62 Xen., *An.* 1.1.6: … καὶ γὰρ ἦσαν αἱ Ἰωνικαὶ πόλεις Τισσαφέρνους τὸ ἀρχῖον ἐκ βασιλέως δεδομέαι, …

63 See Sekunda, "Achaemenid Colonization," 10–11.

64 Compare also for Babylonia Briant, *A History of the Persian Empire*, 442–446.
65 Translation of Walter Miller, *Xenophon, Cyropaedia, Vol. 2, Books V–VIII*, Loeb Classical Library 52 (Cambridge: Harvard University Press, 1914), 413–417.
66 See Clarisse Herrenschmidt, "Une lecture iranisante du poème de Symmachos dédié à Arbinas, dynaste de Xanthos," *Revue des Études Anciennes* 87 (1985): 125–133. See also Xenophon's entire description of the satrapal court Xen., *Cyr.* 8.6.10–14.
67 Xen., *Cyr.* 8.6.10.
68 See for example at Lydian Sardis: Dusinberre, *Aspects of Empire*, 196–203. This does not mean that Persians in the satrapies actively practise a kind of "Persianization" in a sense of "cultural imperialism". See therefore ibid., 203–216.
69 See for example for Sardis: Dusinberre, *Aspects of Empire*, 145–156. Especially on the Persian aristocracy, see Briant, *A History of the Persian Empire*, 352–354; on Persian aristocracy in the Achaemenid empire, see Hilmar Klinkott, "Zum persischen Adel im Achaimenidenreich," in *Die Macht der Wenigen: Aristokratische Herrschaftspraxis, Kommunikation und "edler" Lebensstil in Antike und Früher Neuzeit*, ed. Hans Beck, Peter Scholz, and Uwe Walter (München: Oldenbourg Verlag, 2008), 207–251.
70 Xen., *Cyr.* 8.6.10. "πρῶτον μὲν ἱππέας καθιστάναι ἐκ τῶν συνεπιστομένων Περσῶν καὶ συμμάχων καὶ ἁρματηλάτας". Translation: Miller, *Xenophon, Cyropaedia*, 414.
71 See Florian Knauss, Iulon Gagoshidze, and Ilias Babaev, "A Persian Propyleion in Azerbaijan: Excavations at Karacamirli," in *Achaemenid Impact in the Black Sea: Communication of Powers*, ed. Jens Nieling and Ellen Rehm, Black Sea Studies 11 (Aarhus: Aarhus University Press, 2010), 111–122; Maria Brosius, "Pax Persica and the People of the Black Sea Region: Extent and Limits of Achaemenid Imperial Ideology," in ibid., 29–40; Florian Knauss, "Caucasus," in *L'archéologie de l'empire achéménide: Nouvelle recherches; Actes du colloque organize au Collège de France par le Réseau international d'études et de recherches achéménides, 21–22 novembre 2003*, ed. Pierre Briant and Rémy Bourchalat, Persika 6 (Paris: De Boccard, 2005), 197–210, at 204.
72 See Bruno Jacobs, *Griechische und persische Elemente in der Grabkunst Lykiens zur Zeit der Achaimenidenherrschaft* (Göteborg: Åström, 1987), 15–23.
73 In my opinion, claims that the palace was the residence of a satrap remain speculation. It is also possible that it could have been the possessions of the king, his family, Persian generals (*strategoi*), or aristocratic court members.
74 Beyond the garrisons, see for example also the **kāra-tanu-ka*—the "bodyguards" known from the Aramaic documents of Bactria: Naveh and Shaked, *Aramaic Documents*, 29.
75 For the fortress, see in detail Alain Davesne and Françoise Laroche-Traunecker, *Gülnar I: Le site de Meydancıkkale* (Paris: Éd. Recherche sur les Civilisations, 1998); Alain Davesne and George Le Rider, *Gülnar II: Le trésor de Meydancıkkale (Cilicie Trachée, 1980)* (Paris: Éd. Recherche sur les Civilisations, 1999).
76 Even the definition as an outpost is part of a subjective perspective. From a Persian point of view, the Caspian region may have been perceived in a quite different manner.

References

Ahn, Gregor. *Religiöse Herrscherlegitimation im Achämenidischen Iran*. Acta Iranica 31. Leiden: E. J. Brill & Peeters Press, 1992.

Balcer, Jack Martin. "Fifth Century B.C. Ionia: A Frontier Redefined." *Revue des Études Anciennes* 87 (1985): 31–42.

Baslez, Marie-Françoise. "Présence et traditions Iraniennes dans les cités de l'Égée." *Revue des Études Anciennes* 87 (1985): 137–155.

Bockisch, Gabriele. "Ἁρμοσταί." *Klio* 46 (1965): 129–239.

Bongenaar, Arminius Cornelius Valentinus Maria. *The Neo-Babylonian Ebbabar Temple at Sippar: Its Administration and Its Prosopography*. Istanbul: Nederlands Historisch-Archaeologisch Instituut, 1997.

Briant, Pierre. "Dons de terre et de ville: L'Asie Mineure dans le context Achéménide." *Revue des Études Anciennes* 87 (1985): 53–72.

Briant, Pierre. "De Sardes à Suse." In *Asia Minor and Egypt: Old Cultures in a New Empire*, edited by Heleen Sancisi-Weerdenburg and Amélie Kuhrt, 67–82. Achaemenid History 6. Leiden: NINO, 1991.

Briant, Pierre. "Une curieuse affaire à Éléphantine en 419 av. n.é.: Widranga, le sanctuaire de Khnum et le temple de Yahweh." *Méditerranées* 6/7 (1996): 115–131.

Briant, Pierre. *From Cyrus to Alexander: A History of the Persian Empire*. Winona Lake: Eisenbrauns, 2002.

Brosius, Maria. "Pax Persica and the People of the Black Sea Region: Extent and Limits of Achaemenid Imperial Ideology." In *Achaemenid Impact in the Black Sea: Communication of Powers*, edited by Jens Nieling and Ellen Rehm, 29–40. Black Sea Studies 11. Aarhus: Aarhus University Press, 2010.

Busolt, Georg, Adolf Bauer, and Iwan Müller. *Die griechischen Staats-, Kriegs- und Privataltertümer*. Nördlingen: C. H. Beck'sche Verlagsbuchhandlung, 1887.

Canepa, Matthew. *The Iranian Expanse: Transforming Royal Identity Through Architecture, Landscape, and the Built Environment, 550 BCE–642 CE*. Oakland: University of California Press, 2018.

Davesne, Alain, and Françoise Laroche-Traunecker. *Gülnar I: Le site de Meydancıkkale*. Paris: Éd. Recherche sur les Civilisations, 1998.

Davesne, Alain, and George Le Rider. *Gülnar II: Le trésor de Meydancıkkale (Cilicie Trachée, 1980)*. Paris: Éd. Recherche sur les Civilisations, 1999.

Dusinberre, Elspeth R. M. *Aspects of Empire in Achaemenid Sardis*. Cambridge: Cambridge University Press, 2003.

Funke, Peter. "Was ist der Griechen Vaterland? Einige Überlegungen zum Verhältnis von Raum und politischer Identität im antiken Griechenland." *Geographica Antiqua* 18 (2009): 123–131.

Gezgin, Ismail. "Defensive Systems in Aiolis and Ionia Regions in the Achaemenid Period." In *Achaemenid Anatolia: Proceedings of the First International Symposium on Anatolia in the Achaemenid Period, Bandirma 15–18 August 1997*, edited by Tomris Bakır, Heleen Sancisi-Weerdenburg, Gül Gürtekin, Pierre Briant, and Wouter Henkelman, 181–188. Leiden: NINO, 2001.

Graf, David Frank. "The Persian Royal Road System." In *Continuity and Change*, edited by Heleen Sancisi-Weerdenburg, Amélie Kuhrt, and Margaret Cool Root, 167–189. Achaemenid History 8. Leiden: NINO, 1994.

Grelot, Pierre. *Documents Araméens d'Égypte*. Paris: Les Éditions du Cerf, 1972.

Hailer, Ulf. *Einzelgehöfte im Bergland von Yavu (Zentrallykien)*. Bonn: Dr. Rudolf Habelt Verlag, 2008.

Herrenschmidt, Clarisse. "Une lecture iranisante du poème de Symmachos dédié à Arbinas, dynaste de Xanthos." *Revue des Études Anciennes* 87 (1985): 125–133.

Jacobs, Bruno. *Griechische und persische Elemente in der Grabkunst Lykiens zur Zeit der Achaimenidenherrschaft*. Göteborg: Åström, 1987.

Jursa, Michael. *Der Tempelzehnt in Babylonien vom siebenten bis zum dritten Jahrhundert v. Chr.* Alter Orient und Altes Testament 254. Münster: Ugarit Verlag, 1998.

Jursa, Michael. "The Transition of Babylonia from the Neo-Babylonian Empire to Achaemenid Rule." In *Regime Change in the Ancient Near East and Egypt*, edited by Harriet Crawford, 73–94. Oxford: Oxford University Press, 2007.

Kehne, Peter. "Zur Logistik des Xerxesfeldzuges 480 v. Chr." In *Zu Wasser und zu Lande: Verkehrswege in der antiken Welt, Stuttgarter Kolloquium zur historischen Geographie des Altertums 7, 1999*, edited by Eckart Olshausen and Holger Sonnabend, 29–47. Geographica Historica 17. Stuttgart: Franz Steiner Verlag, 1999.Klinkott, Hilmar. *Der Satrap: Ein achaimenidischer Amtsträger und seine Handlungsspielräume*. Frankfurt: Verlag Antike, 2005.

Klinkott, Hilmar. "Zum persischen Adel im Achaimenidenreich." In *Die Macht der Wenigen: Aristokratische Herrschaftspraxis, Kommunikation und "edler" Lebensstil in Antike und Früher Neuzeit*, edited by Hans Beck, Peter Scholz, and Uwe Walter, 207–251. München: Oldenbourg Verlag, 2008.

Klinkott, Hilmar. "Der Großkönig und das Meer: Achaimenidisches Reichsverständnis in einem neuen Weltbild." In *Studien zu Geschichte und Kultur des alten Iran und seiner Nachbarn: Festschrift für R. Schmitt zum 80. Geburtstag*, edited by Hilmar Klinkott, Andreas Luther, and Josef Wiesehöfer, 111–136. Oriens et Occidens 36. Stuttgart: Franz Steiner Verlag, 2021.

Klinkott, Hilmar. "Söldner, Siedler und Soldaten? Zum Status von und Umgang mit multiethnischen Streitkräften im Achaimenidenreich." In *Shaping Politics and Society: Mercenaries in the Greek World*, edited by Patrick Sänger and Sandra Scheuble-Reiter. Stuttgart: Franz Steiner Verlag, forthcoming.

Knauss, Florian. "Caucasus." In *L'archéologie de l'empire achéménide: Nouvelle recherches; Actes du colloque organize au Collège de France par le Réseau international d'études et de recherches achéménides, 21–22 novembre 2003*, edited by Pierre Briant and Rémy Bourchalat, 197–210. Persika 6. Paris: De Boccard, 2005.

Knauss, Florian, Iulon Gagoshidze, and Ilias Babaev. "A Persian Propyleion in Azerbaijan: Excavations at Karacamirli." In *Achaemenid Impact in the Black Sea: Communication of Powers*, edited by Jens Nieling and Ellen Rehm, 111–122. Black Sea Studies 11. Aarhus: Aarhus University Press, 2010.

MacGinnis, John. "The Role of Babylonian Temples in Contributing to the Army in the Early Achaemenid Empire." In *The World of Achaemenid Persia*, edited by John Curtis and St. John Simpson, 495–502. London: Tauris, 2010.

MacGinnis, John. *The Arrow of the Sun: Armed Forces in Sippar in the First Millennium B.C.* Dresden: ISLET-Verlag, 2012.

Masson, Vadim Mihailovic. *Das Land der tausend Städte: Baktrien, Chorasmiern, Margiane, Parthien, Sogdien; Ausgrabungen in der südlichen Sowjetunion*. München: Pfriemer, 1981.

Matarese, Chiara. *Deportationen im Perserreich in teispidisch-achaimenidischer Zeit*. Classica et Orientalia 27. Wiesbaden: Harrassowitz, 2021.

Miller, Walter, trans. *Xenophon, Cyropaedia, Vol. 2, Books V–VIII*, translated by Walter Miller. Loeb Classical Library 52. Cambridge: Harvard University Press, 1914.

Minardi, Michele. *Ancient Chorasmia: A Polity Between the Semi-Nomadic and Sedentary Cultural Areas of Central Asia*. Leuven: Peeters, 2015.

Naveh, Joseph, and Shaul Shaked. "Ritual Texts or Treasury Documents?" *Orientalia* 42 (1973): 445–457.

Naveh, Joseph, and Shaul Shaked. *Aramaic Documents from Bactria (Fourth Century BCE.) from the Khalili Collection*. London: The Khalili Family Trust, 2012.

Pétigny, Amaury. "Des étrangers pour garder l'Égypte." In *L'armée en Égypte aux époches perse, ptolémaïque et romaine*, edited by Anne-Emmanuelle Veïsse and Stéphanie Wackenier, 5–44. Geneva: Librairie Droz, 2014.

Porten, Bezalel. *Archives from Elephantine: The Life of an Ancient Jewish Military Colony*. Berkeley: University of California Press, 1968.

Porten, Bezalel. *The Elephantine Papyri in English: Three Millennia of Cross-Cultural Continuity and Change.* Leiden: Brill, 1996.

Rollinger, Robert. "'Griechen' und 'Perser' im 5. und 4. Jahrhundert v. Chr. im Blickwinkel orientalischer Quellen, oder: Das Mittelmeer als Brücke zwischen Ost und West." In *Grenzen und Entgrenzungen: Historische und kulturwissenschaftliche Überlegungen am Beispiel des Mittelmeerraums*, edited by Beate Burtscher-Bechter, Peter W. Haider, Birgit Mertz-Baumgartner, and Robert Rollinger, 125–153. Saarbrücker Beiträge zur Vergleichenden Literatur- und Kulturwissenschaft 36. Würzburg: Königshausen & Neumann, 2006.

Rollinger, Robert. "Dareios und Xerxes an den Rändern der Welt und die Inszenierung von Weltherrschaft: Altorientalisches bei Herodot." In *Herodots Quellen: Die Quellen Herodots*, edited by Boris Dunsch and Kai Ruffing, 95–116. Classica et Orientalia 6. Wiesbaden: Harrassowitz, 2013 [2014].

Rollinger, Robert. "Between Deportation and Recruitment: Craftsmen and Specialists from the West in Ancient Near Eastern Empires (from Neo-Assyrian Times through Alexander III)." In *Infrastructure and Distribution in Ancient Economies*, edited by Bernhard Woytek, 425–444. Vienna: Austrian Academy of Sciences Press, 2018.

Rollinger, Robert, and Julian Degen. "Conceptualizing Universal Rulership: Considerations on the Persian Achaemenid Worldview and the Saka at 'the End of the World.'" In *Studien zu Geschichte und Kultur des alten Iran und seiner Nachbarn: Festschrift für R. Schmitt zum 80. Geburtstag*, edited by Hilmar Klinkott, Andreas Luther, and Josef Wiesehöfer, 187–224. Oriens et Occidens 36. Stuttgart: Franz Steiner Verlag, 2021.

Rollinger, Robert, and Kai Ruffing. "World View and Perception of Space." In *Aneignung und Abgrenzung: Wechselnde Perspektiven auf die Antithese von "Ost" und "West" in der griechischen Antike*, edited by Nicolas Zenzen, Tonio Hölscher, and Kai Trampedach, 93–161. Heidelberg: Verlag Antike, 2013.

Schmitt, Rüdiger. *The Old Persian Inscriptions of Naqsh-I Rustam and Persepolis (CII vol. I,2).* London: School of Oriental and African Studies, 2000.

Seibert, Jakob. *Die Eroberung des Perserreich durch Alexander d. Gr. auf kartographischer Grundlage.* Tübinger Atlas des Vorderen Orients, Reihe B, 68. Wiesbaden: Dr. Ludwig Reichert Verlag, 1985.

Seibt, Gunter. *Griechische Söldner im Achaimenidenreich.* Bonn: Dr. Rudolf Habelt, 1977.

Sekunda, Nicholas Victor. "Achaemenid Colonization in Lydia." *Revue des Études Anciennes* 87 (1985): 7–29.

Sekunda, Nicholas Victor. "Achaemenid Settlement in Caria, Lycia, and Greater Phrygia." In *Asia Minor and Egypt: Old Cultures in a New Empire*, edited by Heleen Sancisi-Weerdenburg and Amélie Kuhrt, 83–143. Achaemenid History 6. Leiden: NINO, 1991.

Smith, Henry Sydney, and Amélie Kuhrt. "A Letter to a Foreign General." *Journal of Egyptian Archaeology* 68 (1982): 199–209 with pl. XX.

Stolper, Matthew W. *Entrepreneurs and Empire, The Murašu Archive, the Murašu Firm, and Persian Rule in Babylonia.* Istanbul: NINO, 1985.

Thommen, Lukas. *Lakedaimonion politeia: Die Entstehung der spartanischen Verfassung.* Historia Einzelschriften 103. Stuttgart: Franz Steiner Verlag, 1996.

Trundle, Matthew. *Greek Mercenaries from the Late Archaic Period to Alexander.* London: Routledge, 2004.

Tuplin, Christopher. "Persian Garrisons in Xenophon and Other Sources." In *Method and Theory*, edited by Amélie Kuhrt and Heleen Sancisi-Weerdenburg, 67–70. Achaemenid History 3. Leiden: NINO, 1988.

Tuplin, Christopher. "Taxation and Death: Certainties in the Persepolis Fortification Archive?" In *L'archive des fortifications de Persépolis: État des questions et perspectives de recherches*, edited by Pierre Briant, Wouter Henkelman, and Matthew Stolper, 317–386. Persika 12. Paris: De Boccard, 2008.

Tuplin, Christopher. "The Persian Military Establishment in Western Asia Minor." In *Kelainai-Apameia Kibotos: Une métropole achéménide, hellénistique et romaine*, edited by Latife Summerer, Askold Ivantchik, and Alexander von Kienlin, 15–27. Bordeaux: Ausonius éditions, 2016.

Tuplin, Christopher. "Warlords and Mercenaries in the Achaemenid Empire." In *War, Warlord, and Interstate Relations in the Ancient Mediterranean*, edited by Toni Ñaco de Hoya and Fernando López-Sánchez, 17–35. Leiden: Brill, 2017.

Tuplin, Christopher. "Xenophon, Isocrates and the Achaemenid Empire: History, Pedagogy and the Persian Solution to Greek Problems." *Trends in Classics* 10, no. 1 (2018): 13–55.

Tuplin, Christopher. "Mercenaries in the Achaemenid Empire." In *Blackwell Companion to the Achaemenid Empire II*, edited by Bruno Jacobs and Robert Rollinger, 1183–1198. Hoboken: Wiley Blackwell, 2021.

Vandorpe, Katelijn. "Persian Soldiers and Persians of the Epigone: Social Mobility of Soldiers-Herdsmen in Upper Egypt." *Archiv für Papyrusforschung* 54 (2008): 87–108.

Waerzeggers, Caroline. "The Carians of Borsippa." *Iraq* 68 (2006): 1–22.

Wiesehöfer, Josef. "Beobachtungen zum Handel des Achaimenidenreichs." *Münstersche Beiträge zur antiken Handelsgeschichte* 1 (1982): 5–16.

Wiesehöfer, Josef. "PRTRK, RB ḤYL', SGN und MR': Zur Verwaltung Südägyptens in achaimenidischer Zeit." In *Asia Minor and Egypt: Old Cultures in a New Empire*, edited by Heleen Sancisi-Weerdenburg and Amélie Kuhrt, 305–309. Achaemenid History 6. Leiden: NINO, 1991.

Wiesehöfer, Josef. *Das antike Persien*. Zürich: Artemis & Winkler Verlag, 1994.

Wiesehöfer, Josef. "Deportations." In *Blackwell Companion to the Achaemenid Empire II*, edited by Bruno Jacobs and Robert Rollinger, 871–880. Hoboken: Wiley Blackwell, 2021.

2

IMMIGRANT SOLDIERS AND PTOLEMAIC POLICY IN HELLENISTIC EGYPT (LATE FOURTH CENTURY–30 BCE)

Reflections on a Military Diaspora and Its Components

*Patrick Sänger**

In 332 BCE, Alexander III of Macedon, commonly known as Alexander the Great, invaded Egypt and wrested it from the control of the Persians, who had taken it in turn from the Pharaohs. There he founded what would become one of the great cities of the ancient world, Alexandria, modestly named after himself, and then continued to conquer the rest of the Persian Empire. When Alexander died in Babylon in 323 BCE, his general Ptolemy, son of Lagos (Ptolemy I Soter, the founder of the Ptolemaic dynasty), took hold of the conqueror's body and brought it to Egypt, establishing the first of what we call the "successor kingdoms". While Alexander's other generals fought for 20 years in order to determine who was to control the other parts of the conqueror's empire, Ptolemic Egypt was sufficiently strong, and sufficiently difficult to reach with an army, that it was able to participate in the wars of the successors mostly when it wished to. In the following years, the Ptolemies were intermittently able to extend their power north up the coast of the Levant, into the Aegean and the coast of Asia Minor, and west along the coast of Africa.

Ptolemy was Macedonian, and he and his court at Alexandria spoke Greek, as did his descendants: The immigration of Greek speakers to Egypt was encouraged, and Alexandria soon became the most populous Greek-speaking city in the world. Outside of Alexandria was the vast extent—or better, length—of Egypt, an immensely rich serpent of settlements stretching down the Nile Valley, which widened as it approached the delta on the Mediterranean, where Alexandria was situated. The administration of Ptolemaic Egypt has left behind a vast number of records, for beside the Nile grows the papyrus plant, and when its leaves are properly processed, they can be written on, and papyrus was the preferred writing medium of the ancient world. And, because Egypt is so dry, a vast number of these writings have survived over the centuries. So, we can examine the

DOI: 10.4324/9781003245568-3

administration and society of Ptolemaic Egypt far more closely than we can that of any other Hellenistic state.

By virtue of the papyri, Egypt also provides modern historians with the richest documentary evidence of migrations and their impact on the ancient world. Countless papyri mention various ethnic designations derived from foreign (non-Egyptian) cities or regions and, therefore, attest to individuals coming to Egypt from abroad as well as their descendants. This evidence leads to the conclusion that immigration into Egypt after its conquest by Alexander the Great reached dimensions that could not have been imagined in the eras before. But not only Greeks (from mainland Greece, the Peloponnese, the Greek islands, Asia Minor, and Greek towns in other regions) came to Alexandria and Egypt: Thracians from the east of the Balkan Peninsula, for instance, also arrived. Furthermore, there was a great inflow of Jews into Ptolemaic Egypt, and in time Alexandria came to contain the largest concentration of Jews in the world outside of the Holy Land. These processes of immigration not only altered the population of Egypt but also brought about new administrative challenges and caused social, cultural, and religious changes.

In our documentation, soldiers form the greatest single category of immigrants to Ptolemaic Egypt, and so it is natural to wonder whether the papyri from Egypt allow us to trace the formation of a "military diaspora". Kostas Buraselis first applied this term to Ptolemaic Egypt, to describe the whole body of soldiers from Greece and other regions who settled there.[1] The present chapter seeks to investigate whether this is a useful concept by having a closer look at the practical expressions and facets of military immigration. This requires tracking down organized groups whose origins lie in immigrant soldiers and who were also bound by a shared Greek culture or a specific ethnic identity, that is, a socially constructed identity based on cultural markers and "the belief (however fictive) in a shared kinship or common origin".[2] The parts of this question will be addressed in the following order: First, the potential and limits of our source material must be assessed (see "Ethnic Designations and Fiscal/Legal Categories in Ptolemaic Egypt: Traces of Diasporic Groups?"); second, the importance of immigrants to the Ptolemaic army and the emergence of two population groups will be examined, both of them illuminating different military immigration and employment patterns (see "Recruitment Policy and Structure of the Ptolemaic Army: Two Components, Two Strategies?").

Ethnic Designations and Fiscal/Legal Categories in Ptolemaic Egypt: Traces of Diasporic Groups?

Greek (and Demotic) papyri and stone inscriptions from Greco-Roman Egypt provide us with various ethnic designations alluding to foreign (non-Egyptian) cities or regions. Public documents—tax lists, legal documents, and honorary decrees, for example—illustrate the use of ethnic designations in administrative contexts. Private documents—letters on papyrus or private dedicatory inscriptions, for example—show how individuals were identified by others or express their own non-official self-perception.[3] The greatest quantity and variety of such designations occur in Hellenistic Egypt. This evidence was compiled by Csaba La'da who created a catalogue of ethnic designations derived from cities and regions outside of Egypt. For the middle of the third century BCE, when we find the greatest variety, over 170 different terms are attested.[4] Such designations allow us (as will be explained below) to identify persons who, in an administrative context, were neither counted among the indigenous Egyptians nor among the citizens of one of Egypt's three Greek *poleis* (Alexandria, Naukratis, Ptolemais), otherwise their status as citizens would have been indicated instead.

Tax lists reveal the existence of a fiscal category called the *Hellenes*, literally "Greeks", who were exempt from the obol tax: A very modest fiscal privilege.[5] But not all of these so-called tax-*Hellenes* were ethnically Greek or descended from Greeks: Thracians and Jews, for instance, also belonged to the category of tax-*Hellenes* from the outset, and Egyptians could become members of this group, too, as a result of their occupation.[6] In practice, the term "*Hellen*" (Ἕλλην) mostly denoted an "immigrant" or a "foreign settler" who was to be distinguished from "native Egyptians" (*Aigyptioi*),[7] and for reasons that will be explained further below, tax-*Hellenes* of actual Greek origin were much more common in the third than in the second or first centuries BCE. A comparable case of ethnic designations that give a name to a functional category without implying that their holders actually were of the origin the term implied is the second-century BCE military designations *Makedon* and *Perses*, terms that probably denoted status groups within the army, both of which were open to soldiers of Egyptian background.[8]

In addition to the tax-*Hellenes* (not all of whom were Greek), there also existed official "legal ethnic designations".[9] This category includes the just-mentioned military designations *Makedon* and *Perses*. Other examples are "Arab", "Cretan", and "Thracian", and some that referred to individual cities (e.g., "Athenian", "Pergamene", or "Cyrenean"). These designations had their origin in a Ptolemaic governmental requirement that for a document or contract to have legal validity, persons had to indicate their *patris* or place of origin.[10] It seems that the administration tried to enforce consistency in the indication of the name and the *patris* of a person in legal documents,[11] although we know nothing about the actual process of registration of persons to a *patris*.

Both the ethnic designation "*Hellen*" used for marking a certain tax status and the requirement for "legal ethnic designations" show that population categories whose names alluded to ethnic groups played an important role in Ptolemaic administration, presumably reflecting a ruling ideology that attempted to highlight the Greek parts of the Ptolemaic population without discriminating against the indigenous Egyptian population.[12] We have to remember that the Ptolemies were a Macedonian dynasty and keen to preserve that Greco-Macedonian image;[13] and this also explains the existence of the military category of *Makedones*.[14] But if we are interested in specific immigrant groups, we must rely on such designations broadly to sketch the world of such groups as a whole, for in any given instance, an ethnic designation may not describe the actual origin of the individual bearing it. The situation is well described by Dorothy Thompson, one of the leading experts on the society of Hellenistic Egypt, who notes:

> What exactly this *patris* involved—how real it was in all cases—is under debate. But that *patris* is the word used to describe this designation must mean, I think, that at least at some stage it indicated the place of origin of its holder, a term of geographical affiliation or descent.[15]

As this statement reminds us, such ethnic designations were also hereditary, and their survival through many generations of people actually born in Egypt is yet another reason why they might not be, in our terms, descriptive of ethnic or cultural reality.[16] Especially after the third century BCE, most holders of such designations will have been descendants of persons who migrated to Egypt at an earlier date: At the end of the third and in the course of the second century BCE, we can observe a distinct decrease in the variety of ethnic designations in use, a decrease that seems to indicate both that the geographical diversity of immigrants was diminishing and that the total immigration into Egypt was diminishing drastically as well. Such changes mirror the political situation of the Ptolemaic kingdom (see "Recruitment Policy and Structure of the Ptolemaic Army: Two Components, Two Strategies?") and at the same time, reflect the fact that increasingly distant descendants of immigrants were more and more assimilated into local society, with the result that they no longer all used their "legal ethnic designations" but instead referred in legal documents to their place of residence in Egypt.[17]

However inexact in their application, the ethnic designations in the papyri and on stone that tempt and frustrate are the easy part: Far more interesting but even more intractable is how immigrants and immigrant groups felt about and experienced ethnicity in Ptolemaic Egypt. No systematic treatment of the vast literature on ethnicity and ethnic identity in Ptolemaic Egypt—an issue entangled with the question of the fusion or separation of the Greco-Macedonian and indigenous populations, and with scholarly contributions often inspired by colonial or post-colonial experiences and European imperialism—can be attempted here,[18] and we must also leave aside the rich writings discussing the facets and grades of Jewish assimilation or separation in the Greek and Egyptian cultures of Ptolemaic Egypt.[19] It must suffice, for the purpose of this chapter, to point out that recent research on the relations between the Egyptian and Greek milieux in Ptolemaic Egypt, and the separation or non-separation of the two groups, is moving towards the *communis opinio* that the two identities in question were, in large part, a matter of choice and circumstance. As Jane Rowlandson puts it:

> Throughout the Ptolemaic period, Greek and Egyptian identities (expressed through formal status, language, nomenclature, and visual culture such as statues) were at once differentiated yet mutually compatible. Just as the Ptolemies were projected in public in the distinct roles of Pharaoh and Basileus, with only muted and subtle allusions from one to the other, so their subjects could shift between Egyptian and Greek identities depending on their careers and life-choices.[20]

Indeed, at least by the second or first century BCE, the tie between the descendants of the immigrants and the city or region their ancestors came from was (in most cases) attenuated by generations, and intermarriages between people of Greek and Egyptian origin created a culturally mixed population. This situation is well attested—at least in the second century BCE and after—in the group of military settlers or cleruchs (who were granted a plot of land, a *kleros*) to be discussed in the next section of this chapter.[21] Emblematic of the emerging multicultural amalgam are individuals who possessed double names, one Greek and one Egyptian,[22] and even a Greek name alone is no certain indication of the identity of its holder: Jews, for instance, could bear Greek or Hellenized names that nonetheless allowed them to express that they were Jewish, i.e. by adopting certain theophoric names or by Hellenizing Hebrew names.[23] In the Roman period, Roger Bagnall has demonstrated that the Greek names of members of the prestigious status category, which is attested for the administrative district (nome) of Arsinoites and said to include "6,475 Hellenic men", according to its designation, often allude to Egyptian divinities, Macedonian rulers and generals or the Ptolemaic dynasty; this marked the largest part of the members of this group as consisting of persons who had a Greek identity that was rooted in Egypt and its Hellenistic past but not in Greece.[24] Ethnic identity can be multi-faceted, hybrid, and flexible, and so could the ethnic identity of holders of "legal ethnic designations", even in the third century BCE.

Recruitment Policy and Structure of the Ptolemaic Army: Two Components, Two Strategies?

As already indicated above, soldiers formed the largest (visible) migration group in Hellenistic Egypt. According to a recently published estimate, in the third century BCE, approximately 5 percent of the perhaps four million inhabitants of Egypt were Greek, and a little more than half of these Greek migrants, that is, some 2.9 percent of the total population, were members of Greek military families.[25] Data from the Arsinoite nome, a district that was—as will be explained below—drained and resettled in the first half of the third century BCE, also suggest that in this century, the males and females countable among military groups outnumbered the civilian tax-*Hellenes*.[26] Furthermore, in the mid-third century BCE, the descendants of military settlers, the *epigonoi*, could have formed a large part (up to 16 percent) of the civilian tax-*Hellenes* in the Arsinoite nome.[27] In other words, the largest sector numerically of the Greek population resulted from the recruitment policy of the Ptolemaic army. The army itself was divided into two parts: A force of reservist regulars (infantry and cavalry) and a force of mercenaries or professional soldiers (light infantry).

The Regular Army: Preserver of a Greek Cultural Identity

The regular army of the Ptolemies consisted of the so-called cleruchs, who can be described as reservists because they served only when called up, and rather than being paid in coin, they received a plot of land that secured their livelihoods in peacetime[28]—a system whereby the Ptolemaic government drew not only on Macedonian but also Egyptian traditions.[29] Until the end of the third century BCE, this cleruch army was predominantly recruited from immigrants or their descendants. A closer look at the "legal ethnic designations" used by the cleruchs makes clear that their origins lay for the most part in regions that were not under the control of the Ptolemies;[30] of particular importance were Macedonia, mainland Greece, and Thrace. Recently published studies have argued that cleruchs were recruited in these regions at least until the end of the third century BCE, when the Ptolemies lost all of their possessions on the coasts of the North Aegean and Asia Minor as well as those in the Levant.[31] That the Ptolemies continued, as long as they could, to recruit cleruchs from their now-distant "homeland" Macedonia is due to the ideological importance these recruits had for the regime (and no doubt the Ptolemies valued their military quality), which also explains why they were offered land as an inducement to come to Egypt and stay; had they been employed as mercenaries for cash, the government might have lost access to them and their sons after the end of their military service, because they could return to where they came from, lands not directly controlled by the Ptolemies.[32] The same applies to recruits from mainland Greece, whom the Ptolemies also attempted to bind into long-term military service by grants of land.

The most important settlements of cleruchs were located in the Arsinoite nome. Probably as early as the first Ptolemaic king, Ptolemy I Soter (305–293 BCE), and certainly under his successor Ptolemy II Philadelphus (285–246 BCE), this region was drained and resettled.[33] In this as well as in other regions, cleruchs were settled in newly founded or already existing villages. Occasionally, however, settlements of cleruchs are also attested in nome or district capitals.[34] These military settlers, who, in term of their socio-economic situation, could be described as rural middling class,[35] were followed by civilian immigrants coming from Greece and neighbouring regions. Both groups worked in a broad variety of businesses and official capacities. In the mid-third century BCE, papyrological evidence suggests that in the Arsinoite nome, the new settlers could have made up 29 percent of the adult population.[36] The presence of these immigrants is also evidenced by the numerous *gymnasia* that they founded in villages and even in the nome capital Krokodilopolis/Ptolemais Euergetis. This custom, however, was not restricted to the Arsinoite nome and observable throughout Egypt.[37] As institutions borrowed from the Greek city states or *poleis*, the *gymnasia* existed specifically for the preservation of Greek culture. Elsewhere in the Greek-speaking world, *gymnasia* were institutions for Greek education, and for physical and military training. Whether they performed all these functions in the countryside of Ptolemaic Egypt is uncertain, but Ptolemaic *gymnasia* were certainly places of physical training. Given the fact that most rural *gymnasia* were founded as private foundations by soldiers, and that the majority of their members were cleruchs or military settlers, they are also likely to have been places of military training.[38]

Should we classify the cleruchs as part of a single broad Greek military diaspora or do we have hints that some of them formed specific military immigrant communities with different ethnic identities? The *gymnasia* suggest the first conclusion. Although we cannot prove that all cleruchs were members of the *gymnasia* nor that all *gymnasium* members were soldiers, the rural *gymnasia* especially were characterized by a strong presence of military personnel.[39] This suggests that the Greco-Macedonian cleruchs in Egypt wished to preserve a common Greek identity and create focal points of social networks where a common Greek lifestyle was manifested in a formal institution.[40] Therefore, the military diaspora reflected by the *gymnasia* should be understood as part of a Greek diaspora whose identity was not ethnic but cultural.[41] This conclusion is supported by the later history of the *gymnasia* in Egypt: Although they faced social and cultural transformations over time, the *gymnasia* never stopped representing Greek culture.[42] And from the second century BCE, military recruits of Egyptian or Greco-Egyptian origin came to be admitted as members,[43] so much so that Christelle Fischer-Bovet has argued convincingly that "the gymnasium became an engine of integration".[44]

But can we detect specific ethnic identities among the military settlers? In some villages, there lived substantial groups of cleruchs sharing the same origin; in already existing Egyptian settlements, cleruchs could, indifferently, live close to each other, or have Egyptian neighbours.[45] Occasionally even whole settlements or quarters within a nome capital seem to have been named after a foreign region, a suggestive fact, although we know nothing about the actual population of these neighbourhoods.[46] Thanks to onomastics and the use of "legal ethnic designations", concentrations of cleruchs with a common origin can be identified in the following locations: At Pitos (Memphite nome), we meet a group of Thracian cleruchs in the first half of the third century BCE; in the lower Oxyrhynchite toparchy, at the villages of Tholthis and Takona, Cyreneans formed the majority of the Greek military settlers in the second half of the third century BCE; and the same probably applies to those Jewish inhabitants of Samareia (Arsinoite nome) who are attested from the mid-third to the mid-second century BCE and served in the Ptolemaic army (among them several cleruchs).[47] Had these military groups a communal character and a sense of their ethnicity? It is possible, yet in none of the cases are structures of internal governance and shared worship known to us, but that may merely be owed to the lack of evidence;[48] and the Cyreneans, at least, continued to use their Greek dialect.

Community-building along ethnic lines and the existence of ethnic neighbourhoods would of course hardly be surprising among cleruchs,[49] if not yet amenable to proof. The question is rather how long such posited ethnic groups lasted after the first generation of settlers. Generally, our evidence about the settlement of cleruchs in the Egyptian countryside does not show systematic ethnic clustering. Rather, the evidence for the *gymnasia* implies that it was cultural "Greekness" and not city or region of origin based "ethnicity" that mattered from a social and occupational perspective.[50] The state perhaps took notice of the weakness of ethnic feeling: At the beginning of the second century BCE, the cleruch cavalry were no longer divided into both ethnic and numbered subdivisions (*hipparchiai*), the former categories being dissolved and incorporated into the latter,[51] albeit with the preservation, no doubt for ideological reasons, of the single special category of *Makedones*. To find groups that were both organized as associations or communities and seemed to have preserved some kind of ethnic identity, we have to turn to the second pillar of the Ptolemaic army, the mercenaries, or professional soldiers.

The Mercenaries or Professional Soldiers

The Ptolemies recruited full-time mercenary soldiers to use in war, but who also functioned in peacetime to garrison strategically significant points.[52] A significant proportion of such military bases were in larger or urban settlements. The roots of this system lay in late Pharaonic times and can be traced back to the seventh century BCE.[53] In general, it seems that the great majority of soldiers in garrisons were professionals and not cleruchs.[54] In the third century BCE, these professional soldiers were (similarly to cleruchs) immigrants or the sons of immigrants. Statistics show that in this period, the Ptolemies recruited mercenaries—in contrast to cleruchs—by preference in regions where they had possessions or influence, as in Asia Minor, Crete, and the Levant,[55] a practice that is likely explained by the fact that mercenaries recruited from within the Ptolemaic empire would not vanish after the end of their service because they would return to areas controlled by the Ptolemies from which, if necessary, they could be rehired.[56] This pattern of recruiting perhaps also explains why there are few signs of official attempts to integrate mercenaries who had come to Egypt to serve there into local life. Nevertheless, evidence survives of a small number of mercenaries who were apparently given grants of Egyptian land, albeit smaller than the plots given to cleruchs—perhaps because these mercenaries continued to receive pay.[57] Furthermore, there is evidence for the institution of the *politeuma* ("polity"), a kind of association that was probably tailored to specific segments of the population whose origins lay in groups of immigrant mercenaries of the same provenance.[58] This institution is of prime interest to our investigation and deserves a closer look.

Politeumata were described by ethnic designations that pointed to foreign ethnic groups.[59] In Egypt, a *politeuma* of Cilicians (named after the region of Cilicia in the southeast of Asia Minor),[60] one of Boeotians (named after Boeotia in central Greece),[61] one of Cretans,[62] one of Jews,[63] and one of Idumaeans (named after Idumaea south of Judaea)[64] are attested. We come across all these *politeumata* in the second or first century BCE.[65] For their locations in Egypt, we know only that the Boeotian *politeuma* was based in the nome capital of Xois in the north of the Nile delta, the Idumaean in Memphis (at the southern tip of the Nile delta), and the Jewish in Heracleopolis in Middle Egypt. The Cilician and the Cretan *politeuma* cannot be located exactly, but it appears likely that they were in the Fayum or the Arsinoite nome. Other *politeumata* are only attested after Egypt fell under Roman rule and became a Roman province—in the year 30 BCE, after Octavian defeated Antony and the last Ptolemaic monarch Cleopatra at the Battle of Actium in 31 BCE—but they are probably older, originating in the Ptolemaic period. At the end of the first century BCE, we come across a *politeuma* of Phrygians (named after the region Phrygia in the west of Asia Minor), whose location in Egypt is unknown,[66] and many years later, in 120 CE, we encounter a *politeuma* of Lycians (named after the region Lycia in southwestern Asia Minor), which existed in Alexandria.[67]

The link between *politeumata* and foreign mercenaries serving the Ptolemies seems secure. The texts illuminating a *politeuma* of Cilicians, Boeotians, Cretans, and Idumaeans indicate that these groups had close links with military dignitaries or consisted partly of professional soldiers.[68] Furthermore, an inscription that dates from the year 112/111 or 76/75 BCE refers to a *politeuma* of soldiers of unspecified ethnicity stationed in Alexandria (*SEG* 20.499). Outside Egypt, the three *politeumata* at Sidon (now in Lebanon), when it was still under Ptolemaic control, are known from gravestones of their members—gravestones that depict armed men.[69] The Jewish *politeuma* of Heracleopolis was located in the harbour district of that nome capital: In the fifties of the second century BCE, shortly before the *politeuma* is attested, a fortress was built in this same area, and it seems most natural to conclude that the original membership of the Jewish *politeuma* would have consisted of Jewish soldiers residing near the strongpoint they garrisoned.[70]

Like the Jewish *politeuma* of Heracleopolis, moreover, hitherto in Egypt *politeumata* are securely attested only in nome capitals, a fact that itself suggests a connection between the *politeumata* and troops of mercenaries or professional soldiers who were characteristically garrisoned in such towns. That the origins of the known *politeumata* are to be found in bodies of mercenaries (and their civilian staff and families) is further confirmed by the ethnic designations they bore. Most of these refer—Boeotians and Phrygians excluded—to regions (Lycia, Cilicia, Judaea, Idumaea) that were temporarily in the possession of the Ptolemies or where, as in Crete, they had a military presence,[71] regions where—as already indicated—the Ptolemies tended to recruit mercenaries in the third century BCE. These patterns of recruitment may imply that most of the *politeuma* go back to the third century BCE, because afterwards the Ptolemies lost their large extra-Egyptian possessions in Asia Minor and the Levant.[72] There is no actual evidence for a *politeuma* dated to the third century BCE, nor for that matter, for the date of foundation of any of the *politeumata* in Egypt. But evidence from outside Egypt could lend some support to the hypothesis of third-century origin: As already indicated, *politeumata* are attested for Ptolemaic Sidon at the end of the third century BCE.[73] Nothing, however, excludes the possibility of either the foundation of *politeumata* in Egypt, or the migration of their members to Egypt, in the second (or even first) century BCE: Even after the territory of the Ptolemaic kingdom had been reduced to Egypt, Cyprus, and the Cyrenaica, the Ptolemies were still eager and able to recruit soldiers from other regions.[74] From lands, once Ptolemaic but now under hostile control, powerful political refugees and their existing forces or retainers were natural recruits, a fact illustrated by the Ptolemaic reception of the Judaean Onias, member of the Oniad family (descendants of Zadok, high priest under Solomon, whose ancestors had held the office of the high priest at Jerusalem since Onias I [ca. 320–280 BCE]).[75] Political confusion in Judaea, a consequence of the revolt of the Maccabees, drove Onias—accompanied by fellow Jews—to Egypt, and he was allowed by Ptolemy VI to found a Jewish temple and form a military colony in Leontopolis (southeast of the Nile Delta).[76] The start of construction can, depending on our interpretation of Josephus, be dated between 164 and 150 BCE.[77] Some years later, Idumaens possibly took refuge in Egypt after Idumea had been captured and annexed by the Jewish leader John Hyrcanus in ca. 125 BCE.[78] In short, even in a period of declining Ptolemaic power, there is no reason to think the influx of outside soldiers into Egypt ever came to an abrupt end. It rather continued to a lesser degree even in an altered geopolitical context.[79] Therefore, although the Ptolemies started to recruit professional soldiers primarily within Egypt at the turn of the second century BCE,[80] they seem also to have tried—as far as possible—to maintain the recruitment patterns they used in the third century BCE when the kingdom ruled the sea and had far-flung possessions.

Apart from its military character, the decisive characteristic of the *politeuma* is that it was an administrative unit sanctioned by the Ptolemaic authorities that was based on a (semi-autonomous) community or association and its territorial base.[81] This conclusion is drawn from *P.Polit.Iud.*, the archive of twenty papyri texts (dated between 144/143 and 133/132 BCE) attesting the Jewish *politeuma* at Heracleopolis.[82] This archive provides the first definite attestation of a Jewish *politeuma* in the Hellenistic period. The existence of a *politeuma* in Alexandria is not proven, nor is the supposed Jewish *politeuma* of Leontopolis originating in Onias' military colony;[83] and the documents illuminating the Jewish *politeuma* of Berenice in Cyrenaica are dated to Roman, not Ptolemaic, times.[84] *P.Polit.Iud.* suggests that the Jewish *politeuma* of Herakleopolis actually governed its own quarter of the city, an area that was located in the harbour district (with its new fort, which it is likely the Jews garrisoned), which was about a mile removed from the town and located on the Bahr Yusuf, the western branch of the Nile. There the officials of the Jewish *politeuma*, the archons, under a higher official called the politarch, seemed to act (at least in judicial matters) like state functionaries and were supported by lesser officials. Like Ptolemaic officials, the officials of the *politeuma* were approached by means of petitions from their subjects, ordinarily in private legal disputes between Jews, but sometimes also in disputes between Jews and non-Jews. The petitioners appear always to be Jewish. What petitioners expected of the archons was not that they should summon a court that would generate a judicial verdict (as might be rendered by a Greek court in Egypt like the *dikasterion* or the court of the *chrematistai*) but rather the judgement of cases by the archons themselves, by virtue of their own authority, and the enforcement of legal claims that had been granted by the archons of the *politeuma*, by virtue of the authority that inhered in their position. The procedure, therefore, followed the same patterns as the justice of Ptolemaic officials, who gave justice in their own right as magistrates. The petitions show that Jewish beliefs, particularly the ancestral Jewish law, here called the *patrios nomos*, flow into the argumentation and the structure of the petitions to the archons. The allusions and explicit references to Jewish belief seem to be a strategy of argumentation directed at specifically Jewish officials, who would understand the religious considerations adduced by the petitioners, and so be vulnerable to persuasion and influenced thereby.[85] The jurisdiction and significance of the Jewish *politeuma*, moreover, was not restricted only to Herakleopolis or its harbour district. For the papyri attest that Jews living in villages outside Herakleopolis petitioned the archons, and that rural Jewish communities or associations seem to have had links to them—an unmistakable sign of the wide sphere of influence of the Jewish *politeuma* of Herakleopolis, even if we are not exactly sure of the sources and nature of that influence outside the *politeuma*'s formal boundaries.

There is no reason to regard the Jewish *politeuma* of Herakleopolis as unique, or distinct from the *politeumata* of other ethnic groups. In the section "Ethnic Designations and Fiscal/Legal Categories in Ptolemaic Egypt: Traces of Diasporic Groups?" of this chapter, it was already pointed out that Jews in general were classified among the tax-*Hellenes* and this also applied to Boeotians, Cilicians, Cretans, Lycians, Phrygians, or Idumaeans—other ethnic groups that were also organized as *politeumata* and only some of whom had claims to real Greek ancestry. Viewed constitutionally and socio-politically, therefore, Jews did not form a separate class of population in the Ptolemaic kingdom, and there is no reason to consider the Jewish *politeuma* of Herakleoplis a special case.[86] Rather, we should consider—as a working hypothesis—the likelihood that all the *politeumata* listed above held the same position in the Ptolemaic state.

This does not mean that all *politeumata* were organized identically: To be sure, a council of archons, which presided over the Jewish *politeumata* of Heracleopolis and Berenike, is well known from Jewish associations or synagogue communities.[87] But non-Jewish ethnic *politeumata* seem to have employed different officials. In the case of the *politeuma* of soldiers stationed in Alexandria, one encounters a *prostates* (president) and a *grammateus* (scribe); for the Phrygian and Boeotian *politeuma* a priest is attested. Furthermore, we are informed that the Boeotian, Cilician, and Idumaean *politeuma* each had its own sanctuary or temple district; it can, therefore, be assumed that in the last two *politeumata*, as well as in the first, a priest presided over the cult of each group. In the case of the Phrygians, the Boeotians, and the Idumaeans, it is unquestionable that their religious identities were strongly connected to the homelands to which their respective ethnic designations alluded: The Phrygians worshiped Zeus Phrygios, the Boeotians Zeus Basileus, a particularly Boeotian aspect of Zeus,[88] and the Idumaeans (as their sanctuary, called an Apollonieion, reveals) Apollo, who is to be identified with Qos, the main god of the Idumaeans before they converted to Judaism.[89] The cult of the Cilicians is less specifically directed at a homeland god but has at least a strong Greek connotation: It is devoted to Zeus and his doughter, Athena. In the case of the Jewish *politeuma* of Heracleopolis, Jewish belief becomes apparent in the petitions addressed to the archons, and the titles of these officials may suggest that behind the *politeuma* is hidden a synagogue community.

Given the fact that *politeumata* formed cult associations that carried on the rites of the "homeland" indicated by their ethnic designation and had their own administration, which—if the Jewish *politeuma* of Herakleopolis is anything to go by—seems to have a territorial character (a feature which, by the way, fits the most common Greek sense of the word *politeuma*), they cannot be categorized merely as "ethnic networks" or "ethnic associations", but should be regarded as "ethnic communities" according to the terminology of social science.[90] Furthermore, the location of the *politeumata* and the ethnic designations they bore suggest that these communities were the outcome of the settlement of ethnically defined mercenary groups whose units had been stationed—as far as we can see—in nome capitals, where most of these professional soldiers lived in the same neighbourhood and probably in the vicinity of their garrison. The *politeumata* are without doubt the best example of a process described by Dorothy Thompson:

> Local ethnic communities in the Ptolemaic period often derived in origin from military groups; [but] in their developed form they were total communities, consisting of far more that just the military.[91]

In other words, *politeumata* were founded as an aspect of Ptolemaic "military policy" but over time, may have lost much of their military character: We cannot know how many of the members of a *politeuma* chose military careers after the first generation, although our sources suggest that some did or that new members of the same ethnic group were imported to do so (the 500 men who are said to have reinforced the Cretan *politeuma* could have well been soldiers recruited in Crete[92]), if only because *politeumata* do not appear to have multiplied in cities, as would have happened if most or all the descendants of the original mercenaries chose civilian careers and the Ptolemies had to bring in new mercenaries to perform the military functions they abandoned. That said, we have no indication that *politeumata* themselves mainly served military functions. Rather, the transformation of ethnic communities, consisting of soldiers and their families, into administrative units seems to have been a civil and social measure:[93] This is certainly the case with the Jewish *politeuma* in Heracleopolis we witness in *P.Polit.Iud.*

For mercenaries or professional soldiers and their existence in the military diaspora of Hellenistic Egypt, evidence is not restricted to those groups just mentioned. In Memphis, where the *politeuma* of the Idumaeans was located, we know that the so-called Hellenomemphites and Karomemphites—descendants of Ionian and Carian mercenaries settled in Memphis in the sixth century BCE—inhabited their own quarters, had a cult centre and—as far as the Hellenomemphites are concerned—their own leaders, the *timouchoi* or "honourables".[94] Much older even than these groups were the Phoenico-Egyptians of Memphis. Possibly originating as Canaanite merchants, migrating to Memphis as early as the fifteenth century BCE, and as Phoenician mercenaries, settled (like the Ionians and Carians) in the sixth century BCE, in Ptolemaic times, they still had their own priests and a temple.[95] Traces of comparable groups with Semitic and Jewish backgrounds and connected with Persian garrisons of the fifth century BCE can be found in Memphis and Syene/Elephantine.[96] These groups may provide us with more or less clear examples of ethnic communities that were rooted in migrating mercenaries or soldiers and survived under the Pharaohs or Persian domination into the age of the Ptolemies.

Besides the *politeumata* and the Jewish military colony in Leontopolis, the evidence for organized ethnic groups whose emergence is arguably linked to Ptolemaic military policy is limited: First, a single papyrus from the third century BCE indicates that in the Arsinoite village of Philadelphia, a group of Arabs (who, as a category, held the same fiscal privilege as the tax-*Hellenes*[97]) was represented by elders and officials called *dekadarchai*, while other documents suggest that in the Arsinoite nome these Arabs often served as guards or formed some kind of special police force.[98] Second, a group of *xenoi*, mercenaries, who call themselves *Apollonia(s)tai* are attested in two fragmentary inscriptions dated to the first century BCE in the nome capital of Hermoupolis.[99] They and those sharing the ceremonial act (*sympoliteuomenoi*) dedicated a sanctuary to Apollo, Zeus, and related gods. An onomastic analysis of the dedicants, whose names are inscribed beneath the main text and broken down by military units, seems to indicate that most but not all of them were Idumaeans.[100] Because some of the members of the *Apollonia(s)tai* have cult titles, it seems that we are dealing with a cult association that probably consisted mainly of Idumaean mercenaries.[101] But neither about the Arsinoite Arabs nor the *Apollonia(s)tai* do we have enough information to draw conclusions about the experience of migrant soldiers in Egypt that go beyond those we have already reached, other than to confirm that an organizational structure and joint religious observance seem to have been important to them. Finally, still in the Ptolemaic realm but outside Egypt, there are the mysterious ethnic *koina* in Cyprus. These are associations or assemblies—the word *koinon* can have both meanings[102]—of Achaeans and other Greeks, as well as of Cretans, Ionians, Thracians, Lycians, and Cilicians. Once again, these are groups of mercenaries or professional soldiers, but all that we know about them is that they met or gathered to honour high officials, predominantly the governor of the island, but sometimes also other dignitaries.[103]

Résumé

Despite the vast number of documentary sources (inscriptions, papyri) providing evidence of the use of ethnic designations in Hellenistic Egypt, experiences of diaspora groups in this region are difficult to deduce: Although the holders of ethnic designations may have been immigrants or descendants of immigrants, these designations cannot be taken as simple or one-sided indication of a person's actual identity.

Due to the recruitment policy of the Ptolemaic army, soldiers formed the largest migrant group in Egypt. To understand this military diaspora, two different kinds of Ptolemaic soldiers need to be differentiated: The military settlers, or cleruchs, representing the regular army and the mercenaries or professional soldiers. Statistical analyses of ethnic designations show that, at least in the third century BCE, both groups were recruited mainly from outside Egypt. Apparently, the Ptolemies even tried—as far as they could—to channel migration from certain extra-Egyptian regions into the two different military "job profiles": Cleruchs were recruited by preference from Macedon, mainland Greece, and Thrace—regions that were not controlled by the Ptolemies—and mercenaries or professional soldiers from the Ptolemaic outer possessions, especially Asia Minor and the Levant, even when the Ptolemies no longer controlled these areas. The distribution of migrants in two different military occupational groups is also reflected in the strategies employed to retain these immigrants in Egypt. On the one hand, cleruchs, who were intended for long-term employment, were granted plots of land for cultivation. On the other hand, there were the *politeumata*, which appear in the second century BCE and—because their number seems to have been limited—probably bear witness to the selective promotion of certain ethnic communities of particular importance for the Ptolemaic government that originated in contingents of mercenaries or professional soldiers. By incorporating communities of valuable mercenary warriors into the administrative structure of Ptolemaic Egypt, the *politeuma* can be regarded as the urban counterpart of the cleruchic settlements that were created with land grants: Both testify to how the Ptolemies tried to strengthen the ties between them and their army.[104]

Both military groups illuminate different aspects of the military diaspora in Hellenistic Egypt. By investigating the underlying identities of the soldiers, two main patterns appear. First, the evidence suggests that the emergence and adoption of a common Greek identity is an important feature of the milieux of the cleruchs, a phenomenon of which the *gymnasia* are emblematic. What we see is, therefore, a military diaspora that was part of a culturally defined Greek diaspora. Second, as far as specific ethnic identities are concerned, our information is most instructive in the case of mercenaries or professional soldiers. Apart from the Arabs, all the relevant groups appear in urban contexts, and the question arises whether it was the milieu of active (or once active) military men and/or the urban environment that fostered the emergence of ethnic associations or communities.[105] The clearest examples of these "ethnic components" of the military diaspora are without a doubt the *politeumata*, and, more generally, it is the *politeumata* that provide the best evidence for authentic ethnic communities in Hellenistic Egypt. These communities could be regarded as single, ethnically defined military diasporas of their own,[106] each of which belonged, at the same time, to the category of *Hellenes* or the "Hellenic sector" of Ptolemaic society.[107]

Abbreviations

BL	*Berichtigungsliste der griechischen Papyrusurkunden aus Ägypten.*
C.Pap.Jud.	*Corpus Papyrorum Judaicarum.* Cambridge, MA.
CIG	*Corpus Inscriptionum Graecarum.*
IG	*Inscriptiones Graecae.*
IGR	*Inscriptiones graecae ad res romanas pertinentes.*
Josephus, *AJ*	Josephus. *Antiquitates Iudaicae.*
Josephus, *BJ*	Josephus. *Bellum Iudaicum.*
P.Cair.Zen.	*Zenon Papyri, Catalogue général des antiquités égyptiennes du Musée du Caire*, edited by C. C. Edgar. Cairo.
P.Giss.	*Griechische Papyri zu Giessen.*
P.Polit.Iud.	*Urkunden des Politeuma der Juden von Herakleopolis (144/3, 133/2 v. Chr.)*, edited by Klaus Maresch and James M. S. Cowey. Wiesbaden: Westdeutscher Verlag, 2001.
P.Tebt.	*The Tebtunis Papyri.* London.
Polyb.	Polybius.
OGIS	*Orientis Graeci Inscriptiones Selectae.*
SB	*Sammelbuch griechischer Urkunden aus Ägypten.*
SEG	*Supplementum Epigraphicum Graecum.*
W.Chr.	Ludwig Mitteis, and Ulrich Wilcken, *Grundzüge und Chrestomathie der Papyruskunde, I. Band Historischer Teil, II. Hälfte Chrestomathie.* Leipzig: B. G. Teubner, 1912.

Notes

* I thank Georg Christ and Jon E. Lendon for their criticism and help with the English style. The present text is a longer version of my article "Military Immigration and the Emergence of Cultural or Ethnic Identities: The Case of Ptolemaic Egypt," *Journal of Juristic Papyrology* 45 (2015): 229–253; see also my book *Die ptolemäische Organisationsform* politeuma*: Ein Herrschaftsinstrument zugunsten jüdischer und anderer hellenischer Gemeinschaften* (Tübingen: Mohr Siebeck, 2019), esp. 141–156.

1 See Kostas Buraselis, "Ambivalent Roles of Centre and Periphery: Remarks on the Relation of the Cities of Greece with the Ptolemies until the End of Philometor's Age," in *Centre and Periphery in the Hellenistic World*, ed. Per Bilde et al. (Aarhus: Aarhus University Press, 1993), 251–270, at 258.

2 For a definition of "ethnic identity" in the context of Ptolemaic Egypt, see Jane Rowlandson, "Dissing the Egyptians: Legal, Ethnic, and Cultural Identities in Roman Egypt," in *Creating Ethnicities & Identities in the Roman World*, ed. Andrew Gardner, Edward Herring, and Kathryn Lomas (London: Institute of Classical Studies, 2013), 213–247, at 215–216 (quotation from p. 216) (drawing on Jonathan M. Hall, *Ethnic Identity in Greek Antiquity* [Cambridge: Cambridge University Press, 1997]; id., *Hellenicity between Ethnicity and Culture* [Chicago: The University of Chicago Press, 2002]); Christelle Fischer-Bovet, *Army and Society in Ptolemaic Egypt* (Cambridge: Cambridge University Press, 2014), 172–173. For an exact definition of "military diaspora", see the chapter of Nathanael Andrade in this volume, and for that of "ethnic minority" in Ptolemaic Egypt, see n. 90, below.

3 See Fischer-Bovet, *Army and Society in Ptolemaic Egypt*, 171.

4 Csaba La'da, *Foreign Ethnics in Hellenistic Egypt*, Prosopographia Ptolemaica 10; Studia Hellenistica 38 (Leuven: Peeters, 2002); Katja Müller, *Settlements of the Ptolemies: City Foundations and New Settlement in the Hellenistic World* (Leuven: Peeters, 2006), 168.

5 See Dorothy J. Thompson, "Hellenistic Hellenes: The Case of Ptolemaic Egypt," in *Ancient Perceptions of Greek Ethnicity*, ed. Irad Malkin (Cambridge, MA: Harvard University Press, 2001), 301–322, at 307–310, and Willy Clarysse and Dorothy J. Thompson, *Counting the People in Hellenistic Egypt,* vol. 2, *Historical Studies* (Cambridge: Cambridge University Press, 2006), 138–147.

6 Thracians and Jews: See Joseph Mélèze Modrzejewski, "Le statut des Hellènes dans l'Égypte lagide: Bilan et perspectives des recherches," *Revue des Études Grecques* 96 (1983): 241–268, at 265–266; and Clarysse and Thompson, *Counting the People*, 145, 147–148. Egyptians: See Thompson, "Hellenistic Hellenes," 310–312; Clarysse and Thompson, *Counting the People*, 142–145.

7 See Roger S. Bagnall, "The People of the Roman Fayum," in *Hellenistic and Roman Egypt: Sources and Approaches*, ed. id. (Aldershot: Ashgate, 2006), chap. 14, 1–19, at 3 (originally published in *Portraits and Masks: Burial Customs in Roman Egypt*, ed. Morris L. Bierbrier [London: British Museum Press, 1997], 7–15); Clarysse and Thompson, *Counting the People*, 155.

8 See Thompson, "Hellenistic Hellenes," 306; Katelijn Vandorpe, "Persian Soldiers and Persians of the Epigone: Social Mobility of Soldiers-Herdsmen in Upper Egypt," *Archiv für Papyrusforschung* 54 (2008): 87–108; and Fischer-Bovet, *Army and Society in Ptolemaic Egypt*, 177–191.

9 The terminology "legal ethnic designations" is taken from Vandorpe, "Persian Soldiers and Persians of the Epigone," 87.

10 See Thompson, "Hellenistic Hellenes," 304–306.

11 For the evidence, see Sandra Scheuble, "Griechen und Ägypter im ptolemäischen Heer—Bemerkungen zum Phänomen der Doppelnamen im ptolemäischen Ägypten," in *Interkulturalität in der Alten Welt: Vorderasien, Hellas, Ägypten und die vielfältgen Ebenen des Kontakts*, ed. Robert Rollinger et al. (Wiesbaden: Harrassowitz, 2010), 551–560, at 551 (with further bibliographical references in n. 1).

12 See Csaba A. La'da, "*Ethnic Designations in Hellenistic Egypt*" (PhD diss., University of Cambridge, 1996), 173–189; Thompson, "Hellenistic Hellenes," 307; Clarysse and Thompson, *Counting the People*, 124–147, 155; Ian S. Moyer, *Egypt and the Limits of Hellenism* (Cambridge: Cambridge University Press, 2014), 30, n. 110.

13 See Buraselis, "Ambivalent Roles of Centre and Periphery," 159; La'da, "Encounters with Ancient Egypt: The Hellenistic Greek Experience," in *Ancient Perspectives on Egypt: Encounters with Ancient Egypt*, ed. Roger Matthews and Cornelia Roemer (London: UCL Press, 2003), 157–169, at 166–167; Tony Spawforth, "'Macedonian Times': Hellenistic Memories in the Provinces of the Roman Near East," in *Greeks on Greekness: Viewing the Past under the Roman Empire*, ed. David Konstan and Suzanne Saïd (Cambridge: Cambridge Philological Society, 2006), 1–26, at 5–7.

14 As a result, despite the fact that actual immigration to Egypt declined from the late third century BCE on, at least until the middle of the second century BCE

there was no reduction in the proportion of military settlers or cleruchs designated as *Makedones*; see Sandra Scheuble-Reiter, *Die Katökenreiter im ptolemäischen Ägypten* (Munich: C. H. Beck, 2012), 114–115; Mary Stefanou, "Waterborne Recruits: The Military Settlers of Ptolemaic Egypt," in *The Ptolemies, the Sea and the Nile: Studies in Waterborne Power*, ed. Kostas Buraselis, Mary Stefanou, and Dorothy J. Thompson (Cambridge: Cambridge University Press, 2013), 108–131, at 123–124.

15 Thompson, "Hellenistic Hellenes," 305–306.

16 Corresponding to the results of recent research, official identity displayed by ethnic designations has to be distinguished from cultural or ethnic identity (the underlying concept of which was explained in the introduction); see Bernard Legras, *L'Égypte grecque et romain* (Paris: Colin, 2004), 60 and especially Fischer-Bovet, *Army and Society in Ptolemaic Egypt*, 170–173.

17 See La'da, "*Ethnic Designations in Hellenistic Egypt*," 87–91; Thompson, "Hellenistic Hellenes," 304; Müller, *Settlements of the Ptolemies*, 168; Fischer-Bovet, *Army and Society in Ptolemaic Egypt*, 171. Many years ago, Elias Bickermann, "Beiträge zur antiken Urkundengeschichte I.: Der Heimatsvermerk und die staatsrechtliche Stellung der Hellenen im ptolemäischen Ägypten," *Archiv für Papyrusforschung* 8 (1927): 216–239, at 234–238, had already noticed that the decrease in the use of ethnic designations in documents is paralleled by an increase in the use of Egyptian geographical designations.

18 For the research history, see the recent overviews of Moyer, *Egypt and the Limits of Hellenism*, 11–36 and (focusing the Ptolemaic army and its social role) Fischer-Bovet, *Army and Society in Ptolemaic Egypt*, 4–6. Among studies of ethnicity in Ptolemaic Egypt, a colonialist interpretation—driven by the idea that the native Egyptians were ruled by a superior class of Macedonian and Greek conquerors and that, in the course of time, contact or intermarriage between these two parts of society eventually led to the decline of Hellenism and the Ptolemaic kingdom—is detectable in Bickermann, "Beiträge zur antiken Urkundengeschichte." Although, according to him, there was no legal difference between a common *Hellen* and an Egyptian (238–239), he argued, based on his observations mentioned in n. 17 (above), for an Egyptianization of the *Hellenes* so that Hellenism was extinct at the end of the Ptolemaic era. A post-colonial view is represented Mélèze Modrzejewski, "Le statut des Hellènes," esp. 252–258 and recently id., *Loi et coutume dans l'Égypte grecque et romaine* (Warsaw: Faculty of Law and Administration of the University of Warsaw, 2014), 108–109, 186–188, 192, 197, 221, 228 who argues that the conquerors, marked by their legal ethnic designations pointing to their belonging to a Greek ethnic, social, and cultural milieu, were sharply separated from the conquered people, the indigenous Egyptians, through the entire course of Ptolemaic rule. Like Bickermann, Modrzejewski, "Le statut des Hellènes," 256–258, pointed out that, in the legal sphere at least, there was no discrimination against Egyptians. Also arguing against this form of ethnic discrimination—often implicit in separation models favoured by the older scholarship (see, e.g., the handbooks on law and society of Hellenistic Egypt of Ludwig Mitteis, *Grundzüge und Chrestomathie der Papyruskunde, Zweiter Band: Juristischer Teil, Erste Hälfte: Grundzüge* [Leipzig: B. G. Teubner, 1912], xii–xiii; Rafael Taubenschlag, *The Law of Greco-Roman Egypt in the Light of the Papyri: 332 B.C.–640 A.D.*, 2nd ed. [Warsaw: Państwowe wydawnictwo Naukowe, 1955], 8–12; Erwin Seidl, *Ptolemäische Rechtsgeschichte*, 2nd ed. [Glückstadt: Augustin, 1962], 1–2) and influential in major studies on ancient Jewry (see n. 19 below)—in Ptolemaic policy, are the works of Koen Goudriaan, *Ethnicity in Ptolemaic Egypt* (Amsterdam: J. C. Gieben, 1988) and La'da, *Ethnic Designations* (see also id., "Encounters with Ancient Egypt"), who, although their focuses are different, both re-evaluate the evidence related to the ethnic designations. Goudriaan (drawing on Frederik Barth, *Ethnic Groups and Boundaries: The Social Organization of Culture Difference* [Boston: Little, Brown, 1969]) only investigated the (unofficial) use of the designations *Hellen* and *Aigyptos* aiming at self-description, but on the whole the study is too limited to be satisfactory and sometimes even misleading (see Jane

Rowlandson, Review of *Ethnicity in Ptolemaic Egypt*, by Koen Goudriaan, *The Classical Review*, n.s., 40 [1990]: 370–371; Csaba A. La'da, Review of *Ethnicity in Hellenistic Egypt*, ed. Per Bilde et al., *Topoi* 4 [1994]: 341–350, at 343). La'da offers a comprehensive discussion of ethnic designations and divides them into designations that reflect real ethnic affiliations and those that are fictitious (like the so-called tax-*Hellenes*). Considerations of the relationship between "real" ethnic designations and ethnic identity are, however, lacking.

19 A widespread assumption in the older scholarship was that a community of pious Jews required self-governing associations that allowed them a degree of independence and certain legal privileges so as to preserve their alien customs and laws. With some variations, we find these ideas—especially that a Jewish community in a pagan environment could only survive under the protection of a *politeuma* (a kind of association dealt with below), as a state within a state—in the influential works of Victor A. Tcherikover (*Hellenistic Civilisation and the Jews* [Philadelphia, Jerusalem: Jewish Publication Society of America, 1959], 298–305), Mary Smallwood (*The Jews under Roman Rule from Pompey to Diocletian* [Leiden: Brill, 1976], 225–226), Shimon Applebaum ("The Legal Status of the Jewish Communities in the Diaspora" and "The Organization of the Jewish Communities in the Diaspora," both articles in *The Jewish People in the First Century: Historical Geography, Political History, Social, Cultural and Religious Life and Institutions*, ed. Shmuel Safrai and Menahem Stern [Assen: Van Gorcum, 1974], 1:420–463 and 464–503, at 430, 452, 465), and Aryeh Kasher (*The Jews in Hellenistic and Roman Egypt: The Struggle for Equal Rights* [Tübingen: J. C. B. Mohr (Paul Siebeck), 1985], with a summary on 356–357). However, this approach was challenged with good cause by Joseph Mélèze Modrzejewski (*The Jews of Egypt: From Ramses II to Emperor Hadrian* [Princeton: Princeton University Press, 1997], 80–83; *Loi et coutume dans l'Égypte grecque et romaine*, 153–157), Constantine Zuckerman ("Hellenistic *politeumata* and the Jews: A Reconsideration," *Scripta Classica Israelica* 8/9 [1985–1988]: 171–185), and Sylvie Honigman ("Philon, Flavius Josèphe, et la citoyenneté alexandrine: Vers une utopie politique," *Journal of Jewish Studies* 48 [1997]: 62–90, at 62–65, 89–90), and is also not confirmed by the new evidence provided by *P.Polit.Iud.* attesting a Jewish *politeuma* in Herakleopolis: see Sylvie Honigman, "The Jewish *Politeuma* at Heracleopolis," *Scripta Classica Israelica* 21 (2002): 251–266, at 264–265; ead., "*Politeumata* and Ethnicity in Ptolemaic Egypt," *Ancient Society* 33 (2003): 61–102, at 93–95; Patrick Sänger, "Heracleopolis, Jewish *politeuma*," in *The Oxford Classical Dictionary*, ed. Sander Goldberg (New York, 2016), accessed 24 March 2022, https://doi.org/10.1093/acrefore/9780199381135.013.8036.

20 Rowlandson, "Dissing the Egyptians," 218. On these "two faces" of Ptolemaic society, see ibid., 218–219; Moyer, *Egypt and the Limits of Hellenism*, 30–32.

21 On intermarriage within the milieu of cleruchs, see Fischer-Bovet, *Army and Society in Ptolemaic Egypt*, 247–250; Scheuble-Reiter, *Die Katökenreiter*, 140.

22 See Willy Clarysse, "Greeks and Egyptians in the Ptolemaic Army and Administration," *Aegyptus* 65 (1985): 57–66; id., "Some Greeks in Egypt," in *Life in a Multi-Cultural Society: Egypt from Cambyses to Constantine and Beyond*, ed. Janet H. Johnson (Chicago: The Oriental Institute of the University of Chicago, 1992), 51–56, at 52–53; La'da, "Encounters with Ancient Egypt," 167–168; Sandra Scheuble, "Griechen und Ägypter im ptolemäischen Heer."

23 For a summary overview, Mélèze Modrzejewski, *The Jews of Egypt*, 76–80; James M. S. Cowey and Klaus Maresch, *Urkunden des Politeuma der Juden von Herakleopolis (144/3–133/2 v. Chr.) (P.Polit.Iud.)* (Wiesbaden: Westdeutscher Verlag, 2001), 30–32 (with further literature at n. 98).

24 See Bagnall, "The People of the Roman Fayum," 7–9.

25 Christelle Fischer-Bovet, "Counting the Greeks in Egypt: Immigration in the First Century of Ptolemaic Rule," in *Demography and the Graeco-Roman World: New Insights and Approaches*, ed. Claire Holleran and April Pudsey (Cambridge: Cambridge University Press, 2011), 135–154.

26 See Clarysse and Thompson, *Counting the People*, 94 (table 4:1 and 4:2), 139–140. At least in the third century BCE, the military registration was separate from the civilian one; see ibid., 62, 139–140, 155.
27 See Clarysse and Thompson, *Counting the People*, 154; Fischer-Bovet, *Army and Society in Ptolemaic Egypt*, 183–186.
28 See Fischer-Bovet, *Army and Society in Ptolemaic Egypt*, 118–123.
29 See Scheuble-Reiter, *Die Katökenreiter*, 24; Fischer-Bovet, *Army and Society in Ptolemaic Egypt*, 199–200.
30 See Roger S. Bagnall, "The Origins of Ptolemaic Cleruchs," *Bulletin of the American Society of Papyrologists* 21 (1984): 7–20; Scheuble-Reiter, *Die Katökenreiter*, 114–118.
31 Scheuble-Reiter, *Die Katökenreiter*, 18–23, 117 with n. 20; Stefanou, "Waterborne Recruits," 108–131.
32 See Scheuble-Reiter, *Die Katökenreiter*, 25.
33 See Dorothy J. Crawford, *Kerkeosiris: An Egyptian Village in the Ptolemaic Period* (Cambridge: Cambridge University Press, 1971), 55; Müller, *Settlements of the Ptolemies*, 149–151; Fischer-Bovet, *Army and Society in Ptolemaic Egypt*, 201.
34 On the settlement of the cleruchs, see Scheuble-Reiter, *Die Katökenreiter*, 27–32; on the residence of the cleruchs in the villages, see ibid., 33–38; Fischer-Bovet, *Army and Society in Ptolemaic Egypt*, 239–242.
35 See Willy Clarysse, "Egyptian Estate-Holders in the Ptolemaic Period," in *State and Temple Economy in the Ancient Near East*, ed. Edward Lipiński (Leuven: Department Oriëntalistiek, 1979), 2:731–743, at 735; see also Clarysse and Thompson, *Counting the People*, 151; Scheuble-Reiter, *Die Katökenreiter*, 285.
36 See Clarysse and Thompson, *Counting the People*, 139–140 and Fischer-Bovet, "Counting the Greeks in Egypt," 151, n. 62.
37 On the diffusion of *gymnasia* in Ptolemaic Egypt, see the map provided by Wolfgang Habermann, "Gymnasien im ptolemäischen Ägypten—eine Skizze," in *Das hellenistische Gymnasion*, ed. Daniel Kah and Peter Scholz (Berlin: Akadmie Verlag, 2004), 335–348, at 337.
38 On the relationship between the army and the *gymnasia*, see most recently Scheuble-Reiter, *Die Katökenreiter*, 309–315; Fischer-Bovet, *Army and Society in Ptolemaic Egypt*, 280–290.
39 See also Clarysse and Thompson, *Counting the People*, 133–134.
40 See Fischer-Bovet, *Army and Society in Ptolemaic Egypt*, 279–280.
41 On an Egyptian or national identity that was opposed to a Greek or cultural identity, see Rowlandson, "Dissing the Egyptians," 216–217.
42 See Thompson, "Hellenistic Hellenes," 312.
43 See Fischer-Bovet, *Army and Society in Ptolemaic Egypt*, 281–282, 283–284, 289–290, and also Scheuble-Reiter, *Die Katökenreiter*, 313–314.
44 Fischer-Bovet, *Army and Society in Ptolemaic Egypt*, 289.
45 See Willy Clarysse, "Ethnic Diversity and Dialect among the Greeks of Hellenistic Egypt," in *The Two Faces of Graeco-Roman Egypt*, ed. Pieter W. Pestman and Arthur M. F. W. Verhoogt (Leiden: Brill, 1998), 1–13, at 1–2; Clarysse and Thompson, *Counting the People*, 151; Scheuble-Reiter, *Die Katökenreiter*, 27–32; Fischer-Bovet, *Army and Society in Ptolemaic Egypt*, 247.
46 See Scheuble-Reiter, *Die Katökenreiter*, 27–29; Fischer-Bovet, *Army and Society in Ptolemaic Egypt*, 202.
47 Thracians: *P.Cair.Zen.* 1.59001 (274 BCE); Scheuble-Reiter, *Die Katökenreiter*, 27–28. Cyreneans: Clarysse, "Ethnic Diversity," 2–6; Sylvie Honigman, "The Jewish *Politeuma* at Heracleopolis," 265; ead., "*Politeumata* and Ethnicity in Ptolemaic Egypt," 99; Clarysse and Thompson, *Counting the People*, 320–321; Jews: Clemens Kuhs, "Das Dorf Samareia im griechisch-römischen Ägypten" (MA thesis, University of Heidelberg), 1996, 85–91, 107–110.
48 In the case of the Jews in Samareia, the absence of any kind of organizational structure (already noted by Kuhs, "Das Dorf Samareia," 110–111) is perhaps even more

astonishing given the fact that there are several examples of Jewish associations or synagogue communities scattered over Ptolemaic Egypt; relevant source texts are collected by Anders Runesson, Donald D. Binder, and Birger Olsson, *The Ancient Synagogue from its Origins to 200 C.E.: A Source Book* (Leiden: Brill, 2008), 171–217.

49 For the use of the term "ethnic neighborhood", see Clarysse, "Ethnic Diversity," 4–5 who assumed that the Cyreneans settled in the lower Oxyrhynchite toparchy formed such concentrated or closed communities ("By sticking together they were able to fend off the disappearance of their dialect for several generations", p. 5); cf. Clarysse and Thompson, *Counting the People*, 151 ("cleruchs were resident in the villages of the Egyptian countryside, sometimes living among other villagers but more often forming their own community within a village").

50 Cf. Scheuble-Reiter, *Die Katökenreiter*, 326–329.

51 See ibid., 60–71; Fischer-Bovet, *Army and Society in Ptolemaic Egypt*, 132–133.

52 See Sandra Scheuble, "Bemerkungen zu den μισθοφόφοι und τακτόμισθοι im ptolemäischen Ägypten," in "*... vor dem Papyrus sind alle gleich!" Papyrologische Beiträge zu Ehren von Bärbel Kramer*, ed. Raimar Eberhard et al. (Berlin: Walter de Gruyter, 2009), 213–222, at 214–215; Fischer-Bovet, *Army and Society in Ptolemaic Egypt*, 261–263, 269–279.

53 See Fischer-Bovet, *Army and Society in Ptolemaic Egypt*, 18–37.

54 Scheuble, "Bemerkungen zu den μισθοφόφοι und τακτόμισθοι," 218–220; ead., *Die Katökenreiter*, 240; Fischer-Bovet, *Army and Society in Ptolemaic Egypt*, 262.

55 See Bagnall, "The Origins of Ptolemaic Cleruchs," 16; Stefanou, "Waterborne Recruits," 127–131.

56 Marcel Launey, *Recherches sur les armées hellénistique*, 2nd ed. (Paris: De Boccard, 1987), 276–280 and Stefanou, "Waterborne Recruits," 127 explained the low number of Cretan cleruchs by assuming that Cretans soldiers preferred to be hired as mercenaries and returned home after military service.

57 See Scheuble, "Bemerkungen zu den μισθοφόφοι und τακτόμισθοι," 218.

58 See Patrick Sänger, "The *Politeuma* in the Hellenistic World (Third to First Century B.C.): A Form of Organisation to Integrate Minorities," in *Migration und Integration—wissenschaftliche Perspektiven aus Österreich. Jahrbuch 2/2013*, ed. Julia Dahlvik, Christoph Reinprecht, and Wiebke Sievers (Göttingen: V&R Unipress, 2014), 51–68, at 57–58.

59 On the evidence for the *politeumata*, see most recently Sänger, "The *Politeuma* in the Hellenistic World," 53–55.

60 *SB* 4.7270 = *SEG* 8.573 = Étienne Bernand, *Recueil des inscriptions grecques du Fayoum, vol. 1, La 'méris' d'Hérakleides* (Leiden: E. J. Brill, 1975), no. 15 = id., *Inscriptions grecques d'Égypte et de Nubie au musée du Louvre* (Paris: Centre National de la Recherche Scientifique, 1992), no. 22.

61 *SEG* 2.871 = *SB* 3.6664.

62 *P.Tebt.* 1.32 = *W.Chr.* 448.

63 *P.Polit.Iud.* 1–20. Against Bradley Ritter, "On the 'πολίτευμα in Heracleopolis'," *Scripta Classica Israelica* 30 (2011): 9–37, who rejects the commonly accepted existence of a Jewish *politeuma* in Herakleopolis, see Sänger, "The *Politeuma* in the Hellenistic World," 54, n. 7.

64 *OGIS* 737 = Joseph G. Milne, *Greek Inscriptions* (Oxford: Oxford University Press, 1905), 18–19, no. 33027 = *SB* 5.8929 = André Bernand, *La prose sur pierre dans l'Égypte hellénistique et romaine, 2 vols.* (Paris: Centre National de la Recherche Scientifique, 1992), no. 25. On the identification of the Idumaean *politeuma*, see Dorothy J. Thompson Crawford, "The Idumaeans of Memphis and the Ptolemaic *Politeumata*," in *Atti del XVII Congresso Internazionale di Papirologia*, vol. 3 (Naples: Centro Internazionale per lo Studio dei Papiri Ercolanesi, 1984), 1069–1075; ead., *Memphis under the Ptolemies*, 2nd ed. (Princeton: Princeton University Press, 2012), 93–96.

65 The testimony for the Cilician *politeuma* mentioned above could also be dated to the third century BCE. Bernand, *Inscriptions grecques*, no. 22, p. 65 summarized the various dating suggestions (from the third to the first century BCE) and favoured,

following Leon Mooren, *The Aulic Titulature in Ptolemaic Egypt: Introduction and Prosopography* (Brussels: Paleis der Academiën, 1975), no. 281, p. 173, a dating to the first century BCE.

66 *IG* 14.701 = *OGIS* 658 = *SB* 5.7875 = *IGR* 1.458 = François Kayser, *Recueil des inscriptions grecques et latines (non funéraires) d'Alexandrie impériale (i[er]-iiie s. apr. J.-C.)* (Cairo: Institut Français d'Archéologie Orientale, 1994), no. 74. On the provenance of the inscription, see also Werner Huß, *Die Verwaltung des ptolemaiischen Reiches* (Munich: Verlag C. H. Beck, 2011), 299 with further bibliographical references in n. 232.

67 *SB* 3.6025 = V 8757 = *IGR* 1.1078 = *SEG* 2.848 = Bernand, *La prose sur pierre*, no. 61 = Kayser, *Recueil des inscriptions*, no. 24.

68 The Boeotian *politeuma*, whose priest was *strategos* (the highest nome official), consisted of a group of soldiers and a group of civilians; see Zuckerman, "Hellenistic *politeumata* and the Jews," 175; Dorothy J. Thompson, "Ethnic Minorities in Hellenistic Egypt," in *Political Culture in the Greek City after the Classical Age*, ed. Onno M. van Nijf and Richard Alston (Leuven: Peeters, 2011), 101–117, at 110. In the case of the Cilician *politeuma*, we encounter a high-ranking military officer of *machairophoroi* (a troop of professional soldiers, literally "saber-bearers") acting as a benefactor of the community concerned. In the case of the Idumaean *politeuma*, a *strategos*, who simultaneously held the position of a priest of *machairophoroi*, was honoured by the Idumaeans. Given the position of both the benefactor of the Cilician *politeuma* and the honouree of the Idumaean *politeuma*, it is natural to assume that some members of these *politeumata* served as *machairophoroi*. Regarding the Cretan *politeuma*, it is documented that two representatives of the community were involved in the administrative processing of the promotion of a member of the *politeuma* to a higher rank within the military hierarchy.

69 For the Sidonian *politeumata*, Theodor Macridy, "À travers les nécropoles sidoniennes," *Revue biblique* 13 (= n.s. 1) (1904): 547–572 (p. 549: stele A; p. 551: stele 2; 551–552: stele 3). A *politeuma* is also mentioned in stele 8 (553–554); however, the name of the city from which the members of this *politeuma* came is lost. The Sidonian *politeumata*, consisting of persons from three cities of Kaunos (in Caria), Termessos Minor near Oinoanda, and Pinara (both in Lycia)—situated in the south of Asia Minor—thus differ from the *politeumata* in Egypt because they are associated with a home city rather than a region. For the Sidonian *politeumata* being Ptolemaic and not Seleucid, see Sänger, "The *Politeuma* in the Hellenistic World," 61–62.

70 Thomas Kruse, "Das jüdische *politeuma* von Herakleopolis und die Integration fremder Ethnien im Ptolemäerreich," in *Volk und Demokratie im Altertum*, ed. Vera V. Dement'eva and Tassilo Schmitt (Göttingen: Edition Ruprecht, 2010), 93–105, at 100–101 and id., "Die Festung in Herakleopolis und der Zwist im Ptolemäerhaus," in *Ägypten zwischen innerem Zwist und äußerem Druck: Die Zeit Ptolemaios' VI. bis VIII., Internationales Symposion Heidelberg 16.–19.9.2007*, ed. Andrea Jördens and Joachim F. Quack (Wiesbaden: Harrassowitz Verlag, 2011), 255–267, at 261.

71 See n. 74, below.

72 *Pace* the widespread assumption that there is no evidence for *politeumata* dating before the reign of Ptolemy VI (180–145 BCE) and that the form of organization in question was therefore introduced by this king: see Launey, *Recherches sur les armées hellénistiques*, 1077; Honigman, "*Politeumata* and Ethnicity in Ptolemaic Egypt," 67; Dorothy J. Thompson, "The Sons of Ptolemy V in a Post-Secession World," in *Ägypten zwischen innerem Zwist und äußerem Druck: Die Zeit Ptolemaios' VI. bis VIII.*, ed. Andrea Jördens and Joachim F. Quack (Wiesbaden: Harrassowitz Verlag, 2011), 10–23, at 21–22 with further bibliographical references at n. 47; cf. also Fischer-Bovet, *Army and Society in Ptolemaic Egypt*, 293–294.

73 See Sänger, "The *Politeuma* in the Hellenistic World," 61–62.

74 Until the reign of Ptolemy VI Philometor (180–145 BCE) an active Ptolemaic policy in the Aegean is attested, and until his reign Ptolemaic garrisons were kept in Itanos (north-eastern Crete), Methana (Eastern Peloponnese on the Saronic Gulf), and on the Aegean island of Thera; see Kostas Buraselis, "A Lively 'Indian Summer': Remarks on the Ptolemaic Role in the Aegean under Philometor," in *Ägypten*

zwischen innerem Zwist und äußerem Druck, ed. Andrea Jördens and Joachim F. Quack (Wiesbaden: Harrassowitz Verlag, 2011), 151–160; Eva Winter, "Formen ptolemäischer Präsenz in der Ägäis zwischen schriftlicher Überlieferung und archäologischem Befund," in *Militärsiedlungen und Territorialherrschaft in der Antike*, ed. Frank Daubner (Berlin: Walter de Gruyter, 2011), 65–77; Scheuble-Reiter, *Die Katökenreiter*, 117–118; Fischer-Bovet, *Army and Society in Ptolemaic Egypt*, 168–169. All these outposts could have assisted recruitment in the surrounding areas. The Ptolemies also employed trusted recruitment officers (*xenologoi*) to hire soldiers outside Egypt (Polyb. 5.63.8–9; 15.25.16–18). Stefanou, "Waterborne Recruits," 118–120 concluded (p. 120) "that individual Macedonians might render their services to the Ptolemies, regardless of Ptolemaic relations with the Antigonids", and see 120–121 for Ptolemaic recruitment of prisoners of war and renegades.

75 It is still not possible to determine with certainty whether Onias should be identified with Onias III or his son, though the second possibility is slightly preferred in the literature: see Aryeh Kasher, *The Jews in Hellenistic and Roman Egypt*, 132–135, for the controversy, but who leaves open whether Onias III or IV is meant. Fausto Parente, "Onias III's Death and the Founding of the Temple of Leontopolis," in *Josephus and the History of the Greco-Roman Period: Essays in Memory of Morton Smith*, ed. Fausto Parente and Joseph Sievers (Leiden: E. J. Brill, 1994), 69–98 argued for Onias III, as did (with more or less conviction), Joan E. Taylor, "A Second Temple in Egypt: The Evidence for the Zadokite Temple of Onias," *Journal for the Study of Judaism* 29 (1998): 297–321, at 298–310 and Walter Ameling, "Die jüdische Gemeinde von Leontopolis nach den Inschriften," in *Die Septuaginta—Texte, Kontexte, Lebenswelten. Internationale Fachtagung veranstaltet von Septuaginta Deutsch (LXX.D), Wuppertal 20.–23. Juli 2006*, ed. Martin Karrer and Wolfgang Kraus (Tübingen: Mohr Siebeck, 2008), 117–133, at 118–119. Mélèze Modrzejewski, *The Jews of Egypt*, 124–125 identifies Onias with Onias IV, an identification also preferred by Erich S. Gruen, "The Origins and Objectives of Onias' Temple," *Scripta Classica Israelica* 16 (1997): 47–70, at 47–57 (n. 26 cites older literature for this position); Livia Capponi, *Il tempio di Leontopoli in Egitto: Identità politica e religiosa dei Giudei di Onia (c. 150 A.C.–73 D.C.)* (Pisa: Edizioni ETS, 2007), 42–53; Peter Nadig, "Zur Rolle der Juden unter Ptolemaios VI. und Ptolemaios VIII.," in *Ägypten zwischen innerem Zwist und äußerem Druck*, ed. Andrea Jördens and Joachim F. Quack (Wiesbaden: Harrassowitz Verlag, 2011), 186–200, at 188–194.

76 Josephus, *BJ* 1.33; 7.427; *AJ* 13.65–66.

77 Capponi, *Il tempio di Leontopoli*, 59; Nadig, "Zur Rolle der Juden," 188, 191–193; see also Gruen, "The Origins and Objectives of Onias' Temple," 69–70 pointing to 159–152 BCE, when the office of high priest was vacant. As to whether the military colony of Onias was organized as a *politeuma*, which seems likely, see Patrick Sänger, "Considerations on the Administrative Organisation of the Jewish Military Colony in Leontopolis: A Case of Generosity and Calculation," in *Expulsion and Diaspora Formation: Religious and Ethnic Identities in Flux from Antiquity to the Seventeenth Century*, ed. John Tolan (Turnhout: Brepols, 2015), 171–194.

78 Uriel Rapaport, "Les Iduméens en Égypte," *Revue de philologie, de littérature et d'histoire anciennes* 43 (1969): 73–82, at 78–79, 81–82; Thompson Crawford, "The Idumaeans of Memphis," 1071–1072; ead., *Memphis under the Ptolemies*, 79–80; Honigman, "*Politeumata* and Ethnicity in Ptolemaic Egypt," 66, n. 22, 83–84.

79 See Fischer-Bovet, *Army and Society in Ptolemaic Egypt*, 293: "Indeed, the reorganization of the army during the period of crisis (Period B) [c. 220 and c. 160 BCE] favored the use of professional soldiers in garrisons. Even if recruitment was mainly internal to Egypt, foreigners were also hired at times".

80 See the preceding n. and Fischer-Bovet, *Army and Society in Ptolemaic Egypt*, 269–271, 273–279.

81 Patrick Sänger, "Das *politeuma* in der hellenistischen Staatenwelt: Eine Organisationsform zur Systemintegration von Minderheiten," in *Minderheiten und Migration*

in der griechisch-römischen Welt: Politische, rechtliche, religiöse und kulturelle Aspekte, ed. id. (Paderborn: Ferdinand Schöningh, 2016), 25–45, at 35–38, 44; id., "Heracleopolis, Jewish *politeuma*"; Kruse, "Das jüdische *politeuma* von Herakleopolis," 95, 97, 99–100.

82 On the Jewish *politeuma* of Heracleopolis, see, in general, Cowey and Maresch, *Urkunden des Politeuma der Juden von Herakleopolis*, 1–34; Honigman, "The Jewish *Politeuma* at Heracleopolis;" Maria R. Falivene, Review of *Urkunden des Politeuma der Juden von Herakleopolis*, by James M. S. Cowey and Klaus Maresch, *Bibliotheca Orientalis* 59 (2002): cols. 541–550; Aryeh Kasher, Review Essay of *Urkunden des Politeuma der Juden von Herakleopolis*, by James M. S. Cowey and Klaus Maresch, *The Jewish Quarterly Review* 93 (2002): 257–268; Klaus Maresch and James M. S. Cowey, "'A Recurrent Inclination to Isolate the Case of the Jews from their Ptolemaic Environment'? Eine Antwort auf Sylvie Honigman," *Scripta Classica Israelica* 22 (2003): 307–310; James M. S. Cowey, "Das ägyptische Judentum in hellenistischer Zeit: Neue Erkenntnisse aus jüngst veröffentlichten Papyri," in *Im Brennpunkt: Die Septuaginta; Studien zur Entstehung und Bedeutung der Griechischen Bibel; Band 2*, ed. Siegfried Kreuzer and Jürgen P. Lesch (Stuttgart: Verlag W. Kohlhammer, 2004), 24–43; Thomas Kruse, "Das *politeuma* der Juden von Herakleopolis in Ägypten," in *Die Septuaginta—Texte, Kontexte, Lebenswelten*, ed. Martin Karrer and Wolfgang Kraus (Tübingen: Mohr Siebeck, 2008), 166–175; id., "Das jüdische *politeuma* von Herakleopolis;" Peter Arzt-Grabner, "Die Stellung des Judentums in neutestamentlicher Zeit anhand der Politeuma-Papyri und anderer Texte," in *Papyrologie und Exegese: Die Auslegung des Neuen Testaments im Licht der Papyri*, ed. Jens Herzer (Tübingen: Mohr Siebeck, 2012), 127–158.

83 The questionable sources are, for Alexandria, Aristeas 310 [= Josephus, *AJ* 12.108] and, for Leontopolis, *SB* 1.5765 = *C.Pap.Jud.* III 1530A = Étienne Bernand, *Inscriptions métriques de l'Égypte gréco-romaine: Recherches sur la poésie épigrammatique des Grecs en Égypte* (Paris: Les Belles Lettres, 1969), no. 16 = William Horbury and David Noy, *Jewish Inscriptions of Graeco-Roman Egypt* (Cambridge: Cambridge University Press, 1992), no. 39 (Augustan times to early second century?). For scepticism, Zuckerman, "Hellenistic *politeumata* and the Jews," 181–184; Gert Lüderitz, "What is the Politeuma?," in *Studies in Early Jewish Epigraphy*, ed. Jan W. van Henten and Pieter W. van der Horst (Leiden: E. J. Brill, 1994), 183–225, at 204–210; Walter Ameling, "'Market-Place' und Gewalt: Die Juden in Alexandrien 38 n. Chr.," *Würzburger Jahrbücher für die Altertumswissenschaft, n.s.*, 27 (2003): 71–123, at 88–98 (with n. 112); id., Die jüdische Gemeinde von Leontopolis," 128–129.

84 *CIG* 3.5362 = *SEG* 16.931 = Gert Lüderitz, *Corpus jüdischer Zeugnisse aus der Cyrenaika mit einem Anhang von Joyce M. Reynolds* (Wiesbaden: Reichert, 1983), no. 70 (Augustan times?) and *CIG* 3.5361 = Lüderitz, *Corpus jüdischer Zeugnisse aus der Cyrenaika*, no. 71 (24/25 CE).

85 For treatment of individual petitions, their contents and legal reasoning, see Joseph Mélèze Modrzejewski, "La fiancée adultère: À propos de la pratique matrimonial du judaïsme hellénisé à la lumière du dossier du *politeuma* juif d'Hérakléopolis (144/3–133/2 av. n.è.)," in *Marriage: Ideal—Law—Practice; Proceedings of a Conference Held in Memory of Henryk Kupiszewski*, ed. Zuzanna Służewska and Jakub Urbanik (Warsaw: Warsaw University, Faculty of Law and Administration, Chair of Roman and Antique Law, 2005), 141–160; Robert Kugler, "Dorotheos Petitions for the Return of Philippa (P. Polit. Jud. 7): A Case Study in the Jews and their Law in Ptolemaic Egypt," in *Proceedings of the 25th International Congress of Papyrology*, ed. Traianos Gagos and Adam Hyatt (Ann Arbor: University of Michigan Library, Scholarly Publ. Office, 2010), 387–395; id., "Dispelling an Illusion of Otherness? A First Look at Judicial Practice in the Heracleopolis Papyri," in *The "Other" in Second Temple Judaism: Essays in Honor of John J. Collins*, ed. Daniel C. Harlow et al.(Grand Rapids, MI: Wm. B. Eerdmans, 2011), 457–470; id., "Uncovering New Dimensions of Early Judean Inter-

pretation of the Greek Torah: Ptolemaic Law Interpreted by its own Rhetoric," in *Changes in Scripture: Rewriting and Interpreting Authoritative Traditions in the Second Temple Period*, ed. Hanne von Weissenberg, Juha Pakkala, and Marko Martilla (Berlin: Walter de Gruyter, 2011), 165–175; id., "Peton Contests Paying Double Rent on Farmland (P.Heid.Inv. G 5100): A Slice of Judean Experience in the Second Century BCE Herakleopolite Nome," in *A Teacher for All Generations: Essays in Honor of James C. VanderKam*, ed. Eric Mason et al.(Leiden: Brill, 2012), 537–551; id., "Judean Marriage Custom and Law in Second-Century BCE Egypt: A Case of Migrating Ideas and a Fixed Ethnic Minority," in *Minderheiten und Migration in der griechisch-römischen Welt: Politische, rechtliche, religiöse und kulturelle Aspekte*, ed. Patrick Sänger (Paderborn: Ferdinand Schöningh, 2016), 123–139.

86 See Thompson, "Ethnic Minorities in Hellenistic Egypt," 113; ead., "The Sons of Ptolemy V in a Post-Secession World," 22; Sänger, "The *Politeuma* in the Hellenistic World," 60.

87 See Carsten Claußen, *Versammlung, Gemeinde, Synagoge: Das hellenistisch-jüdische Umfeld der frühchristlichen Gemeinden* (Göttingen: Vandenhoeck & Ruprecht, 2002), 273–278; Daniel Stökl Ben Ezra, "A Jewish 'Archontesse': Remarks on an Epitaph from Byblos," *Zeitschrift für Papyrologie und Epigraphik* 169 (2009): 287–293, at 291.

88 See Launey, *Recherches sur les armées hellénistique*, 954–955, 1067.

89 See Rapaport, "Les Iduméens en Égypte," 73; Thompson Crawford, "The Idumaeans of Memphis," 1071; ead., *Memphis under the Ptolemies*, 92–93.

90 For this definition, see Anthony D. Smith, *The Ethnic Origins of Nations* (Oxford: Basil Blackwell, 1986), 22–31; Gerard Delanty and Krishan Kumar, *The SAGE Handbook of Nations and Nationalism*, (London: SAGE, 2006), 171–172; Thomas H. Eriksen, *Ethnicity and Nationalism: Anthropological Perspectives*, 3rd ed. (London: Pluto Press, 2010), 48–53 (based on Don Handelman, "The Organization of Ethnicity," *Ethnic Groups: An International Periodical of Ethnic Studies* 1 [1977]: 187–200). See also Thompson, "Ethnic Minorities in Hellenistic Egypt," 108–109 summarizing her view of features by which members of an ethnic group can be identified: "Whereas many of these factors [ethnic designation, language, nomenclature, a person's appearance, cultural practices, occupation] serve to identify individuals rather than communities, in the case of the last four features—temples, the existence of ethnic quarters, of ethnic leaders and local responsibility for some degree of legal control—we have features which may define communities".

91 Thompson, "Ethnic Minorities in Hellenistic Egypt," 112–113.

92 *P.Tebt.* 1.32 = *W.Chr.* 448.16–17.

93 Along these lines but with varying nuances, Launey, *Recherches sur les armées hellénistique*, 1078–1079; Honigman, "*Politeumata* and Ethnicity in Ptolemaic Egypt," 94–95; Thompson Crawford, "The Idumaeans of Memphis," 1074–1075; ead., "Ethnic Minorities in Hellenistic Egypt," 109–113; ead., "The Sons of Ptolemy V in a Post-Secession World," 22; Fischer-Bovet, *Army and Society in Ptolemaic Egypt*, 290–295.

94 See Thompson, "Ethnic Minorities in Hellenistic Egypt," 107; ead., *Memphis under the Ptolemies*, 77–78, 87–90.

95 Thompson, "Ethnic Minorities in Hellenistic Egypt," 108; ead., *Memphis under the Ptolemies*, 76–77, 81–87.

96 Thompson, "Ethnic Minorities in Hellenistic Egypt," 101; ead., *Memphis under the Ptolemies*, 90–92.

97 See Clarysse and Thompson, *Counting the People*, 159–161.

98 See Sylvie Honigman, "Les divers sens de l'ethnique Ἄραψ dans les sources documentaires grecques d'Égypte," *Ancient Society* 32 (2002): 43–72, at 61–69; Clarysse and Thompson, *Counting the People*, 159–161, 175–176; John Bauschatz, *Law and Enforcement in Ptolemaic Egypt* (Cambridge: Cambridge University Press, 2013), 156–157.

99 Étienne Bernand, *Les inscriptions grecques d'Hermoupolis Magna et de la nécropole (Cairo:* Institut Français d'Archéologie Orientale, 1999), no. 5 = *SB* 1.4206 (80/79 BCE); Bernand, *Les inscriptions grecques*, no. 6 = *SB* 5.8066 (78 BCE).

100 As it is only *Apoll*[that survives on one of the inscriptions, scholars made two suggestions as to how to complete the word: *Apolloniatai* (see Friedrich Zucker, *Doppelinschrift spätptolemäischer Zeit aus der Garnison von Hermopolis Magna* [Berlin: Akademie der Wissenschaften, 1938]), Idumaeans from the city of Apollonia (located in Palestine between Jaffa and Caesarea Maritima), or *Apolloniastai* (Rapaport, "Les Iduméens en Égypte," 74–77), worshippers of Apollo/Qos.

101 See Launey, *Recherches sur les armées hellénistique*, 974–975, 1024–1025, 1031, 1034, 1080–1081; Fischer-Bovet, *Army and Society in Ptolemaic Egypt*, 292. Against Thompson, *Memphis under the Ptolemies*, 94 and Bernand, *Les inscriptions grecques*, no. 5, p. 48, there is no reason to suppose that the term *sympoliteuomenoi* would indicate that the *Apollonia(s)tai* were organized as *politeuma* because in Cyprus we find this word usage also associated with groups of soldiers describing themselves as *koinon* ("association" or "gathering"); see Launey, *Recherches sur les armées hellénistique*, 1031–1035, 1080–1081, and further below. *P.Giss.* 99 (with *BL* 6.43), a fragmentary papyrus from Hermoupolis dated to the second or third century CE, could suggest that the *Apollonia(s)tai* continued to exist until Roman times; see Thompson Crawford, "The Idumaeans of Memphis," 1071; Launey, *Recherches sur les armées hellénistique*, 1025.

102 See Franz Poland, *Geschichte des griechischen Vereinswesens* (Leipzig: B. G. Teubner, 1909), 164–165; Jacek Rzepka, "*Ethnos, Koinon, Sympoliteia*, and Greek Federal States," in *Euergasias Charin: Studies Presented to Benedetto Bravo and Ewa Wipszycka by Their Disciples*, ed. Tomasz Derda, Jakub Urbanik, and Marek Wecowski (Warsaw: Fund. im. Rafała Taubenschlaga, 2002), 225–247, at 227–234; Roland Oetjen, *Athen im dritten Jahrhundert v. Chr.: Politik und Gesellschaft in den Garnisonsdemen auf der Grundlage der inschriftlichen Überlieferung* (Duisburg: Wellem, 2014), 148–149.

103 Roger S. Bagnall, *The Administration of the Ptolemaic Possessions Outside Egypt* (Leiden: Brill, 1976), 56–57 and Appendix B, 263–266; Launey, *Recherches sur les armées hellénistique*, 1032–1034; Mariano San Nicolò, *Ägyptisches Vereinswesen zur Zeit der Ptolemäer und Römer. Erster Teil: Die Vereinsarten*, 2nd ed. (Munich: C. H. Beck'sche Verlagsbuchhandlung, 1972), 198–200.

104 See also Thompson Crawford, "The Idumaeans of Memphis," 1074–1075; ead., "Ethnic Minorities in Hellenistic Egypt," 109–113; ead., "The Sons of Ptolemy V in a post-secession World," 21–22 who argued that *politeumata* should be treated as an expression of military and related immigration policies the Ptolemies pursued in the middle of the second century BCE as an alternative to granting land to military immigrants as they did in the previous century.

105 Cf. Thompson, "Ethnic Minorities in Hellenistic Egypt," 107: "Such ethnic quarters, however, would appear to have been a feature of well-established cities rather than of a rural setting. They may even serve as an urban indicator".

106 For this concept, see the chapter of Andrade in this volume.

107 For the "Hellenic sector" of Ptolemaic society, see Clarysse and Thompson, *Counting the People*, 154–157.

References

Ameling, Walter. "'Market-Place' und Gewalt. Die Juden in Alexandrien 38 n. Chr." *Würzburger Jahrbücher für die Altertumswissenschaft*, n.s., 27 (2003): 71–123.

Ameling, Walter. "Die jüdische Gemeinde von Leontopolis nach den Inschriften." In *Die Septuaginta—Texte, Kontexte, Lebenswelten. Internationale Fachtagung veranstaltet von Septuaginta Deutsch (LXX.D), Wuppertal 20.–23. Juli 2006*, edited by Martin Karrer and Wolfgang Kraus, 117–133. Tübingen: Mohr Siebeck, 2008.

Applebaum, Shimon. "The Legal Status of the Jewish Communities in the Diaspora." In *The Jewish People in the First Century: Historical Geography, Political History, Social, Cultural and Religious Life and Institutions*, edited by Shmuel Safrai and Menahem Stern, 420–463. Vol. 1. Assen: Van Gorcum, 1974.

Applebaum, Shimon. "The Organization of the Jewish Communities in the Diaspora." In *The Jewish People in the First Century: Historical Geography, Political History, Social, Cultural and Religious Life and Institutions*, edited by Shmuel Safrai and Menahem Stern, 464–503. Vol. 1. Assen: Van Gorcum, 1974.

Arzt-Grabner, Peter. "Die Stellung des Judentums in neutestamentlicher Zeit anhand der Politeuma-Papyri und anderer Texte." In *Papyrologie und Exegese: Die Auslegung des Neuen Testaments im Licht der Papyri*, edited by Jens Herzer, 127–158. Tübingen: Mohr Siebeck, 2012.

Bagnall, Roger S. *The Administration of the Ptolemaic Possessions Outside Egypt*. Leiden: Brill, 1976.

Bagnall, Roger S. "The Origins of Ptolemaic Cleruchs." *Bulletin of the American Society of Papyrologists* 21 (1984): 7–20.

Bagnall, Roger S. "The People of the Roman Fayum." In *Portraits and Masks: Burial Customs in Roman Egypt*, edited by Morris L. Bierbrier, 7–15. London: British Museum Press, 1997.

Bagnall, Roger S. "The People of the Roman Fayum." In *Hellenistic and Roman Egypt. Sources and Approaches*, edited by Roger S. Bagnall, chap. 14, 1–19. Aldershot: Ashgate, 2006.

Barth, Frederik. *Ethnic Groups and Boundaries: The Social Organization of Culture Difference*. Boston: Little, Brown, 1969.

Bauschatz, John. *Law and Enforcement in Ptolemaic Egypt*. Cambridge: Cambridge University Press, 2013.

Bernand, André. *La prose sur pierre dans l'Égypte hellénistique et romaine*. 2 vols. Paris: Centre National de la Recherche Scientifique, 1992.

Bernand, Étienne. *Inscriptions métriques de l'Égypte gréco-romaine: Recherches sur la poésie épigrammatique des Grecs en Égypte*. Paris: Les Belles Lettres, 1969.

Bernand, Étienne. *Recueil des inscriptions grecques du Fayoum*. Vol. 1, *La 'méris' d'Hérakleides*. Leiden: E. J. Brill, 1975.

Bernand, Étienne. *Inscriptions grecques d'Égypte et de Nubie au musée du Louvre*. Paris: Centre National de la Recherche Scientifique, 1992.

Bernand, Étienne. *Les inscriptions grecques d'Hermoupolis Magna et de la nécropole*. Cairo: Institut Français d'Archéologie Orientale, 1999.

Bickermann, Elias. "Beiträge zur antiken Urkundengeschichte I. Der Heimatsvermerk und die staatsrechtliche Stellung der Hellenen im ptolemäischen Ägypten." *Archiv für Papyrusforschung* 8 (1927): 216–239.

Buraselis, Kostas. "Ambivalent Roles of Centre and Periphery: Remarks on the Relation of the Cities of Greece with the Ptolemies until the End of Philometor's Age." In *Centre and Periphery in the Hellenistic World*, edited by Per Bilde, Troels Endberg-Pedersen, Lise Hannestad, Jan Zahle, and Klavs Randsborg, 251–270. Aarhus: Aarhus University Press, 1993.

Buraselis, Kostas. "A Lively 'Indian Summer': Remarks on the Ptolemaic Role in the Aegean under Philometor." In *Ägypten zwischen innerem Zwist und äußerem Druck*, edited by Andrea Jördens and Joachim F. Quack, 151–160. Wiesbaden: Harrassowitz Verlag, 2011.

Capponi, Livia. *Il tempio di Leontopoli in Egitto: Identità politica e religiosa dei Giudei di Onia (c. 150 A.C.–73 D.C.)*. Pisa: Edizioni ETS, 2007.

Clarysse, Willy. "Egyptian Estate-Holders in the Ptolemaic Period." In *State and Temple Economy in the Ancient Near East*, edited by Edward Lipiński, 731–743. Vol. 2. Leuven: Department Oriëntalistiek, 1979.

Clarysse, Willy. "Greeks and Egyptians in the Ptolemaic Army and Administration." *Aegyptus* 65 (1985): 57–66.

Clarysse, Willy. "Some Greeks in Egypt." In *Life in a Multi-Cultural Society: Egypt from Cambyses to Constantine and Beyond*, edited by Janet H. Johnson, 51–56. Chicago: The Oriental Institute of the University of Chicago, 1992.

Clarysse, Willy. "Ethnic Diversity and Dialect among the Greeks of Hellenistic Egypt." In *The Two Faces of Graeco-Roman Egypt*, edited by Pieter W. Pestman and Arthur M. F. W. Verhoogt, 1–13. Leiden: Brill, 1998.

Clarysse, Willy, and Dorothy J. Thompson. *Counting the People in Hellenistic Egypt*. Vol. 2, *Historical Studies*. Cambridge: Cambridge University Press, 2006.

Claußen, Carsten. *Versammlung, Gemeinde, Synagoge. Das hellenistisch-jüdische Umfeld der frühchristlichen Gemeinden*. Göttingen: Vandenhoeck & Ruprecht, 2002.

Cowey, James M. S. "Das ägyptische Judentum in hellenistischer Zeit: Neue Erkenntnisse aus jüngst veröffentlichten Papyri." In *Im Brennpunkt: Die Septuaginta; Studien zur Entstehung und Bedeutung der Griechischen Bibel; Band 2*, edited by Siegfried Kreuzer and Jürgen P. Lesch, 24–43. Stuttgart: Verlag W. Kohlhammer, 2004.

Cowey, James M. S., and Klaus Maresch. *Urkunden des Politeuma der Juden von Herakleopolis (144/3–133/2 v. Chr.) (P.Polit.Iud.)*. Wiesbaden: Westdeutscher Verlag, 2001.

Crawford, Dorothy J. *Kerkeosiris: An Egyptian Village in the Ptolemaic Period*. Cambridge: Cambridge University Press, 1971.

Delanty, Gerard, and Krishan Kumar. *The SAGE Handbook of Nations and Nationalism*. London: SAGE, 2006.

Eriksen, Thomas H. *Ethnicity and Nationalism. Anthropological Perspectives*. 3rd ed. London: Pluto Press, 2010.

Falivene, Maria R. Review of *Urkunden des Politeuma der Juden von Herakleopolis*, by James M. S. Cowey and Klaus Maresch. *Bibliotheca Orientalis* 59 (2002): cols. 541–550.

Fischer-Bovet, Christelle. "Counting the Greeks in Egypt: Immigration in the First Century of Ptolemaic Rule." In *Demography and the Graeco-Roman World: New Insights and Approaches*, edited by Claire Holleran and April Pudsey, 135–154. Cambridge: Cambridge University Press, 2011.

Fischer-Bovet, Christelle. *Army and Society in Ptolemaic Egypt*. Cambridge: Cambridge University Press, 2014.

Goudriaan, Koen. *Ethnicity in Ptolemaic Egypt*. Amsterdam: J. C. Gieben, 1988.

Gruen, Erich S. "The Origins and Objectives of Onias' Temple." *Scripta Classica Israelica* 16 (1997): 47–70.

Habermann, Wolfgang. "Gymnasien im ptolemäischen Ägypten—eine Skizze." In *Das hellenistische Gymnasion*, edited by Daniel Kah and Peter Scholz, 335–348. Berlin: Akademie Verlag, 2004.

Hall, Jonathan M. *Ethnic Identity in Greek Antiquity*. Cambridge: Cambridge University Press, 1997.

Hall, Jonathan M. *Hellenicity between Ethnicity and Culture*. Chicago: The University of Chicago Press, 2002.

Handelman, Don. "The Organization of Ethnicity." *Ethnic Groups: An International Periodical of Ethnic Studies* 1 (1977): 187–200.

Honigman, Sylvie. "Philon, Flavius Josèphe, et la citoyenneté alexandrine: Vers une utopie politique." *Journal of Jewish Studies* 48 (1997): 62–90.

Honigman, Sylvie. "Les divers sens de l'ethnique Ἄραψ dans les sources documentaires grecques d'Égypte." *Ancient Society* 32 (2002): 43–72.

Honigman, Sylvie. "The Jewish *Politeuma* at Heracleopolis." *Scripta Classica Israelica* 21 (2002): 251–266.

Honigman, Sylvie. "*Politeumata* and Ethnicity in Ptolemaic Egypt." *Ancient Society* 33 (2003): 61–102.

Horbury, William, and David Noy. *Jewish Inscriptions of Graeco-Roman Egypt*. Cambridge: Cambridge University Press, 1992.

Huß, Werner. *Die Verwaltung des ptolemaiischen Reiches*. Munich: Verlag C. H. Beck, 2011.

Kasher, Aryeh. *The Jews in Hellenistic and Roman Egypt: The Struggle for Equal Rights*. Tübingen: J. C. B. Mohr (Paul Siebeck), 1985.

Kasher, Aryeh. Review Essay of "*Urkunden des Politeuma der Juden von Herakleopolis*, by James M. S. Cowey and Klaus Maresch." *The Jewish Quarterly Review* 93 (2002): 257–268.

Kayser, François. *Recueil des inscriptions grecques et latines (non funéraires) d'Alexandrie impériale (Ier–IIIe s. apr. J.-C.)*. Cairo: Institut Français d'Archéologie Orientale, 1994.

Kruse, Thomas. "Das *politeuma* der Juden von Herakleopolis in Ägypten." In *Die Septuaginta—Texte, Kontexte, Lebenswelten*, edited by Martin Karrer and Wolfgang Kraus, 166–175. Tübingen: Mohr Siebeck, 2008.

Kruse, Thomas. "Das jüdische *politeuma* von Herakleopolis und die Integration fremder Ethnien im Ptolemäerreich." In *Volk und Demokratie im Altertum*, edited by Vera V. Dement'eva and Tassilo Schmitt, 93–105. Göttingen: Edition Ruprecht, 2010.

Kruse, Thomas. "Die Festung in Herakleopolis und der Zwist im Ptolemäerhaus." In *Ägypten zwischen innerem Zwist und äußerem Druck: Die Zeit Ptolemaios' VI. bis VIII., Internationales Symposion Heidelberg 16.–19.9.2007*, edited by Andrea Jördens and Joachim F. Quack, 255–267. Wiesbaden: Harrassowitz Verlag, 2011.

Kugler, Robert. "Dorotheos Petitions for the Return of Philippa (P. Polit. Jud. 7): A Case Study in the Jews and Their Law in Ptolemaic Egypt." In *Proceedings of the 25th International Congress of Papyrology*, edited by Traianos Gagos and Adam Hyatt, 387–395. Ann Arbor: University of Michigan Library, Scholarly Publ. Office, 2010.

Kugler, Robert. "Dispelling an Illusion of Otherness? A First Look at Judicial Practice in the Heracleopolis Papyri." In *The "Other" in Second Temple Judaism: Essays in Honor of John J. Collins*, edited by Daniel C. Harlow, Karina M. Hogan, Matthew Goff, and Joel S. Kaminsky, 457–470. Grand Rapids: Wm. B. Eerdmans, 2011.

Kugler, Robert. "Uncovering New Dimensions of Early Judean Interpretation of the Greek Torah: Ptolemaic Law Interpreted by its own Rhetoric." In *Changes in Scripture: Rewriting and Interpreting Authoritative Traditions in the Second Temple Period*, edited by Hanne von Weissenberg, Juha Pakkala, and Marko Martilla, 165–175. Berlin: Walter de Gruyter, 2011.

Kugler, Robert. "Peton Contests Paying Double Rent on Farmland (P.Heid.Inv. G 5100): A Slice of Judean Experience in the Second Century BCE Herakleopolite Nome." In *A Teacher for All Generations: Essays in Honor of James C. VanderKam*, edited by Eric Mason, Samuel I. Thomas, Alison Schofield, and Eugene Ulrich, 537–551. Leiden: Brill, 2012.

Kugler, Robert. "Judean Marriage Custom and Law in Second-Century BCE Egypt: A Case of Migrating Ideas and a Fixed Ethnic Minority." In *Minderheiten und Migration in der griechisch-römischen Welt: Politische, rechtliche, religiöse und kulturelle Aspekte*, edited by Patrick Sänger, 123–139. Paderborn: Ferdinand Schöningh, 2016.

Kuhs, Clemens. "Das Dorf Samareia im griechisch-römischen Ägypten." MA thesis, University of Heidelberg, 1996.

La'da, Csaba A. "Review of Ethnicity in '*Hellenistic Egypt*,' edited by Per Bilde, Troels Engberg-Pedersen, Lise Hannestad, and Jan Zahle." *Topoi* 4 (1994): 341–350.

La'da, Csaba A. "Ethnic Designations in Hellenistic Egypt." PhD diss., University of Cambridge, 1996.

La'da, Csaba A. *Foreign Ethnics in Hellenistic Egypt*. Prosopographia Ptolemaica 10; Studia Hellenistica 38. Leuven: Peeters, 2002.

La'da, Csaba A. "Encounters with Ancient Egypt: The Hellenistic Greek Experience." In *Ancient Perspectives on Egypt: Encounters with Ancient Egypt*, edited by Roger Matthews and Cornelia Roemer, 157–169. London: UCL Press, 2003.

Launey, Marcel. *Recherches sur les armées hellénistique*. 2nd ed. Paris: De Boccard, 1987.

Legras, Bernard. *L'Égypte grecque et romain*. Paris: Colin, 2004.

Lüderitz, Gert. *Corpus jüdischer Zeugnisse aus der Cyrenaika mit einem Anhang von Joyce M. Reynolds*. Wiesbaden: Reichert, 1983.

Lüderitz, Gert. "What Is the Politeuma?" In *Studies in Early Jewish Epigraphy*, edited by Jan W. van Henten and Pieter W. van der Horst, 183–225. Leiden: E. J. Brill, 1994.

Macridy, Theodor. "À travers les nécropoles sidoniennes." *Revue biblique* 13 (= n.s. 1) (1904): 547–572.

Maresch, Klaus, and James M. S. Cowey. "'A Recurrent Inclination to Isolate the Case of the Jews from their Ptolemaic Environment'? Eine Antwort auf Sylvie Honigman." *Scripta Classica Israelica* 22 (2003): 307–310.

Mélèze Modrzejewski, Joseph. "Le statut des Hellènes dans l'Égypte lagide: Bilan et perspectives des recherches." *Revue des Études Grecques* 96 (1983): 241–268.

Mélèze Modrzejewski, Joseph. *The Jews of Egypt: From Ramses II to Emperor Hadrian*. Princeton: Princeton University Press, 1997.

Mélèze Modrzejewski, Joseph. "La fiancée adultère: À propos de la pratique matrimonial du judaïsme hellénisé à la lumière du dossier du *politeuma* juif d'Hérakléopolis (144/3–133/2 av. n.è.)." In *Marriage: Ideal—Law—Practice; Proceedings of a Conference Held in Memory of Henryk Kupiszewski*, edited by Zuzanna Służewska and Jakub Urbanik, 141–160. Warsaw: Warsaw University, Faculty of Law and Administration, Chair of Roman and Antique Law, 2005.

Mélèze Modrzejewski, Joseph. *Loi et coutume dans l'Égypte grecque et romaine*. Warsaw: Faculty of Law and Administration of the University of Warsaw, 2014.

Milne, Joseph G. *Greek Inscriptions*. Oxford: Oxford University Press, 1905.

Mitteis, Ludwig. *Grundzüge und Chrestomathie der Papyruskunde, II. Band: Juristischer Teil, I. Hälfte: Grundzüge*. Leipzig: B. G. Teubner, 1912.

Mooren, Leon. *The Aulic Titulature in Ptolemaic Egypt: Introduction and Prosopography*. Brussels: Paleis der Academiën, 1975.

Moyer, Ian S. *Egypt and the Limits of Hellenism*. Cambridge: Cambridge University Press, 2014.

Müller, Katja. *Settlements of the Ptolemies: City Foundations and New Settlement in the Hellenistic World*. Leuven: Peeters, 2006.

Nadig, Peter. "Zur Rolle der Juden unter Ptolemaios VI. und Ptolemaios VIII." In *Ägypten zwischen innerem Zwist und äußerem Druck*, edited by Andrea Jördens and Joachim F. Quack, 186–200. Wiesbaden: Harrassowitz Verlag, 2011.

Oetjen, Roland. *Athen im dritten Jahrhundert v. Chr.: Politik und Gesellschaft in den Garnisonsdemen auf der Grundlage der inschriftlichen Überlieferung*. Duisburg: Wellem, 2014.

Parente, Fausto. "Onias III's Death and the Founding of the Temple of Leontopolis." In *Josephus and the History of the Greco-Roman Period. Essays in Memory of Morton Smith*, edited by Fausto Parente and Joseph Sievers, 69–98. Leiden: E. J. Brill, 1994.

Poland, Fran. *Geschichte des griechischen Vereinswesens*. Leipzig: B. G. Teubner, 1909.

Rapaport, Uriel. "Les Iduméens en Égypte." *Revue de philologie, de littérature et d'histoire anciennes* 43 (1969): 73–82.

Ritter, Bradley. "On the 'πολίτευμα in Heracleopolis'." *Scripta Classica Israelica* 30 (2011): 9–37.

Rowlandson, Jane. "Review of Ethnicity in Ptolemaic Egypt, by Koen Goudriaan." *The Classical Review*, n.s., 40 (1990): 370–371.

Rowlandson, Jane. "Dissing the Egyptians: Legal, Ethnic, and Cultural Identities in Roman Egypt." In *Creating Ethnicities & Identities in the Roman World*, edited by Andrew Gardner, Edward Herring, and Kathryn Lomas, 213–247. London: Institute of Classical Studies, 2013.

Runesson, Anders, Donald D. Binder, and Birger Olsson. *The Ancient Synagogue from Its Origins to 200 C.E.: A Source Book*. Leiden: Brill, 2008.

Rzepka, Jacek. "*Ethnos, Koinon, Sympoliteia*, and Greek Federal States." In *Euergasias Charin: Studies Presented to Benedetto Bravo and Ewa Wipszycka by Their Disciples*, edited by Tomasz Derda, Jakub Urbanik, and Marek Wecowski, 225–247. Warsaw: Fund. im. Rafała Taubenschlaga, 2002.

San Nicolò, Mariano. *Ägyptisches Vereinswesen zur Zeit der Ptolemäer und Römer. Erster Teil: Die Vereinsarten*. 2nd ed. Munich: C. H. Beck'sche Verlagsbuchhandlung, 1972.

Sänger, Patrick. "The *Politeuma* in the Hellenistic World (Third to First Century B.C.): A Form of Organisation to Integrate Minorities." In *Migration und Integration—wissenschaftliche Perspektiven aus Österreich. Jahrbuch 2/2013*, edited by Julia Dahlvik, Christoph Reinprecht, and Wiebke Sievers, 51–68. Göttingen: V&R Unipress, 2014.

Sänger, Patrick. "Considerations on the Administrative Organisation of the Jewish Military Colony in Leontopolis: A Case of Generosity and Calculation." In *Expulsion and Diaspora Formation: Religious and Ethnic Identities in Flux from Antiquity to the Seventeenth Century*, edited by John Tolan, 171–194. Turnhout: Brepols, 2015.

Sänger, Patrick. "Military Immigration and the Emergence of Cultural or Ethnic Identities: The Case of Ptolemaic Egypt." *Journal of Juristic Papyrology* 45 (2015): 229–253.

Sänger, Patrick. "Das *politeuma* in der hellenistischen Staatenwelt: Eine Organisationsform zur Systemintegration von Minderheiten." In *Minderheiten und Migration in der griechisch-römischen Welt: Politische, rechtliche, religiöse und kulturelle Aspekte*, edited by Patrick Sänger, 25–45. Paderborn: Ferdinand Schöningh, 2016.

Sänger, Patrick. *Die ptolemäische Organisationsform politeuma: Ein Herrschaftsinstrument zugunsten jüdischer und anderer hellenischer Gemeinschaften*. Tübingen: Mohr Siebeck, 2019.

Sänger, Patrick. "Heracleopolis, Jewish *politeuma*." In *The Oxford Classical Dictionary*, edited by Sander Goldberg. New York, 2016. Accessed 24 March 2022. https://doi.org/10.1093/acrefore/9780199381135.013.8036.

Scheuble, Sandra. "Bemerkungen zu den μισθοφόφοι und τακτόμισθοι im ptolemäischen Ägypten." In *"... vor dem Papyrus sind alle gleich!" Papyrologische Beiträge zu Ehren von Bärbel Kramer*, edited by Raimar Eberhard, Holger Kockelmann, Stefan Pfeiffer, and Maren Schentuleit, 213–222. Berlin: Walter de Gruyter, 2009.

Scheuble, Sandra. "Griechen und Ägypter im ptolemäischen Heer—Bemerkungen zum Phänomen der Doppelnamen im ptolemäischen Ägypten." In *Interkulturalität in der Alten Welt: Vorderasien, Hellas, Ägypten und die vielfältgen Ebenen des Kontakts*, edited by Robert Rollinger, Birgit Gufler, Martin Lang, and Irene Madreiter, 551–560. Wiesbaden: Harrassowitz, 2010.

Scheuble-Reiter, Sandra. *Die Katökenreiter im ptolemäischen Ägypten*. Munich: C. H. Beck, 2012.

Seidl, Erwin. *Ptolemäische Rechtsgeschichte*. 2nd ed. Glückstadt: Augustin, 1962.

Smallwood, Mary. *The Jews under Roman Rule from Pompey to Diocletian*. Leiden: Brill, 1976.

Smith, Anthony D. *The Ethnic Origins of Nations*. Oxford: Basil Blackwell, 1986.

Spawforth, Tony. "'Macedonian Times': Hellenistic Memories in the Provinces of the Roman Near East." In *Greeks on Greekness: Viewing the Past under the Roman Empire*, edited by David Konstan and Suzanne Saïd, 1–26. Cambridge: Cambridge Philological Society, 2006.

Stefanou, Mary. "Waterborne Recruits: The Military Settlers of Ptolemaic Egypt." In *The Ptolemies, the Sea and the Nile: Studies in Waterborne Power*, edited by Kostas Buraselis, Mary Stefanou, and Dorothy J. Thompson, 108–131. Cambridge: Cambridge University Press, 2013.

Stökl Ben Ezra, Daniel. "A Jewish 'Archontesse': Remarks on an Epitaph from Byblos." *Zeitschrift für Papyrologie und Epigraphik* 169 (2009): 287–293.

Taubenschlag, Rafael. *The Law of Greco-Roman Egypt in the Light of the Papyri: 332 B.C.–640 A.D.* 2nd ed. Warsaw: Państwowe wydawnictwo Naukowe, 1955.

Taylor, Joan E. "A Second Temple in Egypt: The Evidence for the Zadokite Temple of Onias." *Journal for the Study of Judaism* 29 (1998): 297–321.

Tcherikover, Victor A. *Hellenistic Civilisation and the Jews*. Philadelphia, Jerusalem: Jewish Publication Society of America, 1959.

Thompson Crawford, Dorothy J. "The Idumaeans of Memphis and the Ptolemaic *Politeumata*." In *Atti del XVII Congresso Internazionale di Papirologia*, 1069–1075. Vol. 3. Naples: Centro Internazionale per lo Studio dei Papiri Ercolanesi, 1984.

Thompson, Dorothy J. "Hellenistic Hellenes: The Case of Ptolemaic Egypt." In *Ancient Perceptions of Greek Ethnicity*, edited by Irad Malkin, 301–322. Cambridge, MA: Harvard University Press, 2001.

Thompson, Dorothy J. "Ethnic Minorities in Hellenistic Egypt." In *Political Culture in the Greek City after the Classical Age*, edited by Onno M. van Nijf and Richard Alston, 101–117. Leuven: Peeters, 2011.

Thompson, Dorothy J. "The Sons of Ptolemy V in a Post-Secession World." In *Ägypten zwischen innerem Zwist und äußerem Druck: Die Zeit Ptolemaios' VI. bis VIII.*, edited by Andrea Jördens and Joachim F. Quack, 10–23. Wiesbaden: Harrassowitz Verlag, 2011.

Thompson, Dorothy J. *Memphis under the Ptolemies*. 2nd ed. Princeton: Princeton University Press, 2012.

Vandorpe, Katelijn. "Persian Soldiers and Persians of the Epigone: Social Mobility of Soldiers-Herdsmen in Upper Egypt." *Archiv für Papyrusforschung* 54 (2008): 87–108.

Winter, Eva. "Formen ptolemäischer Präsenz in der Ägäis zwischen schriftlicher Überlieferung und archäologischem Befund." In *Militärsiedlungen und Territorialherrschaft in der Antike*, edited by Frank Daubner, 65–77. Berlin: Walter de Gruyter, 2011.

Zucker, Friedrich. *Doppelinschrift spätptolemäischer Zeit aus der Garnison von Hermopolis Magna*. Berlin: Akademie der Wissenschaften, 1938.

Zuckerman, Constantine. "Hellenistic *politeumata* and the Jews: A Reconsideration." *Scripta Classica Israelica* 8/9 (1985–1988): 171–185.

3

SYRIAN RECRUITS AND UNITS IN THE ROMAN ARMY

A Military Diaspora?

*Nathanael Andrade**

Sometime in the second or third century, the *Cohors II Flavia Commagenorum equitata sagittariorum*, stationed at Micia in Dacia (now Romania), dedicated an altar bearing a Latin inscription to Jupiter Turmazgades.[1] The cohort had originated from Commagene, a region of Roman north Syria. A recently discovered inscription places the divinity in Commagene, and units raised or stationed there left dedications to the god elsewhere. The god's Commagenian origin is therefore secure, and the dedication celebrated a god "indigenous" to the region in which the *Cohors II Flavia Commagenorum* was first formed.[2]

In a certain sense, the cohort's dedication to Jupiter Turmazgades could be understood to express a distinctive ethnic or regional identification. It could even be construed as produced by a diaspora of north Syrians. If the unit consisted of soldiers from a relatively homogenous social background who worshipped the "ancestral" divinity of their homeland, then this interpretation is valid. The dedication would reflect how north Syrians enacted and expressed a shared identification, cognition of community, and a meaningful connection to an absent homeland, even after they had been dispersed throughout the Roman military structure and imperial geography. It would illustrate how north Syrian soldiers formed communities or settlements that were "foreign" to those of local inhabitants. As such, it would emblematize the phenomenon of military diaspora.

But certain problems complicate this interpretation too. First, it is apparent that Roman military units often accepted local recruits and did not maintain their initial ethnic or regional composition. Second, officers of the Roman army experienced frequent transfers and were often foreign to the soldiers whom they commanded. Third, the oft-cited concept of "diaspora" is fraught with methodological and theoretical difficulties. Due to such considerations, the dedication to Jupiter Turmazgades could be understood as a unit tradition through which soldiers expressed their occupational or institutional identification.[3] If

DOI: 10.4324/9781003245568-4

so, it articulated the military status of soldiers who maintained diverse social backgrounds, a collective occupational identity, and many of the cultural forms common to Roman soldiers, including the use of Latin in inscriptions. It did not express ethnicity, shared regional background, or above all diaspora.

Whether the concept of "diaspora" captures the experiences of expatriate soldiers in the Roman army is therefore a complex issue. The organizing principles of Roman imperialism and its military apparatus do not afford simple answers. The Roman army notably integrated personnel from diverse provincial societies, and soldiers of varied ethnicity, regional origin, or social background often laboured within common military institutions. Likewise, the army dispersed soldiers of common ethnic or regional origins throughout the Roman imperial landscape, thereby recreating the cultural contexts into which it situated them. In this respect, the Roman empire constituted a "state of becoming" and not a static entity,[4] and diverse factors exerted social pressure on soldiers to assimilate. As soldiers faced varied degrees of dispersal and isolation from their ethnic or regional compatriots, they also experienced integration into military institutions, quotidian occupational practices, social bonds with comrades of diverse backgrounds, and relationships with local civilians. These could have superseded the value of their ethnic or regional origins. Soldiers even witnessed their ancestral customs being adopted by fellow soldiers, military collectives or networks, and local populations in ways that denuded them of their ethnic or regional significances. For such reasons, soldiers did not necessarily forge and experience "diaspora" amid such phenomena.

As this chapter maintains, an examination of the social, cultural, and religious practices that Syrian soldiers enacted can help clarify this issue. The Roman army recruited heavily from Syrian and Upper Mesopotamian populations, and a recent discovery of early Syriac graffiti from Germany provides just one example of how the Roman army moved Syrians far from their ancestral regions.[5] What needs analysis is whether these recruits meaningfully constituted a diaspora (or diasporas) and, if so, how their acts of diasporic expression are to be identified. To explore these issues, the chapter first identifies why the concept of "diaspora" among Syrians in the Roman army raises certain analytical problems ("Methodological Criteria and Issues"). By doing so, it establishes the types of practices that can be defined as diasporic. Subsequently, it analyzes the diasporic expressions of Syrians enrolled in legions and regular *auxilia* and how the social organization and recruiting of the different unit types shaped them ("Syrians in the Legions: A Nexus of Diasporic and Occupational Expression" and "Syrians in the Auxilia: Creating Unit Identities from Diasporic Expressions"). But as it does so, the chapter also scrutinizes how Syrian diasporic practices were reconstituted as unit traditions by various soldiers, appropriated and redefined by military and civilian networks of non-Syrians, or integrated into local municipal cultures. Similar patterns in the famous Palmyrene *numeri* of Dacia and north Africa and the notable example of the Palmyrene auxiliary Publius Aelius Theimes are relevant to the main arguments of the chapter. But, as I have treated them in detail in another forthcoming publication, they will only be summarized here ("The Significance of Syrian *Numeri* and the Career of Publius Aelius Theimes").[6]

Methodological Criteria and Issues

Due to the issues raised so far, the criteria and scope through which this chapter defines and analyzes military diasporas must first be established. The complexities of Syrian identification in the Roman empire; the instability of the term "diaspora" in modern usages; the impact of military recruitment on cultural assimilation and occupational solidarities; and the evidence at disposal all merit clarification. In this regard, the following points should be noted.

First, in recent years, the term "diaspora" has assumed increasingly expansive definitions.[7] A word derived from how ancient Greeks described their own dispersal throughout the Mediterranean,[8] it has traditionally been linked to the historical and migratory experiences of Jews, whose activity in the Roman army is difficult to assess, of modern Armenians, and of the descendants of enslaved Africans in the Atlantic world.[9] But "diaspora" is now often invoked to describe any population that has experienced migration and dispersal throughout an increasingly globalized landscape, and this inclusivity arguably strips the terms of its heuristic utility. As a result, certain scholars have posited that populations only practice diaspora when their members experience a meaningful connection to an absent "ancestral" homeland and articulate it through certain of their cultural or religious practices.[10]

According to such scholars, another key feature of diaspora is intergenerational transmission. For a population to be truly diasporic, the descendants of its original expatriates must implement practices that mark their ancestors' migration and their current separation from their absent homeland. Immigrants do often perpetuate their ancestral traditions, but a population is only uniquely diasporic when the descendants of expatriates replicate such traditions too and thereby commemorate a migration that they did not physically undertake themselves.[11] By this logic, the term "diaspora" foremost reflects the practices and experiences of migratory peoples who engage in intergenerational processes of becoming, reinvention, and transformation as they negotiate the demands of new or shifting social contexts. Through certain key continuities in ancestral practice and their attachment to their absent homeland, such peoples enact cognition of commonality and difference from "others" amid their navigation of assimilation and dispersal. A "diaspora" therefore only exists through the enactment of diasporic practices and experiences by expatriates and their descendants. The treatment of a "diaspora" as a discrete group or bounded object of analytical study, which often reflects the tendency for scholars to see a "diaspora" in any migratory population, has less merit.[12] Accordingly, this chapter invokes "diaspora" principally in reference to how Syrian expatriates maintained a meaningful conceptual connection to the "ancestral" territory of their Syrian subgroup, expressed it through shared cultural or religious practices that ensured cognition of community among their dispersed settlements, and transmitted such practices to descendants.

Second, in this chapter, "Syrian" refers to people with familial origins in the territories of the Levant and Mesopotamia that had by the third century been integrated into the provincial system under the names Coele Syria, Syria Phoenice, Syria Palestina, Osrhoene, and Mesopotamia. Roman authorities identified these regions as inhabited by Syrian peoples, and in turn, their inhabitants identified themselves as Syrian.[13] In the Roman empire, self-ascription as Syrian encapsulated a meaningful cognition of social commonality. Inscriptions clarify that Syrians throughout the empire expressed Syrian identifications and maintained vast social or commercial networks.[14] Nonetheless, Syrians were also regionally diverse, ethnically striated, or citizens of varied civic polities. Not sharing a single common language, they could variously speak Greek, Aramaic dialects, Latin, or a host of other languages, with bilingualism or multilingualism being common. Expressions of Syrianness also occupied a vast spectrum of cultural interweaving informed by Roman, Greek, or Near Eastern precedents or even reflected new cultural forms generated in the Roman imperial context.[15]

Accordingly, despite their cognition of common Syrianness, diasporic experiences and practices among Syrians were structured foremost by specific civic, ethnic, or regional frameworks. Through these, expatriates maintained their densest social networks, shared political ideologies, or common cultural idioms. Such frameworks often informed the composition of Syrian auxiliary units and *numeri* (Emesenes, Ituraeans, Palmyrenes, Commagenians, and so forth), and they defined how Syrians experienced and practiced diaspora. As a consequence, rather than adopting the loose concept of a broad Syrian diaspora, one should approach the issue by addressing diasporic practices or diasporic experiences of Syrian communities, or more specifically Emesene, Palmyrene, and Commagenian/north Syrian diasporas.

A third consideration pertains to the concept of "military diaspora". This can refer to how recruits of shared ethnic or regional origin are organized into units, expatriated, and then stationed in regions in which they form communities or settlements that are foreign to the local inhabitants. The categorization of similar "occupational" diasporas has in fact become increasingly common in scholarly literature, especially when they constitute the subset of a broader national or ethnic diaspora.[16] Since many auxiliary units raised in Syria consisted initially of a specific ethnic or regional core, the definition just expressed is at first glance valid for the experiences of certain Syrian recruits. Nonetheless, if one adheres to it, then to speak of enduring Syrian military diasporas in the Roman army is problematic. Over time, expatriated units of Syrians integrated local recruits who in turn adopted Syrian practices. Likewise, Syrian soldiers and their descendants intermarried, procreated, and worshipped common divinities with civilians, and they adopted Latin for speech and epigraphic display. All told, Syrian auxiliary units did not form endogamic communities whose ethnic or regional foreignness separated them from the local communities in which they were deployed. As units became heterogeneous, they also transformed into occupational communities and ceased to be ethnic, regional, "foreign", or indeed diasporic collectives.

But one need not address "military diasporas" solely as military units that are continuously marked by collective ethnicity or regional origins. One can also define them as cognitively experienced communities created and maintained by expatriated soldiers and their descendants through shared cultural practices. If so, the concept becomes more meaningful for analyzing the situation in the Roman empire, and it becomes inclusive of Syrian civilians and women whose diasporic experiences were shaped by an initial military migration, even if they were not strictly soldiers. According to this understanding, "military diasporas" refer to how expatriated Syrian soldiers and their military or civilian descendants enacted forms of cognition and practice through which they defined community boundaries and marked their differences from other soldiers (even those in their units) and local civilians. Syrian soldiers in various places did in fact maintain a diasporic consciousness of community that distinguished them from the other military and civilian personnel among whom they lived, even as their units integrated diverse new recruits and even as they otherwise underwent cultural assimilation. In such instances, their diasporic identifications and practices did not encapsulate the unit or garrison in which they served, and over time their cognitively experienced communities included civilian (and female) descendants. Nonetheless, even as fellow soldiers adopted their ancestral practices and reconstituted them as unit traditions, diasporic Syrians still perceived such practices as uniquely "ancestral" to them and not for their non-Syrian comrades. In this sense, Syrian soldiers and their descendants enacted military diasporas that were constantly "becoming", that transformed to meet the demands of shifting contexts, and that maintained social boundaries amid various forms of cultural assimilation and interchange.[17]

Even so, it is often difficult to distinguish the diasporic practices of expatriate Syrian soldiers and their descendants from unit traditions, expressions of occupational identity, or activities conducted in municipal life. Soldiers differentiated themselves from civilians in various ways, but relationships and cultural transfers occurred between soldiers and the inhabitants of regions in which units were stationed.[18] Soldiers and local civilians sometimes worshipped the same divinities.[19] Similarly, as non-Syrians enlisted in what had originally been homogenous Syrian units or otherwise formed social bonds with Syrian soldiers, they often worshipped Syrian "ancestral" divinities and reconstituted Syrian diasporic practices as unit traditions or expressions of occupational status. As a result, certain cults of Syrian origin travelled independently from the ethnic backgrounds or regional identifications of their worshippers. Such cults instead travelled when officers who worshipped Syrian gods were transferred among units; when descendants of Syrian expatriates joined local legions or auxiliary cohorts; or when social contacts and bonds were forged between members of Syrian and non-Syrian units stationed in the same place. The worship of Jupiter Dolichenus famously emblematizes the transmission of such a "military" cult.[20] Nonetheless, such cults were not necessarily devoid of diasporic value for Syrians, and we will encounter how Syrians could still deem them "ancestral".

Fourth, given the differences in legal status, weaponry, military tactics, and organization among unit types (at least until the third century), this chapter treats Syrian soldiers in the legions[21] and regular *auxilia* in separate categories,[22] and it directs its readers to the treatment of irregular *numeri*, particularly Palmyrene ones, in another publication.[23] However, it should be noted that this separation is somewhat arbitrary. Members of the different unit types interacted, transmitted culture to one another, or engaged in common practices. As we will see, Syrians soldiers in the *auxilia* or *numeri* sometimes produced offspring who in turn enrolled in local legions or auxiliary units, and they frequently transmitted their cultural practices from one type of unit to another through such means. Finally, it must be stated that the legions, *auxilia*, and *numeri* did not form homogenous entities; particular units representing them varied due to their regional contexts of deployment, their internal ethnic compositions and influences, their patterns of recruitment and transfer, and their social networks.[24]

The final consideration pertains to the value of religious dedications. While epitaphs, military diplomas, and papyrus documents provide important contextual information regarding soldiers, religious dedications are valuable for establishing diasporic practices and experiences. Of the varying epigraphic types, these most articulately expressed a cultural or religious attachment to an "ancestral" homeland and commemorated the expatriation that separated worshippers from it. Of course, many such religious dedications did not reflect diasporic experience at all; soldiers also venerated divinities that putatively guarded the wellbeing of the Roman state (including the emperor), the army, specific units, and localities of deployment. But, significantly, they also worshipped what they experienced as the gods of their homelands and ancestors.[25] Syrians were no different, and their gods ranged from the quintessentially Roman Jupiter Optimus Maximus to Near Eastern divinities named Elagabal and Manawat. Despite the forms of assimilation that they underwent, expatriate Syrian soldiers and their descendants who dedicated temples or altars to their "ancestral" divinities were implementing practices that concretized their ethnic or regional identifications and oriented them towards a sacred homeland where their "ancestral" gods putatively resided. They were doing so even as they cultivated Latin in epigraphic display, assumed various practices and cultural markers of Roman veterans, and formed new social bonds with other soldiers or civilians. In this way, expatriate Syrian soldiers and their descendants truly enacted and perpetuated shared cognition of diasporic community. It is to their example that we now turn.

Syrians in the Legions: A Nexus of Diasporic and Occupational Expression

First- and second-century legionaries were typically Roman citizens.[26] Exceptions, however, can be noted. Given a relative lack of Roman citizens and Italian colonists, legions stationed in Syria perhaps sometimes recruited non-citizen Syrians,[27] and after the *Constitutio Antoniniana*, the civic differences between legionary and auxiliary recruits had largely become moot.[28] Still, the link between legionary enrolment and Roman citizenship informed certain trends. Syrian enlistments, initially sparse, increased over time, and parts of Syria where preeminent Greek *poleis* and Roman *coloniae* abounded produced the greatest numbers.[29] Since units drew recruits from regions in which they were deployed, legions that had been stationed in Syria invariably possessed greater concentrations of Syrians than others. A legion stationed in Syria and then transferred to North Africa brought many Syrian soldiers with it.[30]

Syrians in legions certainly worshipped prototypically Roman divinities and gods associated with their units,[31] but they also venerated the divinities of their homelands. In one lucid example, a third-century legionary named Lucius Trebonius Sossianus raised a dedication at Rome for Jupiter Heliopolitanus, the patron divinity of "the *colonia* Heliopolis", his home city.[32] But significantly, not all Syrians who served in legions while worshipping "ancestral" gods were first-generation expatriates. As the section "Syrians in the Auxilia: Creating Unit Identities from Diasporic Expressions" discusses in more detail, descendants of Emesene auxiliary soldiers in Pannonia joined legions like the *Legio II Adiutrix* and introduced their divinities to comrades.

For such reasons, a second-century dedication raised at Apulum in Dacia for the north Syrian god Jupiter Dolichenus merits attention. As its inscription states:

> To Jupiter Optimus Maximus Dolichenus and the Syrian Goddess (*Dea Syria*), great *Caelestis*, for the safety of the everlasting Roman empire and the *Legio XIII Gemina*, Flavius, son of Barhadadus, priest of Jupiter Dolichenus, for the legion written above. Willingly he raised the offering for him who deserves it.[33]

The name of Jupiter Dolichenus reflects the origins of his cult at Doliche in north Syrian Commagene,[34] and his priests in the Roman army are sometimes believed to have travelled from Syria to administer the cult.[35] Indeed, Flavius' father Barhadadus possessed a distinctly north Syrian name.[36] But unfortunately, we know little about Flavius other than his Syrian descent and his priestly duties. It is not even clear whether he was a soldier in the *Legio XIII Gemina* or simply serving as a priest for a north Syrian god that the legionaries worshipped.[37]

In fact, north Syrian auxiliary soldiers, especially those of the *Cohors II Flavia Commagenorum*, had been transferred from Moesia to Dacia, including the vicinity of Apulum, during or shortly after its conquest in the early second century. One of them also made a dedication for Jupiter Dolichenus at Micia.[38] Such soldiers may have even been responsible for initially transporting the cult of Jupiter Dolichenus to Dacia during the second century, from where networks of non-Syrian Roman officers transmitted it throughout the empire.[39] Yet, many of the north Syrians attested by inscriptions in Dacia as worshipping Jupiter Dolichenus were probably not immigrants themselves. Instead, they likely belonged to longstanding Syrian communities descended from expatriate soldiers and consequently continued to worship Jupiter Dolichenus.[40] Noticeably, some members of the *Legio XIII Gemina* bear names widely attested among Syrians and were probably recruited from such local Syrian communities.[41] This may explain how the worship of Jupiter Dolichenus became popular in the legion. Flavius, a member of a local north Syrian community, then came to serve in the legionary camp as a priest of north Syrian gods whose rites he knew.

To be sure, many non-Syrian soldiers of the *Legio XIII Gemina* who worshipped Jupiter Dolichenus did not deem him an "ancestral" divinity. They had instead adopted a "foreign" god from their Syrian comrades and reoriented his cult in ways that expressed their occupational status as soldiers, not ancestral practice.[42] But we can expect that north Syrians in the legion still embraced the worship of Jupiter Dolichenus as a uniquely "ancestral" diasporic practice through which they expressed shared north Syrian identifications. An example of this practice is provided by a soldier of a non-legionary unit. He served in the *classis praetoria Misenensis* at Rome and made a dedication in which he identified himself as Syrian and referred to Jupiter Dolichenus as his "ancestral" (*paternus*) and Commagenian (*Comogenus*) god.[43] Civilians who bore north Syrian names likewise raised dedications for Jupiter Dolichenus as an explicitly Commagenian god or a god of Commagenians at Apulum and Ampelum of Dacia.[44] Such persons perceived their divinity to be distinctly north Syrian or "Commagenian" and the patron of a cognitively experienced, if dispersed, community of north Syrian expatriates in the Roman empire, even after the inhabitants of Commagene no longer called themselves Commagenian.[45]

The dedication of Flavius should be understood in such terms. Alongside Jupiter Dolichenus, he also worshipped "the Syrian Goddess" (*Dea Syria*), whose primary cult site was at Hierapolis-Manbog in north Syria.[46] As her name indicates, she too was an explicitly "ancestral" divinity for north Syrian soldiers and their descendants in Dacia. Accordingly, even if his dedication exemplifies how Roman military institutions often transformed expressions of Syrian diaspora into unit traditions or markers of military status, this did not prevent Syrians from endowing these expressions with ancestral value. Flavius in fact embodied the transmission of diasporic cult practices through which north Syrians venerated their ancestral divinities Jupiter Dolichenus, "the Syrian Goddess", or "the Commagenian god". Similarly, in Pannonia, members of the *Legio II Adiutrix* adopted the worship of Elagabal from Emesene auxiliary soldiers. But whether they were soldiers or civilians, Emesene expatriates and their descendants in the region still conceived of Elagabal as their ancestral divinity.[47]

Syrians in the Auxilia: Creating Unit Identities from Diasporic Expressions

We have already encountered how Syrian auxiliary soldiers established long-standing communities in Dacia and transferred their cults to others in the region. This was in fact a recurring pattern. Since peregrines without Roman citizen status typically enrolled in the *auxilia*, the evidence for Syrian recruitment is quite ample. Many auxiliary units of Syrian origin were formed during the first through third centuries.[48]

While the units of the *auxilia* maintained ethnic or regional labels, their recruits became increasingly diverse in origin over time. When first formed in a given locality, auxiliary units typically consisted of recruits who possessed a relatively homogenous ethnic or regional origin that was reflected by their titles. But such units invariably accepted recruits from other regions in which they were stationed. Despite their titles, their recruits were diverse in origin and bonded with one another due to their occupational or institutional commonality, not an ethnic or regional one. The histories of the *Cohors I Ituraeorum*, *Ala I Augusta Ituraeorum*, and various Parthian *alae* confirm this development.[49] Significantly religious dedications indicate that soldiers in units of Syrian origin engaged in the individual or collective veneration of Syrian ancestral divinities, and this reflects the diasporic practices of the units' original recruits. These expatriates in turn transmitted such practices to their descendants, whether they were civilians or had joined the army. Such diasporic practices were over time reconstituted by non-Syrian officers and recruits as unit traditions, and through these traditions, heterogeneous recruits expressed their regimental affiliations, articulated their broader occupational identity as Roman soldiers, and distinguished themselves from civilians.[50]

The *Cohors I Hemesenorum* provides one key example of such a phenomenon. Formed in the mid- to late second century, it was primarily composed of Emesenes. By 180 CE or so, the unit was stationed at Intercisa in Pannonia (Hungary), where the remains and inscriptions of many auxiliary soldiers have survived. The latter often refer to the origins of the unit's soldiers, or they otherwise record names that can be suggestive of ethnic or regional background. On this basis, it is clear that by the early third century, the unit, which had apparently earned a collective grant of Roman citizenship, had integrated non-Emesene Syrians and Pannonians. But it still included Emesene recruits, some of whom may have been descended from veterans who belonged to the Emesene community at Intercisa.[51] Moreover, descendants of the cohort's Syrian soldiers, whom inscriptions could describe as Emesene (*domo Hemesa*), enrolled in the nearby *Legio II Adiutrix* or formed social bonds with its members, who in turn worshipped the cohort's *genius*.[52]

Given such internal heterogeneity, members of the cohort noticeably offered collective dedications to the solar divinity Elagabal. Elagabal was the divine patron of Emesa, where the god's most notable temple and aniconic cult statue were located. Elagabal is most famous for being the primary divinity worshipped by Marcus Aurelius Antoninus "Elagabalus", the notorious emperor with an ancestral connection to Emesa.[53] But the veneration of Elagabal by the unit's members took shape independently of "Elagabalus'" reign.[54] As early as the reign of Septimius Severus, the unit collectively raised a temple and Latin inscription for "the divine Sol Aelagabal" at Intercisa, as well as a dedication in its vicinity.[55] By the reign of "Elagabalus", when the unit raised an inscribed altar for its "ancestral god (*deus patrius*) Sol Elagabal", the worship of Elagabal was already established.[56]

All told, the cohort's collective veneration of Elagabal reflects a unit tradition and an expression of occupational identification. But the inscription dedicated during "Elagabalus'" reign adds an intriguing complication; it defines Elagabal as an "ancestral god". Apparently, some soldiers recognized that the god of their unit was distinctively ancestral to people descended from the Emesenes that had constituted its original core. This phenomenon bears relevance for the heterogeneous members of the *Cohors I Hamiorum* in Britain. Originally formed at the north Syrian city of Epiphaneia (Hama), the unit's members worshipped "the Syrian Goddess" of Hierapolis in a similar vein.[57]

One can make similar observations regarding the internal dynamics of the *Cohors XX Palmyrenorum*, stationed at Dura-Europos after 208 (and perhaps long before). A concentration of papyri from Dura-Europos, supplemented by inscriptions, provides ample documentation for the unit; some of the papyri even consist of rosters that list dates of recruitment. Initially consisting solely of Palmyrene recruits, the unit increasingly integrated Syrians from Dura-Europos and the Middle Euphrates during the early third century, as onomastic patterns among its soldiers demonstrate. But it also clearly continued to draw recruits that belonged to the Palmyrene community at Dura-Europos and were even descended from its previous members.[58] Even if the composition of the unit remained largely Syrian, it was nonetheless diverse and integrated Syrians from different subgroups who had their own unique "ancestral" homelands. But since Dura-Europos was located quite close to Palmyra and maintained social connections with it, it falls beyond the strict parameters of this study. I am exploring the unit traditions of the *Cohors XX Palmyrenorum* and the ancestral practices of its soldiers elsewhere.[59]

Moreover, due to the frequency of transfer, officers in units of Syrian origin were often not Syrian or, if Syrian, not of the same Syrian subgroup as the regulars. Likewise, Syrian officers could serve in many different units, whether of Syrian or non-Syrian origin. The careers of the Osrhoenian Barsemis, son of Abbeus,[60] and the Palmyrene Agrippa, son of Themus,[61] demonstrate such trends. Being Syrians, they remarkably served as officers in units of their specific Syrian subgroups, of other Syrian subgroups, and of non-Syrians. Such officers often worshipped the divinities of their home cities and more mainstream "Roman" gods commonly venerated in the army. While the officer named Barsemis mentioned above erected a surviving dedication for Jupiter Optimus Maximus at Intercisa in Pannonia Inferior (Hungary), an official from the Roman *colonia* of Berytus (Beirut) worshipped Jupiter Heliopolitanus and other divinities of Heliopolis/Baalbek while serving as prefect of the *Cohors I Aquitanorum* in Upper Germany during the mid-third century.[62]

The Significance of Syrian *Numeri* and the Career of Publius Aelius Theimes

As described in the introduction, this chapter will not examine in detail the importance of Syrian *numeri* or the units famously recruited at Palmyra and stationed in Dacia and north Africa, since I am treating these in greater detail in another study.[63] Even so, it can be stated in brief that the same social dynamics that we have witnessed in the legions and auxiliary units pertain to the Syrian *numeri* as well, and these are particularly well documented for the Palmyrene units and the contexts in which they were stationed. The descendants of Palmyrene veterans from *numeri* who settled in Dacia and North Africa during the second century maintained a diasporic connection to Syrian Palmyra, as shown by their worship of Palmyrene gods like Bel, Malakbel, Iarhibol, or other gods that they called *dii patrii* (ancestral gods) in Latin.[64] At the same time, as locals joined the Palmyrene units, they reconfigured the worship of Palmyrene divinities as their unit traditions.[65]

Moreover, as the descendants of Palmyrene expatriates joined local auxiliary units or legions, served on local civic councils, or took over municipal or local priesthoods, the worship of Palmyrene divinities also made the transition to these units and local civilian life in ways that facilitated their worship among people who were not Palmyrene.[66] The veteran Publius Aelius Theimes embodies this phenomenon. His father had arrived in Dacia and earned Roman citizenship through his membership in a Palmyrene *numerus*. Theimes in turn had joined an auxiliary unit stationed near Tibiscum, thus bringing his gods to that unit. While subsequently serving as a civic councillor in the municipal community at Sarmizegetusa, he established a temple to his ancestral Palmyrene divinities just outside the city.[67]

Summary

Syrian soldiers transported many cults to diverse regions of the empire and transmitted them to the following generations who were born abroad. In many instances, these cults were adopted by non-Syrian recruits of Syrian units, by non-Syrian soldiers in other units, by mobile officers, and by local civilians. Frequently, Syrian soldiers or their children transported these cults as they joined other military units, became members of local municipal communities or governments, or fashioned organic bonds with soldiers or civilians whom they befriended. But the fact that non-Syrians in a variety of ways came to worship Syrian gods and reconstituted their significance for their own occupational expressions does not mean that these cults were stripped of their diasporic value for the Syrian expatriates who worshipped them from one generation to the next. For Emesene Syrian auxiliary soldiers, legionaries, and civilians who resided in third-century Intercisa, the worship of their "ancestral" god Elagabal quite meaningfully connected them to the sacred homeland that constituted the god's most central cult site. For certain north Syrians who were descended from immigrant soldiers in Dacia and who served in its legions, Jupiter Dolichenus was a "Commagenian god" or their ancestral divinity, even as his worship became widespread among non-Syrian personnel. For Palmyrenes of Dacia who had perhaps never seen the city from which their expatriate ancestors in the *numeri* had hailed, Palmyra was a sacred homeland. They accordingly worshipped its gods, their ancestral divinities, even as others adopted them in their military or municipal life.

Such expatriate Syrian soldiers and their descendants in fact cultivated meaningful religious expressions that enabled them to generate and maintain cognition of their diasporic identifications. Remarkably, they did so even as they interacted with diverse populations, endured separation from Syrian compatriots, adopted Latin, worshipped new divinities, and served in municipal governments. Even if no single and broad "Syrian military diaspora" can be observed, our sources illuminate how certain Syrians created an Emesene military diaspora, how others enacted a Palmyrene one, and how others forged a north Syrian/Commagenian one. In this sense, military diasporas of Syrians were complex and dynamic social phenomena, but they were constant features of the Roman empire in the second and third centuries CE.

Abbreviations

AE	*L'Année Épigraphique.*
Castellum Dimmidi	Gilbert Charles-Picard, *Castellum Dimmidi*. Paris: Boccard, 1947.
CBFIR	*Corpus der griechischen und lateinischen Beneficiarier-Inschriften des Römischen Reiches.* Stuttgart: Kommissionsverlag, 1990.
CCID	Monika Hörig and Elmar Schwertheim, *Corpus Cultus Iovis Dolicheni*. Leiden: Brill, 1987.
CIL	*Corpus Inscriptionum Latinarum.*
EDH	*Epigraphic Database Heidelberg.*
EDR	*Epigraphic Database Roma.*
IDR	*Inscriptiones Daciae Romanae/Inscripţille Daciei Romane.*
IG	*Inscriptiones Graecae*
ILD	Constantine Petolescu, *Inscriptiones Latinae Daciae/Inscripţii latine din Dacia*. Bucharest: Editura Academiei Române, 2005.
ILS	*Inscriptiones Latinae Selectae.*
P. Dura	C. Bradford Welles, *The Excavations at Dura-Europos, Final Report*. Vol. 5, pt. 1: *The Parchments and Papyri*. New Haven: Yale University Press, 1959.
RIB	*Roman Inscriptions of Britain.*
RIU	*Die römischen Inschriften Ungarns.*
TEAD	*The Excavations at Dura-Europos, Preliminary Reports.* 9 vols. New Haven: Yale University Press, 1929–1952.
Triade	Youssef Hajjar, *La triade d'Héliopolis-Baalbek*. 3 vols. Leiden: Brill, 1977–1985.

Notes

* The topic and perspective formulated in this chapter are intimately linked with those of a parallel study of mine: Nathanael Andrade, "Palmyrene Military Expatriation and its Religious Transfer at Roman Dura-Europos, Dacia, and Numidia," in *Palmyra*

and the Mediterranean, ed. Rubina Raja (Cambridge: Cambridge University Press, forthcoming), which treats related social phenomena in the Palmyrene and other Syrian *numeri*.

The research and writing for this chapter were overwhelmingly completed in 2015, before the publication of many relevant works, particularly John D. Grainger, *Syrian Influences in the Roman Empire to AD 300* (Abingdon: Routledge, 2018); Jean-Baptiste Yon, *L'histoire par les noms: Histoire et onomastique de la Palmyrène à la Haute Mésopotamie romaines* (Beirut: IFPO, 2018); Csaba Szabó, *Sanctuaries in Roman Dacia: Materiality and Religious Experience* (Oxford: Archaeopress, 2018); Nadežda Gavrilović Vitas, *Ex Asia et Syria: Oriental Religions in the Roman Central Balkans* (Oxford: Archaeopress, 2021). The overall narrative of my chapter thus cannot reflect their contributions, though I have updated footnotes for Grainger and Yon in particular as appropriate.

1 *IDR* 3.3.138=*ILS* 9273.

2 James F. Gilliam, "Jupiter Turmesgades," in *Actes du IXe congrès international d'études sur les frontières romaines, Mamaïa, 6–13 septembre 1972*, ed. Dionisie M. Pippidi (Bucharest: Editura Academiei, 1974), 309–314; Ian Haynes, *Blood of the Provinces: The Roman Auxilia and the Making of Provincial Society from Augustus to the Severans* (Oxford: Oxford University Press, 2013), 229–230; Oliver Stoll, "Religions of the Roman Armies," in *A Companion to the Roman Army*, ed. Paul Erdkamp (Oxford: Blackwell, 2007), 452–476, at 470. The *Legio XVI Flavia Firma*, which had been stationed in Samosata of Commagene, worshipped this god in the temple of Jupiter Dolichenus at Dura-Europos. See *TEAD* 9.1, no. 973–974. A new altar for Turmazgades has been found at Doliche. See Michael Blömer and Margherita Facella, "A New Altar for the God Turmasgade from Dülük Baba Tepesi," in *Vom eisenzeitlichen Heiligtum zum christlichen Kloster: Neue Forschungen auf dem Dülük Baba Tepesi*, ed. Engelbert Winter (Bonn: Habelt, 2017), 123–146. For similar continuities in ethnic patterns of worship in military units, also see Tatiana Ivleva, "Peasants into Soldiers: Recruitment and Military Mobility in the Early Roman Empire," in *Migration and Mobility in the Early Roman Empire*, ed Luuk de Ligt and Laurens E. Tacoma (Brill: Leiden, 2016), 158–175.

3 Haynes, *Blood of the Provinces*, treats occupational identity in the Roman *auxilia*; Nigel Pollard, *Soldiers, Cities, and Civilians in Roman Syria* (Ann Arbor: University of Michigan Press, 2000) examines the institutional identities of soldiers stationed in Roman Syria; Oliver Stoll, *Zwischen Integration und Abgrenzung: Die Religion des römischen Heeres im Nahen Osten: Studien zum Verhältnis von Armee und Zivilbevölkerung im römischen Syrien und den Nachbargebieten* (St. Katharinen: Scripta Mercaturae, 2001) examines unit and army traditions.

4 Ann L. Stoler and Carole McGranahan, "Introduction: Refiguring Imperial Terrains," in *Imperial Formations*, ed. Ann L. Stoler, Carole McGranahan, and Peter C. Perdue (Sante Fe: SAR, 2007), 3–45.

5 Andreas Luther, "Osrhoener am Niederrhein: Drei altsyrische Graffiti aus Krefeld-Gellep (und andere frühe altsyrische Schriftzeugnisse)," *Marburger Beiträge zur antiken Handels-, Wirtschafts-, und Sozialgeschichte* 27 (2009): 11–30. For patterns of soldiers settling in places where they were stationed or elsewhere, see Saskia T. Roselaar, "State-Organised Mobility in the Roman Empire: Legionaries and Auxiliaries," in *Migration and Mobility in the Early Roman Empire*, ed. Luuk de Ligt and Laurens E. Tacoma (Leiden: Brill, 2016), 138–157.

6 Andrade, "Palmyrene Military Expatriation."

7 See the Introduction to this volume, esp. 4 and 9-12.

8 Robert Garland, *Wandering Greeks: The Ancient Greek Diaspora from the Age of Homer to the Death of Alexander the Great* (Princeton: Princeton University Press, 2014), esp. xviii–xx.

9 Raúl González Salinero, "El servicio militario de los judíos en el ejército romano," *Aquila Legionis* 4 (2003): 45–89; Jonathan Roth, "Jews and the Roman Army: Perceptions and Realities," in *The Impact of the Roman Army (200 B.C.–A.D. 476): Economic,*

Social, Political, Religious and Cultural Aspects, Proceedings of the Sixth Workshop of the International Network Impact of Empire (Roman Empire, 200 B.C.–A.D. 476), Capri, Italy, March 29–April 2, 2005, ed. Lukas de Blois and Elio Lo Cascio (Brill: Leiden, 2007), 409–420 provide treatment.

10 Martin Baumann, "Diaspora: Genealogies of Semantics and Transcultural Comparison," *Numen* 47 (2000): 314–337; Rogers Brubaker, "The 'Diaspora' Diaspora," *Ethnic and Racial Studies* 28, no. 2 (2005): 1–19; Kevin Kenny, *Diaspora: A Very Short Introduction* (Oxford: Oxford University Press, 2013), esp. 6–15; Ian Lilley, "Diaspora and Identity in Archaeology: Moving beyond the Black Atlantic," in *A Companion to Social Archaeology*, ed. Lynn Meskell and Robert W. Preucel (Malden, MA: Blackwell, 2004), 287–312, at 290–292; Robin Cohen, *Global Diasporas: An Introduction*, 2nd ed. (London: Routledge, 2008), 1–19; Hella Eckhardt, "A Long Way from Home: Diaspora Communities in Roman Britain," in *Roman Diasporas: Archaeological Approaches to Mobility and Diversity in the Roman Empire*, ed. Hella Eckhardt (Portsmouth, RI: JRA, 2010), 99–130, at 107–108. Also, see the Introduction to this volume.

11 Brubaker, "'Diaspora' Diaspora"; Sandra So Hee Chi Kim, "Redefining Diaspora through a Phenomenology of Postmemory," *Diaspora* 16, no. 3 (2007): 337–352.

12 Eckhardt, "A Long Way from Home," and Kristina Killgrove, "Identifying Immigrants to Imperial Rome Using Strontium Isotope Analysis," both in *Roman Diasporas*, along with Hella Eckhardt, Gundula Müldner, and Mary Lewis, "People on the Move in Roman Britain," *World Archaeology* 46, no. 4 (2014): 534–550, outline the theoretical issues underlying the analysis of diaspora as practice (98–99 and 107–109; 157–159), even if their case studies often establish migratory activity and cultural difference, rather than diaspora (109–125; 159–171). Indeed, many articles of *Roman Diasporas* focus on migration and expatriation (including the violence of slavery).

13 Nathanael Andrade, *Syrian Identity in the Greco-Roman World* (Cambridge: Cambridge University Press, 2013) and id., "Assyrians, Syrians, and the Greek Language in the Late Hellenistic and Roman Imperial Periods," *Journal of Near Eastern Studies* 73, no. 2 (2014): 299–317 provides treatment.

14 Andrade, *Syrian Identity*, treats Roman Syria. David Noy, *Foreigners at Rome: Citizens and Strangers* (London: Duckworth, 2000), 318–321, documents Syrians at Rome. Heikki Solin, "Juden und Syrer im westlichen Teil der römischen Welt: Eine ethnisch-demographische Studie mit besonderer Berücksichtigung der sprachlichen Zustände," *ANRW* 2.29.2 (1983): 587–789 and 1222–1249, provides comprehensive coverage of the material known at the time of publication. For social networks among Syrian traders in Rome, for instance, see Taco Terpstra, *Trading Communities in the Roman World: A Micro-Economic and Institutional Perspective* (Leiden: Brill, 2013), 152–164. For ethnicity as cognition and identification, see the articles in Rogers Brubaker, *Ethnicity without Groups* (Cambridge, MA: Harvard University Press, 2004).

15 Andrade, *Syrian Identity* provides analysis.

16 Cohen, *Global Diasporas*, 15–19; Brubaker, "'Diaspora' Diaspora," 2–3.

17 This cultural interchange certainly involved cults and divinities. For the example of Roman Syria, see Stoll, *Zwischen Integration und Abgrenzung*, 160–209.

18 Ibid., esp. 1–132 and 418–440; Pollard, *Soldiers, Cities, and Civilians*; Haynes, *Blood of the Provinces* all have demonstrated that the army was not a "total institution".

19 Stoll, *Zwischen Integration und Abgrenzung*, esp. 210–379.

20 Anna Collar, *Religious Networks in the Roman Empire: The Spread of New Ideas* (Cambridge: Cambridge University Press, 2013), 79–145; ead., "Military Networks and the Cult of Jupiter Dolichenus," in *Von Kummuh nach Telouch: Historische und archäologische Untersuchungen in Kommagene*, ed. Engelbert Winter (Bonn: Habelt, 2011), 217–246; and ead., "Commagene, Communication, and the Cult of Jupiter Dolichenus," in *Iuppiter Dolichenus: Vom Lokalkult zur Reichsreligion*, ed. Michael Blömer and Engelbert Winter (Tübingen: Mohr Siebeck, 2012), 99–110. Haynes, *Blood of the Provinces*, 229 notes the problem.

21 Principal publications are John C. Mann, *Legionary Recruitment and Veteran Settlement during the Principate* (London: Institute of Archaeology, 1983), with esp. 70–160, with tables 1–33 featuring assembled epigraphic data; Yann le Bohec and Catherine Wolff, eds., *Les légions de Rome sous le Haut-Empire: Actes du congrès de Lyon (17–19 septembre 1998) (Colloque "les légions de Rome sous le Haut-Empire")*, 3 vols. (Paris: Boccard, 2000–2003).

22 Principal publications are Paul Holder, *Studies in the Auxilia of the Roman Army from Augustus to Trajan* (Oxford: BAR International Series, 1980), with 127–139 and 169–334 assembling epigraphic data; Haynes, *Blood of the Provinces,* which is current and essential; David L. Kennedy, "The Auxilia and Numeri Raised in the Roman Province of Syria" (PhD diss., Oxford, 1980); and Ovidiu Tentea, *Ex Oriente ad Danubium: The Syrian Units on the Danube Frontier of the Roman Empire* (Bucharest: Mega, 2010). Also, the collected articles of David L. Kennedy, *Settlement and Soldiers in the Roman Near East* (Ashgate: Variorum, 2013).

23 Principal publications are Pat Southern, "The *Numeri* of the Roman Imperial Army," *Britannia* 20 (1989): 81–140; Kennedy, *Auxilia and Numeri.* See Andrade, "Palmyrene Military Expatriation" for my separate treatment of Palmyrene *numeri.*

24 For the complexities of culture(s) in the Roman *auxilia*, Haynes, *Blood of the Provinces*, esp. 145–382. For military networks and cultural transmission (as exemplified by the cult of Jupiter Dolichenus), see Collar, *Religious Networks*, 79–145, with Collar, "Military Networks," 217–246 and ead., "Commagene, Communication, and Cult," 99–110.

25 Stoll, *Zwischen Integration und Abgrenzung*, 210–379 captures this diversity in Roman Syria. Michael Speidel, "Recruitment and Identity: Exploring the Meanings of Soldiers' Origins," *Revue internationale d'histoire militaire ancienne* 6 (2017): 35–50 stresses the limitations of mere references to soldiers' origins in the documents.

26 Haynes, *Blood of the Provinces*, 1–84, both recognizes the principle and explores its complications and complexities over time.

27 David L. Kennedy, "The Military Contribution of Syria to the Roman Imperial Army," in *The Eastern Frontier of the Roman Empire: Proceedings of a Colloquium Held at Ankara in September 1988*, ed. David H. French and Chris S. Lightfoot (Oxford: BAR, 1989), 235–237, repr. in Kennedy, *Settlement and Soldiers.*

28 Haynes, *Blood of the Provinces,* 85–94.

29 Antioch, Apamea, Sidon, Berytus, and Caesarea Maritima are the leading attested contributors. Kennedy, "Military Contribution," 245.

30 For the apparent concentration of Syrians in the *Legio III Augusta*, the key document is *CIL* 8.18084 (mid-second century), but other examples are compiled by Yann le Bohec, "Les Syriens dans l'Afrique romaine: Civils ou militaires?," *Karthago* 21 (1987): 82–92, at 85–86, repr. in *L'Armée romaine en Afrique et en Gaule* (Stuttgart: Steiner, 2007), 457–58.

31 *CIL* 6.210, a marble altar in Rome. The dedicator, a member of the *Legio VI Ferrata* and a praetorian guard, was from Capitolias in south Syria, where the *Legio VI Ferrata* of Jerusalem recruited. His dedication was to Hercules the Defender and the *genius* of his century. Hannah M. Cotton, "The *Legio VI Ferrata*" in *Les légions de Rome sous le Haut-Empire: Actes du congrès de Lyon (17–19 septembre 1998) (Colloque "les légions de Rome sous le Haut-Empire")*, ed. Yann le Bohec and Catherine Wolff (Paris: Boccard, 2000), 1:351–357.

32 *CIL* 6.423=*Triade*, no. 290. For photo, see *EDR* F121720 © Roma-Musei Capitolini. Grainger, *Syrian Influences*, 158–162 (esp. 162).

33 *AE* 1965, no. 30 and 1975, no. 719=*CCID* 154=*IDR* 3.5.221=*AE* 2006, no. 1125. For photo, see *EDH* F005143 and 036381 © Ioan Piso. A date before 212 CE is suggested by Piso in *IDR* 3.5.221, whose reading is adopted here. See Constantin C. Petolescu, "Prêtres de Jupiter Dolichenus dans l'armée romaine de Dacie," in *Pouvoir et religion dans le monde romain*, ed. Annie Vigourt et al. (Paris: PUPS, 2006), 461–471

and Rudolf Haensch, "Pagane Priester des römischen Heeres im 3 Jahrhundert nach Christus," in *The Impact of Imperial Rome on Religious, Ritual, and Religion in the Roman Empire*, ed. Lukas de Blois, Peter Funke, and Johannes Hahn (Leiden: Brill, 2006), 208–218, at 209 on sacred personnel. Also, Szabó, *Sanctuaries,* 66–67.

34 For recent work on the cult and the site of Doliche (in modern Turkey), along with the cult elsewhere in the empire, see the articles in Winter, ed., *Von Kummuh nach Telouch* and Blömer and Winter, eds., *Iuppiter Dolichenus.*

35 Collar, *Religious Networks*, 137–144. Priests whose names suggest Syrian origins mostly appear in regions where Syrian troops were heavily deployed, though contrary to some scholarly interpretations, *CCID* 152=*CIL* 3.7791=*IDR* 3.5.223 refers to a priest named Antiochus, not a priest from Antioch. For photo, see *EDH* F006313 © I. Piso.

36 The Aramaic name Barhadad is attested in Greek at Dura-Europos and in Upper Mesopotamia. Giulia F. Grassi, *Semitic Onomastics from Dura-Europos: The Names in Greek Script and from Latin Epigraphs* (Padua: SARGON, 2012), 37 and 156–157; Han J. W. Drijvers and John F. Healey, *The Old Syriac Inscriptions of Edessa and Osrhoene: Texts, Translations, and Commentary* (Leiden: Brill, 1999), Am 11; Yon, *Histoire par les noms*, 153–156.

37 Ioan Piso, "Die soziale und ethnische Zusammensetzung der Bevölkerung in Sarmizegetusa und in Apulum," in *Prosopographie und Sozialgeschichte: Studien zur Methodik und Erkenntnismöglichkeit der kaiserzeitlischen Prosopographie*, ed. Werner Eck (Vienna: Böhlau, 1993), 315–337, at 320, repr. in id., *An der Nordgrenze des Römischen Reiches*: *Ausgewählte Studien (1972–2003)* (Stuttgart: Steiner, 2005), 214 argues that Flavius was merely a priest who lived in the *canabae.*

38 *AE* 1911, no. 35=*IDR* 3.3.67=*CCID* 159. Kennedy, *Auxilia and Numeri*, 97–101 and Tentea, *Ex Oriente*, 41–45, and Grainger, *Syrian Influences*, 84–87 treat the cohort, which was stationed in Dacia by 110 and headquartered at Micia by 130. For the legion, Ioan Piso, "Les legions dans la province de Dacie," in *Les légions de Rome sous le Haut-Empire: Actes du congrès de Lyon (17–19 septembre 1998) (Colloque "les légions de Rome sous le Haut-Empire")*, ed. Yann le Bohec and Catherine Wolff (Paris: Boccard, 2000), 1:220–224, repr. in *An der Nordgrenze*, 422–427.

39 See, recently, Collar, *Religious Networks*, 79–145 (esp. 109–111) and ead. "Commagene, Communication, and the Cult of Jupiter Dolichenus." But also see Grainger, *Syrian Influences*, 172–190.

40 A noteworthy example of north Syrian activity is reflected by the persons who dedicated a third-century temple at Sarmizegetusa. Amid their dedicatory inscription's fragmentary state, these appear to have borne Roman *nomina* and *cognomina* popular among north Syrians. Ádám Szabó, "Die Bauinschrift des Dolichenums von Sarmizegetusa," *Hungarian Polis Studies* 11 (2004): 139–161 has reconstructed the inscription as a dedication for Jupiter Dolichenus. See *EDH* HD024913, with photo: *EDH* F005452 ©I. Piso.

41 Legionaries of the late second or early third century that bear *cognomina* commonly attested in Syria and Dura-Europos are, for example, Lucius Aurelius Marinus and Marcus Aurelius Marinus. *IDR* 3.5.381=*CIL* 3.14476 (photo: *EDH* F005073 © I. Piso); *IDR* 3.5.30=*CIL* 3.989. On the Latin name Marinus and its Semitic homonyms, Yon, *Histoire par les noms*, 99–101.

42 *CIL* 3.1614 and 8004=*CCID* 158=*IDR* 3.5.220; *CCID* 161=*IDR* 3.3.15 (photo: *EDH* F005975).

43 *AE* 1953, no. 26=*CCID* 433. Blair Fowlkes-Childs, "The Cult of Jupiter Dolichenus in the City of Rome: Syrian Connections and Local Contexts," in *Iuppiter Dolichenus: Vom Lokalkult zur Reichsreligion*, ed. Michael Blömer and Engelbert Winter (Tübingen: Mohr Siebeck 2012), 211–230, at 215–217. The inscription does not cite Jupiter Dolichenus by name, but the traces of feet and a tail for the god's accompanying bull still remain on the base, and the *classis Misenensis* and the *classis Ravennas* are

notable for their members' dedications to Jupiter Dolichenus at Rome. For description and image, see Emaneula Zappata, "Les divinités dolichéniennes et les sources épigraphiques latines," in *Orientalia Sacra Urbis Rome Dolichena et Heliopolitana: Recueil d'études archéologiques et historico-religieuses sur les cultes cosmopolites d'origine commagénienne et syrienne*, ed. Gloria Bellelli and Ugo Bianchi (Rome: Bretschneider, 1996), 87–255, at 193–196, no. 46, pl. 32 (image). In the same volume, Ugo Bianchi, "I.O.M.D. et Deo Paterno Comageno," 599–603 argues that the "Commagenian god" is Jupiter Dolichenus in all attestations.

44 *CIL* 3.1301a and 7834=*ILS* 4298=*IDR* 3.3.298=*CCID* 147 (Ampelum); *CIL* 3.1301b and 7835=*ILS* 4299=*IDR* 3.3.299=*CCID* 148 (Ampelum). On the names, Yon, *Histoire par les noms*, 47–52, 67–69, 99–101. For "Suri negotiatores" and Jupiter Dolichenus, see *CIL* 3.7761=*ILS* 4304=*IDR* 3.5.218=*CCID* 153 (Apulum; photo at *EDH* F005095 © I. Piso) and *CIL* 3.7915=*IDR* 3.2.203=*CCID* 169 (Sarmizegetusa; photo at *EDH* F005544 © I. Piso).

45 Miguel J. Versluys, "Cultural Responses from Kingdom to Province: The Romanisation of Commagene, Local Identities, and the Mara bar Sarapion Letter," in *The Letter of Mara bar Sarapion in Context*, ed. Annette Merz and Teun Tieleman (Leiden: Brill, 2012), 43–66, at 56–59. Rare examples for people called "Commagenian" are *IG* 14.885; *RIB* 758.

46 For the cult of the Syrian Goddess (Atargatis, but sometimes conceived of as Holy Aphrodite or the Assyrian Hera), see Jane L. Lightfoot, *Lucian on the Syrian Goddess* (Oxford: Oxford University Press, 2003), esp. 1–83.

47 *RIU* 2.473=*CIL* 3.4300.

48 Holder, *Studies in the Auxilia*, provides a catalogue for auxiliary soldiers raised between the reigns of Augustus and Trajan, as the state of the evidence facilitated c. 1980, with Paul Holder, "Auxiliary Deployment in the Reign of Hadrian," in *Documenting the Roman Army: Essays in Honour of Margaret Roxan*, ed. John J. Wilkes (London: Institute of Classical Studies, 2003), 101–145 being relevant for the reign of Hadrian. The important analysis of Haynes, *Blood of the Provinces*, represents recent developments in the epigraphy and archaeology, and it extends into the third century. Reference works are John Spaul, *Cohors 2: The Evidence for and a Short History of the Auxiliary Infantry Units of the Roman Imperial Army* (Oxford: Archaeopress, 2000) and id., *Ala 2: The Auxiliary Cavalry Units of the Pre-Diocletianic Roman Imperial Army* (Nectoreca: Andover, 1994), with Ovidiu Tentea and Florian Matei-Popescu, "Alae et cohortes Daciae et Moesiae," *Acta Musei Napocensis* 39–40, no. 1 (2002–2003): 259–296 and Tentea, *Ex Oriente ad Danubium* providing corrections and additions for units along the Danube. Other key works are later cited as appropriate.

49 Haynes, *Blood of the Provinces*, 285–300; David L. Kennedy, "Parthian Regiments in the Roman Army," in *Limes: Akten des XI Internationalen Limeskongresses*, ed. Jenö Fitz (Budapest: Akadémiai Kiadó, 1977), 521–531 and id., *Auxilia and Numeri*; Holder, *Studies in the Auxilia* (passim); Grainger, *Syrian Influences*, 77–83 and 96–98; Everett Wheeler, "Parthian Auxilia in the Roman Army, Part I: From the Late Republic to 70 AD," in *Les auxiliaires de l'armée romaine: Des allies aux fédéres*, ed. Catherine Wolff and Patrice Faure (Paris: Boccard, 2016), 171–222 and "Parthian Auxilia in the Roman Army, Part II: From the Flavians to the Late Empire," *Revue internationale d'histoire militaire ancienne* 5 (2017): 103–137, who nonetheless notes possibilities for ethnic "restocking" (117–118).

50 For occupational/institution identity and the distinction between military and civilian identities, see Haynes, *Blood of the Provinces* and Pollard, *Soldiers, Cities, and Civilians*. On unit/regimental traditions, Stoll, *Zwischen Integration und Abgrenzung is* valuable.

51 Kennedy, *Auxilia and Numeri*, 120–132; Haynes, *Blood of the Provinces*, 135–142; Tentea, *Ex Oriente*, 48–52; Grainger, *Syrian Influences*, 89–93 are fundamental. Inscriptions in which figures are described as *domo Hemesa* are: *AE* 1906, no. 5=*RIU* 5.1203; *CIL* 3.3334(=10316)=*RIU* 5.1184 (*EDH* F010064 © G. Alföldy); AE 1909, no. 150=*RIU*

5.1194; *CIL* 3.10318=*RIU* 5.1202. Relevant treatment is by Agócs Nándor, "People in Intercisa from the Eastern Part of the Roman Empire," *Specimina Nova* 21–22 (2013): 9–28, at 11. For a reference to Roman citizenship, see *RIU* 6.1478=*AE* 1965, no. 10 (*EDH* F004502 © G. Alföldy). Jenö Fitz, *Les Syriens à Intercisa* (Brussels: Latomus, 1972) is still a standard work, even if some of its identifications of Syrians (who bear relatively generic names) are flawed (141–148).

52 *RIU* 5.1195 (*domo Hemesa*); *RIU* 4.1031=*CIL* 3.3301 (*domo Hemesa*); *RIU* 5.1190 (photo: *EDH* F011714 © G. Alföldy) and *CIL* 3.10315=*RIU* 5.1189, on which see Barnabás Lőrincz, *Die römischen Hilfstruppen in Pannonien während der Prinzipatszeit*, Part 1: *Die Inschriften* (Vienna: Forschungsgesellschaft Wiener Stadtarchäologie, 2003), no. 334; *CIL* 3.3334 (=10316)=*RIU* 5.1184 (photo: *EDH* F011586 © G. Alföldy); *CIL* 3.11076=*RIU* 3.737bis (*domo (Hi)erapuli cives Surus*); perhaps *RIU* 2.365. Also *RIU* 5.1075=*CBFIR* 394. For the legion's history, see Barnabás Lőrincz, "*Legio II Adiutrix*," in *Les légions de Rome sous le Haut-Empire: Actes du congrès de Lyon (17–19 septembre 1998) (Colloque "les légions de Rome sous le Haut-Empire")*, ed. Yann le Bohec and Catherine Wolff (Paris: Boccard, 2000), 1:159–168.

53 Leonardo de Arrizabalaga y Prado, *The Emperor Elagabalus: Fact or Fiction?* (Cambridge: Cambridge University Press, 2010); Martin Icks, *The Crimes of Elagabalus: The Life and Legacy of Rome's Decadent Boy Emperor* (Cambridge, MA: Harvard University Press, 2012).

54 Haynes, *Blood of the Provinces*, 227–229 provides treatment.

55 *ILS* 9155=*AE* 1910, no. 141=*RIU* 5.1104 (updated reading by *EDH* HD029862, which also celebrates Roman citizenship); *RIU* 6.1490=*AE* 1973, no. 437bis=Geza Alföldy, "Die Großen Götter von Gorsium," *Zeitschrift für Papyrologie und Epigraphik* 115 (1997): 238–239, no. 5 (*EDH* F004495 © G. Alföldy)=Lőrincz, *Die römischen Hilfstruppen*, no. 303.

56 *AE* 1910, no. 133=*RIU* 5.1139. The inscription provides a consular date.

57 See Stoll, "Religions of the Roman Armies," 470; *RIB* 1.1792. Also, Grainger, *Syrian Influences*, 113–115, 171. For the Syrian goddess, Lightfoot, *Lucian on the Syrian Goddess*.

58 Principal documents are *P. Dura* 55–128 (esp. 98–102, which are key rosters). David L. Kennedy, "The Cohors XX Palmyrenorum at Dura Europos," in *The Roman and Byzantine Army in the East*, ed. Edward Dąbrowa (Krakow: Uniwersytet Jagiellonski, 1994), 89–98 and id., "*Cohors XX Palmyrenorum:* An Alternative Explanation of the Numeral," *Zeitschrift für Papyrologie und Epigraphik* 53 (1983): 214–216; Pollard, *Soldiers, Cities, and Civilians*, 128–129; Andrew Smith II, *Roman Palmyra: Identity, Community, and State Formation* (Oxford: Oxford University Press, 2012), 147 and 153 (who discusses prior Palmyrene military presences at Dura-Europos). A reassessment of the Roman and by extension Palmyrene military presence at Dura-Europos is Simon James, *The Roman Military Base at Dura-Europos, Syria: An Archaeological Visualization* (Oxford: Oxford University Press, 2019). The onomastics of Dura-Europos' inhabitants have been given vital treatment by Grassi, *Semitic Onomastics*, and Spaul, *Cohors*, 434–435 and Yon, *Histoire par les noms*, 110–118 contain a partial list of names of the unit's soldiers. Given the controversy over whether the *Feriale Duranum* (*P. Dura* 54) constituted a military or civilian calendar, it will not be discussed here. In any event, the calendar, reflecting the "official" worship of Roman emperors and divinities, is not useful for discussing Syrian diasporic practices. Stoll, *Zwischen Integration und Abgrenzung*, 160–175 analyzes the *Feriale Duranum* as a document for official religion in the Roman army. But see Haynes, *Blood of the Provinces*, 198–206.

59 Andrade, "Palmyrene Military Expatriation." Of recent relevance is Jean-Baptiste Yon, "L'onomastique de la garnison 'palmyrénienne' de Doura Europos: La *cohors XX Palmyrenorum* et l'origine des recrues," *Revue internationale d'histoire militaire ancienne* 6 (2017): 143–153 and Yon, *Histoire par les noms*, 75–122.

60 Originally recruited into a *numerus Hosroenorum*, Barsemis was the decurion of the *Ala Firma Katafractaria* and *magister* of the *Cohors I Hemesenorum*. Most probably a

native speaker of Syriac, he left a Latin dedication to Jupiter Optimus Maximus at Intercisa in Hungary. *ILS* 2540=*CIL* 3.10307=*RIU* 5.1073 (*EDH* HD037160). For treatment, see Haynes, *Blood of the Province*, 92 and Wheeler, "Parthian *Auxilia* II," 114–115. For names, Drijvers and Healey, *Old Syriac Inscriptions*, As 4, As 33, Am 6, Am 11, and Bs 3 (for Baršamaš or Baršama) and As 61 (for Abba); Yon, *Histoire par les noms*, 50, 204. Likewise, see the Syriac names listed in Greek in *P. Euphrat.* 10 in Denis Feissel, Jean Gascou, and Javier Teixidor, "Documents d'archives romains inédits du Moyen Euphrate (IIIe siècle après J.-C.)," *Journal des savants* (1997): 3–57, at 45–50. The name Barsemis appears in Greek and Latin among the military registers of Dura-Europos. Grassi, *Semitic Onomastics*, 41 and 168–169; Yon, *Histoire par les noms*, 99.

61 Agrippa, son of Themus, from Palmyra, was centurion of the *Cohors III Thracum Syriaca* and then was transferred to the *Cohors I Chalcidenorum*. He then commanded a *numerus* of Palmyrene archers at el-Kantara (Calceus Herculis) in North Africa. *ILS* 9173=*AE* 1896 no. 35 and 1900 no. 197. For updated reading, *EDH* HD028486. Smith, *Roman Palmyra*, 168 discusses.

62 *CIL* 13.6658=*Triade*, no. 281. For date and location of the unit, see Holder, "Auxiliary Deployment," 124. On the god and consorts, see Andreas J. M. Kropp, "Jupiter, Venus, and Mercury of Heliopolis (Baalbek): The Images of the Triad and Its Alleged Syncretisms," *Syria* 87 (2010): 229–264.

63 The article in question is Andrade, "Pamyrene Military Expatriation." For key information, Smith, *Roman Palmyra*, 166–171; le Bohec, "Syriens dans l'Afrique romaine"; Grainger, *Syrian Influences*; Tentea, *Ex Oriente*, 66–76; Southern, "*Numeri*"; Maria Gorea, "Considérations sur la politisation de la religion à Palmyre et sur la devotion militaire des Palmyréniens en Dacie," *Semitica et Classica* 3 (2010): 125–162 and ead., "Stèle funéraire à banquet d'un Palmyrénien veteran à *Potaissa (Dacia superior)*," *Semitica et Classica* 6 (2013): 291–296; Eugenia Equini Schneider, "Palmireni in Africa: *Calceus Herculis*," in *L'Africa romana: Atti del 5. Convegno di studio*, ed. Attilio Mastino (Sassari, Università degli studi di Sassari, 1988), 383–395; Arbia Hilali, "La mentalité religieuse des soldats de l'armée romaine d'Afrique: L'exemple des dieux syriens et palmyréniens," in *Impact of Imperial Rome on Religions: Ritual and Religious Life in the Roman Empire*, ed. Lukas de Blois and Peter Funke (Leiden: Brill, 2006), 150–168 and Agnès Groslambert, "Les dieux orientaux à Lambèse," in *L'Afrique romaine de 69 à 439: Quéstions d'histoire*, ed. Marie-Pierre Arnaud-Lindet and Bernadatte Cabouret-Laurioux (Nantes: Éditions du Temps, 2005), 192–212.

64 Some examples from the inscriptions are *IDR* 3.1.134 (*EDH* F005525 © I. Piso), *IDR* 3.1.135 (=*AE* 1977, no. 695, *IDR* 136 (=*AE* 1983, no. 795, *EDH* HD000610 which contains preferable reading), and *IDR* 3.1.142 and 149=*ILD* 207 (combining two fragmentary inscriptions, see *EDH* HD000604) for Tibiscum; *ILD* 663 (reading at *EDH* HD005830) for Porolissum; and Eugène Albertini, "Inscriptions de el-Kantara et la région," *Revue Africaine* 72 (1931): 193–243, no. 9=*AE* 1933, no. 43 for Calceus Herculis.

65 Some examples from the inscriptions are: *ILD* 663 (reading at *EDH* HD005830); *CIL* 8.8795=*CIL* 8.18020=*AE* 1940, no. 148=*Castellum Dimmidi* 9.

66 Some examples from the inscriptions are: *IDR* 3.2.366 (*EDH* HD046185); *AE* 1936, no. 33; *IDR* 3.1.137-38 (*EDH* HD020605 and HD046508)=*AE* 1977, no. 697; *AE* 1967, no. 572=*AE* 2000, no. 1775 (*EDH* HD015289); *CIL* 8.2497 and (with a *numerus Syrorum*), Olwen Brogan and Joyce Reynolds, "Seven New Inscriptions from Tripolitania," *Papers of the British School at Rome* 28 (1960): 51–54, at 51; *AE* 1901, no. 114, with Equini Schneider, "Palmireni in Africa," 394; *ILD* 680 (*EDH* HD 044623); *AE* 1962, 0304 (*EDH* HD016963); *IDR* 3.2.263-64 (*EDH* HD019507 and HD028596); *ILS* 4344=*CIL* 3.1108=*IDR* 3.5.103; *IDR* 3.5.102=*AE* 1977, no. 661 (*EDH* HD020515; *IDR* 3.4.30=*CIL* 3.7728, with 12555 (*EDH* HD044880); Ioan Piso and Ovidiu Tentea, "Un nouveau temple Palmyrénien à Sarmizegetusa," *Dacia* 55 (2011): 111–121 (esp. 119–120).

67 Haynes, *Blood of the Provinces*, 379–381. The key inscriptions are *IDR* 3.2.369=*CIL* 3.12587 (photo: *EDH* F005486 © I. Piso); *IDR* 3.2.370=*CIL* 3.1472 (*EDH* HD046188); *IDR*. 3.2.152=*CIL* 3.7896 (photo: *EDH* F004936 © I. Piso); *IDR* 3.2.18=*CIL* 3.7954 (*EDH* HD045680).

References

Albertini, Eugène. "Inscriptions de el-Kantara et la région." *Revue Africaine* 72 (1931): 193–243.

Alföldy, Geza. "Die Großen Götter von Gorsium." *Zeitschrift für Papyrologie und Epigraphik* 115 (1997): 225–241.

Andrade, Nathanael. *Syrian Identity in the Greco-Roman World*. Cambridge: Cambridge University Press, 2013.

Andrade, Nathanael. "Assyrians, Syrians, and the Greek Language in the Late Hellenistic and Roman Imperial Periods." *Journal of Near Eastern Studies* 73, no. 2 (2014): 299–317.

Andrade, Nathanael. "Palmyrene Military Expatriation and Its Religious Transfer at Roman Dura-Europos, Dacia, and Numidia." In *Palmyra and the Mediterranean*, edited by Rubina Raja. Cambridge: Cambridge University Press, forthcoming.

Arrizabalaga y Prado, Leonardo de. *The Emperor Elagabalus: Fact or Fiction?* Cambridge: Cambridge University Press, 2010.

Baumann, Martin. "Diaspora: Genealogies of Semantics and Transcultural Comparison." *Numen* 47 (2000): 314–337.

Bianchi, Ugo. "I.O.M.D. et Deo Paterno Comageno." In *Orientalia Sacra Urbis Rome Dolichena et Heliopolitana: Recueil d'études archéologiques et historico-religieuses sur les cultes cosmopolites d'origine commagénienne et syrienne*, edited by Gloria Bellelli and Ugo Bianchi, 599–603. Rome: Bretschneider, 1996.

Blömer, Michael, and Margherita Facella. "A New Altar for the God Turmasgade from Dülük Baba Tepesi." In *Vom eisenzeitlichen Heiligtum zum christlichen Kloster: Neue Forschungen auf dem Dülük Baba Tepesi*, edited by Engelbert Winter, 123–146. Bonn: Habelt, 2017.

Blömer, Michael, and Engelbert Winter, eds. *Iuppiter Dolichenus: Vom Lokalkult zur Reichsreligion*. Tübingen: Mohr Siebeck, 2012.

Bohec, Yann le. "Les Syriens dans l'Afrique romaine: Civils ou militaires?" *Karthago* 21 (1987): 82–92.

Bohec, Yann le. *L'Armée romaine en Afrique et en Gaule*. Stuttgart: Steiner, 2007.

Bohec, Yann le, and Catherine Wolff, eds. *Les légions de Rome sous le Haut-Empire*, 3 vols. Lyon: Boccard, 1998–2003.

Brogan, Olwen, and Joyce Reynolds, "Seven New Inscriptions from Tripolitania." *Papers of the British School at Rome* 28 (1960): 51–54.

Brubaker, Rogers. *Ethnicity without Groups*. Cambridge, MA: Harvard University Press, 2004.

Brubaker, Rogers. "The 'Diaspora' Diaspora." *Ethnic and Racial Studies* 28, no. 2 (2005): 1–19.

Cohen, Robin. *Global Diasporas: An Introduction*. 2nd ed. London: Routledge, 2008.

Collar, Anna. "Military Networks and the Cult of Jupiter Dolichenus." In *Von Kummuh nach Telouch: Historische und archäologische Untersuchungen in Kommagene*, edited by Engelbert Winter, 217–246. Bonn: Habelt, 2011.

Collar, Anna. "Commagene, Communication, and the Cult of Jupiter Dolichenus." In *Iuppiter Dolichenus: Vom Lokalkult zur Reichsreligion*, edited by Michael Blömer and Engelbert Winter, 99–110. Tübingen: Mohr Siebeck, 2012.

Collar, Anna. *Religious Networks in the Roman Empire: The Spread of New Ideas*. Cambridge: Cambridge University Press, 2013.

Cotton, Hannah. "The *Legio VI Ferrata*." In *Les légions de Rome sous le Haut-Empire*, edited by Yann le Bohec and Catherine Wolff, 351–357. 3 vols. Lyon: Boccard, 1998–2003.

Drijvers, Han J. W., and John F. Healey, *The Old Syriac Inscriptions of Edessa and Osrhoene: Texts, Translations, and Commentary*. Leiden: Brill, 1999.

Eckhardt, Hella. "A Long Way from Home: Diaspora Communities in Roman Britain." In *Roman Diasporas: Archaeological Approaches to Mobility and Diversity in the Roman Empire*, edited by Hella Eckhardt, 99–130. Portsmouth, RI: JRA, 2010.

Eckhardt, Hella, Gundula Müldner, and Mary Lewis, "People on the Move in Roman Britain." *World Archaeology* 46, no. 4 (2014): 534–550.

Equini Schneider, Eugenia. "Palmireni in Africa: *Calceus Herculis*." In *L'Africa romana: Atti del 5. Convegno di studio*, edited by Attilio Mastino, 383–395. Sassari: Università degli studi di Sassari, 1988.

Feissel, Denis, Jean Gascou, and Javier Teixidor, "Documents d'archives romains inédits du Moyen Euphrate (IIIe siècle après J.-C.)." *Journal des savants* (1997): 3–57.

Fitz, Jenö. *Les Syriens à Intercisa*. Brussels: Latomus, 1972.

Fowlkes-Childs, Blair. "The Cult of Jupiter Dolichenus in the City of Rome: Syrian Connections and Local Contexts." In *Iuppiter Dolichenus: Vom Lokalkult zur Reichsreligion*, edited by Michael Blömer and Engelbert Winter, 211–230. Tübingen: Mohr Siebeck, 2012.

Garland, Robert. *Wandering Greeks: The Ancient Greek Diaspora from the Age of Homer to the Death of Alexander the Great*. Princeton: Princeton University Press, 2014.

Gavrilović Vitas, Nadežda. *Ex Asia et Syria: Oriental Religions in the Roman Central Balkans*. Oxford: Archaeopress, 2021.

Gilliam, James F. "Jupiter Turmesgades." In *Actes du IXe congrès international d'études sur les frontières romaines, Mamaïa, 6–13 septembre 1972*, edited by Dionisie M. Pippidi, 309–314. Bucharest: Editura Academiei, 1974.

Gorea, Maria. "Considérations sur la politisation de la religion à Palmyre et sur la dévotion militaire des Palmyréniens en Dacie." *Semitica et Classica* 3 (2010): 125–162.

Gorea, Maria. "Stèle funéraire à banquet d'un Palmyrénien veteran à *Potaissa (Dacia superior)*." *Semitica et Classica* 6 (2013): 291–296.

Grainger, John D. *Syrian Influences in the Roman Empire to AD 300*. Abingdon: Routledge, 2018.

Grassi, Giulia F. *Semitic Onomastics from Dura-Europos: The Names in Greek Script and from Latin Epigraphs*. Padua: SARGON, 2012.

Groslambert, Agnès. "Les dieux orientaux à Lambèse." In *L'Afrique romaine de 69 à 439: Quéstions d'histoire*, edited by Marie-Pierre Arnaud-Lindet and Bernadatte Cabouret-Laurioux, 192–212. Nantes: Éditions du Temps, 2005.

Haensch, Rudolf. "Pagane Priester des römischen Heeres im 3 Jahrhundert nach Christus." In *The Impact of Imperial Rome on Religious, Ritual, and Religion in the Roman Empire*, edited by Lukas de Blois, Peter Funke, and Johannes Hahn, 208–218. Leiden: Brill, 2006.

Haynes, Ian. *Blood of the Provinces: The Roman Auxilia and the Making of Provincial Society from Augustus to the Severans*. Oxford: Oxford University Press, 2013.

Hilali, Arbia. "La mentalité religieuse des soldats de l'armée romaine d'Afrique: L'exemple des dieux syriens et palmyréniens." In *Impact of Imperial Rome on Religions: Ritual and Religious Life in the Roman Empire*, edited by Lukas de Blois and Peter Funke, 150–168. Leiden: Brill, 2006.

Holder, Paul. *Studies in the Auxilia of the Roman Army from Augustus to Trajan*. Oxford: BAR International Series, 1980.

Holder, Paul. "Auxiliary Deployment in the Reign of Hadrian." In *Documenting the Roman Army: Essays in Honour of Margaret Roxan*, edited by John J. Wilkes, 101–145. London: Institute of Classical Studies, 2003.

Icks, Martijn. *The Crimes of Elagabalus: The Life and Legacy of Rome's Decadent Boy Emperor*. Cambridge, MA: Harvard University Press, 2012.

Ivleva, Tatiana. "Peasants into Soldiers: Recruitment and Military Mobility in the Early Roman Empire." In *Migration and Mobility in the Early Roman Empire*, edited by Luuk de Ligt and Laurens E. Tacoma, 158–175. Brill: Leiden, 2016.

James, Simon. *The Roman Military Base at Dura-Europos, Syria: An Archaeological Visualization*. Oxford: Oxford University Press, 2019.

Kennedy, David L. "Parthian Regiments in the Roman Army." In *Limes: Akten des XI Internationalen Limeskongresses*, edited by Jenö Fitz, 521–531. Budapest: Akadémiai Kiadó, 1977.

Kennedy, David L. "The Auxilia and Numeri Raised in the Roman Province of Syria." PhD diss., Oxford, 1980.

Kennedy, David L. "*Cohors XX Palmyrenorum*: An Alternative Explanation of the Numeral." *Zeitschrift für Papyrologie und Epigraphik* 53 (1983): 214–216.

Kennedy, David L. "The Military Contribution of Syria to the Roman Imperial Army." In *The Eastern Frontier of the Roman Empire: Proceedings of a Colloquium Held at Ankara in September 1988*, edited by David H. French and Chris S. Lightfoot, 235–246. Oxford: BAR, 1989.

Kennedy, David L. "The Cohors XX Palmyrenorum at Dura Europos." In *The Roman and Byzantine Army in the East*, edited by Edward Dąbrowa, 89–98. Krakow: Uniwersytet Jagiellonski, 1994.

Kennedy, David L. *Settlement and Soldiers in the Roman Near East*. Ashgate: Variorum, 2013.

Kenny, Kevin. *Diaspora: A Very Short Introduction*. Oxford: Oxford University Press, 2013.

Killgrove, Kristina. "Identifying Immigrants to Imperial Rome Using Strontium Isotope Analysis." In *Roman Diasporas: Archaeological Approaches to Mobility and Diversity in the Roman Empire*, edited by Hella Eckhardt, 157–174. Portsmouth, RI: JRA, 2010.

Kim, Sandra So Hee Chi. "Redefining Diaspora through a Phenomenology of Postmemory." *Diaspora* 16, no. 3 (2007): 337–352.

Kropp, Andreas J. M. "Jupiter, Venus, and Mercury of Heliopolis (Baalbek): The Images of the Triad and Its Alleged Syncretisms." *Syria* 87 (2010): 229–264.

Lightfoot, Jane L. *Lucian on the Syrian Goddess*. Oxford: Oxford University Press, 2003.

Lilley, Ian. "Diaspora and Identity in Archaeology: Moving beyond the Black Atlantic." In *A Companion to Social Archaeology*, edited by Lynn Meskell and Robert W. Preucel, 287–312. Malden, MA: Blackwell, 2004.

Lőrincz, Barnabás. "*Legio II Adiutrix*." In *Les légions de Rome sous le Haut-Empire: Actes du congrès de Lyon (17–19 septembre 1998) (Colloque "les légions de Rome sous le Haut-Empire")*, edited by Yann le Bohec and Catherine Wolff, 159–168. Vol. 1. Paris: Boccard, 2000.

Lőrincz, Barnabás. *Die römischen Hilfstruppen in Pannonien während der Prinzipatszeit, Part 1: Die Inschriften*. Vienna: Forschungsgesellschaft Wiener Stadtarchäologie, 2003.

Luther, Andreas. "Osrhoener am Niederrhein: drei altsyrische Graffiti aus Krefeld-Gellep (und andere frühe altsyrische Schriftzeugnisse)." *Marburger Beiträge zur antiken Handels-, Wirtschafts-, und Sozialgeschichte* 27 (2009): 11–30.

Mann, John C. *Legionary Recruitment and Veteran Settlement during the Principate*. London: Institute of Archaeology, 1983.

Nándor, Agócs. "People in Intercisa from the Eastern Part of the Roman Empire." *Specimina Nova* 21–22 (2013): 9–28.
Noy, David. *Foreigners at Rome: Citizens and Strangers*. London: Duckworth, 2000.
Petolescu, Constantin C. "Prêtres de Jupiter Dolichenus dans l'armée romaine de Dacie." In *Pouvoir et religion dans le monde romain*, edited by Annie Vigourt et al., 461–471. Paris: PUPS, 2006.
Piso, Ioan. "Die soziale und ethnische Zusammensetzung der Bevölkerung in Sarmizegetusa und in Apulum." In *Prosopographie und Sozialgeschichte: Studien zur Methodik und Erkenntnismöglichkeit der kaiserzeitlischen Prosopographie*, edited by Werner Eck, 315–337. Vienna: Böhlau, 1993.
Piso, Ioan. "Les legions dans la province de Dacie." In *Les légions de Rome sous le Haut-Empire: Actes du congrès de Lyon (17–19 septembre 1998) (Colloque "les légions de Rome sous le Haut-Empire")*, edited by Yann le Bohec and Catherine Wolff, 220–224. Vol. 1. Paris: Boccard, 2000.
Piso, Ioan. *An der Nordgrenze des Römischen Reiches: Ausgewählte Studien (1972–2003)*. Stuttgart: Steiner, 2005.
Piso Ioan, and Ovidiu Tentea. "Un nouveau temple Palmyrénien à Sarmizegetusa." *Dacia* 55 (2011): 111–121.
Pollard, Nigel. *Soldiers, Cities, and Civilians in Roman Syria*. Ann Arbor: University of Michigan Press, 2000.
Roselaar, Saskia T. "State-Organised Mobility in the Roman Empire: Legionaries and Auxiliaries." In *Migration and Mobility in the Early Roman Empire*, edited by Luuk de Ligt and Laurens E. Tacoma, 138–157. Brill: Leiden, 2016.
Roth, Jonathan P. "Jews and the Roman Army: Perceptions and Realities." In *The Impact of the Roman Army (200 B.C.–A.D. 476): Economic, Social, Political, Religious and Cultural Aspects, Proceedings of the Sixth Workshop of the International Network Impact of Empire (Roman Empire, 200 B.C.–A.D. 476), Capri, Italy, March 29–April 2, 2005*, edited by Lukas de Blois and Elio Lo Cascio, 409–420. Brill: Leiden, 2007.
Salinero, Raúl González. "El servicio militario de los judíos en el ejército romano." *Aquila Legionis* 4 (2003): 45–89.
Smith, Andrew, II. *Roman Palmyra: Identity, Community, and State Formation*. Oxford: Oxford University Press, 2012.
Solin, Heikki. "Juden und Syrer im westlichen Teil der römischen Welt: Eine ethnisch-demographische Studie mit besonderer Berücksichtigung der sprachlichen Zustände." *ANRW* 2.29.2 (1983): 587–789 and 1222–1249.
Southern, Pat. "The *Numeri* of the Roman Imperial Army." *Britannia* 20 (1989): 81–140.
Spaul, John. *Ala 2: The Auxiliary Cavalry Units of the Pre-Diocletianic Roman Imperial Army*. Nectoreca: Andover, 1994.
Spaul, John. *Cohors 2: The Evidence for and a Short History of the Auxiliary Infantry Units of the Roman Imperial Army*. Oxford: Archaeopress, 2000.
Speidel, Michael. "Recruitment and Identity: Exploring the Meanings of Soldiers' Origins." *Revue internationale d'histoire militaire ancienne* 6 (2017): 35–50.
Stoler, Ann. L., and Carole McGranahan. "Introduction: Refiguring Imperial Terrains." In *Imperial Formations*, edited by Ann L. Stoler, Carole McGranahan, and Peter C. Perdue, 3–45. Sante Fe: SAR, 2007.
Stoll, Oliver. *Zwischen Integration und Abgrenzung: Die Religion des römischen Heeres im Nahen Osten: Studien zum Verhältnis von Armee und Zivilbevölkerung im römischen Syrien und den Nachbargebieten*. St. Katharinen: Scripta Mercaturae, 2001.
Stoll, Oliver. "Religions of the Roman Armies." In *A Companion to the Roman Army*, edited by Paul Erdkamp, 452–476. Oxford: Blackwell, 2007.

Szabó, Ádám. "Die Bauinschrift des Dolichenums von Sarmizegetusa." *Hungarian Polis Studies* 11 (2004): 139–161.

Szabó, Csaba. *Sanctuaries in Roman Dacia: Materiality and Religious Experience*. Oxford: Archaeopress, 2018.

Tentea, Ovidiu, *Ex Oriente ad Danubium: The Syrian Units on the Danube Frontier of the Roman Empire*. Bucharest: Mega, 2010.

Tentea, Ovidiu, and Florian Matei-Popescu. "Alae et cohortes Daciae et Moesiae." *Acta Musei Napocensis* 39–40, no. 1 (2002–2003): 259–296.

Terpstra, Taco. *Trading Communities in the Roman World: A Micro-Economic and Institutional Perspective*. Leiden: Brill, 2013.

Versluys, Miguel J. "Cultural Responses from Kingdom to Province: The Romanisation of Commagene, Local Identities, and the Mara bar Sarapion Letter." In *The Letter of Mara bar Sarapion in Context*, edited by Annette Merz and Teun Tieleman, 43–66. Leiden: Brill, 2012.

Wheeler, Everett. "Parthian Auxilia in the Roman Army, Part I: From the Late Republic to 70 AD." In *Les auxiliaires de l'armée romaine: Des allies aux fédéres*, edited by Catherine Wolff and Patrice Faure, 171–222. Paris: Boccard, 2016.

Wheeler, Everett. "Parthian Auxilia in the Roman Army, Part II: From the Flavians to the Late Empire." *Revue internationale d'histoire militaire ancienne* 5 (2017): 103–137.

Winter, Engelbert, ed. *Von Kummuh nach Telouch: Historische und archäologische Untersuchungen in Kommagene*. Bonn: Habelt, 2011.

Yon, Jean-Baptiste. "L'onomastique de la garnison 'palmyrénienne' de Doura Europos: La *cohors XX Palmyrenorum* et l'origine des recrues." *Revue internationale d'histoire militaire ancienne* 6 (2017): 143–153.

Yon, Jean-Baptiste. *L'histoire par les noms: Histoire et onomastique de la Palmyrène à la Haute Mésopotamie romaines*. Beirut: IFPO, 2018.

Zappata, Emaneula. "Les divinités dolichéniennes et les sources épigraphiques latines." In *Orientalia Sacra Urbis Rome Dolichena et Heliopolitana: Recueil d'études archéologiques et historico-religieuses sur les cultes cosmopolites d'origine commagénienne et syrienne*, edited by Gloria Bellelli and Ugo Bianchi, 87–255. Rome: Bretschneider, 1996.

4

PARTICIPANTS IN THE EMPEROR'S GLORY

The Statues for Generals in Late Antique Rome

*Mariana Bodnaruk**

Following his victory over an enormous "barbarian" confederation led by Radagaisus near Florence in 406, the Roman general Flavius Stilicho, *magister utriusque militiae*, the highest-ranking military official in the Empire, received a gilded bronze statue in the Forum Romanum—a prominent monument erected in the conspicuous site (see Figure 4.1).[1] Since he had twice failed to prevail over Alaric[2] (the non-success misrepresented as a triumph by the "official panegyrist",[3] the court poet Claudian), Stilicho's defeat of Radagasius was a resounding victory for Emperor Honorius' government, an achievement exceedingly extolled by the honorific inscription.[4]

What is more is that Stilicho succeeded in drafting elite troops, estimated at 12,000 men, from the defeated army.[5] The strategy of recruitment of the whole "barbarian" contingents had already been in place: The *magister militum* had acquired his troops in the same manner during his inspection tours of the Rhine in 396 and the Danube in 401. Although this practice did not itself begin with Stilicho, it was first consistently employed by him. It was concerned not purely with an easy procurement of a desirable formidable force because given the prospect of rewarding and sending back home these troops after the end of the hostilities, and thus releasing the state from the necessity to support a standing army, this was an effective time and cost-saving strategy.[6] But these troops would not always be let to go home afterwards: although they were recruited directly from non-Roman communities, Stilicho held on to them. By recruiting "barbarian" contingents from outside of the Roman Empire in times of an emergency, Stilicho created "military diasporas" which in this chapter principally refer to *foederati*, the "barbarian" people of mainly northern origin, who served as more or less homogeneous ethnic formations in the Roman military forces on Roman territory, regardless of rank, duration of service, or success.[7]

Military diasporas certainly served on a large scale, either recruited into regular units, which often carried an "ethnic" unit name, or as *foederati*. *Foederati*,

DOI: 10.4324/9781003245568-5

FIGURE 4.1 Honorific inscription on statue base of Flavius Stilicho (*CIL* 6.1731 = 1195). Front side. From Roman Forum, 405–406. Rome, Gardens of Villa Medici. Photograph by the author, object in public domain.

initially "barbarians" in a treaty (*foedus*) with the Empire and coming entirely from outside the effective imperial borders, were significant not only because they entered the Roman military service (they were steadily recruited already during the third century), but to the extent, they participated in this occupational opportunity in the late fourth and early fifth centuries.[8] Seen as distinct and independent political entities of the *gentes externae*[9] by Romans, the military diasporas formed on Roman territory were initially, as Timo Stickler maintains, not a part of the Roman Empire proper.[10] Notably, Stilicho made no attempt to negotiate with or settle Radagaisus' army at the frontier. Confronting the Goths in 418, the *magister militum* Flavius Constantius, however, imposed on them a treaty by the terms of which they had to provide military service to the emperor in return for residence in Aquitaine (i.e. on the territory of the Roman Empire), while still remaining a homogeneous group with their own military organization. From then on, they appear as a settled military diaspora.

The *foederati* settled within Roman territory indeed lived under overall Roman authority and were subject to regulation by Roman officials, when by the early fifth century, some federate units became partly or wholly integrated into the formal structure of the army. Serving permanently in Roman provinces, they eventually drew on Roman recruits to uphold numbers, changing the composition of manpower.[11] The strategy of accommodating the federates by providing them with land produced, however, an unwanted effect as the resettlements were mostly unsuccessful, military diasporas rebelled, were used by usurpers or other "barbarians", and caused other troubles. Units of even thoroughly Romanized "barbarians" finished as political groups in opposition to central authorities every once in a while. Later in the fifth century, any developments of integration processes in the West slowed down and stopped due to civil wars and the ensuing shift of the frontiers, respectively.

Furthermore, in the 13 years between Constantius' death and general Aetius' reaching of unchallenged dominance, the existence of military diasporas had even more profoundly changed the recruitment practice of the Romans: While troops that were not recruited among the "barbarians" played only a minor role, the Empire made itself militarily dependent on ethnically defined "barbarian" contingents headed by their own leaders, including those settled on the Roman territory forming well-established military diasporas. Regarding the fragility of the military command structure in the late Roman West, the Empire ultimately relied on the senior Roman officers, who soon gained the highest positions in government, establishing bonds of loyalty between diverse ranges of "barbarian" groups serving the Roman army.

What is unusual about the statuary dedications to Stilicho is their location. For although the conspicuous public venues of the city of Rome had been crowded with honorific statuary before, they remained—primarily, if not exclusively—spaces for the commemoration of emperors and senators. The Forum Romanum as well as the Forum Traiani,[12] where senators were commemorated, were the most considerable public settings for the display of imperial statues in the fourth and fifth centuries.[13] The Forum Romanum was designed as a scene to display the emperor's military glory: Its architectural ornamentation conveyed the imagery of military triumph. With Stilicho, a new figure of a military honorand emerged in the late antique representational art and epigraphic culture. Possessing an exclusive political and social position derived from individual military achievements, the late Roman general befitted the representational space open to the public eye. Extant epigraphic evidence from late antique Rome attests nine statues erected for senior military officers (*magistri militum*): One from the modern Via del Corso, another in the Forum Traiani, and the rest in the Forum Romanum.

The first part of this chapter is a case study using the example of Stilicho to explore the epigraphic evidence consisting of statue bases erected in Rome for high-ranking military commanders between the last decade of the fourth and the first half of the fifth centuries. Dedicatory inscriptions articulate military power delegated by the emperor as supreme commander of the army to the generals and converted by them into a power base which is symbolically communicated by the sculpted representation and honorific language of the monuments. The second part shall investigate the changing relationship between the emperor and the military high commanders. How can we define the social standing that derived from the proximity to the imperial family and was monopolized by senior military officers? This allows for a re-examination of innovations in the late imperial government and of the role of the military elite in the late Roman West. Finally, I shall briefly relate and compare the senatorial and military image exhibited symbolically in the representational art of the period under discussion. Apprehended symbolically, it will be argued that different forms of representation, which were appropriated by late Roman army commanders on an exclusive basis and whose formation ultimately relied on the commanders' ability to create military diasporas—raise *foederati* and control their military diaspora—provide a fairly accurate image of the development of the social hierarchy in the late Roman West.

Military Honorands and the Roman Spectator: The Case of Stilicho

The Forum Romanum was the most prominent public venue for the display of honorific statues of distinguished military commanders in the city. Even more significantly, the military honorands came to share this space with emperors in contrast to senators, whose statues were restricted to an aristocratic but civilian zone in the Forum Traiani.[14]

Since all of the statues dedicated to military commanders are now lost, scholars are confined to dealing with fragments of honorific inscriptions carved into the statue bases, which were discovered amid the rubble of the Forum. Therefore, the statue bases and the preserved texts are the only remnants of the no longer extant sculptural representation. The inscriptions were, in fact, "written instructions" on how to read and interpret the images above them.[15] As statuary bases for military honorands were set up in a public and monumental context, they were open to a throng of potential readers on a prime site of Rome. It is against this "official" and imperial background that we should read dedicatory inscriptions as texts.

Chronologically, the earliest honorific statue of Stilicho was erected in the Forum Romanum between late 398 and early 399.[16] Honoured by the senate, the commander-in-chief of the army in the Western Roman Empire received not merely a common life-size statue but an equestrian monument, which was usually restricted exclusively to the members of the imperial family.[17] The text provides a career inscription in thirteen lines, praising Stilicho as a military office holder: He appears as *vir illustrissimus*, a man of illustrious rank (the highest senatorial grade), "master of the cavalry and the infantry" (*magister equitum peditumque*, "commander of the imperial guard" (*comes domesticorum*), and praetorian tribune (*tribunus praetorianus*).[18] Although it emphatically starts and dwells at length on the military career of the honorand by emphasizing his personal participation in military campaigns as the "count (*comes*) of the divine Theodosius Augustus in all wars and victories", it makes an important digression on a highly coveted social experience, namely, Stilicho's relationship to the imperial house: The general is extolled for both advancing "over the passing years through the steps of the most glorious military service rising to the height of eternal glory and carrying it up to royal relationship by marriage as the son-in-law of the deified Theodosius"; furthermore Stilicho was "admitted by Theodosius to a second royal kinship by marriage as the father-in-law of our lord Honorius Augustus".[19]

Stilicho claimed to have been appointed by Theodosius I to be the guardian of Honorius and Arcadius and "advisor" (*consultor*) of the emperors on the basis of his exemplary military career.[20] In his qualities as guardian of Honorius and husband of Serena (niece and formally adopted daughter of Theodosius), who bore him a son, Eucherius, Stilicho was part of the imperial family. To reproduce and further reinforce his lasting relationships with the emperor, he arranged the marriage of his daughters Maria and (after Maria's death) Thermantia to Honorius.[21] His matrimonial strategies brought him extremely close to the innermost circles of power in the Western court. High-ranking men with military achievements now wielded direct access—previously controlled by courtiers with civilian offices—to the emperor.

Therefore, as an able army leader, Stilicho managed to convert his military power into a rare social capital—as it was in late antiquity—namely, the proximity to the emperor. Another fragmentary inscription from the Forum Romanum contains a deliberate erasure of Stilicho's name, a sign of *damnatio memoriae*.[22] In five partially surviving lines, the text honours "the wisest, most victorious leader, advisor of our lords, as well as the protector of the divine family (*divini generis*) and of the Roman name".[23] The aforementioned inscription of 398/399 finishes by turning back one more time to the military services, which Stilicho rendered to the emperor Honorius, pointing out the general as a man "whose counsel and foresight delivered Africa"[24] and thus referring to the imperial victory over the Roman Mauri general Gildo,[25] *magister militum per Africam*, whose rebellion had threatened the corn supply of Rome. The Roman senate was not only responsible for the formal declaration of war against Gildo but also for the subsequent dedication to Stilicho set up by a senatorial decree (*ex senatus consulto*).[26] This is remarkable because until the late fourth century, it had been the emperors whose statues were set up to celebrate the victories gained by them as supreme military commanders—victories (advertised as) essentially gained for the senate and people of Rome; since Stilicho, high generals replaced the emperors in this function, both on the battlefield and partially in the ideological representation.

Another inscription, too, commemorates Stilicho's (counselling) role in the Gildonic war. The text belongs to a second statue erected for Stilicho in 400 by the guilds of barge-owners and fishermen (*caudicarii seu piscatores*) of Rome and focuses on Stilicho's recent military achievements:

> Out of high regard for great virtues, among the other benefits which have been bestowed through him upon the city of Rome [...] because with him, having Gildo the public enemy (*hostis publicus*) vanquished and the food supply of the Romans restored, increased the happiness.[27]

Remarkably, Gildo himself had complemented his force constituted of the Roman forces already present in Africa with a huge cavalry of mercenaries from the pre-Sahara zone, but his Mauri mercenaries fled from Mascezel's (Roman) Gallic army in disarray. Claudian's war account, shaped by imperialist prejudices, describes Gildo's forces as predominantly African tribal warriors who could not match Roman discipline.[28]

What both inscriptions are really about is the transformative power of the inscribed word and the need to construct a report on the war: What actually happened is beside the point, because both inscriptions intend to refashion reality to create a story they need to tell. What makes the entire honorific enterprise symptomatic is, on the one hand, the actual marginal, advisory—rather than leading—role of Stilicho in the imperial military conduct in Africa. Much more important, however, on the other hand, is to be aware of the actual political situation behind what is communicated: Gildo's ambiguous status (who, after all, pledged fidelity to Arcadius, emperor of the Eastern Roman Empire) and Stilicho's uncertain standing (who was denounced as *hostis publicus* by the Eastern court in 398). It is here that ideology enters the text.

Two inscriptions from the gates of the Aurelian wall of Rome dated to 401/402 record "the restoration of the walls, gates, and towers of the Eternal City, with the riddance of immense rubble, due to the suggestion of the count and master of both forces, Stilicho, of *clarissimus* and illustrious rank" (see Figure 4.2).[29] The inscriptions refer to the completion of the strengthening of the Aurelian wall under the reign of Honorius and emphasize—once again—the advisory role played by Stilicho. A similar reference is erased from a third inscription, on the Porta Tiburtina (see Figure 4.3),[30] in all probability because Stilicho fell into disfavour in 408.

In 406, leading a large force with Hunnic and Allan allies, Stilicho defeated Radagasius.[31] It is worth noting that Olympiodorus, stemming from the Eastern Roman Empire, refers to the numbers of recruits drafted by the *magister militum* into the Roman army with some surprise, as something typical of the West only.[32] The victory over Radagasius was celebrated by a lost triumphal arch with statues of the emperors Arcadius, Honorius, and Theodosius II dedicated by the senate in Rome, possibly in the Campus Martius.[33] The related inscription claimed that this victory "extinguished forever the nation of the Goths".[34]

FIGURE 4.2 Dedicatory gate inscription recording completion of Honorian works on Aurelianic wall and advisory role played by Flavius Stilicho (*CIL* 6.1189). Porta Praenestina. 401–402. Rome, outside Porta Maggiore. Photograph by the author, object in public domain.

Two further dedications from the Forum Romanum, set up most probably in 406 and therefore after Stilicho's decisive victory over Radagaisus, attest to the eminent position of Stilicho in early fifth-century Rome. The first is the gilded statue set up by the urban prefect Flavius Pisidius Romulus.[35] Its accompanying inscription mentions Stilicho's military offices—commander of both soldieries (*magister utriusque militiae*), comital commander of the imperial guards and of the sacred stable (*comes domesticorum et stabuli sacri*)—and points out that the general is a "partner" (*socius*) to the emperors "in all wars and victories".[36] With rare precision, the text describes the monument and indicates its site: "The Roman people, due to their exceptional love for him and his foresight, have decreed a statue of bronze and silver to be installed on the Rostra as a memory of his eternal glory".[37] A silver-plated statue evoked comparison with the emperor's images and required both imperial and senatorial consent to be set up in the central area of the Forum—even more because it was paired with a statue of Honorius, suggesting co-ruling in a dynastic tradition.[38] While the statue made the absent honorand appear less remote from Rome, the location of the monument had a highly symbolic significance: Following a military triumph, public addresses were held from the speaker's platform in the Forum Romanum.[39] By appropriating the place from where the emperors made their *orationes*, Stilicho effectively

FIGURE 4.3 Dedicatory gate inscription recording completion of Honorian works on Aurelianic wall and advisory role played by Flavius Stilicho (name erased) (*CIL* 6.1190). Porta Tiburtina. 401–402. Rome, outer façade of Porta Tiburtina. Photograph by the author, object in public domain.

replaced the emperor in this ceremonial context. Positioned on the Rostra, the statue effectively reminded the populace of all previous orators, thereby conveying quasi-imperial honour to the distant general.

The second monument, set up in the name of the senate and people of Rome under the supervision of the same city prefect and found in situ in front of the curia near the arch of Septimius Severus, is dedicated, quite unusually, to the *fides* and *virtus* of the emperor's soldiers (see Figure 4.4):

> To the loyalty (*fides*) and valor (*virtus*) of the most devoted soldiers of our lords Arcadius, Honorius, and Theodosius, everlasting Augusti, after the Gothic war had been brought to an end through the good fortune (*felicitas*) of our eternal emperor and lord Honorius, and by the counsels (*consilia*) and bravery (*fortitudo*) of the count and master of both armies, Flavius Stilicho, of illustrious rank, twice consul.[40]

Even though the identity of the statue that stood above this inscribed base is controversial and uncertain, and the name and titles of Stilicho[41] were erased from the base after his downfall in 408, most scholars regard this inscription, emphasizing recent imperial victories over the Goths, and the related statue as dedicated

FIGURE 4.4 Honorific inscription on statue base dedicated to the Fides and Virtus of the emperor's soldiers, associating their victories with actions of Flavius Stilicho (*CIL* 6.31987). Front side. From Roman Forum.406.Rome, Roman Forum (Sopr. For-Pal., inv. no. 12436). Photograph by the author with permission of the Ministero per i Beni e le Attività Culturali e per il Turismo—Parco archeologico del Colosseo.

to the all-powerful general and courtier.[42] The text of the inscription, however, also celebrates the good fortune of Honorius' reign.[43] Therefore, the statue itself, as Carlos Machado notes, may also have been an image of the emperor himself or a personification of the virtues celebrated. Alternatively, Stilicho himself may have been represented as the living personification of military *fides* and *virtus*.[44]

Remarkably, the "emperor's soldiers" who won the victory over the "barbarian" confederation led by Radagaisus in 406 are emphatically praised for loyalty. According to Zosimus, the Roman troops comprised 30 *numeri*, together with auxiliaries from the Alans and Huns, to form an army that does not exceed a figure between 10,000 and 15,000 against prevailing forces gathered by Radagaisus.[45] Stilicho's *numeri*—a type of regular army units—were thus accompanied by several thousand auxiliaries drafted from Alans and Huns. Leif Petersen stresses that Roman identity, against which various "barbarian", tribal ethnicities have been contrasted, both in ancient historiography and modern accounts, was effectively a political identity in late antiquity:

> [A] Roman was whoever was loyal to the Roman cause, no matter his ethnic, geographic or linguistic background. This meant that disloyalty put "Romans" beyond the pale, while "barbarians" from beyond the borders were commended for spending their whole lives in service dedicated to the Roman state.[46]

The symbolic emphasis of the inscriptions discussed in this section is unmistakably on Stilicho's military commands as an embodiment of bravery and valour, which are catalogued in exhaustive detail. The manner in which a *magister militum* like Stilicho received the exceptional honour of statue dedications is instructive of how such a high-ranking *vir militaris* and the relation between him, the emperor, or the senate were perceived: He was the strong man of the Western Roman Empire. After his victory, Stilicho was able to incorporate 12,000 of Radagaisus' *optimatoi* (ὀπτιμάτοι) into his army as auxiliaries.[47]

Military Honorands and Emperors: The Case of Constantius III

The administrative reforms of Constantine I (306–337) had created a new elite that owed its advancement to military service. The ranks of the new service aristocracy or "aristocracy of office" (composed of the *militia civilis/officialis* and the *militia armata*) were made up of members of the existing provincial oligarchies. Although some senatorial governors may have retained military functions into Constantius II's reign, senior military officers, who bore the titles *magister equitum et peditum*, *magister utriusque militiae* and *magister militum* (which were used variably in the inscriptions), essentially replaced Roman senators in warfare expertise and assured supreme military command, the right to recruit and control the troops: *comitatenses*, field troops of the late Roman army, and *limitanei*, troops stationed permanently on the *limes*,[48] as well as *veterani*.[49] It is emblematic of the late Roman Empire that these generals could become so powerful that they were almost equal to the emperor, leading to their participation in the emperor's glory. After Merobaudes, Bauto, and Arbogastes, Stilicho did not establish a precedent and a pattern for these historical developments, but he certainly did it for the epigraphical representation of military commanders-in-chief. The next all-powerful general who dominated imperial politics in the first half of the fifth century and received a greater share of honorific inscriptions appeared immediately after Stilicho's downfall: Flavius Constantius who even managed to gain more influence and power than Stilicho had.

Two lost inscriptions from Rome, originally located in the Forum Romanum or the Forum Iulium, commemorate the third consulship of Flavius Constantius (also known as Constantius III), *magister utriusque militiae*, *patricius*, and a close associate of the imperial house by his marriage to Galla Placidia (daughter of Theodosius I and granddaughter of Valentinian I)[50] in 417. In 414, the Roman princess was wedded to Athaulf, a Gothic king, inaugurating a new phase in Roman-barbarian relations.[51] The dedicatory text on the statue base with the less fragmentary inscription praises Constantius as the "kinsman (*parens*) of the most unconquered emperors".[52] Therefore, the inscription appropriates Stilicho's quasi-official title *parens* to signify Constantius' relationship to the imperial family. These ideological investments worked at the same time as strategies of legitimation, i.e. the acquisition of titles and matrimonial alliances, which aimed at consolidating the exclusive appropriation of social prestige and reproducing it.

Constantius was the first *magister militum* who received the rank of *patricius*[53] (414), a rather unusual title for a general at that time, and shared the consulship with the emperors.[54] Rather than denoting an actual office or a formal position, the title of patrician was a title of a rank or dignity, and as such a sign of special favour and close relationship to the emperor.[55] Still more conspicuously, he was ordinary consul in the Western Roman Empire three times, which, at least for a civilian official, was an unheard-of honour. In 418, the settlement of the Visigoths, under their king Vallia, as a military diaspora in Aquitania Secunda was negotiated by Constantius.[56] The rare honour of a third consulate fell on him in 420. The second, more fragmentary inscription recapitulates his honours up to this point.[57]

The imperial ideology of military triumph acknowledged the military achievements of a general only insofar as his victories enhanced the emperor's glory: The former handed his victory over to the latter and received due recognition in return. A different picture emerges in the case of Flavius Constantius who is praised as "the restorer of the Commonwealth (*res publica*)",[58] an honorific title usually reserved for emperors. Dedications frequently celebrate emperors as maintainers of earthly order not only by vanquishing "barbarians" but also by their victories over usurpers, effectively assimilating the usurper to a "barbarian".[59] In the Western Roman Empire, the threat of usurpation had many sources of origin, but it were primarily military commanders outside the centre of imperial power, who drew on the loyalty of their soldiers and revolted against the emperor. In the 50 years between 406 and 456 alone, 13 usurpers seized power, 10 under the reign of Honorius.[60] Having defeated usurpers and generals alike,[61] it was Flavius Constantius who was eulogized as "the restorer of the Commonwealth", not the emperor; Constantius' proven loyalty to the imperial cause obviously justified claiming the restoration of the Empire. That such a symbolic role was assigned to high-status military commanders reflects the actual power relationships in the last century of the Western Roman Empire. By listing the *cursus honorum* and phrasing the honorific text in a sound military tone, the inscription depicts Constantius as a senior member of the military hierarchy of the Empire. However, the accumulation of titles points to the ambiguity of Constantius' status, which was clearly emperorlike.[62] The final step in the rise of the omnipotent "generalissimo"[63] was to become emperor, and this was eventually achieved by Constantius in 421 when he was appointed by Honorius as Augustus and co-emperor.

As we have seen so far, military ability could be converted into a durable network of connections, most importantly, of course, into proximity to the emperor providing high military officers with prestige expressed in the public recognition, in every sense of the word. Since the child-emperors Arcadius and Honorius failed to fulfil their military leadership, warfare became, at least in the Western Roman Empire, the virtual monopoly of senior military commanders since the late fourth century. Generals like Stilicho and Flavius Constantius were the only officials who could boast extensive military experience, which allowed them to become widely independent from the emperors. Such a power base made up of military prowess and easily convertible into social prestige—because of proximity to the imperial family through marriage bonds, which will be dealt with in the next section in more detail—led to the exclusive position of its holders, and honorific monuments of the military officers in the Forum Romanum represent this particular unity of military authority and high social status accumulated by the generals. The dedications in the Forum Romanum show that the participation in the emperor's glory existed within a constantly recreated framework, confirmed both by military victories and by matrimonial alliances designed to maintain a close relationship which in this form was inaccessible to the senatorial aristocracy.[64]

Military Honorands and Senators: The Case of Aetius

As already pointed out, the combination of outstanding military authority and proximity to the imperial family yielded power to the generals that extended far beyond military affairs. Interestingly enough, they were not related to the senatorial aristocracy. However, the incorporation of military elites into the *ordo senatorius* and the Roman senatorial *cursus honorum* started under Valentinian I (364–375) and Valens (364–378).[65] This brings one back to the question of integrating imperial elites into a single senatorial class by replacing previous rank distinctions, though clearly separating military and civilian careers. First, beginning with Constantine I, the *magister militum* became *clarissimus* and *comes primi ordinis*.[66] The rank predicate *clarissimus* denotes the lowest senatorial rank, while the honorific title *comes primi ordinis*—with *comes* being a designation for courtiers, palace officials, and other officials in imperial service—distinguishes the closest advisors of the emperor. Initially, below the *prefecti pretorio*—the highest civilian administrators presiding over the largest administrative units, the prefectures—who were ranked as *viri illustres*, i.e. the highest senatorial rank, *magistri militum* were elevated to the same rank under Valentinian I.[67] As the political and military situation in the West became increasingly precarious, military leaders, as we have seen, appropriated a primary relation to imperial power, outdoing in this the senators of Rome.

This situation is further exemplified by another distinguished general: Flavius Aetius. A statue base dedicated to him by the senate and the people of Rome was found in a conspicuous location, namely in the area immediately behind the curia, identified as the late antique *atrium libertatis*, a space by this time associated with the senate, where precisely, according to the inscription, the statue was originally installed (see Figure 4.5).[68] It is significant that the monument should have been erected in the Forum Romanum in close proximity to the curia of the senate. This was the political centre of the late imperial forum; the zone in front of the curia was an important location for the erection of inscriptions honouring

FIGURE 4.5 Honorific inscription on statue base of Flavius Aetius (*CIL* 6.41389). Front side, lower part. From Roman Forum. 437–445. Rome, Roman Forum (Lapidario Forense, inv. no. 12462). Photograph by the author with permission of the Ministero per i Beni e le Attività Culturali e per il Turismo—Parco archeologico del Colosseo.

the emperors. The gilded statue was set up by the Roman senate and people on the imperial command of Theodosius II and Valentinian III, who decided where the base was to be placed.[69] Acting as an awarder of the monument, the senate of Rome recognized Aetius' achievements for the state celebrating his victories over the Goths and Burgundians, which have made Italy secure.

The dedication to Aetius extols the honorand as the "upright in morals, rejecting wealth, the bitterest foe of informers and enemies, guarantor of liberty, avenger of modesty",[70] echoing the language of recent imperial laws and the acclamations by the senators receiving the Theodosian Code in 438. What if Aetius in fact was acting not simply as a senior military officer but as a civilian one responsible and publicly recognized for the issuing of laws? In that case, it would be impossible to distinguish Aetius' role from that of the ruling emperor.[71] Aetius' involvement in matters of civilian jurisdiction in the 440s and beyond, reflected through financial legislation,[72] brought him eventually to a close encounter with the senate.

According to the honorific inscription, Aetius was "not only master of the army in Gaul (*magister militum per Gallias*), which he restored a little while ago to Roman rule through victories sworn in war and peace, and *magister utriusque militiae*".[73] The dedication also commemorates him "for the security of Italy" as a general "outstanding in conquering distant peoples, the Burgundians and the Goths".[74] Indeed, in the course of the 430s, Aetius was able to deploy Hunnic *foederati* in his major campaigns in Gaul, and it was only this reinforcement that enabled the Gallic army under his leadership—in alliance with the Hunnic military diaspora—to defeat the Burgundians and the Visigoths. This episode is emblematic of Aetius' career, which was made possible by his excellent contacts with Hunnic groups and by acquiring a substantial personal following (in his youth, Aetius was a hostage, first of the Visigoths and then of the Huns)[75] and the fact that the strength of the Roman army depended on its *foederati*: The military administration in the West tried to cope with the difficulty in recruiting troops, and, facing opposition of senatorial landowners, began to resort to the deployment of military diasporas already from the end of the fourth century.[76]

Furthermore, the inscription focuses on what was more than his military career: Aetius was "ordinary consul (*consul ordinarius*) for a second time and *patricius*, forever dear to the Commonwealth (*res publica*) and adorned with all military gifts".[77] The availability of a military power base allowed priority access to the highest civilian honours: It was Aetius' excessive importance for the defence and security of the state that turned the general into a consul. Among the *magistri militum*, Flavius Bonosus was the first to achieve a consulate in 344.[78] Flavius Merobaudes, who participated in campaigns against the Alamanni and the Nori, delivered a panegyric on Aetius, probably on the consulship of 432. As *magister militum* Merobaudes himself received statue honours in the Forum of Trajan in 435, with the inscription praising both his military and literary skills.[79] Aetius' third consulship in 446, a highly distinguished and unusual honour, signalled remarkable imperial favour. As senators were rarely able to hold more than one consulship—it was usually the emperors who held several consulships—, Aetius' assumption of three consulships was even more conspicuous. Before Aetius, besides the emperors, the only holder of three consulships was, as was already mentioned, Flavius Constantius. Even more conspicuously, Aetius' Western consulship was recognized in the Eastern Roman Empire. Aside from his office of *magister utriusque militiae*, Aetius, like Constantius, not only was a consul but also a *patricius*, thereby outranking other senatorial magistrates.[80] For a society headed by an emperor, illustrious titles like *patricius* had immense prestige. The fact that Aetius possessed the dignity of *patricius*, a title commensurate with his position, no doubt indicates that he had the emperor's confidence and enjoyed a very close relationship with Valentinian III. Amongst many titles recorded for him, Valentinian first addressed Aetius as *comes et magister utriusque militiae et patricius*; by 452, when he was no longer the only *patricius*, a more complex formula was used: *magnificus vir parens patriciusque noster.*[81]

Aetius' triple ordinary consulship represented the apogee of a career unobtainable for senators of Rome. It was an exceptional honour conferred on the *magister militum* by the emperor. Serving as consuls alongside emperors and other senators, senior military commanders had increasingly monopolized chief state offices by successfully exploiting their new social position. Their marriage ties with the imperial family are symbolically represented in the dynastic-mythological iconography of both the ivory diptych of Stilicho and the consular diptych of Constantius.[82] Aetius, too, like Constantius and Stilicho before him, sought a connection to the imperial house by trying to compel Valentinian III to marry the emperor's daughter Placidia to his son Gaudentius.

The old-established senatorial families of Rome, nonetheless, held on to the vision of an empire in which they still governed the state together with the emperor. The advocacy of the senatorial vision, expressed by Symmachus in his orations to Valentinian I in 369–370 and in an address to Gratian in 376,[83] who hoped for a return to the senate-driven imperial policy at the frontiers, voiced a desire substituting for reality. Symmachus' *Letters*, however, testify to the senatorial elite's rapid drift towards recognition of rising *homines novi* by accepting senior imperial officers in military service as acceptable late Roman aristocrats.[84]

Even if Aetius was facing senatorial—or even imperial—opposition during the later period of his career, and thereby remained unable to achieve the absolute dominance in the court that Stilicho or Constantius had claimed,[85] this posed no threat to his immediate position securely based on his military successes—achievements which no one among those in the senate or at court could boast at that time. He relied on military diasporas beyond imperial control that granted him greater invulnerability than Stilicho or Constantius enjoyed. For the imperial government had neither control over the Western Roman army nor over Aetius' Hunnic *foederati*. The military diaspora at his personal disposal provided a power base that had contributed to the growth of his authority.[86] In this context, it is certainly symptomatic that, while in 425 Aetius led the Huns to support the usurpation of John (against Valentinian III), in 451, it was the whole

Roman army he mobilized in the alliance against Attila, and this army consisted mostly of *foederati*.[87]

Conclusion

The documentary evidence from Rome gives an instructive glimpse into the political and social role of the *magistri militum* Stilicho, Constantius III, and Aetius and illuminates a decisive turning point in Roman history. In the late Roman Empire, senior military officers converted actual or potential resources accumulated in the military sphere into social prestige institutionalized in their relationships with the imperial family, membership in the imperial house, or even by undertaking the leadership of the Empire. Their power was essentially based on their army commands, and their careers overshadowed those of civilian office holders (even of those obtaining the consulate). Furthermore, the fate of the Empire relied more and more on the generals' ability to raise or command troops that were not genuine "Roman" but composed of *foederati*, forming well-established or fresh military diasporas within the borders of the Roman Empire—a decisive indicator for Stilicho's, Constantius' and later Aetius' military and political success.

The change in imperial military leadership in the Roman West in the first half of the fifth century accounts for a paradox because it was not an army composed of Roman soldiers that took advantage of this situation. Theoretically, Stilicho would have been able to set up an efficient Italian army in order to withstand the Gothic attack. However, he chose to employ *foederati* who vanished after his demise in 408 and switched to Alaric.[88] The first half of the fifth century indeed witnessed the marginalization of Roman units,[89] which had become less important than the ever-shifting "barbarian" allies of the Western Roman Empire. Senatorial reluctance to furnish recruits even in periods of military emergency forced the *magistri militum* to establish or deploy military diasporas. Therefore, the generals had neither the resources nor the will to preserve the Roman army from the influx of *foederati*; it was this primary power base that allowed Roman generals swiftly to accumulate the highest imperial distinctions.

The honorific monuments erected in the Forum Romanum preserve memories of the political and military roles exercised by the generals of the Western Roman Empire in the first half of the fifth century. The placement and honorific language of these dedications, acting as mediating structures by which the ruler and aristocracy articulated their interaction, reveal the ways in which these members of the senatorial order constructed their relationship to emperors, civilian elites, as well as the broader public. The textual, material, and space-embedded qualities of the late Roman statue dedications, as well as visuality resulting from them, elucidate the newly generated spatial and political mobility networks of the military elite, based on dealings with the military diasporas, which were restricted to generals. While traditional aristocrats mostly remained excluded from military service, these newcomers to the *ordo senatorius* were holding together the Empire due to their ability to speedily raise, mobilize, and transfer their armies interregionally. Despite their low, non-senatorial origin, the imperial military office holders came to share a distinct aristocratic outlook embodied in the honorific statuary and were able to sustain a supra-regional institutional framework under conditions of increasing fragmentation within the Empire, in effect substituting and occasionally even replacing emperors in whose glory they were partaking.

Abbreviations

Amm. Marc.	Ammianus Marcellinus.
Claud., *Bell. Gild.*	Claudianus. *In Gildonem.*
Claud., *Bell. Goth.*	Claudianus. *Bellum Gothicum.*
Claud., *Eutr.*	Claudianus. *In Eutropium.*
Claud., Stil.	Claudianus. *De consulatu Stilichonis.*
Dig.	*Digesta.*
Jer., *Ep.*	Jerome. *Epistulae.*
Olymp. fr.	Olympiodoros. *Fragmentum.*
Oros.	Orosius.
Symm., *Ep.*	Symmachus. *Epistulae.*
Symm., *Or.*	Symmachus. *Orationes.*
Veg., *Mil.*	Vegetius. *De re militari.*
Zos.	Zosimus.

Notes

* An earlier version of this chapter was originally delivered as a paper at the workshop "Military Diasporas and Diasporic Regimes in East Central Europe and the Eastern Mediterranean 500–1800" at the final meeting of the workshop series "Trans-European Diasporas: Migration, Minorities, and Diasporic Experience in East Central Europe and the Eastern Mediterranean 500–1800" at Ruprecht-Karls-Universität Heidelberg in October 2013. I am most grateful to those scholars who organized and attended the conference or read and made comments on the paper, in particular, Patrick Sänger, Georg Christ, Julia Burkhardt, Katalin Szende, Lajos Berkes, Marianne Sághy, Jeroen W.P. Wijnendaele, Christian Witschel, and Bryan Ward-Perkins, but I am especially indebted to Volker Menze for his enlightening course "Warfare in Late Antiquity" taught at Central European University in Winter 2012.

1 *CIL* 6.1731=1195=*LSA*–1437 (by Carlos Machado). Giovanna Tedeschi Grisanti, "'*Dis Manibus*, Pili, Epitaffi, ed altre cose antiche': Un codice inedito di disegni di Giovannantonio Dosio," *Bollettino d'Arte* 18 (1983): 69–102, at 89.

2 At Pollentia and Verona (402). Michael Kulikowski, "The Failure of Roman Arms," in *The Sack of Rome in 410 AD: The Event, its Context and its Impact*, ed. Johannes Lipps, Carlos Machado, and Philipp von Rummel (Wiesbaden: Reichert, 2013), 77–86, at 78, conjectures that Stilicho either "may simply have found it impossible to defeat Alaric" or, what is more, "like all commanders, had every reason to allow defeated enemies some freedom to recover, since they and their units were potentially useful alive."

3 Claud., *Bell. Goth.*, lines 267–404, on Alaric's attack on Italy. Alan Cameron, *Claudian: Poetry and Propaganda at the Court of Honorius* (Oxford: Clarendon, 1970), 59, on the role of Claudian. Remarkably, Claudian himself received a bronze statue in

the Forum of Trajan dedicated by the senate in 400 after he delivered a panegyric to Stilicho in Rome: *CIL* 6.1710.

4 Although this very unusual late dedication ne *numeri* ither specifies victory in war against the Goths nor records the services of Stilicho, a conjecture of his success against Radagaisus (405–406) rather than against Alaric (402) is more likely.

5 Oros., 7.37.14–16; Olymp. fr. 9.

6 Amm. Marc. 31.4.4; on this particular way of recruiting mercenaries, see John H. W. G. Liebeschuetz, "The End of the Roman Army in the Western Empire," in *War and Society in the Roman World*, ed. John Rich and Graham Shipley (London: Routledge, 1993), 265–276; Maria Cesa, "Römisches Heer und barbarische Föderaten: Bemerkungen zur weströmischen Politik in den Jahren 402–412," in *L'armée romaine et les barbares du IIIe au VIIe siècle*, ed. Françoise Vallet and Michel Kazanski (Rouen: Association Française d'Archéologie Mérovingienne, 1993), 21–29.

7 See Timo Stickler, "The *Foederati*," in *A Companion to the Roman Army*, ed. Paul Erdkamp (Oxford: Blackwell, 2007), 495–515. For the term "military diaspora," see the introduction of this volume.

8 On the nature of Gothic "federate" regiments after Adrianople, see John H. W. G. Liebeschuetz, *Barbarians and Bishops: Army, Church and State in the Age of Arcadius and Chrysostom* (Oxford: Clarendon, 1991), 34–36; Thomas S. Burns, *Barbarians within the Gates of Rome: A Study of Roman Military Policy and the Barbarians, ca. 375–425 A.D.* (Bloomington: Indiana University Press, 1994), 92–111; Peter Heather, "*Foedera* and *Foederati* of the Fourth Century," in *Kingdoms of the Empire: The Integration of Barbarians in Late Antiquity*, ed. Walter Pohl (Leiden: Brill, 1997), 85–97; Guy Halsall, *Barbarian Migrations and the Roman West, 376–568* (Cambridge: Cambridge University Press, 2007), 183–184.

9 On ethnogenesis and ethnicity, see *Strategies of Distinction: The Construction of Ethnic Communities, 300–800*, ed. Walter Pohl and Helmut Reimitz (Leiden: Brill, 1998), with bibliography. For criticism of the "Vienna (Toronto) School," see Patrick Amory, *People and Identity in Ostrogothic Italy, 489–554* (Cambridge: Cambridge University Press, 1999) and Halsall, *Barbarian Migrations*, 15–19, for overview. While the debate on ethnogenesis between various schools has received a great amount of attention in the past generation, it has little bearing on the subject examined here.

10 *Dig.* 49.15.5 and 7. On the literary evidence for Gothic *foederati* in the fourth century as anachronistic, see Heather, "*Foedera* and *Foederati*," 85–97. Stickler, "The *Foederati*," 495, on the status of *foederati* granted to the groups already settled within the empire as only one part of the late integration policy.

11 Leif I. R. Petersen, *Siege Warfare and Military Organization in the Successor States (400–800 AD): Byzantium, the West, and Islam* (Leiden: Brill, 2013), 50.

12 Robert Chenault, "Statues of Senators in the Forum of Trajan and the Roman Forum in Late Antiquity," *Journal of Roman Studies* 102 (2012): 103–132.

13 John Weisweiler, "From Equality to Asymmetry: Honorific Statues, Imperial Power, and Senatorial Identity in Late-antique Rome," *Journal of Roman Archaeology* 25 (2012): 319–350; Heike Niquet, Monumenta virtutum titulique*: Senatorische Selbstdarstellung im spätantiken Rom im Spiegel der epigraphischen Denkmäler* (Stuttgart: Franz Steiner, 2000); Carlos Machado, "Building the Past: Monuments and Memory in the Forum Romanum," in *Social and Political Life in Late Antiquity*, ed. William Bowden, Adam Gutteridge, and Carlos Machado (Leiden: Brill, 2006), 157–192; Carlos Machado, *Urban Space and Aristocratic Power in Late Antique Rome: AD 270–535* (Oxford: Oxford University Press, 2019).

14 Dedications for emperors and imperial family in the Forum Romanum: *CIL* 6.40794=36957 (Valentinian I or II), *CIL* 6.3791a=31413 (Valentinian II, 389), *CIL* 6.36959 (Theodosius I, 389), *CIL* 6.3791b=31414 (Arcadius, 389), *CIL* 6.36960 (Thermantia, mother of Theodosius I, 389–391), *CIL* 6.1187=31256a=*ILS* 794 (Honorius and Arcadius, possibly *quadriga* or large statue group, 398), *CIL* 6.36956b (originally Valens, re-erected 421–439, perhaps still Valens); Forum Iulium: *CIL*

6.40798 (Arcadius, 399/400), *CIL* 6.40804 (Galla Placidia, 417–423); Forum Traiani: *CIL* 6.1186=*ILS* 2945 (Theodosius I, 389), *CIL* 6.40797 (Arcadius? 383–408), *CIL* 6.40813 (*domino nostro*, late fourth to early fifth century?), *CIL* 6.1194 (Honorius, 417/418). See also Chenault, "Statues of Senators," 103–132; Machado, "Building the Past," 157–192.

15 Jaś Elsner, "Inventing Imperium: Texts and the Propaganda of Monuments in Augustan Rome," in *Art and Text in Roman Culture*, ed. Jaś Elsner (Cambridge: Cambridge University Press, 1996), 32–53, at 35.

16 *CIL* 6.1730=*LSA*–1436 (by Carlos Machado); Franz Alto Bauer, *Stadt, Platz und Denkmal in der Spätantike* (Mainz: von Zabern, 1996), 20–21; Brigitte Ruck, *Die Großen dieser Welt: Kolossalporträts im antiken Rom* (Heidelberg: Verlag Archäologie und Geschichte, 2007), 264–265.

17 While the inscription does not mention an equestrian statue, the measurements of the marble base (with the front of a base 123 × 134 cm) clearly show that it must have been for a monument bigger than a standing statue. The honour awarded through this dedication, an equestrian statue in the Forum Romanum, the traditional location of the imperial statuary in Rome, was still exceptional. See Niquet, *Monumenta*, 57–58. For the fourth-century non-imperial equestrian statuary, *CIL* 14.4455 (Manilius Rusticianus, prefect of the *annona* and patron, Ostia); *AE* 2000, 735 (Vicarius Usulenius Prosper(ius?), governor of Baetica, Corduba); *CIL* 2.1972 (Quintus Attius Granius Caelestinus, governor of Baetica, Malaca).

18 *CIL* 6.1730, lines 2–4, *Flavio Stilichoni inlustrissimo viro / magistro equitum peditumque / comiti domesticorum tribuno praetoriano*. In inscriptions dated by his consulship Stilicho is called simply *vir clarissimus*, not *vir illustris*, an appellation to which his office would have entitled him: *CIL* 6.1706=*LSA*–1413, *dedic(ata) e(st) / XII k(a)l(endas) Dic(embres) / Fl(avio) Stili / chone v(iro) c(larissimo) / co(n)s(ule)* (400) and *CIL* 11.3238=*ILCV* 3294, *cons(ulatu) v(iri) c(larissimi) Stiliconis* (400/405). By no means this inconsistency suggests that Stilicho is not given the highest rank to which he is entitled, it rather reflects a common practice at the turn of the century.

19 *CIL* 6.1730, lines 5–12, *et ab ineunte aetate per gradus claris/simae militiae ad columen gloriae / sempiternae et regiae adfinitatis evecto / progenero divi Theodosi, comiti divi / Theodosi Augusti in omnibus bellis / adque victoriis et ab eo in adfinitatem / regiam cooptato itemque socero d(omini) n(ostri) Honorii Augusti*. Niquet, *Monumenta*, 144, n. 122, compares it with senatorial honorific inscriptions from Rome.

20 *CIL* 6.3868=31988=41381, line 3, *[c]onsulṭ[o]rị*.

21 *ILS* 800, *Honori, Maria, Stelicho, Serŋna, vivatis! Stelicho, Serena, Thermantia, Eucheri, vivatis!* on a *bulla* discovered in the probable tomb of Maria, bearing the names of the imperial family. Stilicho and Serena are mentioned twice as part of the imperial family, following Stilicho's *parens principum* rhetoric, see *CIL* 9.4051.

22 For bibliography on *damnatio memoriae*, see: Friedrich Vittinghoff, *Der Staatsfeind in der Römischen Kaiserzeit: Untersuchungen zur* 'Damnatio Memoriae' (Berlin: Junker und Dünnhaupt, 1936); Charles Hedrick, *History and Silence: The Purge and Rehabilitation of Memory in Late Antiquity* (Austin: University of Texas Press, 2000); Harriet Flower, *The Art of Forgetting: Disgrace and Oblivion in Roman Political Culture* (Chapel Hill: University of North Carolina Press, 2006) rejects, however, the concept of *damnatio memoriae*, arguing instead for "sanctions against memory." For the high imperial period, see Florian Krüpe, *Die* Damnatio memoriae*: Über die Vernichtung von Erinnerung; Eine Fallstudie zu Publius Septimius Geta (198–211 n. Chr.)* (Gutenberg: Computus, 2011). For hostile mention of Stilicho's half-Vandal origin, Jer., *Ep.* 123.16; Oros. 7.38.1. Had Stilicho not fallen from emperor's grace, there would be no reason to see him as other than Roman, see Elton, *Warfare in Roman Europe*, 141–142.

23 *CIL* 6.3868=31988=41381=*LSA*–1490 (by Carlos Machado), lines 1–5, *[p]rov[identissimo duci] / victọ[rios]isṣ[imo dominor(um) nn(ostrorum)] / [c]onsulṭ[o]rị etiaṃ [fautori divini] / [ge]neris ac no[minis Romani]/ [[[Fl(avio) Stilicho]ṇị [v(iro) c(larissimo) et inl(ustri)—?]]]*. Niquet, *Monumenta*, 57–58, suggests a possibility of the equestrian statue with the width of ca. 125 cm; Ruck, *Die Großen dieser Welt*, 263.

24 *CIL* 6.1730, lines 12–13, *Africa consiliis eius / et provisione liberata.*
25 *PLRE* 1, 395–396 Gildo; on revolt and defeat, see Claud., *Bell. Gild*; Oros. 36. 2–13; Zos. 5.11; Yves Modéran, "Gildon, les Maures et l'Afrique," *Mélanges de l'École française de Rome. Antiquité* 101, no. 2 (1989): 821–872; Andy Blackhurst, "The House of Nubel: Rebels or Players?," in *New Perspectives on Late Antique North Africa*, ed. Andrew Merrills (Aldershot: Ashgate, 2004), 59–75.
26 *CIL* 6.1730, line 13; Claud., *Stil.* 1.325–332, *hoc quoque non parva fas est cum laude relinqui / quod non ante fretis exercitus adstitit ultor / ordine quam prisco censeret bella senatus. / neglectum Stilicho per tot iam saecula morem / rettulit ut ducibus mandarent proelia patres / decretoque togae felix legionibus iret / tessera.*
27 *CIL* 6.41382=*LSA*–1587 (by Carlos Machado), lines 5–6, *pro virtutum veneratione inter cetera / beneficia quae per eum urbi Romae delata s[unt]*, lines 10–12, *qu]od Gildone hoṣṭẹ p[ublico de]/[victo et ali]moniis Romạ[norum resti]/[tutis felicitat]ẹm aụ[xerit—].* On Gildo, see Cameron, *Claudian*, 93–123.
28 Claud., *Eutr.* II. *Praef.* 69–70; *Stil.* 1.354–357.
29 *CIL* 6.1188=*LSA*–1306, lines 3–4 and *CIL* 6.1189=*LSA*–1307, lines 3–4, *ob instauratos urbi(s) aeternae muros portas ac turres egestis inmensis ruderibus ex suggestione v(iri) c(larissimi) et inlustris / comitis et magistri utriusq(ue) militiae Stilichonis.* Since Constantine the title of *comes* as a mark of status had been formalized and regularized; this distinction is recorded in fragmentary inscriptions from Rome honouring Stilicho: *CIL* 6.1732=31914=15. 7134=*ILCV* 65, *Fl(avi) Stilichonis / v(iri) c(larissimi) et inl(ustris) com(itis) / et mag(istri) utrius / que militiae / f(undus) Fimbrianus*; *CIL* 6.1733=15.7133, *in his predies Fl(avi) Sti/lic(honis) co(mitis) magist(ri) utri/usq(ue) militae*; CIL 6.1734=15.7136=*ILCV* 65 adn., *Fl(avi) Still/iconis / v(iri) c(larissimi), com(itis)*; *CIL* 6. 31989=15. 7135=*ILCV* 65 adn., *Fl(avi) Stili/chonis vi/ri inl(ustris) co/mitis.*
30 *CIL* 6.1190=*LSA*–1308, lines 3–4, *ob instauratos urbi(s) aeternae muros portas ac turres egestis [inmensis] ruderibus [[—]] / [[—]].*
31 Zos. 5.26.3, Oros. 7.37.12, on allies. On Radagaisus' attack as a revolt of "barbarian" recruits, see Burns, *Barbarians within the Gates of Rome*, 198.
32 Olymp. fr. 9, on draft.
33 *CIL* 6.1199=*ILS* 798=*LSA*–1311 (by Carlos Machado), Niquet, *Monumenta*, 206, n. 35.
34 *CIL* 6.1196 = *LS* 798, line 3, *quod Getarum nationem in omne aevum doc[u]ere exti[ngui].*
35 *PLRE* 1, 771–772, Romulus 5.
36 *CIL* 6.1371=1195, lines 9–10, *socio / bellorum omnium.*
37 *CIL* 6.1371=1195, lines 15–22, *populus Romanus / pro singulari eius / circa se amore / adque providentia / statuam ex aere argentoque / in rostris ad memoriam / gloriae sempiteranae / conlocandam decrevit.* Note an equally rare dedication by the *populus Romanus*, the old "Republican" source of the sovereignty of the senatorial government.
38 Gregor Kalas, *The Restoration of the Roman Forum in Late Antiquity: Transforming Public Space* (Austin: University of Texas Press, 2015), 90–91. See *CIL* 6.1195.
39 *CIL* 6.1184a=*LSA*–1294 (Gratian, Valentinian II, and Theodosius I) installed on top of the late antique Rostra; see Franz Alto Bauer, "Das Denkmal der Kaiser Gratian, Valentinian II. und Theodosius am Forum Romanum," *Mitteilungen des Deutschen Archäologischen Instituts, Römische Abteilung* 106 (1999): 213–234.
40 *CIL* 6.31987=*LSA*–1363 (by Carlos Machado), lines 1–11, *fidei virtutiq(ue) devotissimorum / militum dom(i)norum nostrorum / Arcadi Honori et Theodosi / perennium Augustorum / post confectum Gothicum / bellum felicitate aeterni / principis dom(i)ni nostri Honori / consiliis et fortitudine / inlustris viri comitis et / [[[magistri utriusq(ue) militiae]]] / [[[Fl(avi) Stilichonis bis co(n)s(ulis) ord(inarii)]]].*
41 *CIL* 6.31987, lines 10–11.
42 Geza Alföldy and Christian Witschel argue in favour of its identification as Stilicho in *Addenda et corrigenda*, *CIL* 6.8.3, 4800, no. 31987.
43 Rudolf H. W. Stichel, *Die römische Kaiserstatue am Ausgang der Antike: Untersuchungen zum plastischen Kaiserporträt seit Valentinian I. (364–375 v. Chr.)* (Rome: Bretschneider, 1982), 94, no. 88; Bauer, *Stadt, Platz und Denkmal*, 20–22; and Wolfgang

Messerschmidt, "Die statuarische Repräsentation des theodosianischen Kaiserhauses in Rom," *Mitteilungen des Deutschen Archäologischen Instituts, Römische Abteilung* 111 (2004): 555–568, at 559, n. 4 attribute the monument to one of the three ruling emperors, most probably Honorius.

44 *LSA*–1363 (by Carlos Machado).

45 Burns, *Barbarians within the Gates of Rome*, 356 n. 24 and Hugh Elton, *Warfare in Roman Europe, AD 350–425* (Oxford: Clarendon Press, 1996), 211. See Jeroen W. P. Wijnendaele, "Stilicho, Radagaisus and the So-Called 'Battle of Faesulae' (406 CE)," *Journal of Late Antiquity* 9, no. 1 (2016): 267–284, at 270–271.

46 Petersen, *Siege Warfare and Military Organization in the Successor States*, 18.

47 Olymp. fr. 9.

48 On the organization of the late Roman army and its *foederati* regiments, see Elton, *Warfare in Roman Europe*, 89–117, 200–214 with table 1. On the merely administrative distinction between *limitanei* and *comitatenses*, see Benjamin Isaac, "The Meaning of the Terms *Limes* and *Limitanei*," *Journal of Roman Studies* 78 (1988): 125–147. The false distinction between *limitanei* and *comitatenses* is also treated in Petersen, *Siege Warfare and Military Organization in the Successor States*, 50, 154.

49 On recruitment and the issue of "barbarization" of the Roman army in late antiquity, see Elton, *Warfare in Roman Europe*, 128–154 contra Liebeschuetz, *Barbarians and Bishops*, 7–47; Constantin Zuckermann, "Two Reforms of the 370s: Recruiting Soldiers and Senators in the Divided Empire," *Revue des Études Byzantines* 56 (1998): 79–139.

50 Olymp. fr. 8, 20, 34.

51 Olymp. fr. 22, 24 transmits western hopes of a new era in the relationship between Romans and barbarians.

52 *CIL* 6.1719=*LSA*–1423 (by Carlos Machado), lines 2–3, *parenti invictissimo[rum] / principum*.

53 *CIL* 6.1719, *CIL* 6.1720=*LSA*–1424 (by Carlos Machado). Niquet, *Monumenta*, 50, 72.

54 In 417 with Honorius, in 420 with Theodosius II; see Roger S. Bagnall et al., *Consuls of the Later Roman Empire* (Atlanta: Scholars Press, 1987).

55 Gratian's law of 382 considered the patriciate only as an appendage of the consulate, see Ralph W. Mathisen, "Emperors, Consuls and Patricians: Some Problems of Personal Preference, Precedence and Protocol," *Byzantinische Forschungen* 17 (1991): 173–190.

56 *PLRE* 2, 321–325 Fl. Constantius 17.

57 *CIL* 6.1720, lines 3–5, *[c]omiti et magistro utriusq(ue) / [m]ilitiae patricio consuli / [ordina] rio [t]er omnes*.

58 *CIL* 6.1719, line 1, *reparatori rei publicae*. Compare *CIL* 3.12333, *CIL* 3.13714, *CIL* 3.13715, *reparatori conservatori patriae* (Aurelian?); *CIL* 3.22, *CIL* 3.5326, *reparatori orbis sui* (Diocletian); *CIL* 10.516=*LSA*–1846, *reparatori orbis sui* (Constantine); *CIL* 11.4781, *reparatores orbis adque urbium restiltutores* (Constantius II and Julian); *CIL* 9.417=*LSA*–1697, *reparatori orbis Romani* (Julian); *AE* 1986, 631=*LSA*–52, *reparatori Romanae rei* (Theodosius I, Arcadius, and Honorius). On the rhetoric of pacification, security and liberty in the honorific inscriptions, see André Chastagnol, *Le pouvoir impérial à Rome: Figures et commemorations, scripta varia IV*, ed. Stéphane Benoist and Ségolène Demougin (Geneva: Droz, 2008), 143–145.

59 *CIL* 6.40768: *[tyrannis extinctis] liberatoribus [atque rei] [publicae] restitutori[bus]*, *CIL* 6.40768a: *[co]nservatori Romani [no]minis propagatori [or]bis sui factionum [ty]rannicarum extinctori [dom]itori gentium barbarum*, *AE* 2003, 2014: *[divi]nae virtutis [principi?] [extinctori? ty]rannicae factionis et v[ictori? defensori?] [pro]vinciarum suarum atque urb[ium?—] d(omino) n(ostro) Flavio Valerio Constantino* (Constantine I), *CIL* 6.1154=36958: *[Im]peratoribus ae[te]rnae urbis sua[e defensoribus] [saevoru]m tyranno[r]um domination[is depulsoribus] [di]gnitatis honorumque [exemplis] [domini]s nostris Fl(avio) Val[entiniano et] [Fl(avio) Theodo] sio* (Valentinian II and Theodosius I), *CIL* 6.36959: *extinctori tyrannorum ac publicae securitati(s) auctori d(omino) n(ostro) Theodosio*, *CIL* 3.737: *difficilis quondam dominis parere serenis*

iussus et extinctis palmam portare tyrannis (Theodosius I), *CIL* 3.735: *haec loca Theudosius decorat post fata tyranni* (Theodosius II). On the rhetoric of tyranny, see Valerio Neri, "L'usurpatore come tiranno nel lessico politico della tarda antichità," in *Usurpationen in der Spätantike*, ed. François Paschoud and Joachim Szidat (Stuttgart: Franz Steiner, 1997), 71–86 contra Timothy Barnes, "Oppressor, Persecutor, Usurper. The Meaning of 'Tyrannus' in the Fourth Century," in *Historiae Augustae Colloquium Barcinonense*, ed. Giorgio Bonamente and Marc Mayer (Bari: Epuglia, 1996), 55–65.

60 Marcus (406–407), Gratianus (407, four months), Constantinus III (407–411), Constans (408–411), Maximus (409–411), Attalus (409/410), Iovinus (411–413), Sebastianus (412–413), Attalus (414–416), Maximus (419–421), Iohannes (423–425), Petronius Maximus (455), Avitus (455–456). See Joachim Szidat, Usurpator tanti nominis*: Kaiser und Usurpator in der Spätantike (337–476 n. Chr.)* (Stuttgart: Franz Steiner, 2010), 415.

61 Constantius defeated Gerontius, the general of usurper Maximus, and Edobichus, the general of Constantine III (411), expelled from Italy the Visigoths under Ataulf (412), captured usurper Priscus Attalus (415). On Constantius' career, see Werner Lütkenhaus, *Constantius III: Studien zu seiner Tätigkeit und Stellung im Westreich 411–421* (Bonn: Habelt, 1998).

62 Meaghan A. McEvoy, *Child Emperor Rule in the Late Roman West, AD 367–455* (Oxford: Oxford University Press, 2013), 282.

63 John M. O'Flynn, *Generalissimos of the Western Roman Empire* (Edmonton: University of Alberta Press, 1983).

64 For an exception: *PLRE* 1, 789 Sabinianus 3 (*magister equitum* (*per Orientem*), 359–360); Joachim Szidat, "Sabinianus: Ein Heermeister senatorischer Abkunft im 4. Jh.," *Historia* 40 (1991): 494–500.

65 André Chastagnol, *Le Sénat romain à l'époque imperiale: Recherches sur la composition de l'Assemblée et le statut de ses members* (Paris: Les Belles Lettres, 1992), 294–295.

66 Alexander Demandt, "*Magister militum*," in *Paulys Realencyclopädie der classischen Altertumswissenschaft, Supplementband XII*, ed. Konrat Ziegler (Stuttgart: Alfred Druckenmüller, 1970), 560–565 (from Constantine I until 353).

67 Men often of non-Roman origin in the lower levels of military service—*comites, duces*, and *tribunes*—could also access the senatorial ranks *clarissimi* and *spectabiles*.

68 Machado, "Building the Past," 162–163. On the identification of the *atrium libertatis*, see Augusto Fraschetti, *La conversion da Roma pagana a Roma cristiana* (Bari: Laterza, 1999), 210.

69 *CIL* 6.41389, lines 10–11, *iussu principum dd(ominorum) nn(ostrorum) Theodosi et Plaçịḍị [Valenti]/[n]iani pp(erpetuorum) Augg(Augustorum) in atrio libertatịs.*

70 *CIL* 6.41389, lines 13–15, *mọrum probo opum refugo delato/rum ut hostium inimicissimo vindici libertatis / pudoris ultor.*

71 See Chenault, "Statues of Senators," 128.

72 Roland Delmaire, "Flauius Aëtius, *delatorum inimicissimus, uindex libertatis, pudoris ultor* (*CIL* VI 41389)," *Zeitschrift für Papyrologie und Epigraphik* 166 (2008): 291–294.

73 *CIL* 6.41389=*LSA*-1434 (by Carlos Machado), lines 2–4, *[n]ec non et magistro militum per Gallias quas dudum / [o]b iuratas bello pace victorias Romano imperio / reddidit magistro utriusq(ue) militiae*. Terézia Olajos, "L'inscription de la statue d'Aétius et Merobaudes," in *Acta of the Fifth International Congress of Greek and Latin Epigraphy Cambridge 1967* (Oxford: Blackwell, 1971), 469–472.

74 *CIL* 6.41389, line 7, *ob Italiae securitatem*; lines 8–9, *quam procul domitis gentib(us) peremptisque / [B]urgundionib(us) et Gotis oppressis vincendo praestit[it]*. The statue for Aetius was set up in the Forum Romanum between 435, as the date for his appointment as *patricius*, and 445, as the inscription only mentions his double consulship, recording his holding of high offices in the empire.

75 Liebeschuetz, "The End of the Roman Army," 269–270. Hunnic bodyguards are already attested with Stilicho: Zos. 5.34.1.

76 Veg., *Mil.* 1.28; Symm., *Ep.* 6.64; it continued to be a subject of lawmaking, see Zuckermann, "Two Reforms of the 370s," 79–139, patterns of recruitment.

77 *CIL* 6.41389, lines 4–6, *secundo / consuli ordinario atq(ue) patricio semper rei publicae / [i] npenso omnibusq(ue) donis militarib(us) ornato.*
78 Demandt, "*Magister militum*," 564.
79 *LSA*–319=*CIL* 6.1724. *PLRE* 2, 756–758 Fl. Merobaudes.
80 On the nature of the patriciate in the first half of the fifth century, see Timothy D. Barnes, "'Patricii' under Valentinian III," *Phoenix* 29 (1975): 155–170.
81 *NVal.* 36 (29 June 452). As *patricius* among other *patricii*, as for Valentinian III's reign, the higher number of *patricii* is attested, Aetius was consistently referred to in the laws as *parens*. See Barnes, "'Patricii' under Valentinian III," 166.
82 On the Monza diptych: Delbrück, *Die Consulardiptychen*, 242–248, no. 63; Bente Kiilerich and Hjalmar Torp, "*Hic est: hic Stilicho*: The Date and Interpretation of a Notable Diptych," *Jahrbuch des Deutschen Archäologischen Instituts* 104 (1989): 319–371; Bente Kiilerich, *Late Fourth Century Classicism in the Plastic Arts: Studies in the so-called Theodosian Renaissance* (Odense: Odense University Press, 1993), 137–141, figs. 78–79. On the Halberstadt diptych: Delbrück, *Die Consulardiptychen*, 87–93, no. 2 (Fl. Constantius, 417); cf. Alan Cameron, "Consular Diptychs in Their Social Context: New Eastern Evidence," *Journal of Roman Archaeology* 11 (1998), 384–403 (Fl. Constans, 414). See also Gudrun Bühl, *Constantinopolis und Roma: Stadtpersonifikationen der Spätantike* (Zürich: Akanthus, 1995), 151–164 (following Delbrück); and Gudrun Bühl, "Eastern or Western?—That is the Question: Some Notes on the New Evidence Concerning the Eastern Origin of the Halberstädter Diptych," *Acta ad archaeologiam et artium historiam pertinentia* 15 (2001): 193–203. On the Trier diptych: *CIL* 13.3674.
83 Symm., *Or.* 2.1–2, 4.6.25–27.
84 To Stilicho (4.1–4.14), to Richomeres (3.54–3.69), and to Bauto (4.15–4.16). See also Michele Renee Salzman, "Symmachus and the 'Barbarian' Generals," *Historia* 55 (2006): 352–367.
85 Briggs L. Twyman, "Aetius and the Aristocracy," *Historia* 19 (1970): 480–503.
86 Personal military followings of "barbarians" later known as *bucellarii* played an important role for generals of the early fifth century and appeared to have become a regular feature of armies at this time witnessed in the case of Stilicho. But Olymp. fr. 7.4 states that both Romans and non-Romans served as *buccellarii* and *foederati*.
87 Franks, Sarmatians, Armoricans, Liticians, Burgundians, Saxons, Riparians, Olibriones (once Roman soldiers, now *foederati*). What cannot but strike the eye of anyone well-versed in the Roman military history is the conspicuous absence of any reference to Roman soldiers: Liebeschuetz, "The End of the Roman Army," 272.
88 Zos. 5.35.
89 Liebeschuetz, "The End of the Roman Army," 267, argues that "in the course of the first half of the fifth century, the regular army, that is the class of units listed in the *Notitia*, became unimportant as compared with the federates."

References

Standard Editions, Databases, Reference Works

AE = *L'Année épigraphique*. Paris, 1888–.

CIL = Mommsen, Theodor, et al., eds. *Corpus Inscriptionum Latinarum*. Vols. 1–17. Berlin, 1853–.

ILCV = Diehl, Ernst, ed. *Inscriptiones Latinae Christianae Veteres*. Vols. 1–3. Berlin, 1924–1931, repr. 1970.

ILS = Dessau, Hermann, ed. *Inscriptiones Latinae Selectae*. Vols. 1–5. Berlin, 1892–1916.

LSA = *Last Statues of Antiquity* database, http://laststatues.classics.ox.ac.uk/.

PLRE 1 = Jones, Arnold H. M., et al., eds. *The Prosopography of the Later Roman Empire: AD 260–395*. Vol. 1. Cambridge, 1971.

PLRE 2 = Martindale, John Morris, ed. *The Prosopography of the Later Roman Empire: AD 395–527.* Vol. 2. Cambridge, 1980.

Secondary Literature

Amory, Patrick. *People and Identity in Ostrogothic Italy, 489–554.* Cambridge: Cambridge University Press, 1999.

Bagnall, Roger S., Alan Cameron, Seth R. Schwartz, and Klaas Anthony Worp. *Consuls of the Later Roman Empire.* Atlanta: Scholars Press, 1987.

Barnes, Timothy D. "'*Patricii*' under Valentinian III." *Phoenix* 29 (1975): 155–170.

Barnes, Timothy D. "Oppressor, Persecutor, Usurper: The Meaning of 'Tyrannus' in the Fourth Century." In *Historiae Augustae Colloquium Barcinonense*, edited by Giorgio Bonamente and Marc Mayer, 55–65. Bari: Epuglia, 1996.

Bauer, Franz Alto. *Stadt, Platz und Denkmal in der Spätantike.* Mainz: von Zabern, 1996.

Bauer, Franz Alto. "Das Denkmal der Kaiser Gratian, Valentinian II. und Theodosius am Forum Romanum." *Mitteilungen des Deutschen Archäologischen Instituts, Römische Abteilung* 106 (1999): 213–234.

Blackhurst, Andy. "The House of Nubel: Rebels or Players?" In *New Perspectives on Late Antique North Africa*, edited by Andrew Merrills, 59–75. Aldershot: Ashgate, 2004.

Bühl, Gudrun. *Constantinopolis und Roma: Stadtpersonifikationen der Spätantike.* Zürich: Akanthus, 1995.

Bühl, Gudrun. "Eastern or Western?—That is the Question: Some Notes on the New Evidence Concerning the Eastern Origin of the Halberstädter Diptych." *Acta ad archaeologiam et artium historiam pertinentia* 15 (2001): 193–203.

Burns, Thomas S. *Barbarians within the Gates of Rome: A Study of Roman Military Policy and the Barbarians, ca. 375–425 A.D.* Bloomington: Indiana University Press, 1994.

Cameron, Alan. *Claudian: Poetry and Propaganda at the Court of Honorius.* Oxford: Clarendon, 1970.

Cameron, Alan. "Consular Diptychs in Their Social Context: New Eastern Evidence." *Journal of Roman Archaeology* 11 (1998): 384–403.

Cesa, Maria. "Römisches Heer und barbarische Föderaten: Bemerkungen zur weströmischen Politik in den Jahren 402–412." In *L'armée romaine et les barbares du IIIe au VIIe siècle*, edited by Françoise Vallet and Michel Kazanski, 21–29. Rouen: Association Française d'Archéologie Mérovingienne, 1993.

Chastagnol, André. *Le Sénat romain à l'époque imperiale: Recherches sur la composition de l'Assemblée et le statut de ses members.* Paris: Les Belles Lettres, 1992.

Chastagnol, André. *Le pouvoir impérial à Rome: Figures et commemorations, scripta varia IV*, edited by Stéphane Benoist and Ségolène Demougin. Geneva: Droz, 2008.

Chenault, Robert. "Statues of Senators in the Forum of Trajan and the Roman Forum in Late Antiquity." *Journal of Roman Studies* 102 (2012): 103–132.

Delbrück, Richard. *Die Consulardiptychen und verwandte Denkmäler.* Berlin: de Gruyter, 1929.

Delmaire, Roland. "Flauius Aëtius, *delatorum inimicissimus, uindex libertatis, pudoris ultor* (*CIL* VI 41389)." *Zeitschrift für Papyrologie und Epigraphik* 166 (2008): 291–294.

Demandt, Alexander. "*Magister militum.*" In *Paulys Realencyclopädie der classischen Altertumswissenschaft, Supplementband XII*, edited by Konrat Ziegler, 553–790. Stuttgart: Alfred Druckenmüller, 1970.

Elsner, Jaś. "Inventing Imperium: Texts and the Propaganda of Monuments in Augustan Rome." In *Art and Text in Roman Culture*, edited by Jaś Elsner, 32–53. Cambridge: Cambridge University Press, 1996.

Elton, Hugh. *Warfare in Roman Europe, AD 350–425*. Oxford: Clarendon, 1996.
Flower, Harriet. *The Art of Forgetting: Disgrace and Oblivion in Roman Political Culture*. Chapel Hill: University of North Carolina Press, 2006.
Fraschetti, Augusto. *La conversion da Roma pagana a Roma cristiana*. Bari: Laterza, 1999.
Halsall, Guy. *Barbarian Migrations and the Roman West, 376–568*. Cambridge: Cambridge University Press, 2007.
Heather, Peter. "*Foedera* and *Foederati* of the Fourth Century." In *Kingdoms of the Empire: The Integration of Barbarians in Late Antiquity*, edited by Walter Pohl, 85–97. Leiden: Brill, 1997.
Hedrick, Charles. *History and Silence: The Purge and Rehabilitation of Memory in Late Antiquity*. Austin: University of Texas Press, 2000.
Isaac, Benjamin. "The Meaning of the Terms *Limes* and *Limitanei*." *Journal of Roman Studies* 78 (1988): 125–147.
Kalas, Gregor. *The Restoration of the Roman Forum in Late Antiquity: Transforming Public Space*. Austin: University of Texas Press, 2015.
Kiilerich, Bente. *Late Fourth Century Classicism in the Plastic Arts: Studies in the So-called Theodosian Renaissance*. Odense: Odense University Press, 1993.
Kiilerich, Bente, and Hjalmar Torp. "*Hic est: hic Stilicho*: The Date and Interpretation of a Notable Diptych." *Jahrbuch des Deutschen Archäologischen Instituts* 104 (1989): 319–371.
Krüpe, Florian. *Die Damnatio memoriae: Über die Vernichtung von Erinnerung; Eine Fallstudie zu Publius Septimius Geta (198–211 n. Chr.)*. Gutenberg: Computus, 2011.
Kulikowski, Michael. "The Failure of Roman Arms." In *The Sack of Rome in 410 AD: The Event, Its Context and Its Impact*, edited by Johannes Lipps, Carlos Machado, and Philipp von Rummel, 77–86. Wiesbaden: Reichert, 2013.
Liebeschuetz, John H. W. G. *Barbarians and Bishops: Army, Church and State in the Age of Arcadius and Chrysostom*. Oxford: Clarendon, 1991.
Liebeschuetz, John H. W. G. "The End of the Roman Army in the Western Empire." In *War and Society in the Roman World*, edited by John Rich and Graham Shipley, 265–276. London: Routledge, 1993.
Lütkenhaus, Werner. *Constantius III: Studien zu seiner Tätigkeit und Stellung im Westreich 411–421*. Bonn: Habelt, 1998.
Machado, Carlos. "Building the Past: Monuments and Memory in the Forum Romanum." In *Social and Political Life in Late Antiquity*, edited by William Bowden, Adam Gutteridge, and Carlos Machado, 157–192. Leiden: Brill, 2006.
Machado, Carlos. *Urban Space and Aristocratic Power in Late Antique Rome: AD 270–535*. Oxford: Oxford University Press, 2019.
Mathisen, Ralph W. "Emperors, Consuls and Patricians: Some Problems of Personal Preference, Precedence and Protocol." *Byzantinische Forschungen* 17 (1991): 173–190.
McEvoy, Meaghan A. *Child Emperor Rule in the Late Roman West, AD 367–455*. Oxford: Oxford University Press, 2013.
Messerschmidt, Wolfgang. "Die statuarische Repräsentation des theodosianischen Kaiserhauses in Rom." *Mitteilungen des Deutschen Archäologischen Instituts, Römische Abteilung* 111 (2004): 555–568.
Modéran, Yves. "Gildon, les Maures et l'Afrique." *Mélanges de l'École française de Rome. Antiquité* 101, no. 2 (1989): 821–872.
Neri, Valerio. "L'usurpatore come tiranno nel lessico politico della tarda antichità." In *Usurpationen in der Spätantike*, edited by François Paschoud and Joachim Szidat, 71–86. Stuttgart: Franz Steiner, 1997.

Niquet, Heike. Monumenta virtutum titulique: *Senatorische Selbstdarstellung im spätantiken Rom im Spiegel der epigraphischen Denkmäler.* Stuttgart: Franz Steiner, 2000.

O'Flynn, John M. *Generalissimos of the Western Roman Empire.* Edmonton: University of Alberta Press, 1983.

Olajos, Terézia. "L'inscription de la statue d'Aétius et Merobaudes." In *Acta of the Fifth International Congress of Greek and Latin Epigraphy Cambridge 1967*, 469–472. Oxford: Blackwell, 1971.

Petersen, Leif I. R. *Siege Warfare and Military Organization in the Successor States (400–800 AD): Byzantium, the West, and Islam.* Leiden: Brill, 2013.

Pohl, Walter, and Helmut Reimitz, eds. *Strategies of Distinction: The Construction of Ethnic Communities, 300–800.* Leiden: Brill, 1998.

Ruck, Brigitte. *Die Großen dieser Welt: Kolossalporträts im antiken Rom.* Heidelberg: Verlag Archäologie und Geschichte, 2007.

Salzman, Michele. "Symmachus and the 'Barbarian' Generals." *Historia* 55 (2006): 352–367.

Stichel, Rudolf H. W. *Die römische Kaiserstatue am Ausgang der Antike: Untersuchungen zum plastischen Kaiserporträt seit Valentinian I. (364–375 v. Chr.).* Rome: Bretschneider, 1982.

Stickler, Timo. "The *Foederati*." In *A Companion to the Roman Army*, edited by Paul Erdkamp, 495–515. Oxford: Blackwell, 2007.

Szidat, Joachim. "Sabinianus: Ein Heermeister senatorischer Abkunft im 4. Jh." *Historia* 40 (1991): 494–500.

Szidat, Joachim. Usurpator tanti nominis: *Kaiser und Usurpator in der Spätantike (337–476 n. Chr.).* Stuttgart: Franz Steiner, 2010.

Tedeschi Grisanti, Giovanna. "'*Dis Manibus*, Pili, Epitaffi, ed altre cose antiche': Un codice inedito di disegni di Giovannantonio Dosio." *Bollettino d'Arte* 18 (1983): 69–102.

Twyman, Briggs L. "Aetius and the Aristocracy." *Historia* 19 (1970): 480–503.

Vittinghoff, Friedrich. *Der Staatsfeind in der Römischen Kaiserzeit: Untersuchungen zur 'Damnatio Memoriae'.* Berlin: Junker und Dünnhaupt, 1936.

Weisweiler, John. "From Equality to Asymmetry: Honorific Statues, Imperial Power, and Senatorial Identity in Late-antique Rome." *Journal of Roman Archaeology* 25 (2012): 319–350.

Wijnendaele, Jeroen W. P. "Stilicho, Radagaisus and the So-Called 'Battle of Faesulae' (406 CE)." *Journal of Late Antiquity* 9, no. 1 (2016): 267–284.

Zuckermann, Constantin. "Two Reforms of the 370s: Recruiting Soldiers and Senators in the Divided Empire." *Revue des Études Byzantines* 56 (1998): 79–139.

5

THE PERSIAN AND ARAB OCCUPATIONS OF EGYPT IN THE SEVENTH CENTURY

*Lajos Berkes**

The Persian and Arab Occupation in the Papyrological Record of the Seventh Century

As a part of the Roman Empire since 30 BC, Egypt was mostly spared from the violent events of late Roman history. The turbulent seventh century brought invasions and regime changes again: In 619, the Sassanid Persian armies occupied the country, but only ten years later, they departed in accordance with a treaty with the Byzantine emperor Heraclius. The country had only started to recover from the Persian occupation when Arab troops arrived: After occupying most of the Near East, they reached Egypt in 639. Even though the Byzantine troops were defeated in the battle of Heliopolis in 640, the Arabs could not take Alexandria. Finally, an 11-month ceasefire was agreed upon in 641, but it still took several years until the conquerors were able to completely control Middle and Upper Egypt. After an unsuccessful Byzantine attempt to reconquer their former province in 646, Arab rule remained unquestionable, although the southern frontier of the country could only be secured in 652 after a final peace treaty with the Nubians.[1]

The present chapter deals with the Persian and the Arab occupation of Egypt in the seventh century as reflected in the papyrological documentation that offers a contemporary, bottom-up perspective on the interactions of locals with the conquering armies. The dry climate of the country has preserved documents of everyday life written on papyri, potsherds, parchment, and many other writing materials only scarcely found in other regions. These papyrological[2] texts include letters, contracts, accounts, and many other types of documents that offer deeper insights into several areas of everyday life than any other sources from this period. From the seventh century, a rich documentation consisting overwhelmingly of Greek and Coptic but also Persian and Arabic papyri, has been preserved.[3]

DOI: 10.4324/9781003245568-6

The seventh century, a crucial period of not only Egyptian but also Mediterranean history, offers a possibility to study and compare two different military diasporas that were active in the same country hardly more than a decade apart, based on the same type of source material. Identifying the Persian occupying forces as a military diaspora, even if only a short-term one, is straightforward. The diaspora was composed of Persian soldiers and officials living far away from their homeland in an occupied province of the Byzantine Empire, against which their empire had fought wars for hundreds of years. Furthermore, they were Zoroastrians in a Christian country and spoke a language with which the locals were not familiar. Even though our sources do not provide explicit evidence on the matter, these factors must certainly have strengthened their group identity as Persian soldiers in an occupied country.

The question of a military diaspora in the Arab period is a more complicated issue, since the conquering troops included Christian Arabs and non-Arab non-Muslims as well, such as Greeks and Persians.[4] Furthermore, the early Muslim conquests marked the beginning of a new empire, where an imperial identity was still being formed, and tribal affiliations remained of utmost importance to the Arabs. Moreover, it is difficult to grasp how Islam was exactly understood in this period and thus, it constitutes a complicated issue whether the terms "Arab" and "Muslim" can be employed interchangeably, as I will do in this chapter. Nevertheless, we do know that the above-mentioned non-Muslim units in Egypt certainly converted to Islam by 659–661[5] and it is thus reasonable to assume that two decades after the conquest, the majority of occupying troops identified as Muslim and could be regarded as a military diaspora with a shared religious identity. An additional factor to understand the interactions of Egyptians and the Arab troops may be the fact that some of the Arab conquerors were originally Byzantine subjects. Arabs appeared regularly in various parts of Egypt and were thus to a lesser extent, foreigners or enemies than the Persians.[6] It is also worth pointing out that Arab soldiers were present in Egypt as part of the multi-ethnic imperial troops.[7]

Papyri are mostly silent on how the Persians and Arabs perceived their own situation in Egypt. This is hardly surprising, since most of our documents are administrative texts such as official letters, petitions, accounts, and legal documents that are unlikely genres for self-reflective statements. Although documents from both occupying forces, issued in their own language, are preserved, this chapter will focus on the Greek and Coptic documentary texts. This may seem paradoxical at first and thus requires an explanation. While more than a thousand Middle Persian documents written in Pahlavi script are preserved, their interpretation poses formidable challenges.[8] While it is clear that many Pahlavi papyri contain administrative documents, such as letters and accounts, the current state of the research does not allow for building an interpretation on them. This is well illustrated by an account containing toponyms and numbers that could be interpreted as either a recruitment list, a tax account, a register of distances between cities, or a document recording economic activities at checkpoints, such as the number of passing caravans.[9] Arabic papyri, likewise, are preserved only in relatively low numbers from the first decades of Arab rule, since the Muslim presence was very limited and thus, it left a much smaller impact on the documentary record than the thousands of Greek and Coptic papyri many of which attest to interactions between the new rulers and their subjects.

This chapter does not intend to give a comprehensive account of the Sassanid occupation and the first decades of Arab rule, a topic that could easily fill one or more monographs. Instead, it will focus on the interaction of the local population and the Persian and Arab occupying troops and local attitudes towards the new rulers as attested in the papyrological evidence. The nature of these sources means that these aspects will be mostly studied through administrative documents that reflect the bureaucratic framework in which most of the interaction between Egyptians and the occupying forces occurred. The short duration of the Sassanid occupation of only ten years does not allow us to witness similar long-term developments as during the Arab rule that turned out to be permanent. While the period under investigation is *per se* defined for the Persian occupation by its beginning in 619 and ending in 629, we need to be more careful with setting a time limit for the Arab period. The most intuitive solution, in my view, is to limit ourselves to roughly the first half-century of Muslim rule, i.e. ca. 642–700. In this period, as Clive Foss put it, Egypt was "a country under occupation, not yet arrived at a point when there was any assimilation between the new conquering forces and the local population".[10] As I have proposed elsewhere, this period could perhaps be best referred to as post-Byzantine Egypt, since the designations "Arab" and "Islamic" give a very misleading impression of realities in the country.[11] The end of the seventh century, or to be more precise, when the empire-wide reforms of the Marwanids left their imprint on Egypt, constitutes a more natural boundary. A first wave of Arabization and Islamization began and local Christian notables who were in charge of the administration were gradually replaced by Muslim officials sent from outside. Village headmen and leaders lost their previous power to assign tax shares, and new Muslim landholders started to threaten their status. All this resulted in increasing unrest and revolts from the beginning of the eighth century.[12] These and other significant administrative, societal, and linguistic changes initiated by these reforms justify setting the limits of our study to around 700.[13]

The Persian Occupation of Egypt (619–629)

The Persian occupation seems to have followed the same pattern in Egypt as in the whole Roman East: "Evidence from the occupied provinces—from Armenia to Egypt—reveals a consistent pattern: Stability, continuity and tolerance followed an initial period of violence".[14] There is ample evidence from Egypt for the initial violent stage of the Persian occupation. An archaeological case can be made for the monastic complex and pilgrimage centre of Abu Mina that lies 40 km southwest of Alexandria. Several layers of destruction were found at the site that can be linked to the Persian attack on Egypt.[15] A recently published inscription from the well-known Apa Apollos Monastery at Bawit dated to the year 620 commemorates a notary and a monk "whom the Persians have killed".[16] The bad reputation of the Persians made the bishop of Coptos, Pisentius, flee from his seat to the area of the ancient city of Thebes, where he stayed during the whole period of the occupation as documented by his correspondence.[17] Papyrological sources report many atrocities against the local population. A woman writes, for instance, in her petition:

> I am this wretched one, miserable beyond (all) men on earth, and sore oppressed with grief and sadness, and heartbroken for my husband who is dead and for my son whom the Persians beat (?) ... and my cattle which the Persians carried off.[18]

An even more dramatic story emerges from a Greek letter. The sender informs his "good lord" that he was kidnaped and tortured by the Persians: He managed to flee, but his children were taken away.[19] In a similar letter, a certain Esaias asks for the help of "his lord", because he is "in the hands of the Persians".[20] In a Coptic letter from the year 621, the bishop of Hermopolis writes to rebels who seized and plundered the city. He tries to convince them to start negotiations with the Persians and even offers his son as hostage.[21]

Apart from these violent incidents, the new regime did not cause major disruption in the administrative system of the country that constituted the framework for most of the interactions of the invaders with Egyptians. This is also manifest in the fact that apart from *sellarios*, the Romanized version of the Persian title *salār*, no other Persian term seems to have entered the vocabulary of Greek and Coptic documents.[22] A recent study has concluded that the main interest of the Sassanids was to ensure tax incomes and supplies for their troops.[23] Strategically important points were controlled by Persian units and officials, but not much changed in terms of the everyday bureaucratic workflow. The same agents collected the taxes on the local level as before, but now for Persian lords, and most of the local stakeholders maintained their position and status. Church institutions that, as we have seen, initially suffered at the hands of the invaders, began to establish a working relationship with them instead, as evidenced by the institutions' role in supplying the Persian troops.[24] There are, however, exceptions to this continuity, but mostly on the higher level: For instance, the aristocratic family of the Apions, which was influential on an imperial level, disappears entirely from our documentation after the Persian conquest.[25]

The Egyptians conceptually integrated the Persians into their Byzantine world: For instance, they applied the same honorary epithets to Persian officials that they would to Byzantine aristocrats in the same position or rank. This is well illustrated by a Greek letter from the city of Oxyrhynchus dating to the year 623, which was written in the name of the Persian Rasbanas by one of his clerks.

It refers to the high-ranking Sassanid official Saralaneozan and the "king of the kings", the Persian ruler:[26]

> Make haste to send us also, within three days, the balance of the first instalment. For remember, you have also had written instructions about this matter from our master the all-praiseworthy Saralaneozan. Now you yourself make haste immediately to send us the balance of the first instalment, since we wish to make the shipment of the gold to our master the king of the kings.[27]

Saralaneozan is referred to in our letter with the honorific "all-praiseworthy" (*paneuphēmos*), which was reserved for high-status Byzantine aristocrats.[28] Some Greek (and Pahlavi) documents attest to provisions for his household, such as foodstuff and oil for his torches. The term used for his household in Greek, *oikos*, is the same one that designates Byzantine aristocratic houses.[29] In a similar vein, in a Coptic legal document addressed to a Persian official from the year 625, we find an oath taken on "God and the wellbeing of the king of the kings" who is, as we have seen above, the Sassanid ruler.[30] Such oaths used to be taken for the well-being of the Byzantine emperor, but were apparently adapted to new realities at request of the new masters. In private documents, however, the changes in oath formularies are more hesitant: In a Coptic legal document from 627, an oath is taken "on God Almighty and the well-being of those who rule over us".[31]

The hesitance to name the Persians already suggests a lack of enthusiasm for the new regime, but a similar document implies more strongly an unwillingness to accept it. From the late sixth century on, Greek (and Coptic) legal documents were introduced by invocations of Christ or the Holy Trinity, sometimes with addition of Mary and the Saints.[32] In the invocation of a legal document from the year 621, Jesus Christ is specified as the "king of kings".[33] Even if this phrase is commonplace in patristic literature, it is new among highly standardized Egyptian invocation formulae and probably thus implies a statement. Specifying Jesus Christ with the same title as the Persian king emphasizes Christ's rule over all kings of this world, even the Sassanid one. By extension, it may hint at the superiority of Christian rulers, i.e. Byzantines, over pagans, i.e. Persians.[34] Even if this small change in the invocation formula could have been highly provocative for the occupying forces, it is unlikely that they would have ever been confronted with private legal documents. Thus, this statement was mostly meant to be read and understood by Egyptians reinforcing their Christian identity and perhaps Byzantine loyalty under a brutal, foreign, non-Christian occupying force.

A final note is due on the memory of the Persian occupation. It is hardly surprising that this period left a lasting imprint on Egyptians, as we can gather from two papyri that were written shortly after the Sassanids left. In the first one, a Coptic legal document written after the Islamic conquest in 646/647, the "time of the Persians" is mentioned.[35] A Greek petition written after the Sassanid occupation, but before the Arabs arrived, is more explicit: It talks about the "time of the godless Persians".[36] Interestingly, both the Greek and Coptic texts use the same term for time, the Greek *kairos,* which may suggest that this phrase was already a well-established commonplace only a couple of years after the conquest.

Between Persians and Arabs: Methodological Problems

Although the main outlines of the Byzantine administrative system did not change under Persian and the first decades of Arab rule, there are nevertheless some novelties. However, the extent, nature, and role of these novelties is a much-debated topic in current scholarship. One of the main problems is dating administrative reforms between the Persian conquest and the first decades of early Islamic rule. Although there are thousands of papyri preserved from this period, only a few can be dated with certainty, since most documents only refer to indiction years. An indiction cycle consisted of 15 years, and thus the years within each cycle cannot be fixed without additional specifications. It is a further difficulty that it is virtually impossible to date handwriting more precisely than to a period of 50–100 years. Thus, if we detect reforms in the administration in this period, they can be connected with at least three different historical events: The Persian occupation of Egypt, the Byzantine reorganization after the Sassanids left the country, and the early Islamic period.[37] This situation is complicated further if we consider it likely that the Arabs imported administrative practices from other Byzantine provinces.

The dating clause of a Greek legal document eminently illustrates this problem and its significance. It refers to the first year of the "administration" of the Byzantine emperor Heraclius and his son Heraclonas, rather than to their reign, as one would expect. However, there is a lacuna of ca. 40 letters between the word "administration" and the mention of the rulers. The editor of the document interpreted this clause as a reference to Roman emperors as the vassals of the Persian king. However, reconsiderations in the placements of the fragments have led to the conclusion that the indiction year was misread in the *editio princeps*. The new reading does not allow a dating into the period of the Persian occupation but points to the early years of Arab rule. The papyrus therefore refers either to the "administration" of Heraclius' widow Martina, or, as it has been argued by Federico Morelli and seems more likely, to the conqueror and first Muslim governor of Egypt ʿAmr ibn al-ʿĀṣ as the administrator of the Byzantine emperor. This later interpretation could imply far-reaching consequences for the understanding of the character of the early Islamic period since the Arabs would appear as the vassals of the Byzantine emperor, as will be discussed in more detail below.[38]

The Early Islamic Period

The conquering Muslim army that reached Egypt in around 639 counted only a few thousand men. The Arabs, if our literary sources are to be trusted, seem to have negotiated treaties with the local communities. It is conceivable that the

Arab troops marched from city to city and reached an agreement with their leading citizens.[39] Their approach to the country seems to have been very pragmatic: Their main interest was securing tax revenues.[40] They did not introduce radical changes to the local administration: Apart from the top echelons of bureaucracy where officials were replaced with Arabs, the local administration managed by the Christian elite was left untouched. This is hardly surprising: These Egyptian notables possessed the necessary bureaucratic know-how and experience and were also able to guarantee the smooth working of the administrative machinery. Although some later Arab historians claim that there was a mass exodus of Byzantine aristocrats after the conquest, this was certainly not the case, or at least not in the papyrologically attested parts of Egypt which lay south of the Nile Delta: Documentary papyri show very well that many high-ranking Byzantine officials continued to work under the new masters.[41]

One of the novelties that the Arabs introduced was the poll tax that seems to have been levied upon non-Muslim males, but this—and the date of its introduction—is a controversial issue.[42] Another visible change is the foundation of the new capital Fusṭāṭ (modern Old Cairo) near the Roman military fortress of Babylon. This is well-illustrated by a story often repeated in Arab historiography: After the Arabs took Alexandria, their soldiers moved into splendid houses in the city, but the caliph ʿUmar recalled them to live in the modest living quarters of the new funded Muslim capital of Egypt, Fusṭāṭ. As Petra Sijpesteijn put it:

> Separated from the local population in order to preserve their Arab culture and religion and in preparation for further conquests, Arab soldiers were in fact actively discouraged from moving into the countryside and making their own living from agriculture.[43]

Thus, only a handful of Arab soldiers were present in the country, and they were mostly isolated from the local population. This is mirrored in the fact that, as a recent study has shown, travelling to Fusṭāṭ seems to have caused anxiety to many Christian Egyptians in the seventh and eighth centuries. This is eminently clear from the regular appearance of the epistolary formula "God has guided us" that appears almost exclusively in connection with journeys to Fusṭāṭ/Babylon: It seems that the need for divine protection was felt, when travelling to the new Arab capital.[44] Furthermore, in the Greek and Coptic administrative correspondences of this period, Arab officials appear almost exclusively as masters who were persistently demanding tax payments and requisitions: There is hardly any sign of contact in other areas. The conquerors were not interested in converting the local population either: Muslims were almost invisible in the countryside for the first 50 years of Islamic rule.

As in the Persian period, Egyptians conceptually incorporated the Arabs in their Byzantine world. A very characteristic example is the case of the high-ranking official Flavius Atias who is attested in the last years of the seventh century and the very beginning of the eighth, first as pagarch, the chief administrator of a city and its district, but also as *dux*, governor of a whole province.

One of the Greek documents issued in his name introduces him as "Flavius Atias, most famous *dux*".[45] The gentilicium Flavius was employed in late antiquity as a status marker for soldiers and high-ranking public servants.[46] His title "most famous *dux*" consists of a typical Byzantine honorific and the graecicized version of the Latin title *dux* that was also a relic of the Byzantine period. This would fit well so far with a "post-Byzantine" official. However, as we learn from some other documents, his full name was ʿAṭiyya b. Ju'ayd. We do not know whether he was an Arab Muslim or perhaps an Egyptian convert, but it is clear that he was treated as a Byzantine aristocrat.[47]

The example of Flavius Atias is far from unique: For instance, Arab *amīrs* are regularly referred to with the same Byzantine honorific as "most famous" (*eukleestatos*).[48] There is, however, one more example that merits mention for its special interest and possible implications: A Greek letter, probably from the early years of Arab rule that identifies the sender, an Arab, as *patrikios* (= lat. patricius). This was one of the highest titles in the Byzantine court bestowed only upon the most important aristocrats and it is thus curious to find it connected to an Arab official—this is, in fact, the only mention of it in papyri after the Islamic conquest. It could imply, therefore, as it has been recently cautiously hypothesized by Federico Morelli, that the Arabs might have presented themselves as the vassals of the Byzantine emperor at least in the early years after the Islamic conquest. As Morelli has argued, Egypt in these years displays many similarities to the situation of the Barbarian vassal kingdoms of the west such as the Ostrogoth kingdom of Theodoric in the fifth–sixth centuries.[49] This interesting theory, however, still needs further direct support from our sources.

What were the attitudes of Egyptians towards a new occupation only a decade after the "godless Persians" left? Some historical sources claim that the miaphysite "Egyptians" would have welcomed the Arab conquerors as their liberators from the "Greek" Chalcedonian oppression—a claim that is still often repeated in modern handbooks. Recent research, however, has shown that sectarian differences (miaphysites and Chalcedonians) or ethnic identities ("Greeks" and "Egyptians") had no significant influence on daily life or the success of the Arab conquest. Such claims stem from later historiographical sources that anachronistically project problems of their own time on the past.[50]

It is also conspicuous that the first years of Arab rule did not leave a similar impression of violent atrocities as the Sassanid occupation. There is obviously no invasion without violence against the local population, and indeed we find examples of heavy fighting during the conquest,[51] but the Arabs seem to have generally been much more cautious.[52] As we have seen above, literary sources suggest that they concluded treaties with the local communities that regulated the terms of surrender. This could have been a reasonable strategy: The Arab troops were few in number and there was an ongoing war against Byzantium that made it necessary to guarantee a peaceful hinterland from which resources could be extracted.

There is a growing number of documents being identified and published from the first years of Arab rule that displays not only a system of requisitions but also some tensions (see below). A famous example is a bilingual receipt from the year 643 that was issued by an Arabic commander to the pagarchs of the city of Heracleopolis about requisitioned sheep:

> (Greek) In the name of God. [From] ʿAbdallāh, governor, to you, Christophoros and Theodorakios, pagarchs of Herakleopolis. I have received from you for the maintenance of the Saracens stationed with me at Herakleopolis 65 sheep, sixty-five only; and in witness thereof I have issued the present receipt, written by me, Ioannes, notary and deacon, on the 30th of the month of Pharmouthi of the first indiction year.
>
> (Arabic) In the name of God, the most Merciful the Compassionate! This is [to certify] what ʿAbdallāh b. Jābir and his companions have taken of slaughter sheep from Ahnās. We received from the representative of Tidhraq, the younger son of Abū Qīr, and from the representative of Iṣṭufur, the elder son of Abū Qīr, fifty sheep for slaughter and fifteen other sheep. He gave them for slaughter for his naval, cavalry and heavy armed infantry units in the month of Jumādī al-Ūlā of the year twenty two. Written by Ibn Ḥadīdō.
>
> (Verso): Receipt for the sheep delivered to the *Magarites* and others who had come, on account of the taxes of the 1st indiction.[53]

Although the Greek and the Arabic version contain essentially the same information, they both follow their own documentary tradition and are not translations of each other. The Arabic text—unlike the Greek—points out that only the representatives of the pagarchs were present. It also reveals that they were brothers: While the Greek talks only about "Christophoros and Theodorakios", the Arabic identifies them as "Tidhraq, the younger son of Abū Qīr … Iṣṭufur, the elder son of Abū Qīr". Moreover, the Arabic version specifies the purpose of the requisitions: "for slaughter for his naval, cavalry and heavy armed infantry units". The layout of the document shows that the Greek text was written first and the Arabic was added only later.[54] It is not entirely clear at first sight why the pagarchs would have needed a receipt in Arabic for this transaction, since the administrative machinery of Egypt worked in Greek. For the pagarchs Christophoros and Theodorakios a receipt only in Greek—that they could have produced on demand—would have been sufficient. That is why the use of Arabic in this document and some similar cases has been explained by Petra Sijpesteijn as a symbolical or political choice:

> Arabic was considered essential to the Islamic empire's communication with its subjects right from the start, although the "message" it conveyed was often less the immediate content of the text than its symbolic power. Arabic identified the new rulers and their triumphant religion, eventually penetrating into the remotest corners of the country. (…) The use of the *hijra* date in these earliest datable papyri similarly functions as a religious and political symbol.[55]

However, as it has been pointed out by Federico Morelli, there is a simpler solution: The pagarchs might have been required to produce this document to other Arabs.[56] This seems to be a more natural interpretation of the bilingual nature of the receipt, even if it does not exclude that the use of Arabic also conveyed a symbolic meaning. Providing the representatives with a document also written in Arabic might have functioned as an extra guarantee against demands of other Arab troops who might not have understood or accepted a Greek document. Even if we do not find mentions of violent atrocities committed by the Arabs in the papyri, a certain tension and stress in dealing with the conquerors is palpable in the internal correspondence of Egyptian officials. For example, a Greek letter from ca. 660–680 reports about an unknown person in an administrative context: "he himself is being annoyed by the Saracens".[57] In another Greek letter that was probably written in the second part of the seventh century, a certain Viktor reminds his colleague of sending the arrears of several villages, closing his message with the following words: "if not,—may God help!—you will be given and taken to Fusṭāṭ!"[58]

In the Byzantine administrative system, local stakeholders and patronage played a crucial role. The Arabs were not integrated into these networks, and therefore employed violent threats as a means of putting pressure on the local bureaucrats. There are examples of and references to "threatening letters", as our sources identify them, demanding taxes written in the name of Muslim officials already from the first years of Arab rule.[59] The harsh style of these communications is a novelty of the Islamic period: Byzantine administrators seldom wrote in such an extravagantly violent style. A Greek letter, already mentioned above, provides an example of such "threatening letters". It was written in the name of an Arab official styled as *patrikios* who demands arrears from the bishop and other leading citizen of Hermopolis:

> ... save your souls, since the appointed time has passed and if you do not pay it all, you will not have a word nor an excuse for me! Since, when you asked me something, I did it and I endured you. So, look at your task, since, as I said, the appointed time has passed! Lo! I will send a man after this letter who is due to collect the public taxes and, by the name of God, if someone of you does not pay in full, if he owes something, I will take away both his children and his wife and all his belongings! That you know: you cannot say a word to me ...[60]

We do not know how often such threats were fulfilled: It is possible that in some cases, the Arabs implemented drastic measures. However, as a recent study suggested, this rhetoric could be interpreted as a sign of the central government's powerlessness while dealing with distant local officials who were supported by and loyal to their local power base.[61] Furthermore, we need to consider that threats of corporal punishment might have been part of legal language in this period.[62]

As we have seen in the documentation of the Persian period, oath formulae and invocations can provide some subtle hints about local attitudes.[63] As during the Sassanid occupation, we find the replacement of the Byzantine emperor, but now by Arab officials. In a tax declaration submitted by village representatives, we find an oath "by the name of God and the well-being of ʿAmr".[64] The reference to ʿAmr ibn al-ʿĀṣ, the conqueror of Egypt provides a date for the papyrus, since he was governor of the country 641–644 and 658–663/664. Similarly, a handful of Hermopolite tax declarations from the 650s attest to oaths sworn on the well-being of the *amīrs*.[65] Our examples consist so far, however, only of official documents, private ones convey a very different picture. Their oath formulae, as in the Persian period, avoid direct references to the Arab rulers, but simply refer to "the wellbeing of those who rule over us now".[66] A document from 644/645, thus the very first years of Muslim rule, modifies the formula in a telling way: "the well-being of any power and authority who rules over us in any time".[67] Another papyrus from 645 still employs the old oath "on the almighty God and the victory of imperial well-being".[68] While in this case, it remains possible that the scribe simply copied an outdated formula without much consideration, a loyalist attitude is also conceivable, especially in the light of the other examples mentioned above. Since official documents show that oath formulae with direct reference to the Arabs were regularly being used, it is, at least in my view, obvious to interpret these general formulations as deliberate omissions and thus as a sign of unwillingness to accept the permanence of Muslim rule. Similarly, the invocation formula referring to Jesus Christ as the "king of kings", which we have already encountered in the Persian period, turns up again after the Arab conquest and may again be understood as a "political statement".[69] It is also important to highlight that there is much evidence for nostalgia for the Byzantine period well into the eighth century.[70]

As we have seen above, the constant demands of the Arabs produced much tension. A clear example for this is a letter from ca. 660–680 in which a local bureaucrat complains about new requisitions:

> … I received other messages from our lord, the most famous *amīr* through four Saracens of the *Amīr* of the believers [i.e. the caliph] about the purchase of many things and he did not concede anything of them today. May he taste water and, truly, Satan has brought me … because I have never been vexed or distressed, if not now![71]

Even if the letter is fragmentary, it clearly conveys the writer's feelings about "the most famous amīr".

There are, however, examples for a more positive attitude as well. In the first years after the conquest, a certain Hypatios writes to one of his colleagues about requisitions of milk for the Arabs. He begins his short message with his wish to please Kulayb, an Arab who is presumably a military commander or some kind of official: "Truly do I wish to greatly honor Kulayb".[72] However, this could also be interpreted as an instrumental use of Arab authority, since mentioning Kulayb's name might have produced quicker results. There are also examples from the first decades of the Arab regime of Christian officials imitating the epistolary style of the conquerors in their communication with locals. Employing these formulae underlined their close contact with the Arab masters and thus their position in the new establishment.[73]

Similarities and Differences: Military Diasporas and Their Host Societies

Both the Persian occupation and the initial stage of Islamic rule point to a similar pattern. However, the occupation took place—with an initial period of violence or rather with more caution and negotiation—after the situation stabilized and the social status of the local elite was not significantly threatened, the Egyptians accepted their new masters. As long as the equilibrium of power did not change significantly between the new rulers and the local elite, life could continue in the same framework. A famous parallel for this continuity from outside Egypt is the case of the family of St. John of Damascus. His grandfather Manṣūr was in charge of taxation in Damascus. Appointed by the Byzantine emperor Maurice, Manṣūr continuously held his position during the Persian occupation. After the Byzantine re-conquest, he was humiliated by the emperor Heraclius and later took part in betraying the city to the Arabs, while his son, Sergius, became the treasurer to the caliph.[74]

The new masters managed the government "from above" and—at least in the initial phase of their rule—remained separated from indigenous society: Their interaction with the Egyptians was almost exclusively restricted to certain administrative matters. The local population interacted with the Persian and Arab occupying forces and new bureaucrats moving into the country very similar to the Byzantine provincial officials and Byzantine practices still regulated most areas of private and public life. The Egyptians conceptually integrated the invaders into their Byzantine world, designated them with the same titles and epithets as their predecessors. It nevertheless seems that there was a certain reluctance both under Persian and Arab rule to accept the new regime as permanent.

There is one more aspect we need to take into account here: The Persian occupation of the Near East might have facilitated the Arab conquests.[75] Would the Arabs have met more resistance if there had not been a Persian occupation of horrendous memory only a decade before? We cannot know, but it is certain that the citizens of the affected regions realized that they could find a *modus vivendi* with the new rulers and thus members of the elite could more or less maintain their standing. The Persian occupation thus created a pattern with which the locals could deal with the Arab invaders. The inhabitants of the ex-Byzantine provinces might have wanted to avoid a similar bloody occupation as in the case of the Persians. They saw that the Persians left after ten years and could thus reasonably hope for a Byzantine re-conquest.

The papyrological record provides a snapshot of the Persian and Arab occupations and thus of occupying troops from the perspective of the local population. It clearly shows the local attitudes towards occupying military forces with different ethnicity, language, and religion, who lived separated from the population of the conquered country. As always, in such historical situations, we find examples of hostility, indifference, and cooperation. Nevertheless, the example of Egypt in the seventh century highlights that subsequent military occupations created patterns for the local population on how to deal with new rulers and their soldiers. Thus, the memory of previous occupations and earlier contact with foreign soldiers created expectations that, in turn, influenced the interactions with the newly arrived troops.

Abbreviations

BKU	*Ägyptische Urkunden aus den Königlichen Museen zu Berlin: Koptische Urkunden.*
CPR	*Corpus Papyrorum Raineri.*
O.AbuMina	Litinas, Nikos, ed. *Greek Ostraka from Abu Mina.* Archiv für Papyrusforschung und verwandte Gebiete, Beihefte 25. Berlin, New York 2008.
P.Apoll.	*Papyrus grecs d'Apollônos Anô.*
P.Gascou	Mélanges Jean Gascou. *Textes et études papyrologiques* (P.Gascou), edited by Jean-Luc Fournet, Arietta Papaconstantinou. Travaux et Mémoires 20,1. Paris 2016.
P.Heid.	*Veröffentlichungen aus der Heidelberger Papyrussammlung.*
P.Iand.	Kalbfleisch, Carl, et al., eds. *Papyri Iandanae.* Leipzig.
P.Lond.Copt.	*Catalogue of the Coptic Manuscripts in the British Museum.*
P.Mich.	*Michigan Papyri.*
P.Oxy.	*The Oxyrhynchus Papyri.* London.
P.Paramone	*Editionen und Aufsätze von Mitgliedern des Heidelberger Instituts für Papyrologie zwischen 1982 und 2004*, edited by James M. S. Cowey and Bärbel Kramer. Archiv für Papyrusforschung und verwandte Gebiete, Beihefte 16. Munich, Leipzig 2004.
P.Ross.Georg.	*Papyri russischer und georgischer Sammlungen.* Tiflis.
SB	*Sammelbuch griechischer Urkunden aus Ägypten.*
SB Kopt.	*Koptisches Sammelbuch.*
SPP	Wessely, Carl, ed. *Studien zur Palaeographie und Papyruskunde.* Leipzig, 1901–1924.

Notes

* References to Greek and Coptic papyri are provided according to the *Checklist of Editions of Greek, Latin, Demotic and Coptic Papyri, Ostraca and Tablets* (http://library.duke.edu/rubenstein/scriptorium/papyrus/texts/clist_papyri.html). All translations are my own, unless indicated otherwise. I thank Graham W. Claytor for comments on a draft of this paper and correcting my English.

1 On the Islamic conquest of Egypt, see the summary of Petra M. Sijpesteijn, *Shaping a Muslim State: The World of a Mid-Eighth-Century Egyptian Official*, Oxford Studies in Byzantium (Oxford: Oxford University Press, 2013), 49–58. Phil Booth has recently argued that, contrary to the narrative of Muslim historians, the Arab conquerors invaded Egypt both from the north and from the east through the Red Sea, see Phil Booth, "The Muslim Conquest of Egypt Reconsidered," in *Constructing the Seventh Century*, ed. Constantin Zuckerman, Travaux et mémoires 17 (Paris: Association des Amis du Centre d'Histoire et Civilisation de Byzance, 2013), 639–670.

2 The word "papyri" is used in this chapter according to papyrological usage as an umbrella term covering not only papyri but also other perishable writing surfaces such as ostraca, parchment, and wooden tablets.

3 There are of course papyri written in other languages, such as Latin or Syriac that are interesting in their own right, but their number is negligible in comparison to the plethora of Greek and Coptic documents, and they do not provide significant information for the research question of this article.

4 See Wadad al-Qāḍī, "Non-Muslims in the Muslim Conquest Army in Early Islam," in *Christians and Others in the Umayyad State*, ed. Antoine Borrut and Fred M. Donner, Late Antique and Medieval Islamic Near East 1 (Chicago: Oriental Institute of the University of Chicago, 2016), 83–127, esp. 95–96 for Egypt.

5 Ibid., 95.

6 For Arabs in Egypt before the Islamic period, see Janneke De Jong, "Arabia, Arabs, and 'Arabic' in Greek Documents from Egypt," in *New Frontiers of Arabic Papyrology*, ed. Sobhi Bouderbala, Sylvie Denoix, and Matt Malczycki, Islamic History and Civilization 144 (Leiden: Brill, 2017), 1–27.

7 During the Byzantine period certain imperial troops were identified by ethnonyms, such as Franks. They were originally most likely indeed constituted by barbarian units, but as time passed by, their ranks were largely replaced by local Egyptians, see e.g. Roger S. Bagnall and Bernhard Palme, "Franks in Sixth-Century Egypt," *Tyche* 11 (1996): 1–13. However, this does not mean that all ethnic troops stationed in Egypt were completely replaced by locals, see e.g. the recently published tantalizing account *P.Gascou* 32 listing payments to *bucellarii* (a sort of paramilitary unit the exact nature of which is debated) that contains several soldiers identified by their ethnicity. This account lists two Goths, a Slav, an Armenian, a Persian, a Saracen (i.e. Arab), five Romans, a Danubian, a person from Ashkelon, and perhaps also a Cilician, but also many Egyptians, see the analysis of Nikolaos Gonis in the introduction to *P.Gascou* 32, at 180–181. The text was written in the year 612 and thus shows that foreign (para)military groups were well-known to Egyptians before the arrival of the Persians and the Arabs.

8 For a recent overview about the contribution of Pahlavi papyri to the assessment of the Persian period in Egypt, see Dieter Weber, "Die persische Besetzung Ägyptens 619–629 n. Chr.: Fakten und Spekulationen," in *Ägypten und sein Umfeld in der Spätantike: Vom Regierungsantritt Diokletians 284/285 bis zur arabischen Eroberung des Vorderen Orients um 635–646*, ed. Frank Feder and Angelika Lohwasser, Philippika 61 (Wiesbaden: Harrassowitz, 2013), 221–246.

9 Stefanie Schmidt, "*P.Bas.* II 69 and 70: A Look Behind the Text," *Archiv für Papyrusforschung* 64 (2018): 324–342, at 325–330.

10 Clive Foss, "Egypt under Muʿāwiya. Part I: Flavius Papas and Upper Egypt," *Bulletin of the School of Oriental and African Studies* 72 (2009): 1–24, at 3.

11 Lajos Berkes, "Introduction: A Papyrological Perspective on Christians and Muslims in Early Islamic Egypt," in *Christians and Muslims in Early Islamic Egypt*, ed. id., American Studies in Papyrology 56 (Ann Arbor: University of Michigan Press, 2022), 1–9, at 1–2 and 4 with further references.

12 Petra M. Sijpesteijn, "Landholding Patterns in Early Islamic Egypt," *Journal of Agrarian Change* 9 (2009): 120–133, at 130.

13 On these much-discussed reforms, see e.g. the recent summary in Marie Legendre, "Aspects of Umayyad Administration," in *The Umayyad World*, ed. Andrew Marsham (New York: Routledge, 2020), 133–157, at 141–145.
14 Clive Foss, "The Persians in the Roman Near East (602–630 AD)," *Journal of the Royal Asiatic Society* 13 (2003): 149–170, at 167.
15 N. Litinas, O.AbuMina, ix–x.
16 Florence Calament, "Un exceptionnel témoin épigraphique de la dernière occupation perse en Égypte," *Journal of Coptic Studies* 23 (2021): 1–13.
17 On his person, see Jacques van der Vliet, "Pisenthios de Coptos (569–632): Moine, évêque et saint. Autour d'une nouvelle édition de ses archives," in *Autour de Coptos: Actes du colloque organisé au Musée des Beaux-Arts de Lyon (17–18 mars 2000)*, Topoi Supplément 3 (Paris: De Boccard, 2002), 61–72.
18 SB Kopt. 1.295. Translation from the edition: James Drescher, "A Widow's Petition," *Bulletin de la Société d'Archéologie Copte* 10 (1944): 91–96, at 93.
19 *P.Ross.Georg.* 4 appendix.
20 *P.Iand.* 2.22.
21 *BKU* 3.338.
22 Jean-Luc Fournet, "L'impact de la conquête sassanide sur l'Égypte: Notes lexicographiques," in *Mélanges Bernard Flusin*, ed. André Binggeli and Vincent Déroche, Travaux et Mémoires 23 (Paris: Association des Amis du Centre d'Histoire et Civilisation de Byzance, 2019), 287–298, at 288, esp. n. 5 (with further literature).
23 Patrick Sänger, "Saralaneozan und die Verwaltung Ägyptens unter den Sassaniden," *Zeitschrift für Papyrologie und Epigraphik* 164 (2008): 191–201 and id., "The Administration of Sassanian Egypt: New Masters and Byzantine Continuity," *Greek, Roman, and Byzantine Studies* 51 (2011): 653–665.
24 This is well documented by the archive of Theopemptos and Zacharias. However, the documents of this important text group are mostly edited unsatisfactorily, which is also apparent in the fact that they are underutilized in studies of this period, see e.g. Sänger, "Administration," 656. A new edition of the archive is underway by Nikolaos Gonis, whom I thank for sharing his preliminary results with me. For a recent reedition of two texts of this archive, see Fournet, "L'impact."
25 Their household is, however, still attested as an economic unit in the first years of the Persian occupation, see Todd M. Hickey, *Wine, Wealth, and the State in Late Antique Egypt: The House of Apion at Oxyrhynchus* (Ann Arbor: University of Michigan Press, 2012), 17–18.
26 Sänger, "Administration," 655.
27 Translation of John R. Rea from the edition: P.Oxy. 51.3637.14–19.
28 Sänger, "Saralaneozan," 198.
29 For oil for the torches of Saralaneozan and his *oikos*, see *SPP* 10.251.A2; for provisions for his kitchen, Patrick Sänger and Dieter Weber, "Der Lebensmittelhaushalt des Herrn Saralaneozan/Šahr-Ālānyōzān: Neuedition von zwei Speiselisten und einem Geschäftsbrief aus dem sassanidischen Ägypten," *Archiv für Papyrusforschung* 58 (2012): 81–96.
30 *CPR* 4.48.19–20.
31 *SB* 20.14427.30–31.
32 See Roger S. Bagnall and Klaas A. Worp, *Chronological Systems of Byzantine Egypt*, 2nd ed. (Leiden: Brill, 2014), 99–109.
33 SB 1.4483.
34 On this invocation, see the discussions (with further literature) by Nikolaos Gonis and Klaas A. Worp, "*P.Bodl.* I 77: The King of Kings in Arsinoe under Arab Rule," *Zeitschrift für Papyrologie und Epigraphik* 141 (2002): 173–176, at 176 and Fournet, "L'impact," 297, esp. n. 38–39.
35 *SB Kopt.* 1.36.65. A new, much improved edition of this document, also known as *P.Budge* is being prepared by Sebastian Richter to whom I owe the date of the papyrus and who also pointed out to me the importance of the phrase discussed above.

36 *P.Louvre* E 6846.24 which is being prepared for publication by Jean-Luc Fournet, see id., "Papyrologie Grecque," *Annuaires de l'École pratique des hautes études* 139 (2008): 92–93, at 92.

37 Similar problems also emerge in other Roman provinces that were occupied by the Persians and later conquered by the Arabs. See for example the case of the boundaries of greater Syria under early Umayyad rule, in Foss, "The Persians," 162–163.

38 *P.Paramone* 18. See Dieter Hagedorn and Fritz Mitthof, "*P.Paramone* 18: Ein neu platziertes Fragment oder: Cave restauratorem chartarum," *Zeitschrift für Papyrologie und Epigraphik* 149 (2004): 157–158; Nikolaos Gonis, "*P.Paramone* 18: Emperors, Conquerors and Vasalls," *Zeitschrift für Papyrologie und Epigraphik* 173 (2010): 133–135; and Federico Morelli, "ʿAmr e Martina: La reggenza di un'imperatrice o l'amministrazione araba d'Egitto," *Zeitschrift für Papyrologie und Epigraphik* 173 (2010): 136–157. It constitutes a similar problem whether the term *chorion* as "rural tax unit" appears as a novelty in the Islamic period, or already before the conquest. See Jean Gascou, "Notes critiques: *P.Prag.* I 87, *P.Mon. Apollo* 27, *P.Stras.* VII 660," *Zeitschrift für Papyrologie und Epigraphik* 177 (2011): 243–253, at 247: "(...) le terme de χωρίον est propre à l'administration arabe qui, dans sa pratique fiscale, l'a surimposé, au plus tard avant le 8 janvier 643, à la toponomastique traditionnelle, κώμη et parfois ἐποίκιον. Il ne s'agit pas d'un simple changement de vocabulaire. Les Arabes concevaient le χωρίον comme l'unité d'assignation de leurs impôts et réquisitions. À la différence des Byzantins, qui ne regardaient guère plus bas que la cité, leur fiscalité opérait au ras du sol. De ce fait, dans les régions de l'Égypte où les papyrus fiscaux d'époque arabe abondent, comme l'Arsïnoite, la toponymie recensée est plus riche et variée qu'aux époques antérieurs." See also his recent summary, id., "Arabic Taxation in the Mid-Seventh Century Greek Papyri," in *Constructing the Seventh Century*, ed. Constantin Zuckerman, Travaux et mémoires 17 (Paris: Association des Amis du Centre d'Histoire et Civilisation de Byzance, 2013), 671–677, at 672–673. Cf. also the doubts expressed by Federico Morelli, *CPR* XXX, p. 59. For further examples, see Morelli, "ʿAmr e Martina," 147, esp. n. 25.

39 Petra M. Sijpesteijn, "The Arab Conquest of Egypt and the Beginning of Muslim Rule," in *Egypt in the Byzantine World, 300–700*, ed. Roger S. Bagnall (Cambridge: Cambridge University Press, 2007), 451–452. See also Sijpesteijn, *Shaping a Muslim State*, 52, esp. n. 24.

40 On this, see the recent study of Stefanie Schmidt, "Between Byzantine and Muslim Egypt: Mobilizing Economic Resources for an Embryonic Empire," *Journal of Ancient Civilizations* 35 (2020): 241–266.

41 On the flight of the Byzantine elite, see Sijpesteijn, *Shaping a Muslim State*, 56, esp. n. 56 and 57 for further references. For continuity in administration, see e.g. the case of the *anystes* Senuthius and the *illustris* Athansius in the Hermopolite nome or the *defensor civitatis* Athanasius in the Fayyum: Arietta Papaconstantinou, "Administering the Early Islamic Empire: Insights from the Papyri," in *Money, Power and Politics in Early Islamic Syria. A Review of Current Debates*, ed. John Haldon (Farnham: Ashgate, 2010), 57–74, at 61. The *spectabilis comes* Flavius Calomenas may present a similar case, see Nikolaos Gonis, "Notes on the Aristocracy of Byzantine Fayum," *Zeitschrift für Papyrologie und Epigraphik* 166 (2008): 203–210, at 210.

42 See Papaconstantinou, "Administering," 58–64 and Stefanie Schmidt, "Adopting and Adapting: Zur Kopfsteuer im frühislamischen Ägypten," in *Proceedings of the 28th International Congress of Papyrology, August 1–6, 2016, Barcelona*, ed. Alberto Nodar and Sofía Torallas Tovar (Barcelona: Publicacions de l'Abadia de Monserrat, 2019), 586–593.

43 Sijpesteijn, *Shaping a Muslim State*, 78, esp. n. 213 for the sources.

44 Anne Boud'hors, "Babylone-Fusṭāṭ dans les sources papyrologiques coptes," in *Christians and Muslims in Early Islamic Egypt*, ed. Lajos Berkes, American Studies in Papyrology 56 (Ann Arbor: University of Michigan Press, 2022), 53–64, at 59–60.

45 *CPR* 8.84.1.

46 See James G. Keenan, "The Names Flavius and Aurelius as Status Designations in Later Roman Egypt," *Zeitschrift für Papyrologie und Epigraphik* 11 (1973): 33–63 and 13 (1974): 283–304.
47 On his career, see the recent summary in Jennifer Cromwell, "Coptic Texts in the Archive of Flavius Atias," *Zeitschrift für Papyrologie und Epigraphik* 184 (2013): 280–288.
48 See e.g. *CPR* 22.1.1 (643–644).
49 Morelli, "ʿAmr e Martina," 152–154, esp. 154: "Uguale come e da dove avesse lo 'Abdallâh di *SB* 16.12575 il titolo di patrizio, è un fatto che egli, illegittimamente o legittimamente, svalutato o no, porti lo stesso titolo non solo del suo connazionale del VI secolo Alamoundaros, ma anche di Odoacre o di Teodorico; e che in questo egli non si distingue dai due re e 'funzionari imperiali' barbari del V secolo. Se anche si tratta soltanto di un titolo attribuito a 'Abdallâh dai burocrati di lingua greca del suo ufficio, esso testimonia in ogni caso del modo in cui essi consideravano questo funzionario arabo."
50 Bernhardt Palme, "Political Identity versus Religious Distinction? The Case of Egypt in the Later Roman Empire," in *Visions of Community in the Post-Roman World: The West, Byzantium and the Islamic World, 300–1100*, ed. Walter Pohl, Clemens Gantner, and Richard Payne (Cambridge: Ashgate, 2011), 81–98.
51 See the case of the city of Oxyrhynchus, cf. Sijpesteijn, *Shaping a Muslim State*, 55, esp. n. 50 with further references.
52 See also below n. 72.
53 *SB* 6.9576. Translation from Lejla Demiri and Cornelia Römer, *Texts from the Early Islamic Period of Egypt: Muslims and Christians at their First Encounter, Arabic Papyri from the Erzherzog Rainer Collection, Austrian National library, Vienna*, Nilus Studien zur Kultur Ägyptens und des Vorderen Orients 15 (Vienna: Phoibos Verlag, 2009), 8.
54 On the interpretation of this text, see Papaconstantinou, "Administering," 65–69.
55 Sijpesteijn, *The Arab Conquest of Egypt*, 446. This question is fully discussed in Sijpesteijn, *Shaping a Muslim State*, 217–257.
56 Federico Morelli, "Consiglieri e comandanti: I titoli del governatore arabo d'Egitto symboulos e amîr," *Zeitschrift für Papyrologie und Epigraphik* 173 (2010): 158–166, at 161–162.
57 *P.Apoll.* 36.v16 (ca. 660–680).
58 *P.Heid.* 11.488.8–9.
59 See also the "threatening letters" in *SB* 26.16358.8 (644 [?]) and *P.Apoll.* 38.3 (ca. 660–80).
60 *SB* 16.12575.4–13. The papyrus may well have been written in the early decades of Islamic rule, cf. Morelli, "ʿAmr e Martina," 152–154, esp. 154.
61 Arietta Papaconstantinou, "The Rhetoric of Power and the Voice of Reason: Tensions between Central and Local in the Correspondence of Qurra ibn Sharīk," in *Official Epistolography and the Language(s) of Power: Proceedings of the 1st International Conference of the Research Network Imperium and Officium, Comparative Studies in Ancient Bureaucracy and Officialdom. University of Vienna, 10–12 November 2010*, ed. Stephan Procházka, Lucian Reinfandt, and Sven Tost (Vienna: Österreichische Akademie der Wissenschaften, 2015), 267–281.
62 Naïm Vanthieghem and I will discuss this in more detail in our forthcoming study of the letters of Qurra b. Sharīk.
63 For interpreting oath formulae in a similar vein, see Arietta Papaconstantinou, "'What Remains Behind': Hellenism and Romanitas in Christian Egypt after the Arab Conquest," in *From Hellenism to Islam: Cultural and Linguistic Change in the Roman Near East*, ed. Hannah M. Cotton, Robert G. Hoyland, and Jonathan J. Price (Cambridge: Cambridge University Press, 2009), 447–466, at 456 and also Jean Gascou, "L'Égypte byzantine (284–641)," in *Le Mond Byzantin*, vol. 1, *L'Empire Romain d'Orient*, ed. Cécile Morrisson, Nouvelle Clio (Paris: Presses Universitaires de France, 2004), 403–436, at 436.
64 *P.Lond.Copt.* 1079.14–15.

65 Gascou, "Arabic Taxation," 673–674.
66 See e.g. *CPR* 4.80.14, *CPR* 4.103.8, and *CPR* 4.110.7.
67 *SB* 6.8987.39–40.
68 *P.Mich.* 13.662.12.
69 See *SB* 28.17202.2 from the year (671), cf. also the discussion in Gonis and Worp, "*P. Bodl.* I 77."
70 See the detailed discussion in Papaconstantinou, "'What Remains Behind'," and cf. also Gascou, "L'Égypte byzantine (284–641)," 436.
71 *P.Apoll.* 37.9–11.
72 *P.Mich.inv.* 2102, line 1. See the edition in Lajos Berkes and W. Graham Claytor, "Hypatios, Kulayb, and the Requisition of Milk: A Letter from the Senouthios Archive," *Zeitschrift für Papyrologie und Epigraphik* 203 (2017): 223–226.
73 Lajos Berkes, "'Peace Be upon You': Arabic Greetings in Greek and Coptic Letters Written by Christians in Early Islamic Egypt," in *Ties that Bind: Mechanisms and Structures of Social Dependency in the Early Islamic Empire*, ed. Edmund Hayes and Petra Sijpesteijn (forthcoming).
74 Foss, "The Persians," 158.
75 Ibid., 170; Papaconstantinou, "Administering," 65 argues: "compared to the brutality with which Roman power was re-asserted after the recovery of Egypt from the Persians, which left long-lasting traumatisms among the local population, it is very questionable whether the attitude of the Arabs at their arrival had such a powerful effect. Arguably, if the Arabs did have a deliberate policy on this matter, it was to make themselves as little felt as possible, so as to mark their contrast with Heraclius and come across as much more consensual."

References

al-Qāḍī, Wadad. "Non-Muslims in the Muslim Conquest Army in Early Islam." In *Christians and Others in the Umayyad State*, edited by Antoine Borrut and Fred M. Donner, 83–127. Late Antique and Medieval Islamic Near East 1. Chicago: Oriental Institute of the University of Chicago, 2016.

Bagnall, Roger S., and Bernhard Palme. "Franks in Sixth-Century Egypt." *Tyche* 11 (1996): 1–13.

Bagnall, Roger S., and Klaas A. Worp. *Chronological Systems of Byzantine Egypt*. 2nd ed. Leiden: Brill, 2014.

Berkes, Lajos. "Introduction: A Papyrological Perspective on Christians and Muslims in Early Islamic Egypt." In *Christians and Muslims in Early Islamic Egypt*, edited by Lajos Berkes, 1–9. American Studies in Papyrology 56. Ann Arbor: University of Michigan Press, 2022.

Berkes, Lajos. "'Peace Be upon You': Arabic Greetings in Greek and Coptic Letters Written by Christians in Early Islamic Egypt." In *Ties That Bind: Mechanisms and Structures of Social Dependency in the Early Islamic Empire*, edited by Edmund Hayes and Petra Sijpesteijn, forthcoming.

Berkes, Lajos, and W. Graham Claytor. "Hypatios, Kulayb, and the Requisition of Milk: A Letter from the Senouthios Archive." *Zeitschrift für Papyrologie und Epigraphik* 203 (2017): 223–226.

Booth, Phil. "The Muslim Conquest of Egypt Reconsidered." In *Constructing the Seventh Century*, edited by Constantin Zuckerman, 639–670. Travaux et mémoires 17. Paris: Association des Amis du Centre d'Histoire et Civilisation de Byzance, 2013.

Boud'hors, Anne. "Babylone-Fusṭāṭ dans les sources papyrologiques coptes." In *Christians and Muslims in Early Islamic Egypt*, edited by Lajos Berkes, 53–64. American Studies in Papyrology 56. Ann Arbor: University of Michigan Press, 2022.

Calament, Florence. "Un exceptionnel témoin épigraphique de la dernière occupation perse en Égypte." *Journal of Coptic Studies* 23 (2021): 1–13.
Cromwell, Jennifer. "Coptic Texts in the Archive of Flavius Atias." *Zeitschrift für Papyrologie und Epigraphik* 184 (2013): 280–288.
De Jong, Janneke. "Arabia, Arabs, and 'Arabic' in Greek Documents from Egypt." In *New Frontiers of Arabic Papyrology*, edited by Sobhi Bouderbala, Sylvie Denoix, and Matt Malczycki, 1–27. Islamic History and Civilization 144. Leiden: Brill, 2017.
Demiri, Lejla, and Cornelia Römer. *Texts from the Early Islamic Period of Egypt: Muslims and Christians at their First Encounter, Arabic Papyri from the Erzherzog Rainer Collection, Austrian National library, Vienna*. Nilus Studien zur Kultur Ägyptens und des Vorderen Orients 15. Vienna: Phoibos Verlag, 2009.
Drescher, James. "A Widow's Petition." *Bulletin de la Société d'Archéologie Copte* 10 (1944): 91–96.
Foss, Clive. "The Persians in the Roman Near East (602–630 AD)." *Journal of the Royal Asiatic Society* 13 (2003): 149–170.
Foss, Clive. "Egypt under Mu'āwiya. Part I: Flavius Papas and Upper Egypt." *Bulletin of the School of Oriental and African Studies* 72 (2009): 1–24.
Fournet, Jean-Luc. "Papyrologie Grecque." *Annuaires de l'École pratique des hautes études* 139 (2008): 92–93.
Fournet, Jean-Luc. "L'impact de la conquête sassanide sur l'Égypte: Notes lexicographiques." In *Mélanges Bernard Flusin*, edited by André Binggeli and Vincent Déroche, 287–298. Travaux et Mémoires 23. Paris: Association des Amis du Centre d'Histoire et Civilisation de Byzance, 2019.
Gascou, Jean. "L'Égypte byzantine (284–641)." In *Le Mond Byzantin*. Vol. 1, *L'Empire Romain d'Orient*, edited by Cécile Morrisson, 403–436. Nouvelle Clio. Paris: Presses Universitaires de France, 2004.
Gascou, Jean. "Notes critiques: *P.Prag.* I 87, *P.Mon. Apollo* 27, *P.Stras. VII* 660." *Zeitschrift für Papyrologie und Epigraphik* 177 (2011): 243–253.
Gascou, Jean. "Arabic Taxation in the Mid-Seventh Century Greek Papyri." In *Constructing the Seventh Century*, edited by Constantin Zuckerman, 671–677. Travaux et mémoires 17. Paris: Association des Amis du Centre d'Histoire et Civilisation de Byzance, 2013.
Gonis, Nikolaos. "Notes on the Aristocracy of Byzantine Fayum." *Zeitschrift für Papyrologie und Epigraphik* 166 (2008): 203–210.
Gonis, Nikolaos. "*P.Paramone* 18: Emperors, Conquerors and Vasalls." *Zeitschrift für Papyrologie und Epigraphik* 173 (2010): 133–135.
Gonis, Nikolaos, and Klaas A. Worp. "*P.Bodl.* I 77: The King of Kings in Arsinoe under Arab Rule." *Zeitschrift für Papyrologie und Epigraphik* 141 (2002): 173–176.
Hagedorn, Dieter, and Fritz Mitthof. "*P.Paramone* 18: Ein neu platziertes Fragment oder: Cave restauratorem chartarum." *Zeitschrift für Papyrologie und Epigraphik* 149 (2004): 157–158.
Hickey, Todd M. *Wine, Wealth, and the State in Late Antique Egypt: The House of Apion at Oxyrhynchus*. Ann Arbor: University of Michigan Press, 2012.
Keenan, James G. "The Names Flavius and Aurelius as Status Designations in Later Roman Egypt." *Zeitschrift für Papyrologie und Epigraphik* 11 (1973): 33–63 and 13 (1974): 283–304.
Legendre, Marie. "Aspects of Umayyad Administration." In *The Umayyad World*, edited by Andrew Marsham, 133–157. New York: Routledge, 2020.
Morelli, Federico. "ʿAmr e Martina: La reggenza di un'imperatrice o l'amministrazione araba d'Egitto." *Zeitschrift für Papyrologie und Epigraphik* 173 (2010): 136–157.
Morelli, Federico. "Consiglieri e comandanti: I titoli del governatore arabo d'Egitto *symboulos* e *amîr*." *Zeitschrift für Papyrologie und Epigraphik* 173 (2010): 158–166.

Palme, Bernhardt. "Political Identity versus Religious Distinction? The Case of Egypt in the Later Roman Empire." In *Visions of Community in the Post-Roman World: The West, Byzantium and the Islamic World, 300–1100*, edited by Walter Pohl, Clemens Gantner, and Richard Payne, 81–98. Cambridge: Ashgate, 2011.

Papaconstantinou, Arietta. "'What Remains Behind': Hellenism and Romanitas in Christian Egypt after the Arab Conquest." In *From Hellenism to Islam: Cultural and Linguistic Change in the Roman Near East*, edited by Hannah M. Cotton, Robert G. Hoyland, and Jonathan J. Price, 447–466. Cambridge: Cambridge University Press, 2009.

Papaconstantinou, Arietta. "Administering the Early Islamic Empire: Insights from the Papyri." In *Money, Power and Politics in Early Islamic Syria. A Review of Current Debates*, edited by John Haldon, 57–74. Farnham: Ashgate, 2010.

Papaconstantinou, Arietta. "The Rhetoric of Power and the Voice of Reason: Tensions between Central and Local in the Correspondence of Qurra ibn Sharīk." In *Official Epistolography and the Language(s) of Power: Proceedings of the 1st International Conference of the Research Network Imperium and Officium, Comparative Studies in Ancient Bureaucracy and Officialdom. University of Vienna, 10–12 November 2010*, edited by Stephan Procházka, Lucian Reinfandt, and Sven Tost, 267–281. Vienna: Österreichische Akademie der Wissenschaften, 2015.

Sänger, Patrick. "Saralaneozan und die Verwaltung Ägyptens unter den Sassaniden." *Zeitschrift für Papyrologie und Epigraphik* 164 (2008): 191–201.

Sänger, Patrick. "The Administration of Sassanian Egypt: New Masters and Byzantine Continuity." *Greek, Roman, and Byzantine Studies* 51 (2011): 653–665.

Sänger, Patrick, and Dieter Weber. "Der Lebensmittelhaushalt des Herrn Saralaneozan/Šahr-Ālānyōzān: Neuedition von zwei Speiselisten und einem Geschäftsbrief aus dem sassanidischen Ägypten." *Archiv für Papyrusforschung* 58 (2012): 81–96.

Schmidt, Stefanie. "*P.Bas.* II 69 and 70: A Look Behind the Text." *Archiv für Papyrusforschung* 64 (2018): 324–342.

Schmidt, Stefanie. "Adopting and Adapting: Zur Kopfsteuer im frühislamischen Ägypten." In *Proceedings of the 28th International Congress of Papyrology, August 1–6, 2016, Barcelona*, edited by Alberto Nodar and Sofía Torallas Tovar, 586–593. Barcelona: Publicacions de l'Abadia de Monserrat, 2019.

Schmidt, Stefanie. "Between Byzantine and Muslim Egypt: Mobilizing Economic Resources for an Embryonic Empire." *Journal of Ancient Civilizations* 35 (2020): 241–266.

Sijpesteijn, Petra M. "The Arab Conquest of Egypt and the Beginning of Muslim Rule." In *Egypt in the Byzantine World, 300–700*, edited by Roger S. Bagnall, 437–459. Cambridge: Cambridge University Press, 2007.

Sijpesteijn, Petra M. "Landholding Patterns in Early Islamic Egypt." *Journal of Agrarian Change* 9 (2009): 120–133.

Sijpesteijn, Petra M. *Shaping a Muslim State: The World of a Mid-Eighth-Century Egyptian Official*. Oxford Studies in Byzantium. Oxford: Oxford University Press, 2013.

Van der Vliet, Jacques. "Pisenthios de Coptos (569–632): Moine, évêque et saint. Autour d'une nouvelle édition de ses archives." In *Autour de Coptos: Actes du colloque organisé au Musée des Beaux-Arts de Lyon (17–18 mars 2000)*, 61–72. Topoi Supplément 3. Paris: De Boccard, 2002.

Weber, Dieter. "Die persische Besetzung Ägyptens 619–629 n. Chr.: Fakten und Spekulationen." In *Ägypten und sein Umfeld in der Spätantike: Vom Regierungsantritt Diokletians 284/285 bis zur arabischen Eroberung des Vorderen Orients um 635–646*, edited by Frank Feder and Angelika Lohwasser, 221–246. Philippika 61. Wiesbaden: Harrassowitz, 2013.

6

ALEXIOS, EMPEROR OF THE DIASPORAS?

The Komnenian Revolt of 1081 and Foreign Military Groups in Byzantium

Roman Shliakhtin[1]

In the early morning of 1 April 1081, the citizens of Constantinople heard loud sounds from the area of the Charisius Gate. There, the army of the rebellious general Alexios I Komnenos (1081–1118) had entered the city through the tower guarded by a group of foreigners called *Nemitzoi* and crushed the remnant forces of emperor Nikephoros III Botaneiates (1078–1081).[2] Soldiers of the Komnenoi plundered the city for three subsequent days.[3] The daughter of Alexios, Anna Komnene (1082–before 1155), described this event in some detail:

> The whole army (that was composed of foreign and native troops and had come together from home and faraway lands) knew that the city had for a long time been crammed with all kinds of riches.... They spared not a single house, not a single church .... The natives ... apparently forgetting themselves, changed their manners for the worse and did themselves exactly the same things as the barbarians.[4]

These "barbarians" included the Turks and most probably members of "Frankish" diasporas as well as some soldiers from the Caucasus. This chapter aims to provide some basic information about armed foreigners in Byzantine service and investigate possible reasons behind the attention that Anna Komnene pays to these armed foreigners in the wake of the 1081 revolt.[5] The first part of the chapter will trace the different foreign groups in Byzantine service during the eleventh century, the second will analyze the revolt of Komnenoi in the *Alexiad*, the third will follow the history of different foreign military groups in twelfth-century Byzantium, and the final part will frame the *Alexiad* against the background of the mid-twelfth century in order to reconstruct the message that Anna Komnene wished to convey to her audience.

DOI: 10.4324/9781003245568-7

Foreign Military Groups in Pre-Komnenian Byzantium (1025–1081)

Diasporá is an Ancient Greek word that originally meant groups of the people of Israel living in Egypt and other foreign lands. The Byzantines seldom used this word and, until recently, Byzantinists did not analyze diasporas within Byzantium.[6] For the sake of clarity, I use here the definition of military diasporas given in the introduction to this volume, namely "groups with an ethnic or para-ethnic (imagined ethnic) background serving in foreign land that were more or less far away from their point of origin". At the same time, it is important to note that there are many other definitions of a diaspora. An entry in the *Oxford Bibliographies in Anthropology*, for example, describes a diaspora as a "term used to describe the mass, often involuntary, dispersal of a population from a centre (or homeland) to multiple areas, and the creation of communities and identities based on the histories and consequences of dispersal".[7] While these two definitions are not exhaustive, it seems interesting to use them as analytical frameworks for the different foreign units in the Byzantine army and to check whether they are fitting for the situation in 1081.[8] Before applying the term "military diaspora" to Alexios Komnenos' army of 1081, however, I will introduce several ethnic groups that played an important role in Byzantine military history in the eleventh century.

In the beginning of the eleventh century, the Byzantine Empire effectively solved all of the problems existing on the western and eastern borders. The absence of major military confrontations in the middle of the eleventh century allowed Byzantine emperors to ease pressure on the war chests and outsource defence.[9] This stimulated discontent among the military professionals and, together with other factors, led to the series of military revolts. In 1057, Isaac Komnenos, uncle of Alexios, led a rebellion against Michael VI, whereby he captured Constantinople and ruled as emperor until 1159. The next emperor, Constantine X Doukas ruled from 1059 to 1067. After his death, the throne passed to Eudokia Makrembolitissa who ruled as a regent for a short period of time (1067–1068) and then married the prominent general Romanos Diogenes, who reigned from 1068 to 1071. The disastrous battle of Manzikert (1071) allowed the Doukai to recapture the throne. Michael VII Doukas ruled from 1071 until 1078, when he was deposed by a coalition of Anatolian magnates led by Nikephoros Botaneiates. His rule from 1078 to 1081, in turn, was wracked by rebellions, even for this turbulent period, such as the continuous uprising in the Empire's western domains led first by Nikephoros Bryennios the Elder (1077–1078) and then by Nikephoros Basilakes (1078) and the revolt of Nikephoros Melissenos in the East (1080). The rule of Botaneiates was itself ended by the revolt of Alexios I Komnenos in 1081.

The military rebellions constituted the disintegration of old army units. The disastrous battle of Manzikert (1071) contributed to the decline of standing units. What remained were members of the military aristocracy who either survived

Manzikert or did not participate in the battle like the future emperor Alexios Komnenos. These members of the aristocracy gathered around them in small groups of soldiers. These soldiers and their commanders often choose to wait out the fight for the throne in Constantinople. In the 1070s, Kekaumenos, an experienced officer, gave detailed advice on how to wait out a rebellion without participating in it.[10] This turbulent period, which Speros Vryonis has labelled as a "political and military collapse", coincided with growing external threats.[11] The new simultaneous threats from the Pechenegs, the Seljuks, and the Normans led to a significant rise in the number of "external contractors" in the imperial army. As Vryonis wrote, the Byzantine army at the end of the eleventh century was notable for the presence of a "widening ethnic array" of troops.[12] But unfortunately, information about some of these groups is limited by the imprecision of the sources, especially regarding their names and identities. Thus, precise information about the payment of foreign troops or their specific role in the changing structure of the Byzantine army is often difficult to come by.[13] In the following paragraphs, I will summarize the data about some of these groups relevant to the present chapter.

Nevertheless, it is clear that many soldiers from various ethnic groups served in the imperial bodyguard, known today as the "Varangian Guard". This is a later label that nineteenth-century scholars applied to mercenaries from Norway, England, and Kievan Rus'. These mercenaries performed the functions of honorary guards in the Great Palace and fought for the Empire as heavy infantry or marines when required.[14] Thus, one should not speak of an ethnic group of "Varangians", but rather about the combination of many members of many ethnic groups in one military and social body which performed certain court functions (as bodyguards), carried special weapons (first swords, later axes) and lived in the imperial palace. The superior commander of the guard also acted as an interpreter and participated in the military councils of the emperor.[15] In the beginning of the eleventh century, many guardsmen came from the territories of Rus', but by the middle of the eleventh century, they were mostly from Scandinavia (with Harald Hardrada being the most famous example). After the Norman conquest of England in 1066, many Anglo-Saxons found refuge in Constantinople and joined the guard, while the later Varangians in the early twelfth century mostly came from Denmark and Rus'.[16] It seems likely that they intermarried with local people and in the eleventh century, the Varangians had permission to conduct separate church services at a special shrine dedicated to St. Olaf of Norway in Constantinople, famous for the presence of the saint's sword.[17]

The most important division of foreigners *in stricto sensu* were the people whom the Byzantine *literati* called "Franks" or "Latins". One can hardly call them a coherent group, because Byzantine writers used many different terms for many different groups. In the eleventh century, the term "Latin" or "Frank" denoted a person from the West in general, while separate sub-groups had their own names, like the *Nemitzoi*.[18] Since the 1050s, people from the countries of

Western Europe entered Byzantine service in large numbers.[19] From the 1070s, the Byzantine Emperors, especially Romanos IV Diogenes, used Frankish mercenary companies in their campaigns against the Seljuk Turks, where they became especially renowned for their cavalry charges. As Alexander Kazhdan has noted, Herve Frankopoulos (lit. "the son of a 'Frank'") was one of the leading commanders of the Byzantine army in the 1060s.[20] In the 1070s, another "Frank", Russeil de Ballieuil, with a company of mercenaries, rebelled against the Byzantines and carved out a principality in Asia Minor, with its centre at Ankara, for himself. Despite Ballieuil's rebellion, however, the behaviour of the Franks was generally regarded in a positive manner by Byzantine contemporaries.[21] It was around this time that Emperor Michael IX Doukas introduced another group of "Frankish" mercenaries named *Nemitzoi*, who were stationed in the capital.[22] The contemporary chronicler Michael Attaleiates tells us that the term *Nemitza* referred to a "lower part of Gaul".[23]

Eleventh-century Byzantium also attracted people from the eastern fringes of the Empire, who can be identified in three clear groups.[24] The most prominent group were the Armenians, both Chalcedonian and non-Chalcedonian. Many of these served the Empire in the ninth and tenth centuries, and the prominent eleventh-century Byzantine general and rebel, Katakolon Kekaumenos (not to be mixed up with another Kekaumenos, the author of military treatise), was of Armenian origin.[25] In the middle of the eleventh century, the Byzantine Empire resettled many Armenian princes from the Caucasus into the Balkans and Anatolia, exchanging their historical lands for new estates.[26] In the East, the Byzantines allowed Armenians to serve in specially assigned units primarily employed in the wars against the Seljuk Turks.[27] The Armenians settled in Anatolia, gained prominence in the reign of Romanos IV Diogenes, and supported him even after his defeat at Manzikert.[28] A key figure amongst the Armenian notables was the governor of Antioch, Philaretos Brachamios. He became a de facto independent ruler in 1091 and then later defected to the Turks, attracting disdainful comments from Anna Komnene.[29] Although they were present in the 1070s, the Armenians were absent as a group in the army during the Komnenian revolt.

The second eastern military group was those of Caucasian origin, namely those referred to as "Alans" who served Byzantine nobles as cavalrymen and bodyguards from the late eleventh century.[30] The very same epithet was used to denote Orthodox nobility of Georgian origin, who operated on a very different level of the social hierarchy from the Alan soldiers. The most prominent among them was Maria from the Bagrationi family, whom the Byzantines called Maria of Alania, the wife of Michael VII Doukas (1071–1078).[31] After the dissolution of her marriage with Michael VII, Maria married the next emperor, Nikephoros III Botaneiates (1078–1081), and supported the Komnenoi in their bid to seize the imperial throne.

Besides Alans and Armenians, there was a third clearly identifiable eastern group with military affiliations and something that one can call "self-identification".

These people called themselves "Iberians", after the Iberian theme on the north-eastern frontier of the Empire, which incorporated some regions of present-day Georgia in Caucasus. The most prominent among the Iberians was Gregory Pakourianos (fl. 1064–1086), an important general in the service of Alexios Komnenos with a wide network of family connections in Byzantium, Anatolia, and the Caucasus.[32] Very much like the Armenian magnates of Asia Minor, the Pakourianai served the Empire for some generations before the Komnenoi came to power. During his career, Pakourianos fought for the Empire on both its the eastern and western borders. In the 1080s, he waged regular campaigns against the Pechenegs where he was captured and subsequently released. After obtaining his freedom, Pakourianos was then appointed *megas domestikos* of the West and became third in command of the Byzantine army. He died in 1086 in a battle with the Pechenegs in the vicinity of Adrianople.[33]

Fortunately, the sources provide detailed information about Gregory Pakourianos' life, especially his monastic donations. In the 1070s, Gregory and his brother Aspasios donated sums of money to the Georgian monastery of Iviron at Mount Athos,[34] and in 1083, Gregory founded the monastery of Petrizonitissa in present-day Bachkovo in Bulgaria.[35] One of the aims of the foundation was a provision for his family members. In the *typicon* of this monastery, Gregory Pakourianos stated that he came "from the most illustrious clan of Iberians" (ἐκ τῆς τῶν Ἰβήρων παμφανεστάτης φυλῆς). Pakourianos highlighted the "Georgian" character of the monastery and specially prohibited "Roman" (Greek) priests and monks from entering the coenobitic monastery that was reserved exclusively for the community with which Pakourianos identified himself.[36] In the *Alexiad*, Anna Komnene stated that Pakourianos was a descendant of the "Armenian clan".[37] There is no contradiction here: The Pakourianai were an ancient family of mixed, Armeno-Georgian origin and Chalcedonian religious affiliation whose descendant Gregory Pakourianos identified himself as Iberian.[38] Three years after Pakourianos' death, another member of the same clan donated a significant tract of land to the monastery of Iviron at Mount Athos.[39] This monastery has a long story of extensive connections with the principality of Tao-Klarjeti in the Caucasus. Thus, one can note the presence of a certain group of "Iberians" in eleventh-century Byzantium, who were present in the army as a group of military leaders. Interestingly, Maria of Alania was spatially close to this group in the later part of her life, building her own monastery next to that one of Pakourianos.[40]

As one can see, by the end of the eleventh century, the Iberians were well-rooted in Byzantium. The eleventh century also saw the emergence of two new groups that came from two different communities of Turkic-speaking semi-nomads. The first group were the Pechenegs (often named in Byzantine sources as "Scythians") who appeared in the Byzantine army in the 1060s. They served as light cavalrymen, fought with bow and arrow, and were mostly hired on a seasonal basis.[41] The second group were the Seljuk Turks, who entered Byzantine service in 1069. Byzantine generals, including Alexios Komnenos, used them as light cavalry as well. Both Pechenegs and Turks fought under command of their own officers. A certain Erisgen/Arsigi, a Seljuk "prince" known in Byzantium as Chrysoskoulos, an exile from the sultanate of the Great Seljuks, entered the service of Romanos Diogenes in 1069 and received from him the high rank of *proedros*.[42] He was not the only Turk in Byzantine service. In the 1070s, a group under the princeling Sulaiman ibn Qutlamish migrated to Asia Minor. In the civil war between the Doukai and Nikephoros Botaneiates, Sulaiman supported Botaneiates and gained a foothold in Bithynia. In 1078, the Turks under Sulaiman then fought for Alexios Komnenos against another pretender Nikephoros Bryennios the Elder at Kalavrye.[43]

In the second half of the eleventh century, Byzantium was therefore defended by many foreign military groups of different origin, some of which fit the framework of a military diaspora as outlined in this volume. The rise of these groups caused a certain uneasiness among Byzantine nobles. The Byzantine strategist, Kekaumenos, for example, advised his readers in the 1070s to "Neither raise foreigners who are not from the royal kin of their country into great dignities nor entrust them with great positions of command".[44] His advice, however, fell on deaf ears. After Manzikert and the internal warfare of the 1070s, foreign contingents (*ethnikoi*) became *the substitute* for a standing army. In an age of rising inflation, payment to foreign soldiers became a question of life and death.[45] In the year 1080, emperor Nikephoros III Botaneiates launched a new census in Macedonia that led to a substantial tax increase.[46] He also granted the monastery of Vatopedi at Mt. Athos a privilege, exempting it from any payments which would be used for the hiring of different armed groups. The catalogue of foreign groups in this particular document includes "the Ros', Varangians, Kolpings,[47] Inglinings [Englishmen], Franks, Bulgarians, Saracens".[48]

This list gives an idea about the variety of military groups that existed in Byzantium in the 1080s. Some of these groups are more visible in the sources while others are only alluded to in lists such as the one quoted above. In the eleventh century, at least one group of common geographical origin—the Iberians—had in one decade a combination of a military leader (Gregory Pakourianos), separate religious structures (the monasteries of Iviron and Petrizonitissa), and support in the palace (from Maria of Alania). Other military groups, like the Franks, Turks, and Pechenegs formed less stable communities, but still fulfilled crucial military functions, such as by playing the roles of shock troops and reconnaissance forces. Other groups and actors (such as the Turks of Sulaiman and the Franks of Russeil) used the time of crisis to carve out their domains. Some of them succeeded and created their own domains (Sulaiman) while others lost their position, life, and disappeared from the scene with their followers choosing different sides in the ensuing chain of conflicts. The protagonist of the *Alexiad*, Alexios I Komnenos had to overcome all these groups and unite them under one banner. One of his challenges was the rebellion of 1081.

Foreigners in Komnenian Revolt of 1081 as Seen by Anna Komnene

When comparing the two biographies of Alexios I Komnenos—*The Historical Material* by Nikephoros Bryennios the Younger and *The Alexiad* by Anna Komnene—it is clear that both Anna (Alexios' daughter) and Bryennios (Anna's husband and Alexios' son-in-law) were keen to display Alexios as a good manager of foreigners.[49] Alexios began his military career during the reign of Michael VII Doukas whose cousin, Eirene Doukaina, he married in 1078.[50] In his *Historical Material* Nikephoros Bryennios described the adventures of young Alexios in Anatolia while Anna Komnene focused on her father's ascension to the throne and his rule in Constantinople.

After the military coup of 1079, Alexios Komnenos served Nikephoros III Botaneiates and assisted him in subduing the rebellions of Nikephoros Bryennios the Elder and Nikephoros Basilakes. In both cases, Alexios, with the title of *domestikos* of the West, commanded armies of different origins that included Franks and Turks. In 1080, the relations between Alexios and emperor Nikephoros Botaneiates were strained.[51] In the *Alexiad*, Anna Komnene narrates that two slaves "of Slavic origin" (Σκλαβογενῶν), Borilos and Germanos, who held a grudge against Alexios and Isaak Komnenos, set Nikephoros III Botaneiates against the Komnenoi.[52] Thus, according to Anna, it was the actions of these foreigners that were the starting point of the Komnenoi rebellion.

The first foreigners to support the claim of the Komnenoi were the relatives of the empress Maria of Alania. An unknown *magistros* gained for Alexios some information about the events in Constantinople and advised him not to approach the capital.[53] The first person whom the Komnenoi invited to join the plot was the above-mentioned Gregory Pakourianos, the leader of the Iberians, who Alexios persuaded to support the rebellion. Soon Alexios managed to secure help from another "warlike" man, a Frank called Oumbertopoulos ("the son of Umberto" similar to Frankopoulos mentioned above). To secure this new alliance, Alexios and Oumbertopoulos swore oaths to support one another, after which Alexios returned to the capital to pursue his aims.[54] As Anna Komnene reports, the success of the recruitment was facilitated not only by her fathers' bravery but also by her fathers' generosity, for "they [Pakourianos and Oumbertopoulos] also loved him because he was exceptionally generous and very ready to give, although he had not a great abundance of money".[55] In the time of crisis, leaders of diasporas did not act without some payment as an advance.

After many negotiations, another prominent noble, Gregory Palaiologos, joined the ranks of the rebels together with his mother-in-law, who was of Bulgarian descent.[56] They sent a messenger to recruit an exiled member of the Doukai family and grandfather of Alexios' wife Eirene, Caesar John Doukas, who lived in one of his estates as a monk.[57] The latter, despite his old age, decided to assist the Komnenoi. On his way to the rebels, John Doukas secured funds for the rebellion by stopping and effectively robbing a tax official. In addition,

Doukas also secured the support of another foreign military group, the Seljuk Turks, whom he met on the way to Constantinople. Anna Komnene wrote that these Turks joined him and that "John Doukas demanded that their leaders take an oath, wishing to confirm the deal. They swore an oath according to their custom to be the allies of the Komnenos in battle".[58]

In this case, Anna reiterates the importance of cash and mercenaries for the success of the rebellion. The terms Anna used to describe the promises of John Doukas to the Turks are the same she previously used to describe the generosity of her father. These Turks were the last group that the Komnenoi managed to recruit. After securing the support of several Byzantine nobles, the plotters and their allies from the aforementioned military diasporas advanced to Constantinople and held a military council in close proximity to the walls of the city. According to Anna Komnene, there they realized that the combination of many "ethnic" groups within one army posed a problem: "their forces consisted of foreign troops as well as local ones (ξενικῶν και ἐγχωρίων). Where there is a crowd of people of different origin, there will be a voice of discontent".[59] Indeed, at that point, the rebel army included, at the very least, Iberians, Franks, and Turks.

After the military council, Alexios forced the above-mentioned John Doukas to accompany him on a ride before the walls in his monastic garb in order to learn of the disposition of Botaneiates' regiments on the walls. According to Anna, the army loyal to the ruling emperor was no less diverse than the Komnenoi coalition:

> Caesar John learned that at these points the defenders were the so-called Immortals (this was the emperor's personal bodyguard?), close to them the Varangians from Thule (here I speak about the axe-carrying barbarians), in another place—the Nemitzoi (this is a barbaric nation which for a long time served the empire of the Romans). So he urged Alexios to forget the Varangians, as well as the Immortals. For on the one hand, the Immortals were compatriots of the emperor and, naturally, being very loyal to the emperor, they would better lose their souls or suffer something worse than be persuaded to do something against him. On the other hand, those who carry swords on their shoulders [the Varangians] showed their traditional valour, [namely] the trust and loyalty to the emperors (sic!) as their bodyguards who inherited [this office] from one to another, keeping faithfully their unshaken loyalty. They will hardly hold up to the words of treachery. But had Alexios turned to the Nemitzoi, he would be close to accomplishing his aim, for he might profit from their hold of the tower that gave access to the city.[60]

As one can see, the defenders of the city were not natives of Constantinople. The first group were Immortals, a group of select Anatolian warriors whose history goes back to the tenth century. During the reign of Michael Doukas, they had been turned into a body of elite cavalry, which Nikephoros Botaneiates inherited from his predecessor.[61] In 1078, Botaneiates used Immortals against Nikephoros Bryennios, the Elder and Alexios Komnenos, in his capacity as a general, commanded them in the battle of Kalavrye. During the Komnenian revolt, the Immortals took the side of the emperor, and Anna Komnene had to explain to her audience the political background of this group.[62] The second group were Varangians and Anna again had to explain to her audience in the classicizing way that these Varangians were from Thule, a mystical place in the North.[63] The last group mentioned are *Nemitzoi*, who are openly called "barbaric" and people who had long served the Romans.

The outcome of this confrontation depended on the problematic allegiance of the groups on the walls and the key to the city thus lay in their loyalty or disloyalty to the emperor. In the *Alexiad*, John Doukas advised Alexios to concentrate on his negotiation efforts on the *Nemitzoi*. Alexios followed his advice. The negotiations were quickly concluded: The leader of the barbaric *Nemitzoi* named Gilpraktos "agreed to betray the city soon".[64] The next day Alexios sent his confidante George Palaiologos to the tower of the *Nemitzoi* and after some delay, they opened the city gates. The multi-ethnic army of Alexios Komnenos therefore entered the city and while his warriors were plundering, he proclaimed himself emperor.

Context for the *Alexiad*: Foreigners in the Reigns of Alexios, John, and Manuel

The detailed and minute-by-minute description of the Komnenian revolt must have some explanation. Anna Komnene took pain to list all the different parties that participated in the siege of Constantinople, local and foreign alike. However, the focus of the description is not on the locals, be them the rebels or loyalists, but on the foreign armed groups. So why did Anna introduce the detailed analysis of all these foreign groups (and persons of foreign origin) into her narrative?

The reason may lie exactly in the time gap that separates the events of 1081 from the composition of the *Alexiad*. To understand the context in which the audience of Anna Komnene could read her message, there is a need to write a history of many foreign groups in the Byzantine army in the reigns of Alexios, John, and Manuel Komnenos.

During his long reign, Alexios I reformed and re-created the Byzantine army, mustered many new "native" divisions from Asia Minor and altered the role of foreign groups.[65] The shock cavalry troops of the early Komnenian army consisted of Latins. Alexios I Komnenos advised his generals to use their speedy cavalry charges against the Turks of Bithynia who soon stopped being Alexios' allies and became his enemies.[66] *Tourkopouloi*, the "sons of the Turks", instead filled the role of the Bythinian Turks in the army. They served in the capacity of light cavalrymen in the Byzantine armies, which fought the Turks in Bithynia alongside the armies of the First Crusade and their example probably contributed to the development of analogous forces in the Crusader States.[67] Their exact geographical origins, however, remain obscure. After the disappearance of *Tourkopouloi*, the Komnenoi used the mercenaries of the Black Sea region, the Alans, who served as mounted archers and scouts.[68] At the same time, some of the Turks, especially of noble origin served the Byzantine Empire. A captive from Nicaea, John Axouchos, made a career in the army and became *megas domestikos* (commander-in-chief) of the imperial forces in the 1130s.[69]

A certain fluctuation happened also in another part of the army, in the Varangian Guard. Many of those who stood on the walls against Alexios fell whilst fighting the Normans in the battle of Dyrrhachium (1081) after which Alexios had to rebuild their numbers.[70] After this war, Alexios enlisted some Norman nobles into Byzantine service.[71] The Norman Wars also forced Alexios to restore the damaged Byzantine fleet. Being unable to raise the needed number of ships, Alexios outsourced the navy to the Venetians.[72] In exchange for their assistance, the Venetians received trade privileges and territories in Constantinople, near the southern end of the present-day Galata bridge.[73]

The simultaneous struggle with the Normans in the West, the Turkic polities in the East and the Pechenegs in the North demanded a constant influx of human resources. While older groups like the Immortals disappeared from the scene, Alexios I invited new foreign warriors to his Empire, using every possibility to enlist them to his ranks. According to Peter Frankopan, the consequence of the events that eventually led to the First Crusade began with what was originally an emotional invitation for mercenaries.[74] During the First Crusade, Alexios not only mobilized all available ethnic groups to control the crusaders but also famously made the leaders of the crusade armies swear an oath of fealty to him.[75] After the siege of Nicaea and during the siege of Antioch, some of the crusaders, like the famous Petraliphai brothers, entered Byzantine service.[76] The same holds true for the Turks who entered Byzantine service both before and after the First Crusade. The most notable of them was John Axouchos, who was appointed as *megas domestikos* (commander-in-chief) of the Byzantine military by John II Komnenos.[77]

In contrast to the eleventh-century Franks, members of the Norman and crusader elite integrated into the emerging Empire of the Komnenoi through the network of intermarriages in the higher strata of local nobility.[78] Alexios himself used intermarriages to leverage his alliances and married his son John to princess Piroshka, daughter of the Hungarian king Ladislas I. During the reign of John II (1118–1143), the Byzantine Empire continued to attract many foreigners to its military, who came from both East and West. The attitude of the *literati* to these foreigners remained ambiguous. On the one hand, the foreigners were present in the lists of imperial enemies, on the other, the sources report that some of them served the Empire well. In the beginning of the reign of John II, a notable panegyrist Theodore Prodromos mentioned "trembling Alamans" and the "people of Germans" as possible enemies/partners of the Empire.[79] Later historians of the era praised "Italian" knights who served the Byzantines in the

1130s in Cappadocia against the Seljuk state of the Danishmendides.[80] They specialized in cavalry charges that dispersed the Turks. John II also used the old Byzantine practice of resettling captured enemies and using them as soldiers. According to John Kinnamos (who is favourable to John II in his descriptions) in the 1120s, this emperor defeated the Pechenegs, settled them in Byzantine lands and enlisted them into the ranks of the military.[81] In the 1130s, he allegedly did the same with the Serbs and the Turks of Kastamonu.[82] Very much like his father, John II attracted forces on the basis of short-time treatises such as the one with Mas'ud of Ikonion in the 1130s.[83]

Nobles of foreign origin were present at court in great numbers and occupied important positions. First and foremost, amongst them was a confidant of John II Komnenos, the aforementioned John Axouchos. He was of Seljuk origin, taken captive at Nicaea and grew up in the palace together with the emperor since they were roughly of the same age. If one believes Niketas Choniates, Axouchos played an important role in the succession of John II in 1118. During John II's rule, Axouchos was made *megas domestikos* of East and West and was effectively the commander-in-chief of the armed forces.[84] Surviving seals identify him as a person who tried to position himself as a Byzantine and a man of letters who corresponded with leading intellectuals of the era.[85] John Axouchos preserved his position in court until the 1150s. At around this time, a new monastery with the Seljuk name of Koutloumousiou appeared on Mount Athos testifying to the importance of Seljuk influence at the Komnenian court.[86] While the typicon of the monastery is absent, the very existence of it may be interpreted as a hint for the evidence of the Seljuk "diaspora" in Byzantium. However, the absence of an internal or external label for this group does not allow one to call them diaspora *stricto sensu*. From Anna Komnene, one might deduce that some members of the group were called "Persians".

At the same time, in the 1120s, the palace in Constantinople became a new home for another proto-diaspora. Some of them settled in the city during the reign of Alexios and by the 1130s, reached a peak of their influence. The external name for this group as given by John Kinnamos is "Italians".[87] The Norman exile John Roger was the husband of John II's daughter Anna and with the support of the other officers of "Norman" background constituted a group that was relatively close to the throne. A Norman princess, Eirene, was the wife of imperial son Andronikos Komnenos. She sponsored works of poetry that promoted ideas alternative to Byzantine discourse and maintained contact with Orthodox monks of Italian origin who lived in Bithynia.[88] One can therefore speak about the formation of an "Italian" proto-diaspora (military and spiritual) at the court of John II. This group patronized the same poets and literati that worked under the patronage of Eirene Komnene and Anna Komnene.

In 1143, the death of John II Komnenos altered the position of foreign groups in Constantinople. The "Italians" with John Roger at the helm attempted to usurp the throne, but Manuel I Komnenos (1143–1180) defeated them with the help of his relatives. The emperor then exiled them from the court.[89] Manuel continued his policies in attracting foreigners, with the focus on the group that was relatively close to him—the Latins.[90] By the moment of his coronation, Manuel already had foreigners in the family. In 1143, Manuel was betrothed to the German countess Berta of Sulzbach who was an adopted daughter of Emperor Conrad III.[91] The marriage that became real was a result of long negotiations between Emperor Conrad III and John II and was celebrated by a notable feast in Constantinople, which the court poet Theodore Prodromos duly described. In the poem, Prodromos defined Konrad as a "great king of Old and Ancient Rome" and focused mostly on the union of the two families and beauty of the bride.[92] One year after this marriage, Manuel became an emperor and took over formal command of the military forces, starting an expedition against Ikonion, capital of the sultanate of Rum. According to Kinnamos, during this expedition against the Seljuk sultan of Ikonion Manuel demonstrated great personal bravery. Being around 30 years of age, he was still under a "tyranny of his prime age" and committed feats of courage, including attacking the Turks in person with spear in hand.[93] Manuel also displayed a wish to demonstrate his valour after the wedding, which Kinnamos claimed was a Latin custom to impress his Latin bride. According to Kinnamos (who is sympathetic to Manuel) this aggressive and risky behaviour caused significant doubts among his immediate courtiers and led one of Manuel's relatives—his nephew John—to consider a usurpation.[94] As I will demonstrate later, this particular expedition might form part of the context against which the siege of Constantinople in the *Alexiad* was read.

Manuel's German wife probably assisted her husband during the complex and problematic dealings with the leaders of the Second Crusade when they crossed through Byzantium in 1147. As Magdalino has demonstrated, Manuel Komnenos managed to use tensions between the crusading kings of Germany and France to his benefit, ultimately solidifying his connections with Conrad III. Despite this, the immediate Byzantine impression of the Crusaders was mostly hostile. A contemporary panegyrist labelled the Germans as "Alamans" and depicted Constantinople as a lady, who paints her cheeks with the blood of German Crusaders.[95] Many years later, Kinnamos praised the deeds of Manuel against the Germans and demonstrated their failures in Asia Minor.[96] His later contemporary Niketas Choniates mentioned nefarious acts committed by both the aggressive crusaders and the treacherous Byzantines, praising crusader battles against the Turks in Asia Minor and introducing into his narrative a lengthy panegyric of the Crusader king.[97] One can tentatively say that the attitude of the Byzantine literati towards the foreign soldiers remained ambiguous and contradictory. This contradictory discourse helps one to understand better the complexity present in the allegedly plain narrative of Anna Komnene about the events of 1081.

The Memoir of a Grumpy Aunt? The Komnenian Revolt in the *Alexiad* as a Critique and Advice

During the reigns of John and Manuel Komnenos, the daughter of Alexios, Anna Komnene wrote her chronicle of her father's reign. In the beginning of her chronicle, Anna stated that the aim of her work is to preserve the deeds of her father and his associates for posterity. The *Alexiad* was a huge endeavour of an influential princess that kept her busy from 1137 until 1153. By the time of the Second Crusade, she was 60 and was hardly a novice in Byzantine politics. She observed the First Crusade ravaging Constantinople when she was 14 years old. Her husband, Nikephoros Bryennios the Younger, personally participated in the military actions during the First Crusade and probably knew many military diaspora members. If one is to believe Choniates, in 1118, Anna Komnene participated in the plot aimed at the installation of Nikephoros Bryennios the Younger on the throne. The plot failed, Bryennios returned to the army and John II became the emperor.[98]

It is not clear what position Anna held in the imperial family after the alleged plot during the reigns of John and Manuel. According to Neville, in the 1140s, she held a prominent position within the family, but some points in the narrative allow one to doubt the importance of this position.[99] It seems likely that Anna Komnene was an important, but a side-lined member of the ruling family who reconstructed her authority through history-writing after the death of her husband. She tried to create a historical narrative alternative to the "official" history of her father's reign, which was present on the walls of Blachernae palace.[100] Using her privileged status to gain access to documents and people, Anna struggled to construct a genuine history of her father, a history not free from personal biases of the historian.[101]

The rebellion of Alexios occupies the second book of the *Alexiad*. In the description, Anna vividly demonstrates how dangerous the rebellion was and how easily soldiers became robbers. Anna Komnene does not hide a consequence of her father's rebellion from her audience, demonstrating the disastrous side of the story. On some occasions, however, Anna is careful to describe events in detail, while on others (in the scene of plunder), she does not specify the groups or people who robbed the city. Why is this?

The reason might lie in the twelfth-century context of the *Alexiad*. According to Magdalino, Anna also wrote the *Alexiad* in an advisory tone that might have something to do with her position as the aunt of the ruling emperor Manuel.[102] This advisory position might explain Anna's focus in the description of the Komnenian revolt. In her description of rebellion, Anna presented Alexios as a person who rarely communicates with the diasporas or rank-and-file soldiers. Instead, he is using his elder advisers and intermediaries (like Pakourianos) to mobilize every military group available. When Alexios of the *Alexiad* began the rebellion, he never took command on the front line, but took a position in the rear. In some sense, he acted like the emperor *before* becoming the emperor. This all contrasts with the behaviour of his grandson, Manuel I, who in his first expedition as emperor still acted as a heady young prince under the "tyranny of his prime", by attacking the Turks in person and imitating Latin customs. In some way, Manuel remained a prince *after* becoming an emperor which was not a normal order of things.

With her description of the siege of 1081 and of the role of her great grandfather John Doukas, Anna is emphasizing the value of a different strategy of leadership. This strategy was based not only on personal martial prowess but also on the wise advice of surrounding people, to whom Alexios gave an ear. In the *Alexiad*, the man who solves the puzzle in the narrative is not Alexios himself with a spear in hand, but his elder relative John Doukas. Besides Doukai family pride (which played a certain role in the *Alexiad*), the message of Anna here is about the importance of elder advisors present near the young emperor. Thus, the depiction of Alexios might be a subtle form of *Kaiserkritik* aimed at the young emperor Manuel who did not listen to his advisers, but not at him alone.

The personal agenda of Anna Komnene might also lie behind the special attention that she dedicated to the treacherous *Nemitzoi* and their leader Gilpraktos. I suggest that Anna's focus on the *Nemitzoi's* betrayal and her remarks on their "servile" status might be a barb against Germans who posed a threat to the Empire during the Second Crusade and who became a new and influential group at court in the 1140s. The *Nemitzoi* of the *Alexiad*, with their origin in southern Germany, had an important parallel in Constantinople during the reign of Manuel, namely Empress Eirene-Berta. She was born in Sulzbach in southern Germany in the area that might correspond to Byzantine *Nemitza*. In 1146s, she became the wife of Manuel Komnenos. According to a recent study, Eirene-Berta as a young empress, exercised an important influence over the emperor and was a leading figure in the imperial household.[103] Her privileges in Constantinople included a personal palace, a privilege that Anna Komnene probably did not enjoy. If one is to believe Kinnamos and Choniates, the German princess did not integrate herself into Byzantine society well.[104] Thus, uneasiness at court and competition for influence over the emperor could be the reason behind the general anti-German message that the description in the *Alexiad* conveyed to the educated audience: "Do not trust your security and give much privilege to Western foreigners, especially if they are from Germany". If you do, says the *Alexiad*, Constantinople will fall. The focus of Anna on the greedy *Nemitzoi* might be a criticism directed against Berta of Sulzbach. In addition to *Kaiserkritik*, there is also *Kaiserinkritik*. Bearing in mind the many different attitudes that the Byzantine literati demonstrated towards foreigners, one could argue that Anna's nativist arguments found a ready audience in Constantinople.

To conclude, Byzantium was an empire that constantly attracted foreign warriors. The symbolic capital and material riches allowed Byzantine emperors to find armed servants in times of war and peace. Nearly every war with this or that enemy finished with some representatives of this or that group on the Byzantine side. This led to the kaleidoscope of individuals and groups of foreign origin in Constantinople. Some of those groups served emperors for decades, forming stable social units. The leaders of these groups were intermediaries between the emperor and foreign warriors in times of need. Over time, foreigners of the same origin formed clans that had connections not only in the army but also in the church and court. Some of those social formations reached a stage where they became something close to the modern notion of a diaspora. The Iberians in the eleventh century and the "Italians" in the twelfth were aware of their origin, supported their distinct culture, and were represented as such in the army, church, and at court. On the other side of the spectrum were the Pechenegs and the Cumans, who served in the Byzantine army but did not form any coherent group that attracted any specialized attention from the sources or promoted any notable leaders. There were many cases between these two extremes. With certain caution, one may speak about a "failed" diaspora of Franks who tried to form their own state in Anatolia in the eleventh century but did not succeed. The Byzantine Turks were closer to the "Norman-Iberian" side of the continuum, but the absence of information on the contacts between the leaders (such as John Axouchos) and the rank-and-file soldiers does not allow one to label them clearly as a diaspora. Further studies on sigillography and in the archives on Mount Athos may provide a solution to this conundrum.

The presence of foreigners in matters of war and peace in twelfth-century Byzantium was a normal thing. It supported the prestige of the Empire in the eyes of its inhabitants and assisted in recruiting fresh migrants abroad. According to all Byzantine sources mentioned in this chapter, a good emperor should know how to attract and support groups of useful barbarians and balance their interests with that of the locals. In the twelfth century, emperors of the Komnenoi dynasty adopted different strategies to manage different foreign groups. The literati (many of whom were also members of the ruling elite) had their own opinion on the matter. While court poets praised or blamed foreigners in accordance with the events of the day, members of the imperial family enjoyed a certain freedom of options that allowed them to discuss foreigners in a more subtle way. Anna Komnene, a princess and historian, used her position to create a complex description of her father's life. This description included a vivid story of the Komnenian revolt and subsequent sack of Constantinople. In her description, Anna incorporated a subtle criticism of her nephew Manuel Komnenos in a complex and triumphal image of her father, a victorious emperor over both foreigners and Byzantines.

Notes

1 I would like to thank the DAAD, Central European University, University of Heidelberg, Tijana Krstic, Georg Christ, Patrick Sänger, Katalin Szende, and József Laszlovszky for organizing the original conference on military diasporas and this volume. I would like to thank the ANAMED and GABAM of Koc University and GRK 2304 at the University of Mainz for supporting me during the reworking of this chapter. I would also like to thank the anonymous reviewers for their comments and Dr. Tristan Schmidt with whom I discussed the argument.

2 As Chalandon noted, the gate occupies a dominant position in the landscape and is visible from afar. This probably contributed to the success of the revolt. See Ferdinand Chalandon, *Essai sur le Règne d'Alexis Ier Comnène* (Paris: Picard, 1900), 48–49. For relevant summary of Alexios' biography, see Konstantinos Barzos, *Η γενεαλογία των Κομνηνών* (Thessaloniki: Kentron Byzantinōn Ereunōn, 1984), 83–118.

3 Jean-Claude Cheynet wrote a monograph on the Byzantine rebellions in the eleventh century, listing Alexios as a rebel. For the limited statistics on the successful/unsuccessful attempts to enter Constantinople, see Anthony Kaldellis, *Byzantine Republic* (Cambridge, MA: Harvard University Press, 2015), 136–138; Jean-Claude Cheynet, *Pouvoir et contestations à Byzance (963–1210)* (Paris: Publications de la Sorbonne, 1990).

4 Note the contraposition between the "natives" and the "foreigners". Anna Komnene, *Alexias*, ed. Athanasios Kambylis and Dieter Reinsch (Berlin: De Gruyter, 2001), 81, 94–11. In this chapter, I have used the translation of Sewter with alterations: Anna Komnene, *The Alexiad*, trans. Peter Frankopan and Edgar R. A. Sewter (London: Penguin Books, 2006), 171–172.

5 On a recent summary of the context of Anna, see Samuel Pablo Müller, *Latins in Roman (Byzantine) Histories: Ambivalent Representations in the Long Twelfth Century* (London: Brill, 2021), 46–49.

6 To denote a coherent group of people with separate cultural identity within the Empire, the authors of the only collective volume on the topic, Hélène Ahrweiler and Angeliki Laiou, used the term "ethnic group". See Hélène Ahrweiler, "Byzantine Concepts of the Foreigner," in *Studies on the Internal Diaspora of the Byzantine Empire*, ed. Hélène Ahrweiler and Angeliki E. Laiou (Cambridge, MA: Harvard University Press, 1998), 1–15; Steven Rapp, "Caucasia and Byzantine Culture," in *Byzantine Culture: Papers from the Conference "Byzantine Days of Istanbul" Held on the Occasion of Istanbul Being European Cultural Capital 2010, May 21–23 2010*, ed. Dean Sakel (Ankara: Türk Tarih Kurumu, 2014), 217–234, at 224; Most recently, scholars tend to avoid this term. See Rustam Shukurov, *Byzantine Turks: 1204–1461* (Leiden: Brill, 2016), 493.

7 Jemima Pierre, "Diaspora," in *Oxford Bibliographies* (Oxford: Oxford University Press, 2013), accessed 1 January 2019, http://www.oxfordbibliographies.com/view/document/obo-9780199766567/obo-9780199766567-0091.xml.

8 A separate group were religious dissenters who were treated very much like foreigners. For a short overview of Byzantine attitudes to the religious minority of Paulicians, see Ralph-Johannes Lilie, "Zur Stellung von ethnischen und religiösen Minderheiten in Byzanz," in *Visions of Community in Post-Roman World: The West, Byzantium and Islamic World, 300–1100*, ed. Clemens Gantner, Richard E. Payne, and Walter Pohl (Farnham: Ashgate, 2012), 301–316, at 312–315.

9 The cutting of war finances is widely attested in contemporary sources. See the prominent example in Kekaumenos *Consilia et Narrationes*, where the author explicitly warns the reader against the underfunding of the standing army and thematic armies. Кекавмен [Kekaumenos], *Советы и Рассказы* [*Consilia et Narrationes*], ed. Gennadii G. Litavrin (Saint Petersburg: Aleteia, 2003), 293–295.

10 For detailed instructions on the matter, see Kekaumenos, *Советы и Рассказы* [*Consilia et Narrationes*], 265–285.

11 Speros Vryonis, *The Decline of Medieval Hellenism in Asia Minor and the Process of Islamization from the Eleventh through the Fifteenth Century* (Berkeley: University of California Press, 1971), 69.

12 Ibid., 75.

13 See John Birkenmeier, *The Development of the Komnenian Army: 1081–1180* (London: Brill, 2001), 139–147.

14 The complete absence of the discussion of Varangian guard in the volume on "internal diasporas" is notable. See monograph by Sigfus Blöndal and Benedikt S. Benedikz, *The Varangians of Byzantium: An Aspect of Byzantine Military History* (Cambridge: Cambridge University Press, 1973).

15 Ibid., 143.

16 The question about the "dominant element" in the Varangian guard remains open. See Jonathan Shepard, "Another New England? Anglo-Saxon Settlement on the Black Sea," *Byzantine Studies* 1 (1974): 18–39; Jonathan Shepard, "The English and Byzantium: A Study of Their Role in the Byzantine Army in the Later Eleventh Century," *Traditio* 29 (1973): 53–92.

17 Blöndal and Benedikz, *The Varangians of Byzantium*, 185–186; Krinije N. Cigaar, *Western Travellers to Constantinople: The West and Byzantium, 962–1204* (Leiden: Brill, 1996), 126.

18 Alexander Kazhdan, "Latins and Franks in Byzantium: Perception and Reality from the Eleventh to the Twelfth Century," in *The Crusades from the Perspective of the*

Byzantium and Muslim World, ed. Angeliki E. Laiou (Washington, DC: Dumbarton Oaks Research Library and Collection, 2001), 83–100, at 86.

19 According to Michael Attaleiates, an anonymous Frank in the Byzantine army saved the fortress of Manzikert from the army of the Turks some 20 years before the battle at Manzikert. See Michael Attaleiates, *Historia*, ed. Eudoxos T. Tsolakis (Athens: Acad. Atheniensis, Inst. Litterarum Graecarum et Latinarum Studiis Destinatum, 2011), 37.

20 Kazhdan, "Latins and Franks in Byzantium," 92.

21 Dimitri Krallis notes that the attitude of Attaleiates towards the Latins was very individual. See Dimitri Krallis, *Serving Byzantium Emperors: The Courtly Life and Career of Michael Attaleiates* (London: Palgrave Macmillan, 2019), 155–157. This is hardly "rancorous anti-Latin rhetoric" assumed by some modern scholars. See Jason Roche, "The Byzantine Conception of the Latin Barbarian and Distortion in the Greek Narratives of the Early Crusades," in *Fighting for the Faith—The Many Crusades*, ed. Kurt V. Jensen, Carsten S. Jensen, and Janus M. Jensen (Stockholm: Centre of Mediaeval Studies of Stockholm University, 2018), 143–173, esp. 143–147. For the recent summary and re-evaluation of the image of Russeil, see Müller, *Latins in Roman (Byzantine) Histories*, 213–218.

22 The toponym is present in the famous treatise of Constantine Porphyrogenitus, where it is a synonym of Bavaria Constantine Porphyrogenitus, "De Ceremoniis," in *Constantini Porphyrogeniti imperatoris de cerimoniis aulae Byzantinae libri duo*, ed. Johann J. Reiske (Bonn: Weber, 1829), 1:389.

23 Michael Attaleiates, *Historia*, 170, lines 19–21.

24 For a very short summary of the Armenian diaspora, see Nina G. Garsoian, "Armenian Integration into the Byzantine Empire," in *Studies on the Internal Diaspora*, ed. Hélène Ahrweiler and Angeliki E. Laiou (Washington, DC: Dumbarton Oaks, 1998), 82–85.

25 See В. Арутюнова-Фиданян [V. Arutyunova-Fidanyan] *Армяно-Византийская Конта ктная Зона: Результаты Взаимодействия Культур* [*Armenian-Byzantine Contact Zone: Results of Cultural Interchange*] (Moscow: Nauka, 1994), 82.

26 For short summary of resettlement, see Arutyunova-Fidanyan, *Армяно-Византийская Контактная Зона* [*Armenian-Byzantine Contact Zone*], 81–83.

27 These units were active in the time of Romanos Diogenes. Michael Attaleiates mentions special formations of Armenian infantry that participated in the storm of Manzikert before the ill-fated battle. See Michael Attaleiates, *Historia*, 117.

28 Certain Khatatourios was the last to defect to Romanos. See Michael Attaleiates, *Historia*, 131.

29 On Philaretos Brachamios, see Seta B. Dadoyan, *The Armenians in Medieval Islamic Word: Paradigms of Interactions, Seventh to Fourteenth Centuries* (New Brunswick: Transaction, 2013), 56–59.

30 See Nikephoros Bryennios, *Material for History*, ed. Paul Gautier, 164.

31 The family of Maria of Alania had a long history of service to Byzantium. See Rapp, "Caucasia and Byzantine Culture," 225–226.

32 Anna Komnene, *Alexias*, ed. Kambylis and Reinsch, 2, 4, 63.57–58; Cf. Gérard Dédéyan, ed., *Histoire du peuple arménien* (Toulouse: Éd. Privat, impr., 2007), 317.

33 See Arutyunova-Fidanyan, *Армяно-Византийская Контактная Зона* [*Armenian-Byzantine Contact Zone*], 84.

34 Robert Jordan, "Pakourianos: Typikon of Gregory Pakourianos for the Monastery of the Mother of God Petritzonitissa in Backovo," in *Byzantine Monastic Foundation Documents*, ed. John Tomas and Angela C. Hero (Washington, DC: Dumbarton Oaks Research Library and Collection, 2000), 507–563, at 507.

35 It looks likely that the Byzantine name of the general denoted a greater degree of integration in Byzantine society available for the holder. The lesser commanders were named according to their nickname (see the example of Chrysoskoulos below) or simply by Byzantinization of their "barbaric name". There are a few cases when

Byzantines used a barbaric nickname as a name. See Blöndal and Benedikz, *The Varangians of Byzantium*, 143.

36 See the edition of Gautier and the translation and commentary by Robert Jordan. P. Gautier, "Le typikon du sébaste Grégoire Pakourianos," *Revue des études byzantines* 42 (1984), 5–145, at Proem. 26; Jordan, "Pakourianos," 507–563.

37 Anna Komnene, *Alexias*, ed. Kambylis and Reinsch, 63.

38 According to Jordan, the typicon of the monastery was preserved in three versions—Greek, Armenian, and Georgian. The term "Iberia" could mean here not classical Iberia, but Byzantine province of Iberia from where the grandfather of Pakourianos was from. For families of mixed origin in the space of Caucasia, see Rapp, "Caucasia and Byzantine Culture," 219, for origins of Pakourianos, see Arutyunova-Fidanyan, *Армяно-Византийская Контактная Зона* [*Armenian-Byzantine Contact Zone*], 84.

39 Nun Maria, "Testamentum Mariae Monachae," in *Actes d'Iviron*, vol. 2, *Du milieu du XIe siècle à 1204*, ed. Denise Papachryssantou, Jacques Lefort, and Nicolas Oikonomidès (Paris: Lethielleux, 1990), 179; Yasuhiro Otsuki, "Sacred Dedication in Byzantine Imperial Finance: Maria's Bequest and Iviron Monastery," *Mediterranean World* 16 (2003): 89–99, at 93.

40 On the role of Maria of Alania, see Barbara Hill, "Alexios I Komnenos and the Imperial Women," in *Alexios I Komnenos*, ed. Margaret Mullet and Dion Smyth (Belfast: Belfast Byzantine Enterprises, 1996), 37–54, at 38–39, 45.

41 Michael Attaleiates, *Historia*, 99.

42 Michael Attaleiates, *Historia*, Romanos Diogenes, 110–111; cf. Anthony C. S. Peacock, "From Balkhan-Kuhiyan to Nawakiya: Nomadic Politics and the Foundation of Seljuk Rule in Anatolia," in *Nomad Aristocrats in the World of Empires*, ed. Jürgen Paul (Wiesbaden: Ludwig Reichert Verlag), 56–76.

43 Anna Komnene, *Alexias*, ed. Kambylis and Reinsch, 2,1, 55.19–21.

44 To make things more complicated, Kekaumenos was probably part of Kekaumenoi, powerful Armenian clan. According to the treatise, his grandfather was from Caucasus, while another grandfather served in Bulgaria. See Kekaumenos, *Советы и Рассказы* [*Consilia et Narrationes*], 294, lines 8–10.

45 For eleventh-century inflation, see Michael Hendy, *Studies in Byzantine Monetary Economy* (Cambridge: Cambridge University Press, 1985), 510–515.

46 Konstantinos Smyrlis, "The Fiscal Revolution of Alexios I Komnenos: Timing, Scope and Motives," in *Autour du "Premier humanisme byzantin" et des "Cinq études sur le XIe siècle", quarante ans après Paul Lemerle*, ed. Bernard Flusin and Jean-Claude Cheynet (Paris: Association des Amis du Centre d'Histoire et Civilisation de Byzance, 2017), 593–609, at 595.

47 The identity of this group is not clear.

48 *Actes de Vatopédi*, vol. 1, *Des origines à 1329*, ed. Jacques Bompaire et al. (Paris: P. Lethielleux, 2001), 112, lines 30–37: Ῥῶς, Βαράγγ(ων), Κουλπίγγ(ων), Ἰγγλίγγ(ων), Φράγγ(ων), Νεμιτζ(ῶν), Βουλγ(ά)ρ(ων), Σαρακην(ῶν), Ῥωμ(αίων), Ἀθανάτ(ων) (καὶ) λοιπ(ῶν).

49 For different views on image of Alexios in both works, see Penelope Buckley, *The Alexiad of Anna Komnene: An Artistic Strategy in Making a Myth* (Cambridge: Cambridge University Press, 2014), 41–49; Leonora Neville, *Heroes and Romans in the Twelfth-Century Byzantium* (Cambridge: Cambridge University Press, 2012), 267–268; Leonora Neville, *Anna Komnene: The Life and Work of Medieval Historian* (Oxford: Oxford University Press, 2016), 49–51.

50 Barzos, Ἡ γενεαλογία, 1:98–99.

51 See detailed analysis of the constraint in Buckley, *The Alexiad of Anna Komnene*, 73–75.

52 Anna Komnene, *Alexias*, ed. Kambylis and Reinsch, 2.1, 55; Buckley, *The Alexiad of Anna Komnene*, 75.

53 The anonymity is suspicious here. Anna usually mentions the names of this or that person but omits them in case of uneasiness and personal enmity. See Anna Komnene, *Alexias*, ed. Kambylis and Reinsch, 2.4, 63.46.

54 On the person of Oumbertopoulos, see Müller, *Latins in Roman (Byzantine) Histories*, 209–210.

55 Anna Komnene, *Alexias*, ed. Kambylis and Reinsch, 2.4, 64.83–85: Φιλοδωρότατόν τε ὄντα καὶ τὴν χεῖρα, εἴπερ τις ἄλλος, περὶ τὰς δόσεις εὐκίνητον λίαν ὑπερηγάπων, καίτοι μὴ πάνυ τι πλούτῳ περιρρεόμενον, trans. Frankopan and Sewter.

56 Anna Komnene, *Alexias*, ed. Kambylis and Reinsch, 2.6, 69.46.

57 Demetrios Polemis analyzed his participation in the revolt in his monograph on the Doukai family, while Leonora Neville argued that he might be the author of some historical work that Bryennios later used. See Demetrios Polemis, *The Doukai: A Contribution to the Byzantine Prosopography* (London: Athlone Press, 1968), 39–42; Leonora Neville, "A History of Caesar John Doukas in Nikephoros Bryennios' *Material for History*," *Byzantine and Modern Greek Studies* 32 (2008): 168–188.

58 Anna Komnene, *Alexias*, ed. Kambylis and Reinsch, 2.6, 72.17–25, trans. Frankopan and Sewter, 155.

59 Anna Komnene, *Alexias*, ed. Kambylis and Reinsch, 2.9, 78.3–6.

60 Ibid., 2.9, 79.25–30: ὡς δὲ ἐνταῦθα μὲν ἐφεστάναι τοὺς Ἀθανάτους λεγομένους ἐμάνθανε (στράτευμα δὲ τοῦτο τῆς ῥωμαϊκῆς δυνάμεως ἰδιαίτατον), ἐκεῖσε δὲ τοὺς ἐκ τῆς Θούλης Βαράγγους (τούτους δὴ λέγω τοὺς πελεκοφόρους βαρβάρους), ἀλλαχόσε δὲ τοὺς Νεμίτζους (ἔθνος δὲ καὶ τοῦτο βαρβαρικὸν καὶ τῇ βασιλείᾳ Ῥωμαίων δουλεῦον ἀνέκαθεν), φησὶ πρὸς τὸν Ἀλέξιον παραινῶν μήτε τοῖς Βαράγγοις ἐμβαλεῖν μήτε τοῖς Ἀθανάτοις προσεμβαλεῖν. οἱ μὲν γὰρ αὐτόχθονες ὄντες τῷ βασιλεῖ πολλὴν τὴν εἰς αὐτὸν ἐξ ἀνάγκης ἔχοντες εὔνοιαν θᾶττον ἂν τὰς ψυχὰς προδοῖεν ἢ πονηρόν τι κατ' αὐτοῦ μελετῆσαι πεισθησονται, οἱ δέ γε ἐπὶ τῶν ὤμων τὰ ξίφη κραδαίνοντες πάτριον παράδοσιν καὶ οἷον παρακαταθήκην τινὰ καὶ κλῆρον τὴν εἰς τοὺς αὐτοκράτορας πίστιν καὶ τὴν τῶν σωμάτων αὐτῶν φυλακὴν ἄλλος ἐξ ἄλλου διαδεχόμενοι τὴν πρὸς αὐτὸν πίστιν ἀκράδαντον διατηροῦσι καὶ οὐδὲ ψιλὸν πάντως ἀνέξονται περὶ προδοσίας λόγον.τῶν δέ γε Νεμίτζων ἀποπειρώμενος ἴσως οὐ πόρρω βαλεῖ σκοποῦ, ἀλλ' εὐτυχήσει τὴν εἴσοδον ἀπὸ τοῦ ὑπ' αὐτῶν τηρουμένου πύργου. Trans. Frankopan and Sewter, 167–168.

61 The Immortals are a group with a long story behind it. The name goes back to the tenth century when John Tzymisches created them as a shock cavalry corps. See Nikephoros Bryennios, *Material for History*, ed. Gautier, 164; Birkenmeier, *Komnenian Army*, 57–58.

62 Anna's information complements the information provided by her husband on the matter. While Bryennios focuses on the training of this particular group, Anna underlines their loyalty.

63 Müller, *Latins in Roman (Byzantine) Histories*, 200–207.

64 Note the "barbaric" name of Gilpraktos. Anna Komnene, *Alexias*, ed. Kambylis and Reinsch, 2.9, 79.43–44.

65 For Alexian reforms in the army, see J. Birkenmeier, *Komnenian Army*, 139–155. See special discussion on the financing of unprivileged and "native" infantry. Birkenmeier, ibid., 155–159.

66 Müller, *Latins in Roman (Byzantine) Histories*, 209.

67 On Tourkopouloi, see Alexios Savvides, "Tourkopo(u)loi," in: *Encyclopaedia of Islam*, ed. Peri J. Bearman et al., 2nd ed., accessed 18 October 2021, http://dx.doi.org/10.1163/1573-3912_islam_SIM_7594; Brand, "Turkish Element in Byzantium," 13.

68 On the Alans, see Birkenmeier, *Development of Komnenian Army*, 162–233.

69 Niketas Choniates labels him as "Persian". This is a term that Choniates uses to denote members of Seljuk Elite. See Niketas Choniates, *Historia*, ed. van Dieten, 9; Brand, "Turkish Element in Byzantium," 4–5.

70 For the summary of the battle of Alexios with Normans at Dyrrachium, see George Theotokis, *The Norman Campaigns in the Balkans, 1081–1118* (Woodbridge: Boydell Press, 2014), 155–164.

71 Peter Frankopan, "Turning Latin into Greek: Anna Komnene and the *Gesta Roberti Wiscardi*," *Journal of Medieval History* 39 (2013): 85–86.

72 For the recent overview of the image of the Venetians in Byzantine rhetoric, see Müller, *Latins in Roman (Byzantine) Histories*, 75–85.
73 For complex relations between Alexios and Venice, see Dafni Penna, "The Byzantine Imperial Acts to Venice, Pisa and Genoa, 10th–12th Centuries: A Comparative Legal Study" (PhD diss., University of Groningen, 2013), 28–35.
74 Peter Frankopan, *The First Crusade: The Call from the East* (Cambridge, MA: Harvard University Press, 2012), 87–89.
75 Thomas Asbridge, *The First Crusade: A New History* (Oxford: Oxford University Press, 2004), 111–112.
76 Anna Komnene, *Alexias*, ed. Kambylis and Reinsch, 101. In the twelfth century, Petraliphai married into the imperial family. See Barzos, Ἡ γενεαλογία των Κομνηνών, 2: 138–142.
77 As an example, see the letter of Nikephoros Basilakes to Axouchos.
78 Paul Magdalino, *The Empire of Manuel I Komnenos* (Oxford: Oxford University Press 1993), 201–206. On the early integration of the Normans, see Müller, *Latins in Roman (Byzantine) Histories*, 208–210.
79 Theodore Prodromos, *Historische Gedichte*, ed. Wolfram Hörandner (Vienna: Verlag der Österreichischen Akademie der Wissenschaften, 1974), 179–180.
80 Niketas Choniates, *Historia*, 36; Magdalino, *The Empire of Manuel Komnenos*, 221–225.
81 John Kinnamos, *Epitome rerum ab Ioanne et Alexio Comnenis Gestarum* [*Deeds of John and Manuel Komnenos*], ed. August Meineke (Bonn: Weber, 1836), 8.
82 John Kinnamos, *Epitoma*, 15.
83 Maximilian Lau and Roman Shliakhtin, "Mas'ud I of Ikonion: The Overlooked Victor of the Twelfth-Century Anatolian Game of Thrones," *Byzantinoslavica* 76 (2018): 230–252.
84 It seems unlikely that Axouchosos really exercised much command in the presence of John II, acting more like his assistant than as an independent commander. Niketas Choniates, *Historia*, 83–85; Brand, "Turkish Element in Byzantium," 5–7.
85 Roman Shliakhtin, "From Huns into Persians: The Projected Identity of the Turks in the Byzantine Rhetoric of the Eleventh and Twelfth Century" (PhD diss., Central European University Budapest, 2016), 259–266.
86 It is not clear whether the monastery was founded by Axouchosoi, but this is probable. A new Athonite foundation was certainly a work of the imperial elite, and among the elite, there were no other persons who could be connected with the Koutloumousios who is Kutlumush, father of Sulaiman ibn Kutlumush. Sulaiman was for a short time a ruler in Nicaea from where Axouchosos originated.
87 John Kinnamos, *Epitoma*, 37.
88 Elizabeth Jeffreys, "Sevastokratorissa Eirene as Literary Patroness: The Monk Iakovos," *Jahrbuch der Österreichischen Byzantinistik* 32, no. 3 (1981): 241–256.
89 John Kinnamos, *Epitoma*, 37.
90 Ibid., 15.
91 On marriages, see detailed analysis of Stanković and short notes of Magdalino. Vlada Stanković, "John II Komnenos before the year 1118," in *John II Komnenos, Emperor of Byzantium: In the Shadow of Father and Son*, ed. Alexandro Bucossi and Alex Rodriguez Suarez (London: Routledge, 2016), 11–21, at 18–19; Magdalino, *The Empire of Manuel Komnenos*, 42–43.
92 Theodore Prodromos, *Historische Gedichte*, 321.
93 John Kinnamos, *Epitoma*, 47.
94 Ibid., 51.
95 See Elizabeth Jeffreys and Michael Jeffreys, "The 'Wild Beast from the West': Immediate Literary Reactions in Byzantium to the Second Crusade," in *The Crusades from the Perspective of Byzantium and the Muslim World*, ed. Angeliki E. Laiou (Washington, DC: Dumbarton Oaks, 2001), 101–116, esp. 108.

96 Jason T. Roche, "King Conrad III in the Byzantine Empire: A Foil for Native Imperial Virtue," in *The Second Crusade: Holy War on the Periphery of Latin Christendom*, ed. Jason T. Roche and Janus M. Jensen (Turnhout: Brepols, 2015), 183–216.
97 Nicetas Choniates, *Historia*, 66–71.
98 For the debate on this confrontation, see recent works of Samuel Müller and Leonora Neville. Neville claims that Anna is a victim of biased historians, while Müller suspects that relations between two children of Alexios were far from friendly. He noted that Anna was effectively absent from the family memory list in the typicon of Pantokrator monastery. Neville, *Anna Komnene: Life and Work of a Medieval Historian*, 150–174; Müller, *Latins in Roman (Byzantine) Histories*, 48.
99 Neville, *Anna Komnene: The Life and Work of Medieval Historian*, 141–151.
100 On some elements in the "official" narrative, see Stratis Papaioannou, "Τῇ βασιλίσσῃ μοναχῇ κυρᾷ: An Unedited Letter to Eirene Doukaina (and an Êthopoiia in Verse by Her Son for His Father)," in *After the Text: Byzantine Enquiries in Honour of Margaret Mullett*, ed. Liz James, Oliver Nicholson, and Roger Scott (London: Routledge, 2021), 147–166.
101 Neville, *Anna Komnene: The Life and Work of a Medieval Historian*, 76–86.
102 Magdalino, "The Pen of the Aunt: Echoes of the Mid-Twelfth Century in the 'Alexiad'," in *Anna Komnene and Her Times*, ed. Thalia Ghouma-Peterson and Angeliki E. Laiou (New York: Garland, 2000), 15–44.
103 Müller, *Latins in Roman (Byzantine) Histories*, 148–159.
104 Ibid., 149–151.

References

Sources

Acts of Iviron. *Actes d'Iviron*. Vol. 2, *Du milieu du XIe siècle à 1204*, edited by Denise Papachryssantou, Jacques Lefort, and Nicolas Oikonomidès. Paris: P. Lethielleux, 1990.

Acts of Vatopedi. *Actes de Vatopédi*. Vol. 1, *Des origines à 1329*, edited by Jacques Bompaire, Christophe Giros, Vassiliki Kravari, and Jacques Lefort. Paris: P. Lethielleux, 2001.

Attaleiates, Michael. *Historia*, edited by Eudoxos T. Tsolakis. Athens: Acad. Atheniensis, Inst. Litterarum Graecarum et Latinarum Studiis Destinatum, 2011.

Bryennios, Nikephoros. *Histoire* [*Material for History*], edited by Paul Gautier. Brussels: Byzantion, 1975.

Choniates, Niketas. *Historia*, edited by Johannes A. van Dieten. Berlin: De Gruyter, 1975.

Kekaumenos. *Советы и Рассказы Кекавмена* [*Concilia et Narrationes*], edited by Gennadii G. Litavrin. Saint Petersburg: Aleteia, 2003.

Kinnamos, John. *Epitome rerum ab Ioanne et Alexio Comnenis Gestarum* [Deeds of John and Manuel Komnenos], edited by August Meineke. Bonn: Weber, 1836.

Komnene, Anna. *The Alexias*, edited by Athanasios Kambylis and Diether R. Reinsch. Berlin: De Gruyter, 2001.

Komnene, Anna. *The Alexiad*, translated by Edgar R. A. Sewter and Peter Frankopan. London: Penguin, 2009.

Pakourianos, Gregory. "Le typikon du sébaste Grégoire Pakourianos." *Revue des études byzantines* 42 (1984): 5–145.

Porphyrogenitus, Constantine. *Constantini Porphyrogeniti Imperatoris De cerimoniis aulae Byzantinae libri duo*, edited by Johann J. Reiske. Bonn: Weber, 1830.

Prodromos, Theodore. *Historische Gedichte*, edited by Wolfram Hörandner. Vienna: Verlag der Österreichischen Akademie der Wissenschaften, 1974.

Studies

Ahrweiler, Hélène. "Byzantine Concepts of the Foreigner." In *Studies on the Internal Diaspora of the Byzantine Empire*, edited by Hélène Ahrweiler and Angeliki E. Laiou, 1–15. Cambridge, MA: Harvard University Press, 1998.

Arutyunova-Fidanyan, Viada. *Армяно-Византийская Контактная Зона: Результаты Взаимодействия Культур [Armenian-Byzantine Contact Zone: Results of Cultural Interchange]*. Moscow: Nauka, 1994.

Asbridge, Thomas. *The First Crusade: A New History*. Oxford: Oxford University Press, 2004.

Barzos, Konstantinos. *Η γενεαλογία των Κομνηνών*. Thessaloniki: Kentron Byzantinōn Ereunōn, 1984.

Birkenmeier, John. *The Development of the Komnenian Army: 1081–1180*. London: Brill, 2001.

Blöndal, Sigfus, and Benedikt S. Benedikz. *The Varangians of Byzantium: An Aspect of Byzantine Military History*. Cambridge: Cambridge University Press, 1973.

Brand, Charles. "The Turkish Element in Byzantium, Eleventh-Twelfth Centuries." *Dumbarton Oaks Papers* 43 (1989): 1–25.

Buckley, Penelope. *The Alexiad of Anna Komnene: Artistic Strategy in the Making of a Myth*. Cambridge: Cambridge University Press, 2014.

Chalandon, Ferdinand. *Essai sur le Règne d'Alexis Ier Comnène*. Paris: Picard, 1900.

Cheynet, Jean-Claude. *Pouvoir et Contestations à Byzance (963–1210)*. Paris: Publications de la Sorbonne, 1990.

Cigaar, Krijnie N. *Western Travelers to Constantinople: The West and Byzantium, 962–1204*. Leiden: Brill, 1996.

Dadoyan, Seta. *The Armenians in Medieval Islamic Word: Paradigms of Interactions, Seventh to Fourteenth Centuries*. New Brunswick: Transaction, 2013.

Dédéyan, Gérard, ed. *Histoire du peuple arménien*. Toulouse: Éd. Privat, impr., 2007.

Frankopan, Peter. *The First Crusade: The Call from the East*. Cambridge, MA: Harvard University Press, 2012.

Frankopan, Peter. "Turning Latin into Greek: Anna Komnene and the *Gesta Roberti Wiscardi*." *Journal of Medieval History* 39 (2013): 85–86.

Garsoian, Nina G. "Armenian Integration into the Byzantine Empire." In *Studies on the Internal Diaspora of the Byzantine Empire*, edited by Hélène Ahrweiler and Angeliki E. Laiou, 82–85. Washington, DC: Dumbarton Oaks, 1998.

Hendy, Michael. *Studies in Byzantine Monetary Economy*. Cambridge: Cambridge University Press, 1985.

Hill, Barbara. "Alexios I Komnenos and the Imperial Women." In *Alexios I Komnenos*, edited by Margaret Mullet and Dion Smyth, 37–54. Belfast: Belfast Byzantine Enterprises, 1996.

Jeffreys, Elizabeth. "Sevastokratorissa Eirene as Literary Patroness: The Monk Iakovos." *Jahrbuch der Österreichischen Byzantinistik* 32 (1981): 241–256.

Jeffreys, Elisabeth, and Michael Jeffreys. "The 'Wild Beast from the West': Immediate Literary Reactions in Byzantium to the Second Crusade." In *The Crusades from the Perspective of Byzantium and the Muslim World*, edited by Angeliki E. Laiou, 101–116. Washington, DC: Dumbarton Oaks, 2001.

Jordan, Robert. "Pakourianos: Typikon of Gregory Pakourianos for the Monastery of the Mother of God Petritzonitissa in Backovo." In *Byzantine Monastic Foundation Documents*, edited by John Tomas and Angela C. Hero, 507–563. Washington, DC: Dumbarton Oaks Research Library and Collection, 2000.

Kaldellis, Anthony. *Byzantine Republic*. Cambridge, MA: Harvard University Press, 2015.
Kazhdan, Alexander. "Latins and Franks in Byzantium: Perception and Reality from the Eleventh to the Twelfth Century." In *The Crusades from the Perspective of Byzantium and the Muslim World*, edited by Angeliki E. Laiou, 83–100. Washington, DC: Dumbarton Oaks, 2001.
Krallis, Dimitri. *Serving Byzantium Emperors: The Courtly Life and Career of Michael Attaleiates*. London: Palgrave Macmillan, 2019.
Lau, Maximilian, and Roman Shliakhtin. "Mas'ud I of Ikonion: The Overlooked Victor of the Twelfth-Century Anatolian Game of Thrones." *Byzantinoslavica* 76 (2018): 230–252.
Lilie, Ralph-Johannes. "Zur Stellung von ethnischen und religiösen Minderheiten in Byzanz." In *Visions of Community in Post-Roman World: The West, Byzantium and Islamic World, 300–1100*, edited by Clemens Gantner, Richard E. Payne, and Walter Pohl, 301–316. Farnham: Ashgate, 2012.
Magdalino, Paul. *The Empire of Manuel I Komnenos*. Oxford: Oxford University Press, 1993.
Magdalino, Paul. "The Pen of the Aunt: Echoes of the Mid-Twelfth Century in the 'Alexiad'." In *Anna Komnene and Her Times*, edited by Thalia Gouma-Peterson and Angeliki E. Laiou, 15–44. New York: Garland, 2000.
Müller, Samuel Pablo. *Latins in Roman (Byzantine) Histories: Ambivalent Representations in the Long Twelfth Century*. London: Brill, 2021.
Neville, Leonora. "A History of Caesar John Doukas in Nikephoros Bryennios' *Material for History*." *Byzantine and Modern Greek Studies* 32 (2008): 168–188.
Neville, Leonora. *Heroes and Romans in Twelfth-Century Byzantium: The Material for History of Nikephoros Bryennios*. Cambridge: Cambridge University Press, 2012.
Neville, Leonora. *Anna Komnene: Life and Work of a Medieval Historian*. Oxford: Oxford University Press, 2016.
Otsuki, Yasuhiro. "Sacred Dedication in Byzantine Imperial Finance: Maria's Bequest and Iviron Monastery." *Mediterranean World* 16 (2001): 89–99.
Papaioannou, Stratis. "Τῇ βασιλίσσῃ μοναχῇ κυρᾷ: An Unedited Letter to Eirene Doukaina (and an Êthopoiia in Verse by Her Son for His Father)." In *After the Text: Byzantine Enquiries in Honour of Margaret Mullett*, edited by Liz James, Oliver Nicholson, and Roger Scott, 147–166. London: Routledge, 2021.
Peacock, Anthony. "From Balkhan-Kuhiyan to Nawakiya: Nomadic Politics and the Foundation of Seljuk Rule in Anatolia." In *Nomad Aristocrats in the World of Empires*, edited by Jürgen Paul, 56–76. Wiesbaden: Ludwig Reichert Verlag, 2013.
Penna, Dafni. 2012. "The Byzantine Imperial Acts to Venice, Pisa and Genoa, 10th–12th Centuries: A Comparative Legal Study." PhD diss., University of Groningen, 2012.
Pierre, Jemima. "Diaspora." In *Oxford Bibliographies*. Oxford: Oxford University Press, 2013. Accessed 1 January 2019. http://www.oxfordbibliographies.com/view/document/obo-9780199766567/obo-9780199766567-0091.xml.
Polemis, Demetrios. *The Doukai: A Contribution to the Byzantine Prosopography*. London: Athlone Press, 1968.
Rapp, Steven. "Caucasia and Byzantine Culture." In *Byzantine Culture: Papers from the Conference "Byzantine Days of Istanbul" Held on the Occasion of Istanbul Being European Cultural Capital 2010, Istanbul, May 21–23 2010*, edited by Dean Sakel, 217–234. Ankara: Türk Tarih Kurumu, 2014.
Roche, Jason T. "King Conrad III in the Byzantine Empire: A Foil for Native Imperial Virtue." In *The Second Crusade: Holy War on the Periphery of Latin Christendom*, edited by Jason T. Roche and Janus M. Jensen, 183–216. Turnout: Brepols, 2015.

Roche, Jason T. "The Byzantine Conception of the Latin Barbarian and Distortion in the Greek Narratives of the Early Crusades." *In Fighting for the Faith—The Many Crusades*, edited by Kurt V. Jensen, Carsten S. Jensen, and Janus M. Jensen, 143–173. Stockholm: Center of Medieval Studies of Stockholm University, 2018.

Savvides, Alexios. "Tourkopo(u)loi." In *Encyclopaedia of Islam*, edited by Peri J. Bearman, et al. 2nd ed. Accessed 18 October 2021. http://dx.doi.org/10.1163/1573-3912_islam_SIM_7594.

Shepard, Jonathan. "The English and Byzantium: A Study of Their Role in the Byzantine Army in the Later Eleventh Century." *Traditio* 29 (1973): 53–92.

Shepard, Jonathan. "Another New England? Anglo-Saxon Settlement on the Black Sea." *Byzantine Studies* 1 (1974): 18–39.

Shliakhtin, Roman. "From Huns into Persians: The Projected Identity of the Turks in the Byzantine Rhetoric of the Eleventh and Twelfth Century." PhD diss., Central European University Budapest, 2016.

Shukurov, Rustam. *Byzantine Turks: 1204–1461.* Leiden: Brill, 2016.

Smyrlis, Konstantinos. "The Fiscal Revolution of Alexios I Komnenos: Timing, Scope and Motives." In *Autour du "Premier humanisme byzantin" et des "Cinq études sur le XIe siècle", quarante ans après Paul Lemerle*, edited by Bernard Flusin and Jean-Claude Cheynet, 593–609. Paris: Association des Amis du Centre d'Histoire et Civilisation de Byzance, 2017.

Stanković, Vlada. "John II Komnenos before the year 1118." In *John II Komnenos, Emperor of Byzantium: In the Shadow of Father and Son*, edited by Alessandro Bucossi and Alex Rodriguez Suarez, 11–21. London: Routledge, 2016.

Theotokis, George. *The Norman Campaigns in the Balkans, 1081–1118*. Woodbridge: Boydell Press, 2014.

Vryonis, Speros, Jr. *The Decline of Medieval Hellenism in Asia Minor and the Process of Islamization from the Eleventh through the Fifteenth Century*. Berkeley: University of California, 1971.

7

THE CATALAN COMPANY AS A MILITARY DIASPORIC GROUP IN MEDIEVAL GREECE AND ASIA MINOR

Mike Carr and Alasdair Grant

Introduction

After the conquest of Constantinople by the armies of the Fourth Crusade in 1204, the lands of the Byzantine Empire fragmented into a mosaic of separate polities. Some remained in the hands of Greek-speaking Christians, such as the Empire of Nikaia and the Despotate of Epirus; many others were conquered by crusaders ("Latins"), who established their own lordships in Greece and the Aegean region, an area known as "the Romania"—i.e. "Roman land", the (in many places former) Byzantine Empire. In 1261, Michael VIII Palaiologos Nikaia took Constantinople from its Latin conquerors, thus recentring the Byzantine world on its ancestral capital; this did little, however, to restore unity to the area. Most of the Aegean islands and large parts of Greece remained in Latin hands, while the arrival of the Mongols in eastern Anatolia pushed groups of Turkmen westward, the latter in turn conquering most of the Byzantine territory in the peninsula by the early fourteenth century. The Byzantine Empire's resulting need for help against Turkish military aggression coincided with the availability of Catalan-Aragonese mercenaries who had fought in the war of the Sicilian Vespers (1282–1302). These troops were incorporated into what is now known as the Catalan Company—a mercenary band that fought for the Byzantines against the Turks before turning against their masters and ultimately conquering the Frankish Duchy of Athens, which they ruled until 1388. As a military diaspora, the Catalan Company poses a particularly rich case study: they interacted with their host population both as subjects and (in formerly Byzantine lands) as rulers, and they preserved complex political and cultural links with their ancestral homelands of Aragon, Catalonia, and Sicily.

The historiography of the Catalan Company in the Romania is riven by a break in the year 1311. This is for two reasons, the first of which is historical:

DOI: 10.4324/9781003245568-8

1311 saw the Catalan conquest of the Frankish Duchy of Athens from Walter of Brienne and thus the transformation of what David Jacoby has called an "itinerant army" into a settled polity.[1] The second reason is historiographical: the activities of the Catalans during their time in the Byzantine Empire and Turkish Anatolia are recorded chiefly by the two major narrative histories of Ramon Muntaner[2] and Georgios Pachymeres;[3] the affairs of the duchy, however, are recorded not chiefly by narrative texts but rather in documentary sources preserved in western European archives. The result is two scholarly traditions of very different appearance: first, a grand narrative of Catalan mercenary engagement, conquest, betrayal, and destruction; second, a finer-grained but incomplete picture of the social and legal composition of a Latin Christian society. The first period has been extensively, if not perhaps exhaustively studied, especially in the context of the military achievements of the Company and its relations with Byzantium and the other Mediterranean powers.[4] The main sources and their respective problems are well known, though their interpretation remains up for debate. The second period, still dominated by the pioneering work of the Catalan scholar antoni Rubió i Lluch,[5] and subsequent studies by Kenneth Meyer Setton,[6] lacks an up-to-date critical study of the sort that Venetian Crete, for example, has received.[7] The majority of research since Setton has focussed on how the Catalans contributed to the geopolitical developments in the Aegean,[8] although some recent work has revisited the question of Catalan governance of the duchy.[9] The Catalan Company's status as a military diaspora changed when it became the settled ruling elite of its own duchy, but the question of its relationships with the societies it served and subjected pertains equally to the periods before and after 1311; consequently, this chapter treats the Catalan Company's whole history in the East, from 1303 to 1388.

The wider Aegean region in the period after 1204 provides numerous interesting examples of military diasporas as defined in this volume, of which the Catalans are only one case. But as with the Catalans, few of these have been studied as diasporas as such, and none that we know of has been considered within the specific analytical framework of a military diaspora. The following is a brief survey of some promising comparative examples, with an indication of relevant scholarship. The Latin Empire (1204–1261) and the other Frankish crusader states, such as the Kingdom of Thessaloniki and the Principality of Achaia, shared a common "Frankish" or "crusader" identity, which involved service to God and the papacy, along with dynastic links to the West and to one another, as well as shared governmental systems imported from the West.[10] The Knights Hospitaller, who conquered the island of Rhodes in *c.*1306, bore many similarities to these Frankish states, although as a religious order, their ostensible sense of belonging was to the Order as a whole, with the pope as its head, even though in reality most members had strong connections to their homelands.[11] Further east, the Turks, who had conquered most of Anatolia from Byzantium by the early fourteenth century, were hired as mercenaries by the Catalans and Byzantines and could also be classed as establishing military diasporas of sorts.[12] The Italian-ruled states of the Aegean, although providing an important military function, were primarily economic in purpose and thus fit more into the mould of a commercial rather than military diaspora. Nevertheless, the works of McKee on Venetian Crete and Wright on Genoese Lesbos, despite not being studies of military diasporas *per se*, provide perhaps the best examples of recent scholarly works on the Aegean that touch on the questions of ethnic identity and relations with homelands and host populations that concern us here.[13]

Relations with Byzantium

In the latter four decades of the thirteenth century, Byzantium's position in Asia Minor changed dramatically for the worse. In the earlier thirteenth century, the Byzantine Empire of Nikaia had coexisted with its neighbour, the Seljuk Sultanate of Rūm, in reasonable equilibrium. The arrival of the Mongols in eastern Anatolia, however, created a domino effect of migration among Turkic groups in the region. In 1259/1260, the Byzantine frontier was consequently disrupted, and despite multiple attempts to reassert imperial presence, the next 40 years were characterized overall by a precipitous loss of territory. In 1261, Michael VIII Palaiologos, emperor of Nikaia, took the former Byzantine capital of Constantinople from the Latins; thereafter, he found much of his energy drawn away from Anatolia towards Europe. Michael died in 1282 and was succeeded by his son, Andronikos II. In 1302, the Byzantines suffered a significant defeat at the hands of the burgeoning Ottoman *beylik* (principality) at the Battle of Bapheus. Soon, refugees were streaming across the Bosporus into Thrace; no longer able to rely on the manpower or tax revenue of the East—once the heart of the empire—Andronikos was in dire need of assistance.[14] In spring 1303, the ex-Templar Roger of Flor, who had served Frederick III of Sicily in the War of the Sicilian Vespers (1282–1302), sent ambassadors to Andronikos. He came with the offer of a mercenary force in the form of the Catalan Company, constituted following the end of the Sicilian war; it was an offer the emperor was in no position to refuse. The precise terms of the contract are unclear, but they were almost certainly weighted in the Catalans' favour.[15]

While the Catalans possessed groups of skilled soldiers whose military potential was desirable, Andronikos II engaged the Company primarily because it was available at the right time. That is not to say, however, that it did not come to offer Byzantium something that it otherwise lacked. As a corps of professional soldiers with experience of frontier warfare, the Catalans were *ipso facto* useful to the empire in western Anatolia. Their intended remit was to carry out a traditional role of military conquest, winning back the region from the Turks. The Company consisted of various groups. Frontier warriors known as *almogàvers* formed units of socially low-ranking infantry, perhaps 100 men strong (there remain uncertainties regarding *almogàver* cavalry).[16] *Almogatèns* were commoners in charge of units of *almogàvers*; *adalil*s were commoners identified as scouts or captains; archers were grouped as *mortati, murtat* or *murtadd*, who are sometimes identified as originating in the Cretan capital of Candia.[17] All these terms have Arabic roots, representing their origins as warriors on the Iberian Christian-Muslim frontier: *almogàvers* might derive their name from *al-maghāwīr* (singular: *al-mighwār*), referring to raiders or people of bold spirit; the word *almogatèn* comes from the Arabic *al-muqaddam*, in this context meaning a military captain; back in Iberia, *adalil*s (*al-dalīl*) had guided *almogàver*s through unfamiliar terrain and thus enjoyed high reputations despite being of low social rank; a *murtadd* originally meant a Muslim apostate and in this context referred to the child of a Greek and a Turk.[18] Because the Company did not produce its own provisions (needed for both men and horses) but lived off what it could raid or requisition, it soon exhausted the resources in one place and so needed to move on; this meant that it was unwilling and usually unable to sustain long sieges, something that compromised its ability to achieve its own conquest aims but that ultimately worked to Byzantium's advantage once relationships between the two sides broke down.[19]

In order to keep the Company in check, the Byzantines incorporated the Catalan leaders into their system of imperial court titulature. This practice was intended to create ties of interdependence, honouring and privileging the recipient as a member of the Byzantine elite while simultaneously rendering him indebted to imperial patronage and thus, in theory, less likely to turn against the empire. When the terms of the Company's engagement were agreed, Roger was offered the title *megas doux* ("grand duke") and the hand of Andronikos II's niece, Maria, in marriage. (Maria was probably Roger's second wife; she bore him a posthumous son.)[20] Then, on 25 December 1304, Berenguer of Entença, a close ally of Roger who had recently joined the Company with his own band of followers, was formally awarded the title of *megas doux*. In return, he swore fealty to Andronikos II and was therefore—in theory at least—bound to act in Byzantium's interests.[21]

Andronikos responded to the challenge of keeping Roger under control by rewarding him with yet more dignities, a principle he extended to other Catalan notables too. In April 1305, having ignored the emperor's invitation to come to Constantinople for Epiphany, Roger was offered the dignity of *kaisar* (after *despotes* and *sebastokrator*, the third most senior title) and the specific role of *strategos autokrator* (effectively commander-in-chief) of the Byzantine army in Asia Minor. According to the Byzantine historian Pachymeres, Roger was to hold the rural areas he conquered from the Turks but not the "distinguished" cities (the latter still being in imperial hands). The Catalan chronicler Muntaner, who was a member of the Company from 1302 to 1307, meanwhile, claims that Roger's remit was to include "the kingdom of Anatolia and all the islands of the Romania" and that he would have the right to grant cities, towns, and castles to his own men.[22] These two accounts clearly refer to the same promise but disagree about the scale and independence of Roger's mandate. Whether Andronikos really did propose that Roger should become a feudal lord over the eastern part of the empire, the Catalans chose to believe that he had. Upon Roger's death, the title "lord of Anatolia and the islands of the Romania" was adopted by Berenguer of Entença, a far more open enemy of Byzantium than Roger had ever been.[23] In 1307, the nobleman Ferran Ximenes of Arenós was himself elevated to the position of *megas doux* and married the Byzantine princess Theodora. Eiximenis commanded his own detachment of men, sometimes operating as part of the Company but seceding from it in 1303–1304 and 1307.[24] His case illustrates the fractious nature of the Company, built as it was on the followings of multiple military leaders who might find themselves at odds with one another. From the summer of 1307, Roger of Flor having been assassinated in 1305 (see below), the Catalan Company under its new leader Bernard of Rocafort settled at Kassandreia with the aim, ultimately unfulfilled, of conquering the empire's second city, Thessaloniki. Rocafort proved to be a stubborn opponent of outside influence, ensuring that Ferran of Mallorca, sent by Frederick III of Sicily, should fail to establish himself as the head of the Company.[25]

Andronikos' strategy of controlling foreign mercenary leaders through the conferment of dignities looks at first sight like a well-established Byzantine practice but may in fact have been exceptional. In the eleventh century, the Frankish mercenaries Roussel of Baillou and Hervé Frankopoulos ("son of the Frank") were accorded court titles: Roussel was a *vestes* (a dignity of low rank); Hervé was not only a *vestes*, but also a *magistros* (a dignity of higher rank) and *stratelates* (one term for "commander-in-chief") in the eastern regions.[26] Andronikos was obviously keen to avoid his hiring of the Catalans appearing in a more general sense unprecedented or radical, since he apparently made a speech in 1305 defending the decision on the basis of precedent: the empire had a long tradition of hiring mercenaries and Andronikos's more recent predecessors, John III Vatatzes (of Nikaia) and Michael VIII, had continued this in the post-1204 period.[27] The major prosopographical survey for the thirteenth to fifteenth centuries, however, lists only one Latin who had been honoured with the dignity of *megas doux* before the three Catalans: Licario (Ἰκάριος) of Negroponte (Euboea), who received the dignity from Michael VIII Palaiologos (Andronikos's father) in 1277, something reportedly mentioned by Andronikos in his speech of 1305.[28] While such hiring of mercenaries was nothing new in Byzantium, the level of privilege enjoyed by these Catalan figures, as high imperial dignitaries and consorts of princesses, appears exceptional for the time. The Catalans who held high Byzantine dignities did not participate in court life;[29] for them, the benefit of the privileges was rather the promise of a lordship in Asia Minor. In the end, this longed-for lordship would come to nothing, as relations between Byzantium and the Catalans broke down and the latter moved west across the southern Balkans.

A conspicuous feature of the Catalan Company's relationship with its host polity was the widespread capture and enslavement of Greek Christians. This began following the assassination of Roger of Flor in Adrianople (Edirne) at the hands of Alan mercenaries (30 April 1305). The deed occurred in the palace of Andronikos's son and co-emperor, Michael IX, probably upon Michael's orders.[30] Before this point, the Catalans had captured and enslaved Muslim Turks; the Catalan perception that Michael had broken faith, however, was seen to justify widespread destruction, slaughter, and captivity of Greek Christians, something generally called the "Catalan vengeance". At a time when Andronikos had ceased granting the Catalans largescale pay packages, the Company's members realized the financial potential of selling rather than killing their captives.[31] Venetian notarial records demonstrate the trafficking of Greek Christians from Gallipoli, Kassandreia, Mount Athos, and Thessaloniki to (and probably beyond) the Venetian Cretan capital, Candia. Hagiography offers further evidence of the Company's destruction on Mount Athos, mentioning a successful raid against the monastery of St Panteleimon and an unsuccessful raid against that of Hilandar. The trafficking and exploitation of Greeks continued well beyond 1311: the duchy of Athens, and its capital of Thebes in particular, were home to large-scale slave markets that exported both to other Aegean centres such as Candia and directly to Aragon-Catalonia and the Balearics.[32]

After the Catalan conquest of the Duchy of Athens, the relationship of the Christian Greek-speaking population with the Company became that of a subject group and a ruling elite, respectively. Greeks were generally denied the rights to own, acquire, sell, or bequeath property or to marry women of the Catalan ruling classes. Catalan men are known to have married Greek women; their children would grow up as Latin Christians, but such relationships would nevertheless have contributed to the breaking down of inter-ethnic barriers. A small number of Greeks are known to have obtained enfranchisement (i.e. to have gained the legal privileges of a Catalan). Some of the enfranchised are known to have attained the offices of notary (usually the only office open to Greeks in the Latin dominions) and occasionally, simultaneously, chancellor. By contrast, not one Greek is known to have been enfranchised under the Catalans' predecessors in the duchy, the house of Brienne.[33]

The most notable example of an upwardly-mobile Greek in the duchy is the notary Demetrius Rendi. Enfranchised in July 1362 by Frederick "the Simple" of Sicily (known by both regnal numbers III and IV), he and his progeny were accorded the rights of a born Catalan while also still allowed to worship according to the Greek rite. His property was extended in 1375–1377 as a reward for resisting Nerio Acciajuoli's attack on the town of Megara in 1374. He was, moreover, rewarded by Peter IV of Aragon with the perpetual and heritable office of chancellor in the city of Athens, attached to which was a salary paid from customs and tolls.[34] For the most part, enfranchised Greeks were privileged individuals; the exception to the rule were those inhabitants of Livadia who were rewarded with heritable enfranchisement for having thrown open their gates to the Company when it arrived in 1311.[35] The dynamic thus revealed is typically colonial: conquered inhabitants habitually excluded from the rights and privileges of the ruling class except in cases of (at least outward) self-subordination to the interests of that class.

Relations with Homeland/s

The precise details of the original formation and composition of the Company are not known, but the sources provide enough information about the origin of some of its members that the Company's connection with its homeland (or homelands) may be tentatively traced. Roger of Flor recruited the bulk of the Company between August 1302 and September 1303 from various groups who had fought for the Aragonese in the Sicilian war of 1282–1302. The core of these were men of Catalan or Aragonese origin—either those who had sailed from Iberia to Sicily to fight, or those who had been born in Sicily to Catalan-Aragonese settlers after the island's conquest by King Peter III of Aragon in 1282.[36] It is also likely that Roger recruited a number of Italians and Sicilians who were not born of Aragonese or Catalan parents, as suggested by the Florentine chronicler Giovanni Villani and several Byzantine writers.[37] Later on, once the Company was in the East, other contingents from Catalonia joined, such as those led by Bernard of Rocafort and Berenguer of Entença in 1304. According to Muntaner, these numbered 200 horsemen and 1,000 *almogàvers* and 300 horsemen and 1,000 *almogàvers*, respectively.[38] In the following years, the Company also recruited men from the lands in which they were fighting. Gregoras mentions a group of archers, who were presumably Greek, as they had been captured and then incorporated into the Company by the time of the battle of Halmyros (1311); little else is known of them.[39] The sources also talk of groups of Turks and Turcopoles (light, mounted archers) who joined in the spring and summer of 1305. They remained with the Company in the following years and likewise participated in the battle of Halmyros. We learn from Muntaner that the Turks arrived in two contingents, the first led by "Xemelic" (that is Ishak Melek, a descendant of the Seljuk Sultan Kaykaus II, r. 1246–1262) numbering 800 horsemen, and the second led by Ishak Melek's brother with another 400 horsemen and 200 foot-soldiers. Shortly after, a contingent of 1,000 Turcopoles defected from the emperor and also joined the Company.[40] The Venetian crusade propagandist, Marino Sanudo Torsello, who claims to have seen the Catalan army that fought at the battle of Halmyros, stated that it numbered 1,800 Turk and Turcopole cavalry, which seems plausible given the numbers above.[41] The sources do not elaborate on the specific role of the Turks and Turcopoles within the Company, but it is likely that they formed a corps of light cavalry or cavalry archers, which would have supplemented the Catalans' primarily infantry-based force.

The Company was, therefore, very ethnically and linguistically diverse, which makes it unlikely that a connection to a single land of origin was held by all its members. Nevertheless, excepting the Greeks and Turks, it is still evident that the bulk of the Company displayed a close association with King James II of Aragon (r. 1291–1327) and his younger brother, Frederick III of Sicily (r. 1295–1337), who between them ruled a constellation of territories—although not always united—in Iberia and the Mediterranean.[42] While hailing from diverse regions, the members of the Company could still share a common sense of allegiance to the Aragonese royal family as their "natural lords". This was certainly reflected in the written sources pertaining to the Company, which suggest a strong Catalan-Aragonese identity. Good evidence of this is supplied by Muntaner, who recorded that they used a banner and seal of St George, the patron saint of Catalonia, and went into battle with the war cry "Aragon! Aragon! Saint George! Saint George!".[43] External sources (notably documents from the papacy, Venice, and Genoa) also refer to the Company as being "Catalan" and/or recognize King James of Aragon as being its overlord; this suggests that it was generally perceived in the wider Latin sphere as being Catalan, despite the heterogeneous nature of its membership.[44] After its settlement in Greece, the major Byzantine historians, for their part, mostly employ the straightforward term Κατελάνοι ("Catalans") to refer to the Company, although a number of authors also foregrounded the Catalans' immediate Italian/Sicilian origins, apparently using these latter two locales synonymously.[45]

The ideological affiliation that the Catalans displayed towards the Aragonese royal family did not, however, mean that they wilfully submitted to the wider ambitions of James II or Frederick III in the Mediterranean, even though they were regularly called upon to do so. Disparities between the objectives of the Company's leadership and the Sicilian and Aragonese monarchs can be detected in the earliest months of action in the East, when Berenguer of Entença, who travelled to Constantinople to join Roger of Flor in September 1304, was tasked with making sure the Company followed the instructions of James II and Frederick III.[46] The exact nature of these instructions is unknown, but according to Pachymeres, the Genoese brought intelligence around that time that fleets were being prepared in Sicily and under Sancho of Aragon to arrive the following spring at Constantinople, where the invading forces could be expected to find a ready welcome among the Company.[47] If that was the case, then it seems that these plans were not carried out in any meaningful way: the Company's leadership maintained little communication with Frederick and seems never to have seriously contemplated an attack on Constantinople. Rather, as Angeliki Laiou has claimed, this actually marks the point at which Roger of Flor adopted a more conciliatory tone towards Andronikos—probably aided by the latter's bequests of imperial titles, noted above—and helped to moderate the more restless elements of his army.[48] More evidence of the Company's independence from Frederick III comes from April 1305, when the king's half-brother, Sancho of Aragon, arrived in the Aegean with a flotilla of galleys, under orders to launch naval raids against the empire and investigate the possibility of a Catalan rebellion. Here Roger refused to join Sancho and actively blocked him from attacking the heart of the empire by instructing him not to launch any raids north of the island of Lesbos.[49] Roger's ambivalence towards the Sicilian crown was also shared by others: Even after Roger's assassination, the army could not agree on whether to cooperate fully with Sancho.[50] Thus, as Laiou has argued, conflicts existed not only between the Catalans and the Byzantines but also between the interests of the Company's leadership and the Sicilian monarchy.[51]

After Roger's assassination on 30 April 1305, the Catalans ensconced themselves in Gallipoli and focused on the conquest of Thrace. They suffered a setback a month later when Entença was captured by Byzantium's Genoese allies and carried off to Genoa.[52] At this point, the Company took actions that both further established its independence but also, somewhat contradictorily, confirmed existing informal ties with its homelands and other external powers. Evidence of this is given by Muntaner, who claimed that he (as the Catalan commander of Gallipoli) had four standards made representing the royal arms of Aragon and Sicily, as well as of St George and St Peter of Rome. The first three were carried into battle, while that of St Peter was placed atop the Company's fortress at Gallipoli.[53] At this point, the Company also made several appeals to the kings of Aragon and Sicily, along with the pope and Charles of Valois (the titular Latin emperor), for aid against Byzantium.[54] Whether the Catalans actually considered themselves to be tied in any formal way to these powers seems unlikely, however: the clearest evidence of this is given by the negotiations undertaken between the Company and the Infant Ferran of Majorca (the cousin of both James II of Aragon and Frederick III of Sicily), who had been sent by Frederick to govern the Company in his name. Ferran met the Catalans at Gallipoli in May 1307, but by this time, the leadership was split between Bernard of Rocafort and Berenguer of Entença, the latter having been released by the Genoese and having returned to the Company. Entença, supported by Muntaner and Ferran Ximenes of Arenós, showed a willingness to accept Frederick's overlordship (through the Infant), but Rocafort, who held a grudge against the king, persuaded the greater part of the Company to offer fealty only to the Infant, and not to Frederick.[55] The Infant, unable to accept Rocafort's proposal, returned to Europe without concluding any agreement with the Company. This episode, along with the negotiations with Sancho of Aragon, reinforces the notion that although the Catalans were willing to emphasize links to their homelands, no doubt fostered by a genuine sense of kinship, neither the leadership nor the rank-and-file was committed to forming concrete alliances with the king of Sicily. Indeed, as Burns has argued, although "all the while it [the Company] made specific exception in these allegiances in favour of its immediate overlord Frederick of Sicily", nevertheless "above all, it never forgot that it was essentially Aragonese, proud to be a subject of James of Aragon".[56] The difference between the two monarchs was that James, unlike Frederick, never sought to control the Company directly and, even though he was considered by many outsiders as their natural sovereign, he rarely interfered in their business.[57]

After the departure of the Infant Ferran, the Catalans had little direct contact with their homelands until they defeated Walter of Brienne at the battle of Halmyros in 1311 and occupied the Duchy of Athens. In the Athenian period (1311–1388), the Company consolidated its Catalan-Aragonese identity and took steps to strengthen its ties with the Siculo-Aragonese monarchs in ways it had avoided while acting as an itinerant force before 1311. The lack of high nobles in the Company led it to seek leadership figures from outside its ranks: firstly,

Boniface of Verona in Negroponte (Euboea), who turned down the offer; secondly, Roger Deslaur, a Frank taken captive at Halmyros, who accepted but whose tenure was brief.[58] In 1312, the Company finally agreed to accept the overlordship of Frederick III of Sicily, who appointed his second son, Manfred, as duke of Athens. Manfred was a minor at this point, however, so Berenguer Estañol of Ampurias was appointed as vicar general and sent to Athens to govern in his name. Thereafter, the title of duke of Athens remained in the Siculo-Aragonese royal family, but as none of the dukes ever resided in the duchy itself, the vicar general continued to be the *de facto* governor.[59] By this point, the high-ranking members of the Company had settled in the lands they had conquered and assumed roles of local rulership, which they had usurped from the Frankish aristocrats they had defeated at Halmyros. Although the members of the Company theoretically held an enfeoffment from the duke of Athens (of royal lineage and resident in Sicily), they ruled their domain with little outside interference and reserved the highest ranks in the military and civil administration for themselves. They also continued to use their seal of St George alongside the royal seal.[60] The culture of this administration, however, shows strong links to the Company's ancestral lands: Catalan and Latin were employed as the two official languages of the chancery and municipal courts;[61] Catalan statutes were gradually imported into the duchy, such as the Customs of Barcelona, which replaced the Frankish Assizes and Customs of the Romania; the court of the vicar general, meanwhile, replaced that of the Frankish baronage.[62] Throughout this time, the Catalans also recognized the overarching authority of the king of Aragon, even if their suzerain remained the king of Sicily. An example of this comes from *c.*1328, when the vicar general Alfonso Fadrique, the illegitimate son of Frederick of Sicily, wrote to King Alphonse III of Aragon, asking Alphonse to intercede on his behalf with Frederick to ensure that the fortress of Neopatras be granted to him as a fief. Other examples of the Aragonese kings interceding on behalf of the Catalan duchies with the king of Sicily also exist, and interestingly these intercessions do not seem to have been challenged in either Sicily or Greece. As Setton has stated (in a similar vein to Burns), "Beneath the multiplicity of crowns that comprised the Catalan-Aragonese confederation of the fourteenth century there was a peculiar unity and attachment to the homeland [i.e. Aragon]", which the Catalan Company certainly shared.[63]

Conclusions

The Catalan Company was one of a number of mercenary bands employed by the later Byzantine emperors, but its relations with the empire were marked from the outset by unequal power dynamics, the Catalan Company being significantly larger and more effective than the empire could cope with, and its leading figures volatile and far from united. This mismatch explains why the emperor granted several leading figures honorary imperial titles that were more generous than was customary in the period, and why the Catalan leadership was allowed to marry into the imperial family. Each of these leaders had their own contingents of followers among the Company, thus leading to factionalism and secessions. The Company took several years to find a permanent home: during its itinerant phase, it was unable to sustain long sieges and, not producing provisions of its own, supported itself by exploitation of both the land and the local population wherever it went. In particular, the Company relied heavily on the profits of slavery, something that persisted into its settled phase in the Duchy of Athens from 1311. Despite its heterogeneous membership, throughout its itinerant period, it is clear that the bulk of the Company showed an outward affiliation to the kings of Aragon and Sicily, even though it continually refused to engage with Frederick of Sicily's wider Mediterranean ambitions, preferring instead to remain as autonomous as possible. Once the Company had settled in Athens, however, it sought its figurehead from outside its own ranks and ultimately accepted the overlordship of Frederick and incorporated the Catalan language and statutes into its administration. During this period, the Company developed a colonialist relationship with its Greek subjects, who were allowed to own property, seek (limited) public office or marry Catalan women only if enfranchised in return for service to the Catalan ruling classes. The Catalan Company thus offers a notable example of a military diaspora that may be traced from its engagement by a host polity (Byzantium), through its evolving relationship with that polity and through its multifarious relationships with its ancestral lands and overlords both immediate and ultimate (Sicily and Aragon, respectively), all the way to a period of settlement and polity formation.

Notes

1 David Jacoby, "The Catalan Company in the East: The Evolution of an Itinerant Army (1303–1311)," in *The Medieval Way of War: Studies in Medieval Military History in Honor of Bernard S. Bachrach*, ed. Gregory I. Halfond (London: Routledge, 2015), 153–182.

2 *Les quatre grans cròniques: III. Crònica de Ramon Muntaner*, ed. Ferran Soldevila, Jordi Bruguera, and M. Teresa Ferrer i Mallol (Barcelona, 2011); an up-to-date English translation of the Company's activities from 1303 to 1311 is given by Richard D. Hughes, intr. Jocelyn N. Hillgarth, *The Catalan Expedition to the East: From the Chronicle of Ramon Muntaner* (Barcelona: Barcino, 2006). References will be given to chapters in order to facilitate the use of earlier editions.

3 Georgios Pachymeres, *Relations historiques*, ed. Albert Failler, trans. Vitalien Laurent, 5 vols. (Paris: Institut français d'études byzantines, 1984–1999).

4 For example, Robert Ignatius Burns, "The Catalan Company and the European Powers, 1305–1311," *Speculum* 29 (1954): 751–771; Angeliki E. Laiou, *Constantinople and the Latins: The Foreign Policy of Andronicus II, 1282–1328* (Cambridge, MA: Harvard University Press, 1972), 134–243. Savvas Kyriakidis, "The Employment of Large Groups of Mercenaries in Byzantium in the Period ca. 1290–1305 as Viewed by the Sources," *Byzantion* 79 (2009): 208–230; Scott Jessee and Anatoly Isaenko, "The Military Effectiveness of Alan Mercenaries in Byzantium, 1301–1306," *Journal of Medieval Military History* 11 (2013): 107–132; Nikolaos S. Kanellopoulos and Ioanna K. Lekea, "Prelude to Kephissos (1311): An Analysis of the Battle of Apros (1305)," *Journal of Medieval Military History* 12 (2014): 119–138; Jacoby, "The Catalan

Company in the East"; Mike Carr, "Friend or Foe? The Catalan as Proxy Actors in the Aegean and Asia Minor Vacuum," *Journal of Medieval Military History* 14 (2016): 163–177.

5 For a bibliography of Rubió i Lluch's many works, see *Homenatge a Antoni Rubió i Lluch: Miscellània d'estudis literaris, històrics i linguistics* (Barcelona: Institut d'Estudis Catalans, 1936), 1: ix–xv. Perhaps his most important contribution to scholarship on the Catalans was the collection of sources published posthumously: Antoni Rubió i Lluch, ed., *Diplomatari de l'Orient català, 1301–1409: Collecció de documents per a la història de l'expedició catalana a Orient i dels ducats d'Atenes i Neopàtria* (Barcelona: Institut d'estudis Catalans, 1947) [henceforth *DOC*].

6 Kenneth Meyer Setton, "The Avignonese Papacy and the Catalan Duchy of Athens," *Byzantion* 17 (1944–45): 281–230; id., *Catalan Domination of Athens: 1311–1388* (Cambridge, MA: Mediaeval Academy of America, 1948); id., *Los catalanes en Grecia* (Barcelona: Ed. Aymá, 1975); id., "The Catalans in Greece, 1311–1380," in *A History of the Crusades*, ed. Kenneth Meyer Setton (Madison, WI: University of Wisconsin Press, 1975), 3: 167–224; id. "Catalan Society in Greece in the Fourteenth Century," in *Essays to the Memory of Basil Laourdas*, ed. Louisa Laourdas (Thessalonica: E. Sfakianakis, 1975), 241–284, repr. in id., *Athens in the Middle Ages* (London: Variorum, 1975), Item V; id., *The Papacy and the Levant: 1204–1571* (Philadelphia: Medieval Academy of America, 1976–84), 1: 441–473.

7 Sally McKee, *Uncommon Dominion: Venetian Crete and the Myth of Ethnic Purity* (Philadelphia: University of Pennsylvania Press, 2000).

8 David Jacoby, "Catalans, Turcs et Vénitiens en Romanie (1305–1332): Un nouveau témoignage de Marino Sanudo Torsello," *Studi Medievali* 15 (1974): 217–261; Elizabeth A. Zachariadou, "The Catalans of Athens and the Beginning of Turkish Expansion in the Aegean Area," *Studi Medievali* 21 (1980), 821–838; id., *Trade and Crusade: Venetian Crete and the Emirates of Menteshe and Aydin, 1300–1415* (Venice: Hellenic Institute of Byzantine and Post-Byzantine Studies, 1983), 13–16; Norman Housley, *The Avignon Papacy and the Crusades, 1305–1378* (Oxford: Clarendon, 1986), 16–17, 118–119, 156; Karol Polejowski, "The Hospitallers of Rhodes and Attempts to Recover the Duchy of Athens by the Counts of Brienne after 1311," in *Islands and the Military Orders, c. 1291–1798*, ed. Emmanuel Buttigieg and Simon Phillips (London: Routledge, 2013), 139–145; Mike Carr, *Merchant Crusaders in the Aegean, 1291–1352* (Woodbridge: Boydell and Brewer, 2015), 69–70, 95–96, 112–113.

9 Eusebi Ayensa Prat, "Catalan Domination in Greece during the 14th Century: History, Archaeology, Memory and Myth," *Catalan Historical Review* 13 (2020): 43–58.

10 Peter Topping, *Studies on Latin Greece, A.D. 1205–1715* (Aldershot: Variorum, 1977).

11 See, for example, Anthony T. Luttrell, "Feudal Tenure and Latin Colonization at Rhodes: 1306–1415," *The English Historical Review* 85 (1970): 755–775, and in general: id., *The Hospitallers in Cyprus, Rhodes, Greece, and the West, 1291–1440: Collected Studies* (Aldershot: Variorum, 1978); id., *Latin Greece, the Hospitallers and the Crusades, 1291–1440* (Aldershot: Variorum, 1982); id., *The Hospitallers of Rhodes and Their Mediterranean World* (Aldershot: Variorum, 1992); id., *The Hospitaller State on Rhodes and Its Western Provinces, 1306–1462* (Aldershot: Variorum, 1999); id., *Studies on the Hospitallers after 1306: Rhodes and the West* (Aldershot: Variorum, 2007).

12 For the employment of Turks by the Catalans, see below, nn. 40–41. For the employment of Turks by the Byzantines, see Samuel N. C. Lieu, "Between Byzantium and the Turks—Kallipolis/Gallipoli/Gelibolu (1307–1402)," in *Language, Society, and Religion in the World of the Turks: Festschrift for Larry Clark at Seventy-Five*, ed. Zsuzsanna Gulácsi (Turnhout: Brepols, 2018), 205–247.

13 McKee, *Uncommon Dominion*; Christopher F. Wright, *The Gattilusio Lordships and the Aegean World, 1355–1462* (Leiden: Brill, 2014).

14 In addition to Laiou, *Byzantium and the Latins*, see Dimitri A. Korobeinikov, *Byzantium and the Turks in the Thirteenth Century* (Oxford: Oxford University Press, 2014); id., "The Formation of Turkish Principalities in the Boundary Zone: From the

Emirate of Denizli to the Beylik of Menteşe (1256–1302)," *in Uluslararası Batı Beylikleri Tarih, Kültür ve Medeniyeti Sempozyumu-II: Menteşeoğulları tarihi, 25–27 Nisan 2012, Muğla, bildiriler*, ed. Adnan Çevik and Murat Keçiş (Ankara: Türk Tarih Kurumu, 2016), 65–76; George G. Arnakis, "Byzantium's Anatolian Provinces during the Reign of Michael Palaeologus," in *Actes du XIIe congrès d'études byzantines, Ochride 10–16 septembre 1961*, ed. Georges Ostrogorsky, Đurđe Bošković, Svetozar Radojčić, Franjo Barišić, Jadran Ferluga, Ivanka Nikolajević, Nada Mandić and Borislav Radojčić (Belgrade: Naučno Delo, 1964), 2: 37–44.

15 Jacoby, "The Catalan Company in the East," 153–155. For a discussion of the wages of the Catalans and other mercenaries in Byzantium, see Mark C. Bartusis, *The Late Byzantine Army: Arms and Society, 1204–1453* (Pennsylvania: University of Pennsylvania Press, 1992), chap. 7.

16 For an interesting comparison, see the chapter by Weltert and Christ in this volume.

17 For the *mortati*, see Aldo Cerlini, "Nuovo lettere di Marino Sanudo il vecchio," *La bibliofilia* 42 (1940): 321–359, at 337 and 352, n. 47 (doc. II).

18 Jacoby, "The Catalan Company in the East," 178–180; for *almogatèn* and *adalil*, see too Hillgarth's introduction to Hughes' translation of the *Chronicle of Ramon Muntaner, Catalan Expedition*, p. 42, n. 56 and refs.; for *almogatèns*, see Josep-David Garrido Valls, "Arabismes del vocabulari militar català medieval," in *Miscel·lània Germà Colón* (Barcelona: Abadia de Montserrat, 1995), 4: 5–21 at 10–11 and 13. Cf. the discrete entries in Hans Wehr, *Dictionary of Modern Written Arabic*, ed. J. Milton Cowan, 4th ed. (Wiesbaden: Otto Harrassowitz, 1979).

19 Jacoby, "The Catalan Company in the East," 182; Laiou, *Constantinople and the Latins*, 168.

20 Pachymeres, 4.433, trans. 432; Jacoby, "The Catalan Company in the East," 154; Erich Trapp, Rainer Walther, and Hans-Veit Beyer, eds., *Prosopographisches Lexikon der Palaiologenzeit* (Vienna: Österreichische Akademie der Wissenschaften, 1976–2000) [hereafter *PLP*], #24386.

21 Pachymeres, 4.545, trans. 544; Laiou, *Constantinople and the Latins*, 141.

22 Pachymeres, 4.553, 555, trans. 552, 554; Muntaner, chap. 212 (Hughes trans. chap. 17); Laiou, *Constantinople and the Latins*, 143–144.

23 Laiou, *Constantinople and the Latins*, 136, 154; for Berenguer's title, see e.g. *DOC*, doc. 14 (10 May 1305).

24 *PLP* #27944; Jacoby, "The Catalan Company in the East," 162, 166, 177.

25 See below, n. 55, and Muntaner, chap. 230, 233 (Hughes trans. chap. 34, 36); Jacoby, "The Catalan Company in the East," 156.

26 Jonathan Shepard, "The Uses of the Franks in Eleventh-Century Byzantium," *Anglo-Norman Studies* 15 (1993): 275–305, at 297 and 300; Alexander P. Kazhdan et al., eds., *Oxford Dictionary of Byzantium* (Oxford: Oxford University Press, 1991), s.vv. "Vestes" (p. 2162), "Magistros" (p. 1267), "Stratelates" (p. 1965).

27 Pachymeres, 4.597, trans. 596; Laiou, *Constantinople and the Latins*, 165. For the pre-1204 period, see the chapter by Shliakhtin in this volume.

28 Pachymeres, 4.597, trans. 596; *PLP* #8154.

29 Jacoby, "The Catalan Company in the East," 167.

30 Laiou, *Constantinople and the Latins*, 146; Jacoby, "The Catalan Company in the East," 156.

31 Laiou, *Constantinople and the Latins*, 189: "After Roger de Flor's death, no large sums of money were given to the Catalans, mainly because the two parties could not agree on how dearly peace could be bought."

32 Daniel Duran Duelt, "La Companyia catalana i el comerç d'esclaus abans de l'assentament als Ducats d'Atenes i Neopàtria," in *De l'esclavitud a la llibertat: Esclaus i lliberts a l'edat mitjana, actes del Col·loqui Internacional celebrat a Barcelona, del 27 al 29 de maig de 1999*, ed. María Teresa Ferrer i Mallol and Josefina Mutgé i Vives (Barcelona: CSIC, 2000), 557–571; id., "Los Ducados de Atenas y Neopatria en el comercio regional e internacional durante la dominación catalana (siglo XIV). II: El comercio

de larga distancia a través del observatorio de Barcelona y Mallorca," *Estudios bizantinos* 7 (2019): 85–118; Ernest Marcos Hierro, "The Catalan Company and the Slave Trade," in *Slavery and the Slave Trade in the Eastern Mediterranean (c. 1000–1500 CE)*, ed. Reuven Amitai and Christoph Cluse (Turnhout: Brepols, 2017), 321–352. For an integrated study of Greek Christian captives and their enslavement and trafficking, see Alasdair C. Grant, "Cross-Confessional Captivity in the Later Medieval Eastern Roman World, c. 1280–1450" (PhD diss., University of Edinburgh, 2021).

33 Setton, "Catalan Society," 167–224; id., *Catalan Domination*, 157–221.

34 Setton, *Catalan Domination*, 157–169, 243–245; id., "Catalan Society," 269–270; A. Rubió i Lluch, "Une figure athénienne de l'époque de la domination catalane: Dimitri Rendi," *Byzantion* 2 (1925): 193–229.

35 Setton *Catalan Domination*, 244; *DOC*, doc. 268 (July 1362).

36 Jacoby, "The Catalan Company in the East," 158–159; Muntaner, chap. 200–201 (Hughes trans. chap. 6–7), mentions that the members were all of Catalan or Aragonese origin.

37 Giovanni Villani, *Nuova cronica*, ed. Giuseppe Porta (Parma: U. Guanda, 1990–1), 2: 83–84, mentions Genoese and other Italians. Georgios Pachymeres, Nikephoros Gregoras, Thomas Magistros, and Laonikos Chalkokondyles all allude to their Sicilian background: Burns, "The Catalan Company and the European Powers," 766; cf. Jacoby, "The Catalan Company in the East," 159.

38 Muntaner, chap. 206, 211 (Hughes trans. chap. 12, 16); Jacoby, "The Catalan Company in the East," 166.

39 Nikephoros Gregoras, *Byzantina Historia*, ed. Ludwig Schopen and Immanuel Bekker (Bonn: Weber, 1829–1855), 1:252; Jan L. van Dieten, trans., *Rhomäische Geschichte: Historia Rhomaïke* (Stuttgart: Anton Hiersemann, 1973–2007), 1:194; Jacoby, "The Catalan Company in the East," 159.

40 Jacoby, "The Catalan Company in the East," 159–160; Muntaner, chap. 228 (Hughes trans. chap. 32); Pachymeres, 4.643, trans. 642.

41 Cerlini, "Nuovo lettere di Marino Sanudo il vecchio," 352. Muntaner also mentions that the Turks and Turcopoles fought for the Catalans at Halmyros, but he does not give numbers: Muntaner, chap. 240 (Hughes trans. chap. 45).

42 On the relationship between the Aragonese and Sicilian territories in the Mediterranean, see Jocelyn N. Hillgarth, "The Problem of a Catalan Mediterranean Empire 1229–1327," *English Historical Review Supplement* 8 (1975): 1–54; Burns, "The Catalan Company and the European Powers," 765.

43 Muntaner, chap. 207, 219 (Hughes trans. chap. 19, 22–23); Joan-Pau Rubiés Mirabet, "Rhetoric and Ideology in the Book of Ramon Muntaner," *Mediterranean Historical Review* 26 (2011): 1–29, at 17; Jacoby, "The Catalan Company in the East," 161.

44 *DOC*, docs. 17, 19, 21, 33, 35, 43, 57–58; Jacoby, "The Catalan Company in the East," 161; Laiou, *Constantinople and the Latins*, 222.

45 Discussed in Burns, "The Catalan Company and the European Powers," 766.

46 There is some confusion over the nature of the instructions given to Entença, and exactly which monarch he owed fealty to. This arises from a letter of Entença to James II in which he stated that he was travelling to the Romania to execute James's orders, but where he also pledged to work for "the salvation and prosperity of the great lord, King Frederick III", which implies that he recognized the latter as his sovereign. This is further supported by the remarks of Pachymeres, who suggested that Entença had sworn an oath of fealty to Frederick before leaving Sicily: Pachymeres, 4.547, trans. 546; Laiou, *Constantinople and the Latins*, 139–140.

47 Pachymeres, 4.535, 537, trans. 534, 536; Laiou, *Constantinople and the Latins*, 141–142.

48 Laiou, *Constantinople and the Latins*, 142–143.

49 Pachymeres, 4.571, trans. 570.

50 Ibid., 4.577, trans. 576.

51 Laiou, *Constantinople and the Latins*, 144–145.

52 Muntaner, chap. 218 (Hughes trans. chap. 21); Laiou, *Constantinople and the Latins*, 154.

53 Muntaner, chap. 219 (Hughes trans. chap. 22).
54 Laiou, *Constantinople and the Latins*, 177–179.
55 Carr, "Friend or Foe?" 174–175; Burns, "The Catalan Company and the European Powers," 767; Laiou, *Constantinople and the Latins*, 180–181.
56 Burns, "The Catalan Company and the European Powers," 765.
57 On this, see ibid., 765–767 and Hillgarth, "Problem of a Catalan Mediterranean Empire," 43–44.
58 Jacoby, "The Catalan Company in the East," 163; Setton, *Catalan Domination*, 14–15.
59 Setton, *Catalan Domination*, 15–18.
60 Ibid., 19.
61 Ibid., 216–220; *DOC*, docs. 391 (Articles of Athens), 392 (Articles of Salona).
62 Setton, *Catalan Domination*, 19–20.
63 Ibid., 37.

References

Primary Sources

Gregoras, Nikephoros. *Byzantina Historia*, edited by Ludwig Schopen and Immanuel Bekker. 3 vols. Bonn: Weber, 1829–1855, translated by Jan L. van Dieten. Rhomäische Geschichte: Historia Rhomaïke. 6 vols. Stuttgart: Anton Hiersemann, 1973–2007.

Muntaner, Ramon. *The Catalan Expedition to the East: From the Chronicle of Ramon Muntaner*, translated by Richard D. Hughes, introduction by Jocelyn N. Hillgarth. Barcelona: Barcino, 2006.

Muntaner, Ramon. *Les quatre grans cròniques: III. Crònica de Ramon Muntaner*, edited by Ferran Soldevila, Jordi Bruguera, and M. Teresa Ferrer i Mallol. Barcelona: Institut d'Estudis Catalans, 2011.

Pachymeres, Georgios. *Relations historiques*, edited by Albert Failler, translated by Vitalien Laurent. 5 vols. Paris: Institut français d'études byzantines/Les Belles Lettres, 1984–1999.

Rubió i Lluch, Antoni, ed. *Diplomatari de l'Orient català, 1301–1409: Colleció de documents per a la història de l'expedició catalana a Orient i dels ducats d'Atenes i Neopàtria*. Barcelona: Institut d'estudis Catalans, 1947.

Sanudo Torsello, Marino. "Nuovo lettere di Marino Sanudo il Vecchio", ed. Aldo Cerlini. La bibliofilia 42 (1940): 321–359.

Villani, Giovanni. Nuova cronica, edited by Giuseppe Porta. 3 vols. Parma: U. Guanda, 1990–1991.

Secondary Works

Arnakis, George G. "Byzantium's Anatolian Provinces during the Reign of Michael Palaeologus." In *Actes du XIIe congrès d'études byzantines, Ochride 10–16 septembre 1961*, edited by Georges Ostrogorsky, Đurđe Bošković, Svetozar Radojčić, Franjo Barišić, Jadran Ferluga, Ivanka Nikolajević, Nada Mandić, and Borislav Radojčić, 37–44. Vol. 2. Belgrade: Naučno Delo, 1964.

Ayensa Prat, Eusebi. "Catalan Domination in Greece during the 14th Century: History, Archaeology, Memory and Myth." *Catalan Historical Review* 13 (2020): 43–58.

Bartusis, Mark C. *The Late Byzantine Army: Arms and Society, 1204–1453*. Philadelphia: University of Pennsylvania Press, 1992.

Burns, Robert Ignatius. "The Catalan Company and the European Powers, 1305–1311." *Speculum* 29 (1954): 751–771.

Carr, Mike. *Merchant Crusaders in the Aegean, 1291–1352*. Woodbridge: Boydell and Brewer, 2015.

Carr, Mike. "Friend or Foe? The Catalan as Proxy Actors in the Aegean and Asia Minor Vacuum." *Journal of Medieval Military History* 14 (2016): 163–177.

Duran Duelt, Daniel. "La Companyia catalana i el comerç d'esclaus abans de l'assentament als Ducats d'Atenes i Neopàtria." In *De l'esclavitud a la llibertat: Esclaus i lliberts a l'edat mitjana, actes del Col·loqui Internacional celebrat a Barcelona, del 27 al 29 de maig de 1999*, edited by María Teresa Ferrer i Mallol and Josefina Mutgé i Vives, 557–571. Barcelona: CSIC, 2000.

Duran Duelt, Daniel. "Los Ducados de Atenas y Neopatria en el comercio regional e internacional durante la dominación catalana (siglo XIV). II: El comercio de larga distancia a través del observatorio de Barcelona y Mallorca." *Estudios bizantinos* 7 (2019): 85–118.

Garrido Valls, Josep-David. "Arabismes del vocabulari militar català medieval." In *Miscel·lània Germà Colón*, 5–21. Vol. 4. Barcelona: Abadia de Montserrat, 1995.

Grant, Alasdair C. "Cross-Confessional Captivity in the Later Medieval Eastern Roman World, c. 1280–1450." PhD diss., University of Edinburgh, 2021.

Hillgarth, Jocelyn N. "The Problem of a Catalan Mediterranean Empire 1229–1327." *English Historical Review Supplement* 8 (1975): 1–54.

Homenatge a Antoni Rubió i Lluch: Miscellània d'estudis literaris, històrics i linguistics. 3 vols. Barcelona: Institut d'Estudis Catalans, 1936.

Housley, Norman. *The Avignon Papacy and the Crusades, 1305–1378*. Oxford: Clarendon, 1986.

Jacoby, David. "Catalans, Turcs et Vénitiens en Romanie (1305–1332): Un nouveau témoignage de Marino Sanudo Torsello." *Studi Medievali* 15 (1974): 217–261.

Jacoby, David. "The Catalan Company in the East: The Evolution of an Itinerant Army (1303–1311)." In *The Medieval Way of War: Studies in Medieval Military History in Honor of Bernard S. Bachrach*, edited by Gregory I. Halfond, 153–182. London: Routledge, 2015.

Jessee, Scott, and Anatoly Isaenko. "The Military Effectiveness of Alan Mercenaries in Byzantium, 1301–1306." *Journal of Medieval Military History* 11 (2013): 107–132.

Kanellopoulos, Nikolaos S., and Ioanna K. Lekea. "Prelude to Kephissos (1311): An Analysis of the Battle of Apros (1305)." *Journal of Medieval Military History* 12 (2014): 119–138.

Kazhdan, Alexander P. et al., *eds. Oxford Dictionary of Byzantium*. Oxford: Oxford University Press, 1991.

Korobeinikov, Dimitri A. *Byzantium and the Turks in the Thirteenth Century*. Oxford: Oxford University Press, 2014.

Korobeinikov, Dimitri A. "The Formation of Turkish Principalities in the Boundary Zone: From the Emirate of Denizli to the Beylik of Menteşe (1256–1302)." In *Uluslararası Batı Beylikleri Tarih, Kültür ve Medeniyeti Sempozyumu-II: Menteşeoğulları tarihi, 25–27 Nisan 2012, Muğla, bildiriler*, edited by Adnan Çevik and Murat Keçiş, 65–76. Ankara: Türk Tarih Kurumu, 2016.

Kyriakidis, Savvas. "The Employment of Large Groups of Mercenaries in Byzantium in the Period ca. 1290–1305 as Viewed by the Sources." *Byzantion* 79 (2009): 208–230.

Laiou, Angeliki E. *Constantinople and the Latins: The Foreign Policy of Andronicus II, 1282–1328*. Cambridge, MA: Harvard University Press, 1972.

Lieu, Samuel N. C. "Between Byzantium and the Turks—Kallipolis/Gallipoli/Gelibolu (1307–1402)." In *Language, Society, and Religion in the World of the Turks: Festschrift for Larry Clark at Seventy-Five*, edited by Zsuzsanna Gulácsi, 205–247. Turnhout: Brepols, 2018.

Luttrell, Anthony T. "Feudal Tenure and Latin Colonization at Rhodes: 1306–1415." *The English Historical Review* 85 (1970): 755–775.

Luttrell, Anthony T. *The Hospitallers in Cyprus, Rhodes, Greece, and the West, 1291–1440: Collected Studies*. Aldershot: Variorum, 1978.

Luttrell, Anthony T. *Latin Greece, the Hospitallers and the Crusades, 1291–1440*. Aldershot: Variorum, 1982.

Luttrell, Anthony T. *The Hospitallers of Rhodes and Their Mediterranean World*. Aldershot: Variorum, 1992.

Luttrell, Anthony T. *The Hospitaller State on Rhodes and Its Western Provinces, 1306–1462*. Aldershot: Variorum, 1999.

Luttrell, Anthony T. *Studies on the Hospitallers after 1306: Rhodes and the West*. Aldershot: Variorum, 2007.

Marcos Hierro, Ernest. "The Catalan Company and the Slave Trade." In *Slavery and the Slave Trade in the Eastern Mediterranean (c. 1000–1500 CE)*, edited by Reuven Amitai and Christoph Cluse, 321–352. Turnhout: Brepols, 2017.

McKee, Sally. *Uncommon Dominion: Venetian Crete and the Myth of Ethnic Purity*. Philadelphia: University of Pennsylvania Press, 2000.

Polejowski, Karol. "The Hospitallers of Rhodes and Attempts to Recover the Duchy of Athens by the Counts of Brienne after 1311." In *Islands and the Military Orders, c. 1291–1798*, edited by Emmanuel Buttigieg and Simon Phillips, 139–145. London: Routledge, 2013.

Rubiés Mirabet, Joan-Pau. "Rhetoric and Ideology in the Book of Ramon Muntaner." *Mediterranean Historical Review* 26 (2011): 1–29.

Rubió i Lluch, Antoni. "Une figure athénienne de l'époque de la domination catalane: Dimitri Rendi." *Byzantion* 2 (1925): 193–229.

Setton, Kenneth M. "The Avignonese Papacy and the Catalan Duchy of Athens." *Byzantion* 17 (1944–1945): 281–303.

Setton, Kenneth M. *Catalan Domination of Athens: 1311–1388*. Cambridge, MA: Mediaeval Academy of America, 1948.

Setton, Kenneth M. "Catalan Society in Greece in the Fourteenth Century." In *Essays to the Memory of Basil Laourdas*, edited by Louisa Laourdas, 241–284. Thessalonica: E. Sfakianakis, 1975. Reprinted in id., *Athens in the Middle Ages*. London: Variorum, 1975, Item V.

Setton, Kenneth M. *Los catalanes en Grecia*. Barcelona: Ed. Aymá, 1975.

Setton, Kenneth M. "The Catalans in Greece, 1311–1380." In *A History of the Crusades*, edited by Kenneth M. Setton, 167–224. Vol. 3. Madison, WI: University of Wisconsin Press, 1975.

Setton, Kenneth M. *The Papacy and the Levant: 1204–1571*. 4 vols. Philadelphia: Medieval Academy of America, 1976–1984.

Shepard, Jonathan. "The Uses of the Franks in Eleventh-Century Byzantium." *Anglo-Norman Studies* 15 (1993): 275–305.

Topping, Peter. *Studies on Latin Greece, A.D. 1205–1715*. Aldershot: Variorum, 1977.

Trapp, Erich, Rainer Walther, and Hans-Veit Beyer, eds. *Prosopographisches Lexikon der Palaiologenzeit*. Vienna: Österreichische Akademie der Wissenschaften, 1976–2000.

Wehr, Hans. *Dictionary of Modern Written Arabic*, edited by J. Milton Cowan. 4th ed. Wiesbaden: Otto Harrassowitz, 1979.

Wright, Christopher F. *The Gattilusio Lordships and the Aegean World, 1355–1462*. Leiden: Brill, 2014.

Zachariadou, Elizabeth A. "The Catalans of Athens and the Beginning of Turkish Expansion in the Aegean Area." *Studi Medievali* 21 (1980): 821–838.

Zachariadou, Elizabeth A. *Trade and Crusade: Venetian Crete and the Emirates of Menteshe and Aydin, 1300–1415*. Venice: Hellenic Institute of Byzantine and Post-Byzantine Studies, 1983.

8

CHRISTIAN EXPATRIATES IN MUSLIM LANDS

The Many Roles of Aragonese Mercenaries in Medieval Northern Africa

Nikolas Jaspert[1]

In the Middle Ages, military activities generated mobility in many ways. For one, they could lead to the conquest and occupation of new territories, which in turn stimulated emigration of military personnel in order to hold the newly won zones. Then, armed conflicts also forced involuntary expatriation upon refugees or prisoners of war. Finally, violence could provide a framework in which mercenaries made a military career for themselves in unknown lands—the subject of this chapter. It will deal with an unusual group of such mobile militias: Mercenaries who served masters of other creeds, more precisely, Christian soldiers employed by Muslim rulers from the eleventh century to the fifteenth century.[2] These paid combatants represent a particular case of military diasporas, because they not only shared a specific ethnic background but also followed a different belief system to that of the societies in which they were active. Did this have an impact upon their role within the armed forces of the Islamic polity in which they served? What consequences did their particular background have upon the degree to which they integrated into their host societies? Was their land of origin an important component of their identity politics? What was the specific relationship between Muslim employers and Christian employees, and did the latter feel a conflict of loyalty vis-à-vis the rulers of their home countries? To articulate the aim of this chapter in more general terms: Which insights into processes of transcultural contact and communication can be gained by studying mercenaries as cultural brokers and trans-imperial subjects between Islamic and Christian polities?[3] This chapter will first deal with mercenaries in the Western Mediterranean during the eleventh and thirteenth centuries before focusing on the later Middle Ages. It will attempt to shed light on the social cohesion of these groups, on their involvement in domestic politics, their role as Christian representatives in Islamic lands, and their employment by Muslim rulers as diplomatic envoys. It focusses in particular on notions of identity, loyalty, and belonging.

DOI: 10.4324/9781003245568-9

The mercenary—medieval or modern—is commonly seen as a highly ambiguous individual: He is often considered untrustworthy, an unscrupulous agent not tied to superior moral values who offers his services to those who pay the most.[4] But clearly, from a historical standpoint, a more nuanced view is necessary in order to avoid making moralizing simplifications. Stephen Morillo has outlined two ways of categorizing paid fighters from the perspective of cultural studies: First, according to the degree of their embeddedness into the society in which they work; and second, according to their motivations for serving any particular master.[5] In applying these two categories, mercenaries can be situated along a wide spectrum with blurred borders that encompasses at least four major groups: First, recruited locals and thus, in principle, socially integrated men who fought for money ("social armies", to use Morillo's term); second, paid subjects, that is socially integrated men who were obliged to provide military service, but who were nevertheless paid ("stipendiaries"); third, foreign praetorians who remained more detached from society than the other two groups and whose main task it was to provide security to the potentate ("political armies"); and finally "mercenaries": Mobile professional fighters who rapidly changed from one employer in the pursuit of maximum profit. Most of the mercenaries studied in this chapter belong to the third group, the praetorians.[6] Such men who left their homeland in order to fight for pay in other countries and formed communities of their own which might be termed diasporic. However, diaspora is a difficult or at least ambiguous term, as it elicits associations with the Jewish, Armenian, and Greek diasporas, conjuring up images of involuntary emigration.[7] I prefer to speak of "expatriation"—in this case, military expatriation—as it places focus more on the abandonment of one's homeland than on the reasons for leaving.

Christian Mercenaries Serving Almoravid and Almohad Rulers (Eleventh to Thirteenth Centuries)

Since the ninth century, Christians are known to have fought in al-Andalus in the service of Muslim rulers.[8] Amongst these fighters, one figure stands out, a man whose life was apparently widely narrated in the twelfth century, and who has become very much part of collective memory in modern-day Spain: Rodrigo Díaz de Vivar, the man known to history as El Cid (d. 1099). It is not necessary to recount the details of this Castilian nobleman's life, his rapid ascent in the service of King Sancho II of Castile (d. 1072), his fall from grace under Alphonse VI of León-Castile (d. 1109) and his flight to al-Andalus. All this is well known. We are also well informed of his service to the Muslim ruler of Zaragoza, in whose name he also fought against Christians, before he succeeded in establishing a lordship of his own in Valencia during the turbulent final years of the Taifa period, a lordship that he held until his death in 1099.[9] But a point worth emphasizing is that one of this warrior's most important attributes—at least according to his panegyrists—was his unwavering loyalty towards both his Castilian lord and moral Christian values. But why was it necessary to underline the Cid's fidelity? Perhaps precisely because he also served Muslim potentates? This question is not easy to answer, for we have very few direct sources relating to Rodrigo Díaz de Vivar's activities. Historians are on firmer ground in the case of a lesser-known contemporary, Reverter, Viscount of Barcelona (d.1143/1144).[10]

According to the *Chronica Adefonsi Imperatoris*, an anonymous twelfth-century chronicle, Reverter was captured during a Muslim naval raid along the coast of Catalonia. From there, he was deported to Morocco, where his military abilities were soon noticed, with the result that he was ordered to create and command a militia of captured Christians.[11] Contemporary Arabic sources—the chronicles of Ibn ʿIdārī and Mohammed al-Baydaq—tell a story that slightly differs from the *Chronica Adefonsi Imperatoris*:[12] According to it, around the year 1120 the Almoravid ruler Alī ibn Yūsuf (d. 1143), fearing the growing pressure of the reformist Almohad movement in the Atlas Mountains, ordered the establishment of several militias, amongst them a Christian force. This contingent was put under the charge of a certain Reverter as the *Qā'id ar-Rūm*, that is as "captain of the Romans". Thus, contrary to the Christian source, the Arabic text does not describe this Catalan as a prisoner, but rather suggests that he resided at the Almoravid court voluntarily as an expatriate. It is improbable that he was a hostage of any kind who was sent south, as the 1120s were marked by sharp conflicts between the Almoravids and Count Ramon Berenguer III.[13] Whatever the cause of his commission, Reverter made such a successful career as leader of the Christian militia that he remained in the service of the Almoravids for 20 years until his death in 1142—indeed a prominent case of military expatriation. The life of this viscount in Muslim service raises the question of his relationship with his homeland during his time in the Maghreb. He was clearly faithful to his Almoravid masters, but what was his relationship towards his former liege lord, the Count of Barcelona, and to his land of origin? Luckily, we possess several of Reverter's letters which at least provide tentative answers to this question.

In the 1130s, tensions broke out between count Ramon Berenguer IV of Barcelona (d. 1162) and Reverter's nephew. The viscount saw the need to apologize to his lord for his relative's behaviour in two rather personal letters probably written in 1138.[14] In both of them, he explicitly termed himself Ramon Berenguer's *fidelis*, *amicus*, and *vasallus*.[15] Reverter declared that he strove on behalf of his lord both day and night. In an unmistakable reference to his place of residence beyond the Mediterranean, he wrote that, independently of his present abode, he was still and would always remain his master's vassal.[16] The physical distance between lord and vassal would not keep the latter from carrying out his duty: Whoever turned against his master would be laid in chains and sent to Morocco, where Reverter could deal with the culprit according to count Ramon Berenguer's wishes.[17]

Viscount Reverter's letters appear to provide irrefutable proof that Christian knights could still maintain deep ties with their lands of origin and their masters from a distance even after serving a Muslim ruler loyally for decades. The wording of the two missives also suggests that Reverter still felt and understood the mechanisms and mindsets of his former homeland, despite his long absence from it. However, one must consider the specific cultural and political setting in which he articulated this expression of loyalty: Reverter was interested in maintaining lucrative rights and incomes from landed estates in Catalonia, and he did well in trying to convince the count of his unbroken trust. Indeed, in 1139, the count and viscount signed a *convenientia* which guaranteed Reverter's possession of extensive property in Catalonia in return for his swearing an oath of vassalage to the count.[18] Even 20 years later, one of Reverter's sons named Berenguer (the younger) was still in possession of the family's Castle of Guàrdia. This Berenguer had returned to Catalonia several years earlier, but interestingly, he remained loyal to the Arabic culture in which he had been brought up. When he swore his oath of vassalage to the Count of Barcelona in the very Castle of Guàrdia in 1157, he signed a written agreement not in Latin, but in Arabic: As *Berenguer ibn Reverter* ("Berenguer, son of Reverter").[19] Several years later, he also signed his will in Arabic before leaving Catalonia for Morocco; it is at this point that we lose track of *Berenguer ibn Reverter.*[20] His brother, in contrast, chose a different path: Reverter's second son converted to Islam and forged a long and successful career for himself as a military commander under the name of Abū 'l-Ḥasan 'Alī ibn al-Ruburtayr ("son of Reverter"). In 1187, he was killed in combat against the Banū Ġāniya.[21] Reverter's sons are indeed fascinating examples of the integration of military expatriates in the dār al-Islām.

Reverter was not the only Christian employed by the Almoravids. Some of these soldiers fought against competing Muslim powers, while others served as a protective force for tax collectors in Northern Africa, thus effectively projecting both military and administrative power.[22] According to the *Chronica Latina* of the Castilian Kings, Ibn Hūd of Murcia (d. 1238) took 200 noble Christian knights into his service in order to combat the Almohads.[23] These too depended on the service of Christian militias. One such contingent, probably initially in the service of the Almoravids, arguably formed part of 'Abd al-Mu'min's (d. 1163) army.[24] Other Muslim sources relate that several twelfth-century sultans in Northern and Western Africa maintained foreign soldiers under the leadership of Castilian and Arago-Catalan captains.[25] Unfortunately, the relationship between these mercenaries and their Christian lords in their countries of origin cannot be reconstructed as neatly as in the case of Reverter of Barcelona.[26]

From the little we know about Christian militias of the Amoravid and Almohad periods, their activities were confined to an intra-religious context, that is, they were active only in conflicts waged among Islamic powers.[27] They were clearly used for policing and fulfilled a role in civil enforcement of the respective ruler's power, to whom they were personally affiliated. Like other praetorians before and after them, these foreign fighters stabilized the power of local potentates because—in principle—they were not integrated into local networks of power, and were therefore less easily swayed by court intrigues and dynastic struggles. As we shall see, this was not always the case in practice. Due to their particular military and tactical training, these foreign warriors could effectively complement and strengthen local forces on the battlefield, thus providing their employer with a military edge over his adversaries. Diversifying a military force brought advantages; it was precisely for this reason that Christian rulers on the Iberian Peninsula in turn employed Muslim corps—termed *jenets* in the Crown of Aragon and the *guardia morisca* in Castile—during the Middle Ages.[28] Indeed, dispatching paid fighters can be considered a give-and-take practice: "The Aragonese Kings and Northern African sultans exchanged soldiers, Muslim *jenets* for Christian militias".[29] Mercenaries of diverse creeds thus created a distinct trans-imperial cultural framework. These mercenaries formed specific contingents at Christian courts, where they heightened the king's prestige, but also fulfilled specific military tasks as parts of composite armies and provided a very concrete military edge in moments of crisis. The same can be said of Christian mercenaries in the dār al Islām: Whereas Muslim cavalry contingents were generally noted for their high mobility and speed, but seldom fought in fixed formations, Christian warriors—whether infantry or cavalry—provided a high degree of stability and impact. In several battles of the so-called "Reconquista", the military advantages of Christian heavy cavalry and Muslim light cavalry, respectively, became apparent.[30] This military and tactical difference between Muslim and Christian forces was emphasized by the historian Ibn Khaldun at the end of the fourteenth century. He stated that Northern African rulers created Christian militias because they were used to fighting in fixed formations and claimed that these troops were only employed in conflicts between Arab and Berber armies, but not in wars against Christians, since their loyalty could not be relied upon. For Ibn Khaldun, the expatriates remained dubious outsiders.[31] This however does not diminish their importance for the Almohads, who repeatedly employed them to crush rebellions and fight the growing power of the Merinides in the Western Maghreb.[32]

Foreign Fighters in the Thirteenth to Fifteenth Centuries

It is difficult to discern the specific military culture of these foreign fighters and the ways that they might have shaped Muslim armies (a point raised in the introduction to this volume). But it is clear that they formed an integral part of warfare and civil enforcement in medieval Northern Africa over centuries. The practice of employing Christian militiamen to defend the interests of Muslim rulers against internal and external threats continued until the end of the Middle Ages. These combatants remained an integral part of medieval military culture and were employed alongside slave soldiers, who also fought in Muslim armies.[33]

In the mid-thirteenth century, the Merinides replaced the Almohads as the dominant force in the Maghreb, with the former inheriting the latter's contingents of Christian warriors.[34] These continued to be employed as civil enforcement units, e.g. as tax collectors.[35] Even more often, however, these contingents entered the fray during internal conflicts, usually on the side of the ruler, less so on that of the rebels.[36] Mercenary contingents like these must have been quite considerable, for in 1323, King James II of Aragon requested the Merinid sultan to send him 100 Christian mercenaries stationed in Morocco.[37] In the middle of the fourteenth century, the generally well-informed Syrian geographer Ibn Faḍl Allāh al-ʿUmarī (d. 1349)[38] reported that the army of the Merinids of Fes comprised 4,000 Christian cavalrymen,[39] and Ibn abī Zarʿ speaks of an expedition of 24,000 soldiers undertaken in 1266, a number that included Christian warriors.[40]

Further east, the Abdalwadids/Zayyanids and the Hafsids in turn maintained close relations to the Crown of Aragon from the middle of the thirteenth century onwards, which not only led to the establishment of Aragonese merchant hostels (foundouqs[41]) in Tunis, Tlemcen, and Bugia but also to the creation of a Christian guard under Abū ʿAbd Allāh Muḥammad al-Mustanṣir (d. 1277).[42] When Ibn Abū ʿUmāra (d. 1284) revolted in 1283, 180 Christian fighters were said to have been arrested in Tunis.[43] Even a relatively minor ruler like Abū Bakr ibn Mūsā led "many Christian soldiers", if the testimony of the Catalan chronicler Bernat Desclot is to be believed.[44]

As this chapter will show, some of these soldiers acted as delegates of Christian kings; but there were others who hired for different reasons. Most of them must have been attracted by the considerable economic gains they expected; others had fallen from grace like the Catalans Guillem Ramon de Montcada (d. ca 1275) and Guillem Galcerà (d. 1306) or the Castilian Alonso Pérez de Guzmán.[45] These and other noblemen appear in the sources as captains of the Christian guard, as so-called *qāʾid*, in Catalan: *alcayt*.[46] In some cases, one can even determine the origins of these troops. Some, for instance, can be traced using documents that they prepared before leaving for Northern Africa.[47] In general, rulers from the Western Maghreb, particularly the Merinids, preferred employing Castilians, whereas the Hafsids of Ifíqiya recruited Aragonese soldiers.[48] This distribution mirrors the political relations between the Iberian and Northern African rulers. The competitive relationship between the crowns of Castile and Aragon—dealings marked by both cooperation and conflict—was extended to Northern Africa by these mercenaries. Depending on the political situation and inner-Muslim relations, the Abdalwadids and the other minor rulers in Tripoli, Tlemcen (Tilimsān), and Bugia (Biğāya) tended towards Castile or the Crown of Aragon.

Contingents of Christian militiamen were so common in Northern Africa that several thirteenth-century popes saw the need to intervene, fearing their conversion to Islam.[49] These concerns were not unfounded, as individual cases such as Berenguer Reverter's son mentioned above or the Aragonese nobleman Gil de Alagón alias Mūḥammad, who served the Valencian Ruler Abū Saʿīd, illustrate. General references in the sources to renegade knights underscore that apostasy was not as uncommon as modern images of the Middle Ages would make us believe.[50] In 1290, Nicholas IV wrote to Christian mercenaries stationed in Ifrīqiya and the Maghreb, explicitly urging them to stand firm in their faith, affirming that such steadfastness was not only an exemplar for coreligionists in Northern Africa but also for Muslims.[51] We may very well doubt that proselytizing Muslims was an aim of paid fighters. But some of these figures do seem to have been worried about their spiritual well-being in the dār al-Islām, for in 1307, they wrote to Pope Clement V asking that a bishop be sent to them.[52] Indeed, there are several references in evidence from this era not only to Christian bishops in Morocco but also to other clergymen providing religious services there.[53] Before this backdrop, it is telling that Christian mercenaries were generally not expected to enter combat with coreligionists. Clearly, Muslim rulers not only considered them a fighting force but also saw them as a specific religious diasporic group.[54]

Social Integration and Identity

Christian mercenaries in Northern Africa were often termed *farḥān/īfarḥān* in Arabic (probably from *āfrūḥ*, "young men") and *farfanes* in Castilian.[55] They formed a distinct group among the Christian minorities of Ifrīqiya and the Maghreb, more so if they migrated together with their wives and children, as some sources suggest. In these cases of long-term absence from home, we can reasonably suppose that these expatriates experienced feelings typical to diasporic communities. One can easily suppose that these men and women longed for their places of origin, for faraway Castile, Catalonia, or Aragon. But we do not possess hard evidence for such yearning or "diasporic dwelling", to use the term coined by the American sociologist Thomas Tweed.[56] In contrast, proto-national, ethnically grounded animosities appear to have pervaded under certain circumstances: In 1313, the pope had to intervene in order to resolve conflicts between Castilians and Catalano-Aragonese stationed in Marocco.[57] Ethnic and political elements alike appear to have formed part of the corporate identity held by these groups. Indeed, the rivalry between Castile and the Crown of Aragon, who vied for power in Northern Africa during the first half of the fourteenth century, provided an important backdrop for the mercenaries' activities, although the sources are rather scant in this respect. We know little about the everyday life of these Christians and of the challenges which changing political conditions within the Mediterranean posed for them.

We still have to determine how sultans dealt with their guards when war broke out with Christian powers. To what extent did the popes' fears come true; how many *farfanes* or children of *farfanes* converted to Islam? Ibn Faḍl Allāh al-ʿUmarī relates that apart from the above-mentioned 4,000 Christian cavalrymen, the royal militia also comprised 500 renegade knights.[58] Then again, the chronicler al-Ubbī informs us in his *Ikmāl* that a Christian militiaman caused a scandal in Tunis by hurling insults at Muslims when he heard the call to prayer from a nearby minaret.[59]

There is no doubt that groups of *farfanes* existed up to the end of the Middle Ages. The merchant-traveller Anselm Adorno referred to their quarter in the Qaṣba of Tunis as late as 1470–1471.[60] The rich Catalonian registers even allow us to catch a glimpse of mercenaries migrating south: In the years around 1400, royal licences were extended to groups of militiamen and their female companions. For example, in 1399, the Valencian Gonçalvo Díez embarked with 100 men and 30 women, whilst a certain Simó Safont took 30 men and 10 women;[61] in 1406, Gonçalvo Díez recruited 100 men and 15 women, whilst Pere Eiximenis Baldó led 50 men and 10 "foreign women" (*X fembres de nació stranya*) southwards. The high percentage of women—some of whom were probably prostitutes—is not surprising considering the Islamic (and Christian) prohibition levied on Christians against maintaining sexual relations with Muslim women.

At the same time, other mercenaries sought repatriation to their homeland (or to the homeland of their forefathers).[62] Some of them were formally invited to do so: On 10 October 1386, the town council of Seville wrote to the "honourable, noble Christian *farfanes*, the captains, wives and companions of the noble and pure and great lineage of the Goths (of whom the kings of Castile and noble sires descend), who live in the town of Morocco [Fes] in the kingdom of the Merinids". The council had received one of their relatives, Sancho Rodriguez, as an envoy and now invited the *farfanes* and their families to migrate to Andalusia, where they would be most welcome.[63] What exactly was meant by this reference to the Goths? The Castilian chronicler Pedro López de Ayala provides a hint: He relates how in 1390, 50 knights, descendants of Christians who had migrated to Morocco after the Islamic conquest of the eighth century, returned to Castile together with their wives and children following an invitation by King John.[64] Modern historical research has rejected this alleged Visigothic parentage, suggesting instead that these groups probably descended from Christians exiled or deported to Northern Africa in the first half of the twelfth century.[65] The true length of these migrants' genealogical pedigree is not relevant to us here. However, it is certainly worth highlighting that after the *farfanes* and their families migrated back to Castile at the end of the fourteenth century, they continued to form a distinct kinship group in Christian Andalusia, one that celebrated its singular history, particular traits, and specific privileges for decades to come. The *farfanes* doggedly defended their prerogatives and maintained close cohesion as a social group in Christian lands. Some of them became quite well established. For example, one of them, Antón Farfán de los Godos, became commander of the Hospitaller priory of Barbaldo (Salamanca) at the beginning of the sixteenth century.[66] In 1541, the chronicler of the house of Niebla, Pedro Barrantes Maldonado, would use the story of these military expatriates to boost the pedigree of his master's lineage: 15 chapters of the *Ilustraciones de la Casa de Niebla* relate how Don Alonso Pérez de Guzmán, an ancestor of the book's commissioner Juan Alonso de Guzmán, sixth Duke of Medina Sidonia, served Muslim rulers for many years in Northern Africa.[67] Evidently then, the prolonged diasporic experience of Christian mercenaries became the foundation for identity politics after these groups returned back to their homeland.

Can the same be said about the expatriates during their time in the dār al-Islām? Did they maintain a meaningful exchange between their lands of origin and their lands of service? Did connections to their homelands form an integral part of their identity politics as a diasporic group? Some sources provide a faint glimpse of the hopes and feelings these expatriates and their families might have experienced while living in Muslim territory beyond the sea. In 1399, King Martin of Aragon wrote to the Merinid sultan pleading him to permit two mercenaries to return home, after they had spent many years away from Catalonia, "leaving their parents, friends and possessions behind": Arnau Masquefa wanted to re-migrate with his wife and children, Lluis Vilar with his mother.[68] Between 1403 and 1407, King Martin wrote several other letters of this kind: One on behalf of a certain Bernat Spigol, "who has served you for a long time", so that he might fulfil his parents' and friends' wishes and return home together with his wife;[69] another for male and female companions of Gonçalvo Díez,[70] including a Dominican friar who probably provided the mercenaries with spiritual support;[71] and another for Magdalene, a Catalan who "had spent a long time in Marrocco with her husband".[72] In a compelling study, Roser Salicrú has identified by name a further eight wives of Castilian mercenaries and another 17 men established in Fes.[73] King Ferdinand of Aragon drew particular attention to how loyally fighters who wished to return home had served the Merinid Sultan Abū Saʿīd ʿUṯmān (1399–1420).[74]

The Mercenaries' Roles in Domestic Power Struggles

In other cases, militias and their captains appear as participants in rebellions. In 1308, the Christian *alcayt* rose against the Merinid sultan, who upon his return to Marrakech reportedly killed "all the Christians he could find", lining up their heads on the town wall.[75] Two years later, Gonzalo Sánchez de Troncones joined a vizier in his fight for the Merinid throne, and in 1316, Juan Ruiz de Mendoza sided with the usurper Abū ʿAlī.[76] Although they were foreign fighters and so in theory loyal to their employers, mercenaries were embroiled in court intrigues. Christian militiamen supported Abū ʿInān Fāris when he seized power over Marocco in 1350, and in 1368 all the officers of the Christian militia were executed in public after rumours that they were planning a coup d'état reached the sultan's ears.[77] In Tlemcen, the local captain is said to have conspired against the Abdalwadid sultan in 1254,[78] and in 1318, Abū Tašfīn (d. 1337) succeeded in his coup d'état thanks to a Christian militia and his vizier Hilāl, known as "The Catalan".[79] Finally, in the Hafsid kingdom, another vizier and member of the Royal family, Abū Yaḥyā Zakariyā Ibn al-Liyānī (d. after 1318), usurped power in 1311 with the help of Christian fighters.[80] Thus, foreign fighters seem to correspond to the familiar image of a mercenary: They appear as unscrupulous opportunists, who trim their sail to the wind according to the political climate or the prospect of economic gain.

In reality, however, things were more complex. Both the usurper al-Liyānī[81] as well as the rebels Abū ʿAlī (d. 1351)[82] and Abū ʿInān Fāris (d. 1358) were sons of Christian women who belonged to the sultans' harem.[83] In al-Andalus as well as in the Maghreb, a number of rulers stemmed from such trans-religious unions. We cannot say with any certainty whether religious ties were responsible for Christian fighters taking the side of a Christian woman's son. Neither do we know the extent to which a sultan might have been influenced by his Christian mother's beliefs. But the case of Abū Yaḥyā Zakariyā Ibn al-Liyānī shows that Aragonese kings might hope for the conversion of Islamic rulers to Christianity. For several years, this usurper flirted—or gave the impression of flirting—with converting to Christendom, a development that caused a flurry of diplomatic activity between 1311 and 1316.[84]

We also know that in certain cases, Christian women in the harem could maintain close relations to their coreligionists in the palace guard. A case in point is provided by the mother of the above-mentioned Abū ʿAlī. She was the sister of the local Christian captain. In other words, when that particular mercenary, Juan Ruiz de Mendoza, rose against the sultan in favour of the prince, he was not supporting any pretender but his own nephew.[85]

Such ties could be vitally important to the mercenaries. When the Almohad Caliph Idris al-Maʾmūn died in 1232, his Christian wife Habbaba first informed the Christian militia of their employer's demise before passing the news to the Muslim members of the court, thus giving her coreligionists an important head start to adjust to the new situation.[86] Clearly, these expatriates were better integrated into their host societies than vagrant mercenaries merely in search of payment from the highest bidder.

Without a doubt, Christian militias were a common phenomenon in Northern Africa between the twelfth and fifteenth centuries. But with the collapse of the Almohad Empire in the middle of the thirteenth century and the creation of new lordships in Ifrīqiya and the Maghreb, the role of Christian contingents and their captains changed: Now their radius of action not only transgressed the borders of Islamic polities, but they also crossed the borders of the dār al-Islām. For in this new phase in the history of intercultural military expatriates, mercenaries began to be closely associated with the ruler of their country of origin, and acted at the interface between their homeland and their new place of employment. They therefore were both trans-imperial subjects (in as much as they spanned two political entities) and inter-imperial subjects (on account of their activities as intermediaries). This affected the dynamic, but hazy area of intersection between religious affiliation, feudal ties, social context, and economic interests in which these fighters lived.

Representing the King

In March 1265, King James I of Aragon (d. 1276) appointed Pere de Vilaragut as his captain in Tlemcen, ordering that all his subjects there—soldiers and others (*milites et alii*)—should obey him.[87] Two years later, he repeated a similar appointment, this time installing Guillem Galcerà in the role. In this case, he stipulated that the *alcayt* was to enjoy far-reaching judicial rights, namely "the right to hear and to judge criminal and other cases and to comply with the office of an *alcayt* in every way".[88] In 1286, James's grandson Alphonse III of Aragon imposed a similar regulation on the Abdalwadid ruler Abū Saʿīd ʿUthmān ibn Yaghmurāsan. All foreign Christians in Tlemcen were to be judged according to Aragonese law and by the *alcayt* appointed by the king of Aragon: *Item que todos los christianos que seran en la terra del rey de Tirimçe de qualesquier condiciones et senyorias, que sean jutgados por fuero Daragon por aquel alcayt que el rey don Alfonso ala enbiara.*[89]

By that time, that is, the end of the thirteenth century, Christian rulers had been providing fighters to Muslim potentates for many decades. In 1228, the Christian King Ferdinand of Castile reportedly sent an sizeable contingent to the Almohads for which he received ten castles and the sultan's promise that the Christians in Marrakesh would be allowed to practice their religion freely, which included permission to toll bells.[90] It appears that Christian rulers in general and the kings of Aragon in particular aimed at acquiring more immediate means of exerting political influence than mere territorial gains such as castles.[91] Not only did they support specific factions during inner-Muslim strife, they even declared Christian militias and their captains to be representatives of royal interests, that is, their own.

Looking at other Northern African realms, we find further proof for this new policy. In 1274, James I negotiated a treaty with the Merinid sultan Abū Yūsuf (d. 1286), under whose terms 20 galleys and 500 cavalrymen were to be sent to Morocco, where the fighters would be permitted to establish a church of their own.[92] James's son and namesake strove towards personally appointing the *alcayts* in the town of Fes and Tlemcen when he negotiated with the Merinids in 1308.[93] Aragonese policy towards Tunis and the Hafsid kingdom followed a similar pattern. James I sent an official royal contingent of salaried fighters under the leadership of a noble Catalan named Guillem de Montcada (d. ca. 1275) to Tunis in the middle of the thirteenth century.[94] At this point, the character of Christian praetorians in Ifrīqiya changed: Henceforth, expatriate militias were supposed to be associated with the king of their homeland who had dispatched them south. In the Treaty of Panissars, agreed between the Hafsid ruler and King Peter of Aragon in 1285, the sultan permitted that the Aragonese king appoints the local *alcayt*.[95] King Peter's son Alphonse sought a similar agreement two years later.[96] In 1303, King James II asked the sultan to favour the Christian soldiers who "provide you with good and loyal service (*bon service e leal*), which also compensates us".[97] The King repeated his request a couple of years later in letters to the sultan al-Liyānī, and he even tried to strengthen the captain's position by asking for all Aragonese fighters in the realm to be put under his command.[98]

The treaty signed on 30 July 1287 between Alphonse III and the Almohad prince ʿAbd al-Wāḥid (*Abdelehehit*), the pretender to the sultanate in the Hafsid realm, reflects the most far-reaching list of rights to which a Christian ruler could aspire.[99] The sultan-to-be promised to pay a yearly tribute directly to the king, while each of his Christian knights was to receive two horses and a daily payment of three silver bezants (with each squire receiving one horse and two bezants) plus a number of amenities while in the field. Moreover, all Christian fighters in the Hafsid realm (*tam milites quam scutiferi*) together with their families were to fall under the jurisdiction of the Aragonese *alcayt*—indeed, he would be the only one in the kingdom. The captain would possess the jurisdiction over the expatriates in all economic and criminal matters (*in causis pecuniariis quam criminalibus*). The soldiers and their families were allowed to maintain a church of their own, served by a priest (or other clerics) of their choice, and they would be allowed to toll bells during elevation and use incense holders during obsequies. Finally, a fixed quota of wine was to be allotted to each knight and squire, and the pretender even promised military support against the king's Muslim adversaries.[100] Such privileges would undoubtedly have strengthened social cohesion and marked the mercenaries out as a distinct ethnic and religious group which possessed a unique military culture of their own.

Far-ranging promises like the ones negotiated with ʿAbd al-Wāḥid raise the question as to the reliability of such stipulations. Our main sources for these diplomatic contacts are the royal instructions issued for diplomats as a preparation for their negotiations. However—and this is an important point—the fact that the Aragonese aspirations were laid down in royal documents does not necessarily mean that they were ever realized in practice. Indeed, the treaties that were eventually agreed between both potentates generally did not contain such far-reaching rights for Aragonese captains. Furthermore, in the case of Abd al-Wāḥid, the pretender failed in his attempt to take possession of the throne, meaning that the treaty forced upon him by King Alphonse was never fulfilled.[101] Nevertheless, one should not underestimate the importance of Christian fighters in Tunis as representatives of a foreign king. They formed an influential institution amongst the various Christian communities of Northern Africa, and it is surely no coincidence that King Alphonse of Aragon explicitly named the *alcayt* first and foremost among his subjects in Tunis when, in 1290, he addressed "the captains, consuls and our other subjects in Tunis" (*alcaydis, consuli et universis aliis subditis nostris apud Tunisium*).[102] This phase of direct Aragonese involvement in the Hafsid capital ended with the ascent to power by Sultan Abū Bakr II (d. 1346) in 1318.

A comparison between the dealings of the Crown of Aragon with the various Northern African realms west of the Mamluk Empire in the era between 1270 and 1350 shows that Aragonese rulers followed different goals at different times and in different places. In the realm of the Abdalwadids from Tlemcen (Zayyanids), the *alcayt* ideally was to be the head of all local Christians, a political goal pursued in Morocco as well. In Hafsid Tunis in turn, where the economic interests of the crown were particularly strong, the local consul enjoyed greater rights as a representative of the entire Christian community, while the captain only presided over the militiamen. Although the issue of payment is generally not explicitly mentioned, the sultans probably paid the mercenaries' stipend. In some agreements with Hafsid sultans, the exact sum was indeed stipulated, sometimes including a surprising clause according to which part of this payment was to be ceded to the Aragonese king.[103] This ruling not only emphasized that the fighters remained in a personal relationship with their feudal overlord but can also be interpreted as an indirect tribute paid by the Muslim sultans to the Christian king, one that enabled them to avoid a formal submission of the kind that was frowned upon and actually prohibited by Islamic law.[104] In this period and in certain instances, the presence of official Christian contingents that represented the Christian or Aragonese king and enjoyed far-reaching privileges appears to have reflected a hierarchical relationship and can be understood as a sign of short-term power on the one side and weakness on the other.[105]

For the Aragonese ruler, a Christian expatriate *alcayt* in the dār al-Islām fulfilled several functions simultaneously. He could serve as an informant, as a representative or as an envoy at a distant court who could be contacted as the need arose. Prominent captains at the beginning of the fourteenth century, like Bernat Seguí in Fes, or James of Aragon and Felip de Mora in Tlemcen, represented the interests of all Catalano-Aragonese subjects, not only those of their military comrades.[106] This in turn must have had repercussions on the merchant communities and their representatives (consuls). Cooperation and conflict between mercantile and military diasporic groups as well as between their spokespersons arguably went hand-in-hand, but our sources are silent on this. It is, however, clear that the direct ties to their nominal king at home ensured that their land of origin remained an important factor in the identity politics of this group.

Questions of Loyalty

Over the course of the half-century during which the Aragonese monarchs used the instrument of the Christian militias to further their political ambitions in Ifrīqiya and the Maghreb, that is from around 1270 until ca 1320, the loyalty of these expatriate soldiers vis-à-vis their feudal Christian lords and their Muslim employers represents a particularly interesting object of study.[107] Did these warriors express their relation to their homeland and their king? In his letters to Northern African rulers, James II emphasized that Christian mercenaries were trans-imperial subjects who served two lords simultaneously: In 1314, he described the secretary of the captain of Tunis, a man named Llorenç de Berga, as a member of both households—the Aragonese and the Hafsid: *com lo dit Lorenç sia servidor nostre e vostre*.[108] Similarly, he considered the mercenaries Felip de Mora and Jaume Cervitge his own subjects, but ones who served the sultan—*car son nostres naturals e per que son en vostre servei*—"they are our subjects in your service".[109] The document's cautious wording implies that the king wanted to bypass the question of to whom the mercenaries owed greater loyalty.

In the case of these Christian mercenaries explicitly sent by their king, the close association to their home country and to their nominal ruler should not be underestimated. In contrast to individual men who sought for a living in foreign lands, these soldiers were also—albeit not exclusively—understood as representatives of Christian kings. The monarchs in turn strove to demonstrate this special relationship. When James II appointed Berenguer de Cardona captain of the militia in Tunis on 26 October 1299, he ordered that the new *alcayt* be given a royal flag, which the captain of the town traditionally possessed, and which Berenguer too was to bear as both a sign of his affiliation to the Aragonese king and as a symbol of the monarch's honour.[110] A few years later, the king reached an agreement with the sultan, according to which the *alcayt* was to bear the arms of the Crown of Aragon—the *senyera y senyal del senyor rey*.[111] Further west, the captain of the Aragonese militia in Morocco is known to have fought against rebels under the red and yellow arms—the *senyera*—of the house of Barcelona.[112] Finally, in 1325 James II demanded from the Abdalwadid sultan that not only the captain but also the normal infantrymen of the local militia should bear the arms of Aragon—if they indeed were Aragonese subjects.[113] Clearly, then, the Aragonese monarch construed these warriors more as an expeditionary corps than as a group of foreign mercenaries. These official contingents sent or strongly supported by Christian monarchs should be differentiated from individual *farfanes* who were hired by Muslim sultans in Northern Africa.

All these rulings agreed to by sultans and kings are of a normative type, which of course raises the question of their implementation. It is not easy to ascertain if such rules were indeed put into practice, because very few archival documents have survived from the Muslim realms west of the Mamluk Empire. However, a chance finding provides a comprehensive answer to this question, at least in one instance. On 29 December 1311, the Majorcan mercenary Arnau de Torrella/Torroella died in Fes in Morocco (of a natural death).[114] A copy of the inventory of his belongings—compiled by a Christian notary in Fes and signed by the Christian bishop, the dean, a priest serving in Morocco and five other Christians—is still extant in the Archivo del Reino de Mallorca.[115] This document shows that the Christian soldiers in Fes at that time were under the leadership of a certain Bernat Seguí, and that they lived together in a group of houses. In the document, Arnau de Torrella is described as living in *duaro vocato Iffraguen*, which later on is termed *duaro domini Bernardi Seguini*.[116] This means that in the eyes of contemporaries, the housing complex was named both after the mercenaries (*īfarḥān*) or the "franks" (*al-Ifranǧ*) who inhabited it as well as after Bernat Seguí, the captain of the militia. Still more important for our question concerning the extent to which relations of affiliation were articulated is the contents of the inventory. The document enumerates Arnau de Torrella's possessions—a horse, weapons, a chest, textiles, etc. Arnold's shield is also mentioned. It was apparently decorated with his coat of arms—*et unum scutum cum signo turrarum pernularum*. What is particularly revealing, however, is the reference to "a silk hat with the royal insignia in red and yellow" (*unum sombrerium sive capellum de sol forratum cum sendato regali rubeo et croceo*).[117] This documentary reference proves that Christian militias indeed wore garments bearing the royal colours of Aragon.[118] Correspondingly the fifteenth-century Arabized Christian mercenaries, the above-mentioned *farfanes*, were differentiated mostly by their clothing: They did not wear a turban, but rather a hat which functioned as a distinctive sign of their religious affiliation.[119] The inventory of 1311 strongly implies that in public places in the Maghreb, foreign mercenaries could clearly be assigned to their places of origin and to their Christian lords by their garb. To once again invoke the terminology coined by Stephen Morillo, the "unembeddedness" of these praetorians was thus symbolically driven home.[120] Both visually and physically, the military expatriates remained a group apart. The general prohibition to marry Muslim women will only have heightened the segregation of these Christian foreigners, no matter how closely they might have been tied to the sultan's or emir's court.

Proof of this segregational practice is also provided by the famous *Cantigas de Santa Maria*. This collection of miracle stories in honour of the Virgin Mary, written on behalf of the Castilian King Alphonse X between 1250 and 1284, features one narrative that relates how Mary, under certain circumstances, could grant aid to Muslims. It recounts that when the Merinids laid siege to Marrakesh in 1261/1262,[121] the town's inhabitants sent the Christian militia against the attackers bearing a standard depicting the Virgin, upon which the assailants were overcome by fear and ran away. As the text puts it, "in this way holy Mary helped her friends to conquer their enemies, even if they follow a different law".[122] A miniature of this scene contained in the manuscript of the *Cantigas* produced in the last years of Alphonse X's reign, and now kept at the Escorial, is particularly revealing in this respect. The image shows both the Merenid army, and that of Marrakesh, being led by Christian militias. In this case, the praetorians' loyalty to their employers seems to have been more important than any religious solidarity with their fellow Christians. Such intimate ties between Muslim employers and Christian mercenaries lie at the heart of the last part of this chapter, which will deal with cases in which mercenaries transgressed their initial field of action and in fact served as diplomats.

Christian Mercenaries as Envoys of Muslim Powers

In the decades between 1260 and 1340, one can discern a number of cases in which Christian mercenaries in the service of the sultans were sent as diplomatic envoys to the court of their nominal king—the king of Aragon—in order to represent the interests of a Muslim potentate. This practice is not attested in every realm in the Western Mediterranean. The Nasrids, for example, rarely sent Christian emissaries, but rather chose Muslim negotiators instead; and the same holds true for the Mamluk Empire (which could however employ renegades). We therefore not only need to differentiate between the practices followed in the various realms but also chronologically within every polity. One also needs to keep in mind that diplomatic activities of mercenaries were generally complemented and sometimes supplanted by those of other envoys. Nevertheless, it is important to highlight this particular practice, not least in order to determine the challenges it exerted upon everybody involved, particularly upon the individual mercenaries.

There is a long tradition of rulers in the dār al-Islām falling back upon Christians when negotiating with Christian kingdoms. As is well known, both Muslim-Byzantine and Muslim-Latin relations provide many examples of this practice since the early Middle Ages.[123] In Germany, the embassy sent by the Ommayad Caliph ʿAbd ar-Raḥmān III (d. 961) to Emperor Otto I around the year 953 is particularly famous.[124] It was led by a Latin Christian living under Muslim rule, a so-called Mozarab, a man named Recemundus (d. after 961), who was rewarded for his diplomatic service by being appointed bishop, and who later also undertook missions to the Byzantine Empire in the name of the Caliph. The following examples are not only less well known, but in many respects different in the sense that the envoys were not only Christians, but were dispatched by a Muslim ruler to the diplomats' nominal Christian overlord.

Amongst the Christian mercenaries in Northern Africa, one can discern members of royal European families. Some of these were legitimate, exiled princes,[125] but most of them were illegitimate sons of monarchs, who hoped to make a career for themselves in remote lands. An illegitimate son of King Manfred of Hohenstaufen (d. 1266), for instance, a man tellingly christened with the name of Frederick, can be identified in Ifrīqiya,[126] though most of these royal bastards came from the houses of Castile and Aragon: James Peter (d. 1327) (a son of Peter the Great of Aragon, d. 1285), James (d. 1334) and Napoleon (d. 1338) (sons of James II von Aragón), two of King Ferdinand III of Castile's children named Henry (d. 1304) and Frederick (d. 1277) as well as John and Alfonso Sánchez, sons of King Sancho IV of Castile.[127]

In some cases, the royal offspring were employed as diplomats, probably in order to convey a higher reputation to the embassy. In 1294, Henry, the son of King Ferdinand III of Castile, opened the negotiations between the Hafsid sultan and King James II of Aragon.[128] An Aragonese bastard prince—James, the illegitimate son of James II of Aragon, and captain of sultan Abū Tašfīn of Tlemcen's militia between 1325 and 1329—was particularly active as a diplomat.[129] He was sent to his father's court by the sultan in order to acquire naval support.[130] King James, in turn, tried to have him named captain of all Christian militias.[131] At least three further missions by the illegitimate prince are attested, although they apparently were not successful, probably as a result of political blunders that he committed. There is a letter written by several captured Christians and dispatched from Tlemcen to the king in 1327 in which the authors bitterly complain about the prince's shortcomings as a diplomat.[132] Possibly as a consequence of these setbacks, the prince returned to his homeland in 1331 or 1332.[133]

Turning from the group of illegitimate princes to individual noblemen, a very prominent case is that of Bernat de Fonts, a Catalan *alcayt* of the Christian militia who was sent to the Aragonese court in 1313.[134] This envoy was not only a knight, but was also in fact an Aragonese Templar who had evaded capture in the wake of his Order's dissolution in 1312 by fleeing to the dār al-Islām and taking service with Muslim rulers. Such a case of a member of a Christian military religious order conducting military and diplomatic service for a Muslim ruler is indeed spectacular, though not singular.[135] Bernat de Fonts, who successfully negotiated between James II of Aragon and the sultan, was pardoned from the crimes which his Order had been accused of and returned to Aragon in 1315. One year later, he once again entered the Hafsid sultan's service,[136] but he did so without prior permission from the Aragonese king, with the result that his royal pension was revoked.[137]

The most interesting of all figures to cross political and confessional frontiers in this way is our third case, a man already mentioned in this chapter: Bernat Seguí. This figure was a Catalan who spent over 20 years in the Maghreb and served no less than four sultans as *alcayt*.[138] He was attested in the service of King James II at Murcia in 1297, but then appears to have left for Tlemcen, where he fulfilled military as well as diplomatic missions.[139] First, he was sent by Abū Ya'qūb (d. 1307) to the Crown of Aragon in order to recruit new contingents of mercenaries, then he dispatched letters to King James II of Aragon, who in turn sent him to Morocco in October 1303 as the new Aragonese *alcayt*.[140] The king and sultan both employed him as their mediator in order to sign a significant armistice between them which stipulated—among other points—that an armed contingent should be created in Fes under Bernat's leadership, a contingent that was to bear the arms of Aragon, but be paid by the sultan.[141] In the years that followed, the Aragonese *alcayt* even had a personal secretary, further evidence of his diplomatic activities.[142] Indeed, several letters sent from Barcelona to Fes and vice versa are still extant in the Archivo de la Corona de Aragón in Barcelona.[143] These documents, which include detailed royal instructions to other Aragonese envoys, show that the mercenary Bernat Seguí actually acted as a sort of Aragonese consul in Morocco:[144] He negotiated with the sultan, he received Aragonese ambassadors, and accompanied those representatives during their negotiations at the Merinid court.[145] Moreover, he regularly informed his master, the king of Aragon, about political developments,[146] secretly acted on his behalf,[147] and served both the sultan and the king as messenger and ambassador.[148] The instructions given to Viscount Jaspert of Castellnou (d. 1321) on 3 May 1309 provide an overview of the broad range of his activities: In the narrative opening to this document, no less than eight occasions are mentioned in which Bernat de Seguí acted as informant, envoy or mediator between Aragon and Morocco.[149]

This Catalan mercenary was so well integrated into the Merinid court that he was not only employed by James II, to further royal interests, but also by the town council of Barcelona.[150] Bernat was assisted by his brother Arnau, who also acted as a militiaman and diplomat,[151] as well as by a group of relatives (Berenguer Seguí, Jaume Seguí, Pere Seguí, Ramon Toró)[152] and brothers in arms (Bernat de Claramunt, Pere Martí d'Orta, Guillem de Pujalt).[153] It should be emphasized that these expatriates lived in close proximity to each other and formed a coherent ethnoreligious group whose housing complex was known as the "Douar of Bernat Seguí".[154] Bearing this in mind, it seems justified to interpret the Seguí family as an authentic clan of expatriate *alcayts*, whose members tellingly held the position of captain of the mercenary guard in several Northern African towns, and in exceptional cases such as that of Bernat Seguí, even wielded the authority usually reserved for the local Aragonese consuls.

These three case studies alone—the sons of Christian kings who served as militiamen and diplomats, the Templar Bernat de Fonts, and the arch-*alcayt* Bernat Seguí—should suffice to illustrate that Catalonian *alcayts* could be far more than simply the bearers of messages. They also acted also as autonomous negotiators in the name of the sultan. The direct relationship between these praetorians and the Muslim masters to whom they responded and above all their cultural expertise made them valuable diplomats. It seems that Muslim rulers believed that a deeper understanding of Christian cultural particularities was just as important for successful diplomacy as knowledge of foreign languages. If so, then Islamic potentates had a more accurate notion than their contemporary Christian counterparts of cultural brokerage, and the importance of trust for successful diplomacy. The sultans possibly even acquired a strategic advantage by holding this insight, and it might be no coincidence that Aragonese kings in turn began using Muslim subjects, so-called *mudéjares*, at the end of the fourteenth century when dealing with Muslim powers, favouring them over Christians who were knowledgeable in Arabic.[155]

The examples discussed in this chapter corroborate that Christian mercenaries generally served their Muslim employees more or less loyally when dealing with their nominal feudal lords. This is all the more noteworthy when we consider that in doing so, they were transgressing the religious border between Christendom and Islam. However, such Christian envoys did not enjoy unlimited trust. The mercenaries were generally accompanied by a Muslim envoy, who probably not only acted as a scribe in the production of Arabic documents but might also have controlled and supervised the Christian envoys. The sultan entrusted his mercenaries with embassies, but his trust was not total.

The *alcayts* in turn were fully aware of the dilemma which their activities posed to them. At least, this is the impression that one receives when reading their letters.[156] In 1313, after having served the Hafsid sultan repeatedly in diplomatic service, the above-mentioned Bernat de Fonts wrote a letter to James II of Aragon. In this personal message, Bernat maintained that his priorities were clear: "You should know my lord that I owe my obedience first and foremost to you".[157] When Pere Belot acknowledged receipt of a letter in 1303, he underlined that James II would be his "natural lord" (*senyor natural*) for all time.[158] Bernat Seguí went to great pains to remind his king that he worked day and night on his behalf,[159] prompting James to reassure him that he knew he could trust him as a king trusted his subject,[160] and that he was convinced that Bernat was fully committed to his own, the king's, well-being.[161]

In these documents, the semantic field of loyalty, trust, and obedience is marked with a number of terms, primarily with derivations of the Latin word *legalitas* rendered in the vernacular as *leyal, leyalment*, and similar. They not only imply the moral category of common interests and values, but also include the juridical notion of legally correct behaviour. Mercenaries and their masters did not expect blind obedience, but instead reliable and trustworthy behaviour. Unsurprisingly the semantic field was also employed in direct diplomacy between Christian and Muslim rulers.[162] It was not confined to the hierarchical relation between ruler and subject, or king and mercenary, but also circumscribed that between political partners or negotiators. Formally, these relations between Christian and Muslim rulers were generally fashioned as contracts between equals. As we have seen, in practice, this was not necessarily the case: It is no coincidence that the high point of "official" mercenary involvement in Ifrīqiya coincided with the phase of Aragonese strength and Hafsid weakness. Nevertheless, the relationship between the two powers was in principle not one of vassalage and dominion, but rather one of formal equality.

Conclusion

This study of Christian mercenaries in the dār al-Islām has not been restricted to their multiple fields of action as a fighting force, but has also explored their self-consciousness both during and after the time they spent in foreign lands. The category of loyalty is ambivalent when applied to these cultural brokers. In the Mediterranean, both in the dār al-Islām and in the Christian countries of the Northern Mediterranean, multiple loyalties were not unusual. How these diverse obligations were actually fulfilled across political borders and even across religious boundaries still remains an under-researched field of academic enquiry.[163] Even the fear of apostasy did not lead ecclesiastical or secular powers to undertake serious attempts at curbing the practice of Christians—or for that part Muslims—serving the religious other. Documents issued by figures who straddled this frontier indicate that they practiced distinct identity politics based on shared ethnic backgrounds, common religious affiliation, and joint military and social practices. They upheld a meaningful exchange between their lands of origin and their lands of service, an activity that further strengthened their social cohesion. And they could indeed distinguish different grades of loyalty and aimed at providing their feudal lord with signs of their attachment and allegiance. These statements of loyalty, coupled with the symbolic and ritual behaviour they performed in public space, reveal a clear understanding and differentiated notions of belonging. In these ways, Christian trans-imperial mercenaries at Muslim courts provide fascinating insights into the fluid interfaces between material interests, religious affiliation, and political loyalties in the medieval Mediterranean.

Notes

1 This chapter has profited greatly from presentations and discussions at the University of Bonn at the invitation of Stephan Conermann, at the Cluster of Excellence "Asia and Europe in a Global Context" (University of Heidelberg), at the University of Constance and at the Institute for Historical Research, University of London. My thanks go to Aaron Jochim, Julian Reichert, Sandra Schieweck, and Wolf Zöller (Heidelberg) for further advice and technical assistance as well as to Georg Christ (Manchester) and very particularly Simon A. John (Swansea).

2 Cf. Jean Richard, "An Account of the Battle of Hattin Referring to the Frankish Mercenaries in Oriental Moslem States," Speculum 27 (1952): 168–177; Simon Barton, "Traitors to the Faith? Christian Mercenaries in al-Andalus and the Maghreb, c. 1100–1300," in *Medieval Spain: Culture, Conflict and Coexistence; Studies in Honour of Angus MacKay*, ed. Roger Collins and Anthony Goodman (Basingstoke: Palgrave/ Macmillan, 2002), 23–45.

3 Ella Natalie Rothman, *Brokering Empire: Trans-Imperial Subjects between Venice and Istanbul* (Ithaca: Cornell University Press, 2012); Marc von der Höh, Nikolas Jaspert, and Jenny Rahel Oesterle, "Courts, Brokers and Brokerage in the Medieval Mediterranean," in *Cultural Brokers at Mediterranean Courts in the Middle Ages*, ed. iid. (Paderborn: Fink-Schöningh, 2013), 10–31.

4 *Mercenaries and Paid Men: The Mercenary Identity in the Middle Ages*, ed. John France, History of Warfare 47 (Leiden: Brill, 2008).

5 Stephen Morillo, "Mercenaries, Mamluks and Militia: Towards a Crosscultural Typology of Military Service," in *Mercenaries and Paid Men: The Mercenary Identity in the Middle Ages,* ed. John France (Leiden: Brill. 2008) (cf. note 4), 243–260. On the wide field of medieval mercenaries, see: Kenneth Alan Fowler, *Medieval Mercenaries*, vol. 1, *The Great Companies* (Oxford: Blackwell, 2001); Janin Hunt and Ursula Carlson, *Mercenaries in Medieval and Renaissance Europe* (Jefferson: McFarland, 2013).

6 Morillo, "Mercenaries, Mamluks and Militia" (cf. note 5), especially 252–254.

7 See the discussion of the term in the introduction to this volume by Georg Christ and Patrick Sänger.

8 *The Formation of al-Andalus*, vol. 1, *History and Society*, ed. Manuela Marín, Julio Samsó and María Isabel Fierro (Aldershot: Ashgate, 1998), 394; Hussein Fancy, *The Mercenary Mediterranean: Sovereignty, Religion, and Violence in the Medieval Crown of Aragon* (Chicago: University of Chicago Press, 2016), 86–87.

9 Ramón Menéndez Pidal, *La España del Cid*, vol. 2 (Madrid: Ed. Plutarco, 1929); Richard A. Fletcher, *The Quest for El Cid* (London: Hutchinson, 1989); Javier Peña Pérez, *El Cid: Historia, leyenda y mito* (Burgos: Dossoles, 2000); Diego Catalán, *El Cid en la historia y sus inventores* (Madrid: Fundación Ramón Menéndez Pidal, 2002); critical: Alberto Montaner Frutos, "Dichos y hechos: Tácticas de la obediencia en el Cantar de mio Cid." *Cahiers d'études hispaniques medievales* 34 (2011): 29–39.

10 István Frank, "Reverter, vicomte de Barcelone (vers 1130–1145)," *Boletín de la Real Academia de Buenas Letras de Barcelona* 26 (1954/1956): 195–204; José Enrique Ruiz-Domènec, "Las cartas de Reverter, vizconde de Barcelona," *Boletín de la Real Academia de Buenas Letras de Barcelona* 39 (1983/1984): 93–118; François Clément, "Reverter et son fils, deux officiers catalans au service des sultans de Marrakech," *Medieval Encounters* 9 (2003): 79–107; José Enrique Ruiz-Doménec, *Quan els vescomtes de Barcelona eren: Història, crònica i documents d'una família dels segles X, XI i XII* (Barcelona: Pagès editors, 2006), 173–193; José Enrique Ruiz Domènec, *Atardeceres rojos: Cuatro vidas entre el islam y la cristianidad en la época de las cruzadas* (Barcelona: Ariel, 2007), 91–128, 233–235; Patrick Henriet, "'Ad regem Cordube militandi gratia perrexit': Remarques sur la présence militaire chrétienne en al-Andalus (Xe–XIIIe siècle)," in *Regards croisés sur la guerre sainte: Guerre, idéologie et religion dans l'espace méditerranéen latin (XIe–XIIIe siècle)*, ed. by Daniel Baloup and Philippe Josserand (Toulouse: Presses universitaires du Midi, 2006), 359–380.

11 Luis Sánchez Belda, ed., *Chronica Adefonsi imperatoris*, Escuela de Estudios Medievales, Textos 14 (Madrid: "Diana," 1950), 81–83 (chapters 104–106).

12 Clément, "Reverter et son fils" (cf. note 10), 87–89; Alejandro García Sanjuán, "Mercenarios cristianos al servicio de los musulmanes en el norte de África durante el siglo XIII," in *La Península Ibérica entre el Mediterráneo y el Atlántico: Siglos XIII–XV*, ed. Manuel González Jiménez and Isabel Montes Romero-Camacho (Cádiz: Diputación de Cádiz Servicio de Publicaciones, 2006), 435–448, repr. in Alejandro García Sanjuán: *Coexistencia y conflictos: Minorías religiosas en la Península ibérica durante la Edad Media* (Granada 2015: Universidad de Granada), 171–192, at 176-178.

13 Vincent Lagardère, *Les Almoravides: Le djihâd andalou* (Paris: Éditions l'Harmattan, 1999), 82–89; Carles Vela Aulesa, "L'Andalus en la política de Barcelona i la Corona d'Aragó (segle XI–1213)," in: *Tractats i negociacions diplomàtiques amd els regnes peninsulars i l'Ándalus (segle XI–1213)*, ed. Maria Teresa Ferrer i Mallol and Manuel Riu i Riu, Memòries de la Secció Històrico-Arqueològica 106 (Barcelona: Institut d'Estudis Catalans, 2018), 117–180, at 144–149. On hostages, see Adam J. Kosto, *Hostages in the Middle Ages* (Oxford: Oxford University Press, 2012).

14 Recently dated to 1137–1139: Gaspar Feliú, Josep Maria Salrach Marés, and María Josepa Arnall Juan, eds., *Els pergamins de l'Arxiu Comtal de Barcelona de Ramon Borrell a Ramon Berenguer I*, vol. 3, Col·lecció Diplomataris 20 (Barcelona: Pagès Editors, 1999), 1250, doc. 758; but cf. Ruiz Doménec, *Quan els vescomtes de Barcelona eren* (cf. note 10), 470, doc. 174 dated to 1138 and the explanations 173–201.

15 *In rei veritati sciatis quod ego sum vester fidelis homo et vester fidelis amicus et sum paratus ad vestrum servicium facere nocte et die ubi ego poscham, sine nullo engan*—Ruiz Doménec, *Quan els vescomtes de Barcelona eren* (cf. note 10), 471, doc. 174; Baiges, Feliú and Salrach, *Els pergamins de l'Arxiu Comtal de Barcelona* (cf. note 15), 1251, doc. 758: ... *ego, Reverterius vexconte Barchilonensis, fidele tuo et amico tuo et vasallo tuo, salutem et dilectam amiciciam conmo a domino meo et seniore meo et amico meo*—Ruiz Doménec, *Quan els vescomtes de Barcelona eren* (cf. note 10), 470, doc. 173; Baiges, Feliú and Salrach, *Els pergamins de l'Arxiu Comtal de Barcelona* (cf. note 15), 1252, doc. 759.

16 Ruiz Doménec, *Quan els vescomtes de Barcelona eren* (cf. note 10), 470, doc. 173; Baiges, Feliú and Salrach, *Els pergamins de l'Arxiu Comtal de Barcelona* (cf. note 15), 1252, doc. 759.

17 *Et si in ipsa mea honore est nullus homo qui non voleat facere vestram voluntatem, accipite eum et mitite eum in manicis ferreis et mitite eum ad me ad Marrochs et ibi facio vobis directum que vobis placuerit*—(ibid.).

18 Baiges, Feliú and Salrach, *Els pergamins de l'Arxiu Comtal de Barcelona* (cf. note 15), 1254–1256, doc. 761. About the *convenientiae* cf. Adam J. Kosto, *Making Agreements in Medieval Catalonia: Power, Order, and the Written Word, 1000–1200* (Cambridge: Cambridge University Press, 2001).

19 Ruiz Doménec, *Quan els vescomtes de Barcelona eren* (cf. note 10), 482–483, doc. 182 and 183.

20 Ibid., 491, doc. 192: ... *volens pergere apud Marrocos* ... (1167). On Reverter's son cf. ibid, 195–200. According to other reports, he joined the Templars: Henriet, "'Ad regem Cordube militandi gratia perrexit'" (cf. note 10), 368.

21 José Alemany, "Milicias cristianas al servicio de los sultanes musulmanes de Almagreb," in *Homenaje a D. Francisco Codera en su jubilación del profesorado: Estudios de erudición oriental*, ed. Eduardo Saavedra (Zaragoza: Mariano Escar, 1904), 133–169, at 136–137; Clément, "Reverter et son fils" (cf. note 10), 95–106; Amar Salem Baadj, *Saladin, the Almohads and the Banu Ghaniya: The Contest for North Africa (12th and 13th Centuries)*, Studies in the History and Society of the Maghrib 7 (Leiden: Brill, 2015).

22 Alemany, "Milicias cristianas al servicio de los sultanes" (cf. note 21), 134–137; J. F. P. Hopkins, *Medieval Muslim Government in Barbary: Until the Sixth Century of the Hijra* (London: Luzac and Company, 1958), 54–55; Jamil M. Abun-Nasr, *A History of the Maghrib in the Islamic Period* (Cambridge: Cambridge University Press, 1987), 86, 91; Wiebke Deimann, *Christen, Juden und Muslime im mittelalterlichen Sevilla: Religiöse Minderheiten unter muslimischer und christlicher Dominanz (12. bis 14. Jahrhundert)* (Berlin: Lit-Verlag, 2012), 143–148; García Sanjuán, "Mercenarios cristianos al servicio" (cf. note 12), 176.

23 María Desamparados Cabanes Pecourt, ed., *Crónica latina de los reyes de Castilla*, Textos medievales 11 (Valencia: J. Nácher, 1964), 117; Robert Ignatius Burns, "Renegades, Adventurers, and Sharp Businessmen: The Thirteenth Century Spaniard in the Cause of Islam," *The Catholic Historical Review* 58 (1972): 341–366, at 351.

24 Hopkins, *Medieval Muslim Government in Barbary* (cf. note 22), 76–77.

25 Ibid.; Alemany, "Milicias cristianas al servicio de los sultanes" (cf. note 21), 136–142; Deimann, *Christen, Juden und Muslime im mittelalterlichen Sevilla* (cf. note 22), 144–148.

26 Clément, "Reverter et son fils" (cf. note 10), 81–82.

27 Eva Lapiedra Gutiérrez, "Christian Participation in Almohad Armies and Personal Guards," *Journal of Medieval Iberian* Studies 2 (2010): 235–250; Fancy, *The Mercenary Mediterranean* (cf. note 8), 85. Pedro Fernández de Castro did however lead a contingent on behalf of the Almohads against fellow Christians in 1195 at the battle of Alarcos: Patrick Henriet, "Xénophobie et intégration à Léon au XIIIe siècle: Le discours de Lucas de Túy sur les étrangers," in *L'étranger au Moyen Âge: XXXe Congrès de la SHMES (Göttingen, juin 1999).* (Paris: Publications de la Sorbonne, 2000), 37–58, at 51, Henriet, "'Ad regem Cordube militandi gratia perrexit'" (cf. note 9), 372-373.

28 Andrés Giménez Soler, "Caballeros españoles en Africa y africanos en España," *Revue hispanique* 12 (1905): 299–372, at 348–372; Brian Aivars Catlos, "Mahomet Abenadalill: A Muslim Mercenary in the Service of the Kings of Aragon (1290–1291)," in *Jews, Muslims, and Christians in and around the Crown of Aragon: Essays in Honour of Professor Elena Lourie*, ed. Harvey J. Hames, The Medieval Mediterranean 52 (Leiden: Brill, 2004), 257–302; Hussein Fancy, "The Last Almohads: Universal Sovereignty between North Africa and the Crown of Aragon," in *Spanning the Strait: Studies in Unity in the Western Mediterranean*, ed. Yuen-Gen Liang, Abigail Krasner Balbale, Andrew Devereux, and Camilo Gómez-Rivas, Medieval Encounters 19/1–2 (Leiden: Brill, 2013), 102–136; Fancy, *The Mercenary Mediterranean* (cf. note 8). For other areas: Alexis G. C. Savvides, "Late Byzantine and Western Historiographers on Turkish Mercenaries in Greek and Latin Armies: The Turcoples/Tourkopouloi," in *The Making of Byzantine History: Studies Dedicated to Donald M. Nicol*, ed. Roderick Beaton, Publications of the Centre for Hellenic Studies 1 (Aldershot: Ashgate, 1993), 122–136. Cf. José Enrique López de Coca y Castañer, "Caballeros Moriscos al servicio de Juan II y Enrique IV, reyes de Castilla," *Meridies* 3 (1996): 119–136. Ana Echevarría, *Knights in the Frontier: The Moorish Guard of the Kings of Castile (1410–1467)* (Leiden: Brill, 2009).

29 Fancy, *The Mercenary Mediterranean* (cf. note 8), 85.

30 Garcia Fitz, Francisco and Gouveia Monteiro, Joao, ed., *War in the Iberian Peninsula, 700–1600* (New York: Routledge, 2018).

31 Ibn-Khaldun, *The Muqaddimah: An Introduction to History*, vol. 2, transl. Franz Rosenthal (Princeton: Princeton University Press, 1958), 80–81.

32 García Sanjuán, "Mercenarios cristianos al servicio" (cf. note 11), 179-181; Lapiedra Gutiérrez, "Christian Participation in Almohad Armies" (cf. note 27).

33 Lapiedra Gutiérrez, "Christian Participation in Almohad Armies" (cf. note 27); Fancy, *The Mercenary Mediterranean* (cf. note 8), 94.

34 Ibid., 90; Alemany, "Milicias cristianas al servicio de los sultanes" (cf. note 21), 139, 142–155; Abun-Nasr, *A History of the Maghrib in the Islamic Period* (cf. note 22), 104, 116.

35 According to a fifteenth-century and therefore slightly dubious source, the group of Christian tax collectors led by Alonso Pérez de Guzmán, known as Guzmán el Bueno (d. 1309) was even donned with a cross: Miguel Ángel Ladero Quesada, "Una biografia caballeresca del siglo XV. 'La Coronica del yllustre y muy magnifico cauallero don Alonso Perez de Guzman el Bueno,'" *En la España medieval* 22 (1999): 247–283, at 271: *y mandó que en los capuçes delante y detrás truxesen señal de la cruz, que fuesen de las colores, porque de ellos a los moros oviese diferencia quando el rey los embiase a pedir los tributos que los bárbaros y alaraves y las otras naciones al rey eran obligados.*

36 Alemany, "Milicias cristianas al servicio de los sultanes" (cf. note 21), 146–155.

37 Maximiliano A. Alarcón Santón and Ramón García de Linares, *Los documentos árabes diplomáticos del Archivo de la Corona de Aragón: Editados y traducidos* (Madrid: Estanislao Maestre, 1940), 169–171, doc. 83.

38 On al-ʿUmarī's reliability, see Muḥammad Binšarīfa, "Maṣādir maġribiyya šafawiyya fī masālik al-abṣār li-l-ʿUmarī," *Maǧallat Maǧmaʿ al-Luġat al-ʿArabiyya* 82 (1998): 232–255—Thanks to Amar Baadj (Bonn) for this reference.

39 Clément, "Reverter et son fils" (cf. note 10), 82.

40 Roudh El-Kartas, *Histoire des souverains du Maghreb (Espagne et Maroc) et annales de la ville de Fès*, ed. and trans. Auguste Beaumier (Paris: Duprat, 1860), 371–372, cf. Clara Maillard, *Les papes et le Maghreb aux XIIIème et XIVème siècles: Étude des lettres pontificales de 1199 à 1419* (Turnhout: Brepols, 2014), 233.

41 Olivia Remie Constable, *Housing the Stranger in the Mediterranean World: Lodging, Trade and Travel in Late Antiquity and the Middle Ages* (New York: Cambridge University Press, 2003); Dominique Valérian, "Le fondouk, instrument du contrôle sultanien sur les marchands étrangers dans les ports musulmans (XIIe–XVe siècle)," in: *La mobilité des personnes en Méditerranée de l'Antiquité à l'époque moderne: Procédures de contrôle et documents d'identifications*, ed. Claudia Moatti, Collection de l'École Française de Rome 341 (Rome: École Française de Rome, 2004), 677–698.

42 Charles Emmanuel Dufourcq, *L'Espagne catalane et le Maghrib aux XIIIe et XIVe siècles: De la bataille de Las Navas de Tolosa (1212) à l'avènement du sultan mérinide Abou-I-Hasan (1331)* (Paris: Presses Universitaires de France, 1966), 101–103; Alemany, "Milicias cristianas al servicio de los sultanes" (cf. note 21), 155–169; Abun-Nasr, *A History of the Maghrib in the Islamic Period* (cf. note 22), 121–125; Brahim Jadla, "L'Ifriqiyah hafside et la chrétienté: Le temps des ruptures," in *Les relations diplomatiques entre le monde musulman et l'Occident latin (XIIe–XVIe siècle)*, ed. Denise Aigle and Pascal Buresi, Oriente moderno, n.s., 88 (Rome: Istituto per l'Oriente C. A. Nallino, 2008), 311–321, at 313–317.

43 Robert Brunschvig, *La Berbérie orientale sous les Ḥafsides: Des origines a la fin du XVe siècle*, 2 vols. (Paris: Librairie d'Amérique et d'Orient Adrien-Maisonneuve, 1941–1947), 1:446.

44 *Esdevenc-se que aquest sarraí qui tenia Constantina, qui havia nom Bonbòquer et havia molts soldaders crestians amb ell ....*—Jaume I/Bernat Desclot/Ramon Muntaner/Pere III, *Les quatre grans Cròniques. Revisió del text, pròlegs i notes per Ferran Soldevilla* (Barcelona: Edicions Selecta, 1983), 467; *Les quatre grans Cróniques*, ed. Ferran Soldevila, Jordi Bruguera, and Maria Teresa Ferrer i Mallol, vol. 2, *Crònica de Bernat Desclot*, Memòries de la Secció Historico-Arqueològica 80 (Barcelona: Institut d'Estudis Catalans, 2008), 166–167 (chapt. 77). Cf. Alemany, "Milicias cristianas al servicio de los sultanes" (cf. note 21), 163; Burns, "Renegades, Adventurers, and Sharp Businessmen" (cf. note 23), 353.

45 Alemany, "Milicias cristianas al servicio de los sultanes" (cf. note 21), 144–145; Giménez Soler, "Caballeros españoles en Africa" (cf. note 28), 317–321; Faustino Gazulla, *Jaime I de Aragón y los estados musulmanes* (Barcelona: Real Academia de Buenas Letras, 1919), 63; Ladero Quesada, "Una biografia caballeresca" (cf. note 35). Earlier similar cases from Castile: Henriet, "Xénophobie et intégration à Léon au XIIIe siècle" (cf. note 27), 49–53.

46 Dufourcq, *L'Espagne catalane et le Maghrib aux XIIIe et XIVe siècles* (cf. note 42), 650, s.v. "milices chrétiennes"; Miguel Ángel Ochoa Brun, *Historia de la diplomacia española*, vol. 3 (Madrid: Ministerio de Asuntos Exteriores, 1991), 197; Barton, "Traitors to the Faith?" (cf. note 2).

47 Carmen Batlle Gallart, "Noticias sobre la milicia cristiana en el Norte de África en la segunda mitad del siglo XIII," in *Homenaje al profesor Juan Torres Fontes*, ed. Universidad Murcia (Murcia: Universidad de Murcia, 1987), 1:127–137.

48 Roser Salicrú Lluch, "Mercenaires castillans au Maroc au début du XVe siècle," in *Migrations et diasporas méditerranéennes (Xe–XVIe siècles)*, ed. Michael Ballard and Alain Ducellier (Paris: Éditions de la Sorbonne, 2002), 417–434, especially 419–420.

49 Brett Whalen, "Corresponding with Infidels: Rome, the Almohads, and the Christians of Thirteenth-Century Morocco," *Journal of Medieval and Early Modern Studies* 43 (2011): 487–513; Maillard, *Les papes et le Maghreb* (cf. note 40), particularly 226–244; Michael Lower, "The Papacy and Christian Mercenaries of Thirteenth-Century North Africa," *Speculum* 89 (2014): 601–631; Clara Maillard, "Protection des chrétiens en terre d'Islam et discussion entre papes et souverains musulmans: Le cas singulier des mercenaires du Maroc," in *Religious Minorities in Christian, Jewish and*

Muslim Law (5th–15th Centuries), ed. Nora Berend, Youna Hameau-Masset, and John Tolan (Turnhout: Brepols, 2017), 317–333.

50 Ferran Soldevilla, ed., *Les quatre grans Cròniques* (cf. note 44), 43; *Les quatre grans Cróniques*, ed. by Ferran Soldevila, Jordi Bruguera, and Maria Teresa Ferrer i Mallol, vol. 1, *Llibre dels feits del rei En Jaume*, Memòries de la Secció Historico-Arqueològica 73 (Barcelona: Institut d'Estudis Catalans, 2007), 168–169 (chapt. 75); Henriet, "'Ad regem Cordube militandi gratia perrexit'" (cf. note 9), 369-371.

51 Louis de Mas Latrie, ed., *Traités de paix et de commerce et documents divers concernant les relations des chrétiens avec les Arabes de l'Afrique septentrionale au Moyen âge* (Paris: Plon, 1866; New York: Franklin, 1964), 2:17–18, doc. 18; Santiago Domínguez Sánchez, ed., *Documentos de Nicolás IV (1277–1280), referentes a España* (León: Publicaciones Universidad de León, 1999), 300, doc. 284. Cf. James Muldoon, *Popes, Lawyers, and Infidels: The Church and the Non-Christian World 1250–1550* (Liverpool: Liverpool University Press, 1979), 41–42, 54; Barton, *Traitors to the Faith* (cf. note 2), 37; Whalen, "Corresponding with Infidels" (cf. note 49), 499–503; Maillard, *Les papes et le Maghreb* (cf. note 40), 237.

52 Maillard, *Les papes et le Maghreb* (cf. note 39), 237. On apostasy, see Burns, "Renegades, Adventurers, and Sharp Businessmen" (cf. note 23); José Vicente Cabezuelo Pliego, "Cristiano de Alá, renegado de Cristo: El caso de Abdalla, fill d'en Domingo Vallés, un valenciano al servicio del islam," *Sharq al-Andalus* 13 (1996): 27–46; Roser Salicrú Lluch, "En busca de una liberación alternativa: Fugas y apostasía en la Corona de Aragón bajomedieval," in *La liberazione dei "captivi" tra Cristianità e Islam: Oltre la crociata e il Gihad; Tolleranza e servizio umanitario*, ed. Giulio Cipollone, Collectanea Archivi Vaticani 46 (Vatican City: ASV, 2000), 703–713; Nikolas Jaspert, "Konversion zum Islam im spätmittelalterlichen Mittelmeerraum und die Entscheidungen des Anselm Turmeda alias ʿAbdallāh at-Tarǧumān," in *Der Sultan und der Heilige: Islamisch-christliche Perspektiven auf die Begegnung des hl. Franziskus mit Sultan al-Kamil (1219–2019)*, ed. Amir Dziri, Angelica Hilsebein, Mouhanad Khorchide, and Bernd Schmies (Münster: Aschendorff Verlag, 2021), 367–396.

53 Pierre de Cenival, "L'église chrétienne de Marrakesch au XIIIe siècle," *Hespéris* 7 (1927): 69–83; Salicrú Lluch, "Mercenaires castillans au Maroc" (cf. note 48), 430–432; Maillard, *Les papes et le Maghreb* (cf. note 40).

54 A point correctly underlined by Fancy, *The Mercenary Mediterranean* (cf. note 8), 85, 92–93. This general ruling did not impede Christian militiamen from fighting against coreligionists in an opposing Muslim army on exceptional occasions, as can be seen in the 1271/1272 conflict between the Moroccan Merinids and the lords of Tlemcen or in 1320, during a family dispute amongst the Merinids: Alemany, "Milicias cristianas al servicio de los sultanes" (cf. note 21), 142–143, 150; Gazulla, *Jaime I de Aragón y los estados musulmanes* (cf. note 45), 63.

55 Clément, "Reverter et son fils" (cf. note 10), 81; Cenival, "L'église chrétienne de Marrakesch" (cf. note 53), 76; Maíllo Salgado provides a more critical reading as "bastard" or "coward": Felipe Maíllo Salgado, "Precisiones para la historia de un grupo étnico-religioso: Los farfanes," *Al-Qantara* 4 (1983): 265–282, at 266–267; García Sanjuán, "Mercenarios cristianos al servicio" (cf. note 11), 188-189.

56 Thomas A. Tweed, *Crossing and Dwelling: A Theory of Religion* (Cambridge: Harvard University Press, 2006). Cf. William Safran, "Diasporas in Modern Societies: Myths of Homeland and Return," *Diaspora: A Journal of Transnational Studies* 1 (1991): 83–99.

57 Maillard, *Les papes et le Maghreb* (cf. note 40), 238: … *inter Aragonenses et Cathalanos eosdem et dictos Castellanos discordia debeat exoriri.*

58 Clément, "Reverter et son fils" (cf. note 10), 82.

59 Brunschvig, *La Berbérie orientale* (cf. note 43), 1:448; Roser Salicrú i Lluch, "Entre la praxis y el estereotipo: Vivencias y percepciones de lo islámico-ibérico en fuentes archivísticas y narrativas bajomedievales," in *Ritus infidelium: Miradas interconfesionales sobre las prácticas religiosas en la edad media*, ed. José Martínez Gázquez and John Victor Tolan, Collection de la Casa de Velázquez 138 (Madrid: Casa de Velazquez, 2013), 99–111, at 103.

60 Anselme Adorne, *Itinéraire d'Anselme Adorno en Terre sainte, 1470–1471*, ed. and trans. Jacques Heers and Georgette de Groër (Paris: Éditions du Centre national de la recherche scientifique, 1979), 108.

61 María Teresa Ferrer i Mallol, "Marruecos y la Corona catalano-aragonesa: Mercenarios catalanes al servicio de Marruecos (1396–1410)," in *Homenaje al profesor Eloy Benito Ruano* ed. Sociedad española de estudios medievales (Murcia: Universidad de Murcia, 2010), 1:251–272, at 260–261, doc. 1.

62 Alemany, "Milicias cristianas al servicio de los sultanes" (cf. note 21), 154–155; Hopkins, *Medieval Muslim Government in Barbary* (cf. note 22), 77–78.

63 Rafael Sánchez Saus, "Un linaje hispano-marroquí entre la leyenda y la historia: Los Farfán de los Godos," *Actas del Congreso Internacional "El Estrecho de Gibraltar"* (Madrid: UNED, 1988), 2:323–332, at 331, doc. 1: *A los honrados, nobles varones Farfanes christianos, alcaides, donceles, compañones del noble y limpio y gran linage de los Godos que estades e vivides en la ciudad de Marruecos del reino de Benamarín, del qual linage los Reies de Castilla y nobles señores descienden.*

64 *Estando el Rey Don Juan en Alcalá de Henares* [...] *llegaron a él cinquenta caballeros christianos que avía grand tiempo que vivían en tierra de Marruecos, o eran de linaje de christianos, los quales después que los moros conquistaron a España en tiempos del Rey Don Rodrigo, fincaron en tierra de Marruecos que los envió allá Ulit Noramamolín por ruego del Conde Don Illán, ca eran sus amigos, e llamaban los moros a este linage de christianos que así vivían entre ellos los Farfanes, e troxeron consigo sus mujeres e fijos. E el Rey rescibiólos muy bien, ca él avía enviado por ellos a Marruecos, e prometióles de les dar heredades e bienes en su regno e mantenimiento honrado*—Cayetano Rosell, ed., *Crónicas de los reyes de Castilla: Desde don Alfonso el Sabio hasta los Católicos don Fernando y doña Isabel*, Biblioteca de Autores Españoles 68 (Madrid: Atlas, 1953), 2: 143 and 158. Four years later, these "farfanes de los Godos" received privileges from King Henry III: *Crónicas de los reyes de Castilla* (cf. note 64), 158; Cenival, "L'église chrétienne de Marrakesch" (cf. note 53), 82; Maíllo Salgado, "Precisiones para la historia" (cf. note 55), 278; Salicrú Lluch, "Mercenaires castillans au Maroc" (cf. note 48), 423–425, 427–432.

65 Maíllo Salgado, "Precisiones para la historia" (cf. note 55), 271; Sánchez Saus, "Un linaje hispano-marroquí" (cf. note 63), 325.

66 Sánchez Saus, "Un linaje hispano-marroquí" (cf. note 63), 327.

67 Barrantes Maldonado, Pedro: *Ilustraciones de la Casa de Niebla*, Fuentes para la historia de Cádiz y su provincia 3 (Cádiz: Univérsidad de Cádiz, Servicio de Publicaciones, 1998), 29–71: "Como se fue Don Alonso Pérez de Guzmán a servir al rey Abenyuçaf Rey de Fez y de Marruecos, y porqué causa," "De algunas cosas que le sucedieron a Don Alonso Pérez de Guzmán estando en África", "De la primera batalla que Don Alonso Pérez de Guzmán dió a los moros, que fue principio de su riqueza, yendo a cobrar los tributos de los alárabes, para el rey Abenyuçaf su señor", "De como Don Alonso Pérez de Guzmán hizo con el rey Abenyuçaf que prestase al rey Don Alonso de Castilla sesenta mil doblas", "De como Don Alonso Pérez de Guzmán vino de Africa con la embajada y con sesenta mil doblas al rey Don Alonso de Castilla, y como se desposó con Doña María Alonso Coronel, en Sevilla", etc.

68 *Entès havem per relació d'alscuns servidors nostres, los quals són en cert deute de sanch ab n'Arnau Masquefa e ab en Luys Vilar, naturals nostres, que los dits Arnau e Luys han estat lonch temps e estan vuy en vostre servey, lexant los parents, amichs e heretats que han deçà en nostra terra. On, com los dits servidors nostres hajen a nós humilment supplicat que nós volguéssem scriure e pregar a la vostra real amistat que 'ns enviàssets los damunt nomenats, per ço, molt alt príncep, vos pregam affectuosament que, per esguard de nostres prechs e per honor nostra, nos trametats los dits Arnau, amb la muller e fills seus, e en Luys, ab sa maré, francament e quítia, com pus iverçosmant fer se porà, car cosa serà de què·ns farets gran plaer e'ns obligarets per vós e vostres pregàries fer semblants coses e majors*—Ferrer i Mallol, "Marruecos y la Corona catalano-aragonesa" (cf. note 61), 263, doc. 4.

69 *Moguts per humil suplicació a nós feta per alcuns naturals e sotsmeses nostres molt acostats, parents d'en Bernat Spígol, nostre natural e antich servidor vostre, vos pregam axí affectuosament*

com podem que, per honor e contemplació nostra, per satisfer al desig dels parents e amichs del dit Bernat, que açò tenen molt a cor, vullats donar líbera licència e facultat plenera al dit Bernat e a sa muller, lo qual longament vos ha servit, que pugan sens encorriment de alguna pena venir de les parts deçà ab tots lurs béns, certificant-vos que açò serà cosa que us reputarem a singular honor e plaer e altres nostres sotmeses, sabents vós fer tals gràcies, seran animats a passar liberalment en vostre servey. Si algunes coses vos plaen, molt noble rey e car amich, de les parts deçà, rescrivits-nos ço que·us plàcia—Ferrer i Mallol, "Marruecos y la Corona catalano-aragonesa" (cf. note 61), 264–265, doc. 6—note the king's argument that by showing such munificence, the Sultan would attract further mercenaries.

70 ... *com nos volguéssem granment que vinguessen en nostres regnes e terres per star e habitar en aquells Garcia Biscarra e sa muller Johanna Roiz, Elvira Roiz, muller qui fou de mossèn Bernat de Mirambell, e Garcia Biscarra ... companys de Gonçalvo Díez*—Ferrer i Mallol, "Marruecos y la Corona catalano-aragonesa" (cf. note 61), 265, doc. 7.

71 Ferrer i Mallol, "Marruecos y la Corona catalano-aragonesa" (cf. note 61), 270–271, doc. 4: *Pere Abelló, del orde de prehicadors, qui passà a aqueix regne ab mossèn Gonçalvo Díez.*

72 *Magdalena, de nació cathalana, passà lonch temps ha ab son marit en aqueix regne e ha gran desig de tornar deçà entre sos parents e amich*—Ferrer i Mallol, Marruecos y la Corona catalano-aragonesa (cf. note 61), 271, doc. 15.

73 Salicrú Lluch, "Mercenaires castillans au Maroc" (cf. note 48), 430–434. Cf. the highly stylized description in a fifteenth century text of Alonso Pérez de Guzmán's migration to Fez at the end of the thirteenth century: *El qual, con grandeza de su coraçón, deseando usar dellas, no contento de la aprobación que vibía, vendió lo que tenía con propósito de yr a Verbena a el rey de Benamarín. Y aderçado a sí y a diez escuderos, llebando a su amada muger, en una nave pasando el Estrecho en la ciudad de çeuta fue a desembarcar, y el rey saviendo su venida ovo muy gran plazer y mandole enbiar diez cauallos con sus arreos* [*e*] *ginetes y mandóle dar todas las cosas que para él e su muger e los suyos ovíesen menester, e desque oviesen rreposado se viniesen a su merced*—Ladero Quesada, "Una biografia caballeresca" (cf. note 35), 270.

74 Mariano Arribas Palau, "Cartas de recomendación cursadas al Sultán Abuʿ Sa'i'd Utmān III de Marruecos por el Rey de Aragón, Fernando I, el de Antequera," *Hespéris Tamuda* 1 (1956): 387–407, at 403–404, doc. 13: ... *haiades por recomendados todos los christianos assi alcaydes como mosferrates e companyones e sus muyeres e fillos portando como nos somos stado informado que fueron leyales servidores en el tiempo de vuestro manester* ...; on the wives of mercenaries cf. ibid, 400–401, doc. 9, 10. Cf. Salicrú Lluch, "Mercenaires castillans au Maroc" (cf. note 48), 433.

75 Cenival, "L'église chrétienne de Marrakesch" (cf. note 53), 81; Roudh El-Kartas, *Histoire des souverains du Maghreb* (cf. note 40), 551–552, cf. Maillard, *Les papes et le Maghreb* (cf. note 40), 237.

76 Alemany, "Milicias cristianas al servicio de los sultanes" (cf. note 21), 147–148, 150–151; Dufourcq, *L'Espagne catalane et le Maghrib aux XIIIe et XIVe siècles* (cf. note 42), 456–459; Rudolf Thoden, *Abū 'l-Ḥasan 'Alī: Merinidenpolitik zwischen Nordafrika und Spanien in den Jahren 710–752 H./1310–1351* (Freiburg: Schwarz, 1973), 53, 57–60.

77 Serge Gubert, "Pratiques diplomatiques mérinides (XIIIe–XVe siècle)," in *Les relations diplomatiques entre le monde musulman et l'Occident latin (XIIe–XVIe siècle)*, ed. Denise Aigle and Pascal Buresi, Oriente moderno, n.s., 88 (Rome: Istituto per l'Oriente C. A. Nallino, 2008), 435–468, at 444; Maillard, *Les papes et le Maghreb* (cf. note 40), 239.

78 Alemany, "Milicias cristianas al servicio de los sultanes" (cf. note 21), 155–156; Gazulla, *Jaime I de Aragón y los estados musulmanes* (cf. note 45), 58–61; García Sanjuán, "Mercenarios cristianos al servicio" (cf. note 11), 178.

79 Alemany, "Milicias cristianas al servicio de los sultanes" (cf. note 21), 158; Dufourcq, *L'Espagne catalane et le Maghrib aux XIIIe et XIVe siècles* (cf. note 42), 481–487.

80 Alemany, "Milicias cristianas al servicio de los sultanes" (cf. note 21), 167–169.

81 Abun-Nasr, *A History of the Maghrib in the Islamic Period* (cf. note 22), 124–125.

82 Thoden, *Abū ʾl-Ḥasan ʿAlī: Merinidenpolitik* (cf. note 76), 52–53.
83 Gubert, "Pratiques diplomatiques mérinides" (cf. note 77), 444.
84 Ángeles Masiá de Ros, *La corona de Aragón y los estados del norte de África: Política de Jaime II y Alfonso IV en Egipto, Ifriquía y Tremecén* (Barcelona: Instituto español de estudios mediterráneos, 1951), 487–493; Dufourcq, *L'Espagne catalane et le Maghrib aux XIIIe et XIVe siècles* (cf. note 42), 488–494; Nikolas Jaspert, "Interreligiöse Diplomatie im Mittelmeerraum: Die Krone Aragón und die islamische Welt im 13. und 14. Jahrhundert," in *Aus der Frühzeit europäischer Diplomatie: Zum geistlichen und weltlichen Gesandtschaftswesen vom 12. bis zum 15. Jahrhundert*, ed. Claudia Märtl and Claudia Zey (Zürich: Chronos, 2008), 151–190, at 178; Michael Lower, "Ibn al-Lihyani: Sultan of Tunis and Would-Be Christian Convert (1311–18)," *Mediterranean Historical Review* 24 (2009): 17–27.
85 Alemany, "Milicias cristianas al servicio de los sultanes" (cf. note 21), 150–151; Dufourcq, *L'Espagne catalane et le Maghrib aux XIIIe et XIVe siècles* (cf. note 42), 459; Thoden, *Abū ʾl-Ḥasan ʿAlī: Merinidenpolitik* (cf. note 76), 53; Clément, "Reverter et son fils" (cf. note 10), 82; Gubert, "Pratiques diplomatiques mérinides" (cf. note 77), 444. Another member of the Hafsid dynasty, a son of Abū Isḥāq (d. 1283), indeed converted to Christianity: Fancy, *The Mercenary Mediterranean* (cf. note 8), 76.
86 García Sanjuán, "Mercenarios cristianos al servicio" (cf. note 11), 179.
87 *Dantes et concedentes vobis alcaydiam ejusdem loci, ita quod vos sitis alcaydus omnium christianorum, tam militum quam aliorum, qui vobiscum apud Trinicem ibunt, seu de cetero fuerint ibídem—Traités de paix et de commerce et documents divers concernant les relations des Chrétiens avec les Arabes de l'Afrique septentrionale au moyen âge: Supplément et tables*, ed. Louis de Mas Latrie (Paris: Baur et Détaille, 1872), 38–39, doc. 8; Gazulla, *Jaime I de Aragón y los estados musulmanes* (cf. note 45), 61.
88 Mas Latrie, *Traités de paix et de commerce et documents: Supplément et tables* (cf. note 87), 62 (dated erroneously to 1322); Giménez Soler, "Caballeros españoles en Africa" (cf. note 28), 303, note 1; Dufourcq, *L'Espagne catalane et le Maghrib aux XIIIe et XIVe siècles* (cf. note 42), 149–150, 152: *ita quod vos sitis alcaydus omnium xpianorum tam militum quam aliorum qui vobiscum apud Tirimçe ibunt vel iam sunt seu de cetero fuerint ibídem.*
89 Ludwig Klüpfel, *Die äußere Politik Alfonsos III. von Aragonien bis zu den Verhandlungen von Gaeta, 1285–1299* (Berlin: Rothschild, 1912), 171–173, doc. 15 (April 1286); Fancy, *The Mercenary Mediterranean* (cf. note 8), 78–79. The king demanded freedom of movement and a Christian cleric of their own for the soldiers. Cf. the instalment of Rodrigo Sánchez de Vergays as *alcaydus et caput omnium Christianorum terre et jurisdictionis nostre in terra et jurisdictione vestra commorantium* in 1296: Mas Latrie, *Traités de paix et de commerce et documents: Supplément et tables* (cf. note 87), 46, doc. 19 and 20.
90 Cenival, "L'église chrétienne de Marrakesch" (cf. note 53), 74; Javier Albarrán Iruela, "De la conversión y expulsión al mercenariado: Los cristianos en las fuentes almohades," in *La Península Ibérica en tiempos de Las Navas de Tolosa*, ed. by Carlos Estepa and María Antonia Carmona Ruiz, Monografías de la Sociedad Española de Estudios Medievales 5 (Madrid: Sociedad Española de Estudios Medievales, 2014), 79–91, at 89 (who doubts the veracity of the information). The numbers of fighters given range from 500 and 12.000 cavalrymen: Alemany, "Milicias cristianas al servicio de los sultanes" (cf. note 21), 137–138; Hopkins, *Medieval Muslim Government in Barbary* (cf. note 22), 77; Deimann, *Christen, Juden und Muslime im mittelalterlichen Sevilla* (cf. note 22), 144; Allen James Fromherz, *The Near West: Medieval North Africa, Latin Europe and the Mediterranean in the Second Axial Age* (Edinburgh: Edinburgh University Press, 2016), 169–173.
91 Dufourcq, *L'Espagne catalane et le Maghrib aux XIIIe et XIVe siècles* (cf. note 42), 277–311, 407–449, 487–510; Abun-Nasr, *A History of the Maghrib in the Islamic Period* (cf. note 22), 122–126.
92 Mas Latrie, *Traités de paix et de commerce et documents* (cf. note 51), 2:285–290, doc. 3.
93 Alarcón Santón and García de Linares, *Los documentos árabes diplomáticos* (cf. note 37), 164, doc. 80.

94 Mas Latrie, *Traités de paix et de commerce et documents: Supplément et tables* (cf. note 87), 32–33, doc. 1: ... *quando G. de Montecatano, iturus de mandato nostro apud Tunicium, in Barchinona solveret suis militibus solidatam* ...

95 Mas Latrie, *Traités de paix et de commerce et documents* (cf. note 51), 2:286–290, doc. 4: *Item que tots los cavallers o homens d'armes crestians qui son huy, ne seran d'aquí avant en la senyoria del rey de Tunis, que y sien tots per nos et que nos lus donem cap aquel que nos vulrem, e li mudem e li camiem quan nos vulrem, e qu'el dit Miralmomeni do al cap que nos y metrem per sa persona, e als cavallers e als homens d'armes, aytal sou com prenian el temps del noble en Guillem de Muncada* [...]. On the backdrop Dufourcq, *L'Espagne catalane et le Maghrib aux XIIIe et XIVe siècles* (cf. note 42), 259–274.

96 Giuseppe La Mantia, *Codice diplomatico dei re Aragonesi di Sicilia*, vol. 1, *Anni 128–1290* (Palermo: Società siciliana per la storia patria, 1917), 377–386, doc. 158 and complete new edition of the Latin and Arabic text in Fancy, "The Last Almohads" (cf. note 28), appendix B: 129–136.

97 Charles-Emmanuel Dufourcq, "Documents inédits sur la politique ifrikiyenne de la Couronne d'Aragon," *Analecta Sacra Tarraconensia* 25 (1952): 255–292, at 264: *Com nos ayam entes quen Tunis en lo vostro servici aya molts crestians soldariers, alcayts e cavallers e escuders, los quals a vos fan bon servici e leyal de que nos som molt pagats, pregam a la vostra amistat que aquells vos sien recomanats a que per amor e honor de nos los fassats be e merce.*

98 Masiá de Ros, *La corona de Aragón y los estados del norte de África* (cf. note 84), 389–390, doc. 111; 407, doc. 407. However, these regulations were not included in the peace treaty finally signed in 1314 (ibid., 411–414, doc. 128).

99 Cf. note 96 and Fancy, *The Mercenary Mediterranean* (cf. note 8), 82–84.

100 *Item damus et concedimus vobis quod predictus alcaldus vester et omnes Cristiani qui cum eo erunt possint apud Tun[icium hab]ere Ecclesiam, et quod possint secum habere et tenere sacerdotes et alios clericos qui celebrent eis in ipsa [Ec]clesia officium ecclesiasticum sive domini Ihesu Cristi, et qui possint portare corpus Cristi cum signo campane sive squille prout moris est Cristianorum per omnes alfundi eis et vassalos Cristianorum, et tradere ipsum corpus Cristi Cristianis hoc requirentibus, et quod ipsi clerici possint portare crucem et turibula corporibus Cristianorum decedencium donec fuerint tradi[ti] sepulcure secundum quod hoc consuevit fijiere inter Cristianos, sine omni contradiccione et impedimento nostri et [omnibus] cuiuscumque persone. Item promitimus vobis nos facturos et curaturos quod barrile vini vendatur pred[icto] alcaldo vestro et militibus et scutiferis suis et familie et aliis Cristianis pro duobus bisanciis sine omni encameramento et quod dabitur barrile vini militi <ad> quinque dies et scutifero ad septem dies. Item promitimus vobis bona fijide quod [quo]cienscumque a vobis fuerimus requisiti per literas vestras vel per nuncium vestrum erimus vobis adiutores et de[fen]sores toto posse nostro contra quoscumque homines Cristianos scilicet et Sarracenos et alios quoscumque hom[ines] cuiuscumque condicionis legis vel fijidei existant, et hoc faciemus sine omni dolo malo et fraude*—ed. Fancy, "The Last Almohads" (cf. note 28), 131. Cf. with Arabic version and English translation Olivia Remie Constable, "Ringing Bells in Hafs'id Tunis: Religious Concessions to Christian *Fondacos* in the Later Thirteenth Century," in *Histories of the Middle East: Studies in Middle Eastern Society, Economy and Law in Honor of A. L. Udovitch*, ed. Roxani Eleni Margariti, Adam Sabra, and Petra M. Sijpesteijn, Islamic History and Civilization 79 (Leiden: Brill, 2011), 53–72 (many thanks to Georg Christ for this reference).

101 Fancy, "The Last Almohads" (cf. note 28), 124–125.

102 Dufourcq, "Documents inédits sur la politique ifrikiyenne de la Couronne d'Aragon" (cf. note 97), 257; Dufourcq, *L'Espagne catalane et le Maghrib aux XIIIe et XIVe siècles* (cf. note 42), 291. Further references to the *alcayt* in Tunis from 1299: Mas Latrie, *Traités de paix et de commerce et documents: Supplément et tables* (cf. note 87), 46–48, doc. 21–23.

103 Masiá de Ros, *La corona de Aragón y los estados del norte de África* (cf. note 84), 390, doc. 111 (1294); 407, doc. 125 (1313). Cf. the regulations in King Peter the Ceremonious's instructions for an embassy to the Hafsids: *Primerament demanava lo dit senyor Rey en quel Rey de Tuniç tengues M xpians de sou ço es assaber. cc. cavalers et. Dccc. escuders ab*

alcait del senyor Rey darago. E quel dit alcayt prengues per cascun dia. c. bs. dels quals devia pendre lo senyor Rey darago LXX. bs. e los romanents. XXX. bs. devien esser del dit alcayt. E cascun cavaler devia penre per cascun dia III. br. dels quals devia hauer lo senyor Rey darago. V. milareses. E cascun escuder devia penre per cascun jorn II bs. e mig dels quals deuia hauer lo senyor Rey darago V. milareses–Giménez Soler, "Caballeros españoles en Africa y africanos en España II," *Revue hispanique* 16 (1907): 56–69, at 61 (note 28).

104 During certain phases of weakness on the side of the Hafsids, the Aragonese even went as far as to outright demand tributes: Abun-Nasr, *A History of the Maghrib in the Islamic Period* (cf. note 22), 122–123.

105 On similar power relations in the eastern Mediterranean, see Benjamin Arbel, "Venetian Cyprus and the Muslim Levant, 1473–1570," in *Kypros kai oi Staurophories* [Cyprus and the Crusades], ed. Nicholas S. H. Coureas and Jonathan Riley Smith (Nicosia: Cyprus Research Centre, 1995), 159–185 (many thanks to Georg Christ for this reference).

106 Brunschvig, *La Berbérie orientale* (cf. note 43), 2:443–445; Dufourcq, *L'Espagne catalane et le Maghrib aux XIIIe et XIVe siècles* (cf. note 42), 462, 516–520; "Il fait figure de minister plénipotentiare": Dufourcq, *L'Espagne catalane et le Maghrib aux XIIIe et XIVe siècles* (cf. note 42), 472.

107 This issue is dealt with in more detail in Nikolas Jaspert, "Zur Loyalität interkultureller Makler im Mittelmeerraum: Christliche Söldnerführer (alcayts) im Dienste muslimischer Sultane," in *Loyalty in the Middle Ages: Ideal and Practice of a Cross-Social Value,* ed. Jörg Sonntag and Coralie Zermatten (Turnhout: Brepols, 2016), 235–274.

108 Masiá de Ros, *La corona de Aragón y los estados del norte de África* (cf. note 84), 414–415, doc.129.

109 Ibid., 447, doc. 154. On the legal notion, the "natural" subject cf. Henriet, "'Ad regem Cordube militandi gratia perrexit'" (cf. note 9), 376-377.

110 Mas Latrie, *Traités de paix et de commerce et documents: Supplément et tables* (cf. note 87), 47, doc. 21: ... *ac etiam pendonem nostrum qui per alcaydum nostrum tenetur et consuevit teneri ibidem ... et teneatis pendonem nostrum predictum ad honorem et fidelitatem nostrum.*

111 Mas Latrie, *Traités de paix et de commerce et documents: Supplément et tables* (cf. note 51), 51–53, doc. 35; Masiá de Ros, *La corona de Aragón y los estados del norte de África* (cf. note 84), 407, doc. 125.

112 Giménez Soler, "Caballeros españoles en Africa" (cf. note 28), 315.

113 Ibid., 329 (note 1); Masiá de Ros, *La corona de Aragón y los estados del norte de África* (cf. note 84), 457–458, doc. 161. Cf. the instructions of an Aragonese embassy in 1304, according to which 300–400 riders with their coat of arms/banners (*ab sa senyera*) were promised to the Merinid court: Àngels Masià i de Ros, *Jaume II: Aragó, Granada i Marroc: Aportació documental* (Barcelona: C.S.I.C., 1989), 174.

114 He might possibly be identical with the Templar Arnau de Torroella, former preceptor of Gardeny and commander of Cantavella and Valencia—Josep Maria Sans i Travé, *La fi dels templers catalans,* Col·lecció els ordes militars 10 (Lleida: Pagès, 2008), 210, 215, 221–223, 240, 246–247, 251–253, 268, 412, 435; Enric Guinot Rodríguez, ed., *Pergamins, processos i cartes reials: Documentació dispersa valenciana del segle XIII,* Fonts Històriques Valencianes 46 (València: Universitat de València, 2010), 86, doc. 39.

115 Mariano Gual Torrella, "Milicias cristianas en Berbería," *Boletín de la Sociedad Arquelogica Luliana* 34 (1973): 54–63, edition of the inventory on 54–56.

116 Gual Torrella, "Milicias cristianas en Berbería" (cf. note 115), 54. On the urban setup of medieval Fes, see Eugen Wirth, "Stadtplanung und Stadtgestaltung im islamischen Maghreb 1: Fès Djedid als 'ville royale' der Meriniden (1276 n. Chr.)," *Madrider Mitteilungen* 32 (1991): 213–232; Anton Escher and Eugen Wirth, *Die Medina von Fes: Geographische Beiträge zu Persistenz und Dynamik, Verfall und Erneuerung einer traditionellen islamischen Stadt in handlungstheoretischer Sicht,* Erlanger geographische Arbeiten 53 (Erlangen: Palm Enke, 1992). According to the anonymous

al-Dhakīra al-saniyya, the Merinides had an entire quarter of the Fes's palatine city Dār al-Bayḍā' (Fes al-Jadīd), the Rabaḍ al-Naṣārā (Suburb of the Christians), built for the mercenaries, cf. Amira K. Bennison, "Drums, Banners and Baraka: Symbols of Authority during the First Century of Marinid Rule, 1250–1350," in *The Articulation of Power in Medieval Iberia and the Maghrib*, ed. Amira K. Bennison (Oxford: Oxford University Press, 2014), 195–216, at 206.

117 Gual Torrella, "Milicias cristianas en Berbería" (cf. note 115), 55 (both quotations).

118 All in all, most of the deceased's possessions were red, as far as the colour was mentioned at all (*cotas sive ciprecium panni rubei ... unam gramaleam et unum pelotum de panno sete vocato de diaspe cum sendato rubeo ... espatllerias de sendato rubeo*, etc.). Cf. the list of weapons of Bernat Seguí on his departure to North-Africa in 1297: Batlle Gallart, "Noticias sobre la milicia cristiana" (cf. note 47), 133.

119 Brunschvig, *La Berbérie orientale* (cf. note 43), 1:449.

120 Morillo, "Mercenaries, Mamluks and Militia" (cf. note 5), 247–248 (with references to impediments to marry).

121 Jesús Montoya, "El frustrado cerco de Marrakech (1261–1262)," *Cuadernos de estudios medievales* 8/9 (1980/1981): 183–192.

122 *E assi Santa Maria ajodou a seus amigos, pero que d'outra lei eran, a britar seus eemigos*—Cantiga 181, cf. *Alfonso X. el Sabio: Cantigas de Santa Maria*, ed. Walter Mettmann, Clásicos Castalia (Madrid: Castalia, 1986–1989), trans.: *The Songs of Holy Mary by Alfonso X, the Wise: A Translation of the Cantigas de Santa Maria*, trans. by Kathleen Kulp-Hill, Medieval & Renaissance Texts & Studies (Tempe: Arizona Center for Medieval and Renaissance Studies 2000).

123 Evariste Lévi-Provençal, *La conquête et l'émirat hispano-umaiyade (710–912)* (Paris: Maisonneuve, 1950), 164, 189–191; Andreas Kaplony, *Konstantinopel und Damaskus: Gesandtschaften und Verträge zwischen Kaisern und Kalifen 639–750* (Berlin: Schwarz, 1996); Nicolas Drocourt, "Christian-Muslim Diplomatic Relations: An Overview of the Main Sources and Themes of Encounter (600–1000)," in *Christian Muslim Relations: A Bibliographical History, vol. 1, (600–900)*, ed. David Thomas, Barbara Roggema, and Juan Pedro Monferrer-Sala, History of Christian-Muslim Relations 11 (Leiden: Brill, 2009), 29–72, at 60; Nikolas Jaspert and Sebastian Kolditz, "Christlich-muslimische Außenbeziehungen im Mittelmeerraum: Zur räumlichen und religiösen Dimension mittelalterlicher Diplomatie," *Zeitschrift für Historische Forschung* 41 (2014): 1–88.

124 Johannis abbas S. Arnulfi, "Vita Iohannis Abbatis Gorziensis auctore Iohanne abbate S. Arnulfi," in *Annales, chronica et historiae aevi Carolini et Saxonici*, ed. Georg Heinrich Pertz, Monumenta Germania Historia Scriptores 4 (Hannover: Hahn, 1841), 335–377, at 374 (chapt. 128); Fernando Valdés Fernández, "Die Gesandtschaft des Johannes von Gorze nach Cordoba," in *Otto der Grosse, Magdeburg und Europa*, vol. 1, *Essays* (Mainz: Zabern, 2001), 525–536; Nicolas Drocourt, "Al-Andalus, l'Occident chrétien et Byzance: Liens et réseaux de personnes autour des évêques Recemundo et Liutprand de Crémone; Quelques hypothèses," in *Le Maghreb, al-Andalus et la méditerranée occidentale (VIIIe–XIIIe siècles)*, ed. Philippe Sénac (Zaragoza: Pórtico Librerías, 2007), 57–80.

125 See the cases mentioned by García Sanjuán, "Mercenarios cristianos al servicio" (cf. note 11), 183.

126 Giménez Soler, "Caballeros españoles en Africa II" (cf. note 103), 61–64.

127 Mas Latrie, *Traités de paix et de commerce et documents* (cf. note 51), 1:137; Alemany, "Milicias cristianas al servicio de los sultanes" (cf. note 21), 161, 163; Dufourcq, *L'Espagne catalane et le Maghrib aux XIIIe et XIVe siècles* (cf. note 42), 102, 303–304, 314–316, 472–473; Giménez Soler, "Caballeros españoles en Africa" (cf. note 28), 323–343, 346–347; Giménez Soler, "Caballeros españoles en Africa II" (cf. note 103), 56–57; Brunschvig, *La Berbérie orientale* (cf. note 43), 1:441–443; Maria de las Mercedes Costa Paretas, "El noble Jaume d'Aragó, fin bastard de Jaume II," *Estudis d'història medieval* 1 (1969): 39–60.

128 *Inclitus Infans Enricus filius illustris domini Ferrandi bone memorie Regis Castelle ad nostram accedens presentiam prudenter exposuit coram nobis quod cum vestre intentionis et propositi existat nobiscum pacem firmam habere vobis plurimum esse gratum ut inter nos et vos federa renovarentur amicitie et pacis antique* ...—Masiá de Ros, *La corona de Aragón y los estados del norte de África* (cf. note 84), 389–391, doc. 111; Giménez Soler, "Caballeros españoles en Africa" (cf. note 28), 305; cf. Dufourcq, *L'Espagne catalane et le Maghrib aux XIIIe et XIVe siècles* (cf. note 42), 303–304. Henry had served in Tunis as a mercenary for three decades prior to this (ibid.).

129 Giménez Soler, "Caballeros españoles en Africa" (cf. note 28), 323–342; Costa Paretas, "El noble Jaume d'Aragó" (cf. note 127), 45–52.

130 *Rey fem vos saber que es vengut davant la nostra presencia lo noble en Jacme darago fill nostre quins presenta vostra letra de creença. E nos entesem be e cumplidament ço que era contengut en la letra e ço que ell nos dix de vostra part. E axi creets lo de tot ço queus dira de nostra part axi com aquell en que podets be fiar*–Giménez Soler, "Caballeros españoles en Africa" (cf. note 28), 329 (note 1); Masiá de Ros, *La corona de Aragón y los estados del norte de África* (cf. note 84), 457–458, doc. 161.

131 *Item demanen lo senyor Rey el senyor infant en lo tractament de la pau que en la casa de Tirimçe haia daquiavant alcayt al qual sia provist honradament per lo rey de Tirimçe lo qual ni sia mes per lo senyor Rey darago e port senyera del rey darago la qual guarden tots los soldaders naturals del rey darago. E que ades volen que sia alcayt lo dit noble en Jacme darago por ells*— Giménez Soler, "Caballeros españoles en Africa" (cf. note 28), 330 (note 1).

132 Masiá de Ros, *La corona de Aragón y los estados del norte de África* (cf. note 84), 472–473, doc. 173.

133 Giménez Soler, "Caballeros españoles en Africa" (cf. note 28), 332–342; Masiá de Ros, *La corona de Aragón y los estados del norte de África* (cf. note 84), 463–481, doc. 167–177; Dufourcq, *L'Espagne catalane et le Maghrib aux XIIIe et XIVe siècles* (cf. note 42), 472–473, 483–486. Napoleon of Aragon is also mentioned as an envoy of the Hafsid sultan: Masiá i de Ros, *Jaume II: Aragó, Granada i Marroc* (cf. note 113), 570.

134 Alemany, "Milicias cristianas al servicio de los sultanes" (cf. note 21), 165–166; Andrés Giménez Soler, "Documentos de Túnez, originales o traducidos del Archivo de la Corona de Aragón," *Anuari* 1909–10 (1911): 210–259, at 229–232, doc. 12–13; Masiá de Ros, *La corona de Aragón y los estados del norte de África* (cf. note 84), 490–492, doc. 186.

135 Similar cases from the twelfth and thirteenth centuries are mentioned by Henriet, "'Ad regem Cordube militandi gratia perrexit'" (cf. note 9), 368-369, 374.

136 Mas Latrie, *Traités de paix et de commerce et documents* (cf. note 51), 2:306–310, doc. 14; Alan Forey, *The Fall of the Templars in the Crown of Aragon* (Aldershot: Ashgate, 2001), 216; Sans i Travé, *La fi dels templers catalans* (cf. note 114), 108, 164, 183, 210, 215, 219, 222–223, 281, 359, 382, 384.

137 Forey, *The Fall of the Templars in the Crown of Aragon* (cf. note 136), 223, 358, 367, 380, 385, 393; by 1317, he had already passed away (ibid., 246).

138 Giménez Soler, "Caballeros españoles en Africa" (cf. note 28), 305–317; Charles-Emmanuel Dufourcq, "Nouveaux documents sur la politique africaine de la Couronne d'Aragon," *Analecta Sacra Tarraconensia* 26 (1953): 291–322, at 301–302, 307–310, 314–315; Dufourcq, *L'Espagne catalane et le Maghrib aux XIIIe et XIVe siècles* (cf. note 42), 354–356, 363–365, 372–381, 391–397; Thoden, *Abū 'l-Ḥasan 'Alī: Merinidenpolitik* (cf. note 76), 79; Dominique Valérian, "Les agents de la diplomatie des souverains maghrébins avec le monde chrétien (XIIe–XVe siècles)," *Anuario de estudios medievales* 38 (2008): 886–900, at 888–890; Jaspert, "Interreligiöse Diplomatie im Mittelmeerraum" (cf. note 84), 180. A pertinent collection of sources contains no less than 17 documents for 1302–1309 that refer to Bernat's diplomatic activities—Masiá i de Ros, *Jaume II: Aragó, Granada i Marroc* (cf. note 113), 155, 157, 166, 174, 316, 346–347, 348–349, 353, 360, 370, 404, 405, 408, 409, 411, 412, 414, 416, 417, 419, 423, 424–426, 427, 432.

139 Batlle Gallart, "Noticias sobre la milicia cristiana" (cf. note 47), 132–133.

140 Dufourcq, *L'Espagne catalane et le Maghrib aux XIIIe et XIVe siècles* (cf. note 42), 354–355; Dufourcq, "Nouveaux documents sur la politique africaine de la Couronne d'Aragon" (cf. note 138), 301–302: *Significamus vobis quod volumus et placet nobis ut ad dictum regem eatis et in suo servicio permaneatis*—Giménez Soler, "Caballeros españoles en Africa" (cf. note 28), 306 (note 1). Employed the very next day: *Nos Iacobus etc. Confidentes de fide, legalitate et industria vestri dilecti militis nostri Bernardi Seguini vos in alcaydum prepositum seu capitaneum gentis seu familie militum equitum et peditum vobiscum ad illustrem regem Abenjacob transfretantium et in servicio suo debentium existere et que sunt vel erunt de cetero in servitio dicti Regis providimus et duximus statuendum. Mandantes per presentes universis et singulis militibus equitibus et peditibus supradictis quod vos pro alcaydo preposito et capitaneo habeant et tractent et vobis pareant et obedient in omnibus sicut nobis*—(ibid., note 2). Cf. the similar wording applied for Guillems de Pujalt's employment in 1305: *Confidentes de fide et legalitate ac industria vestri dilecti scutiferi nostri Guillelmi de Podio alto vos in alcaydum prepositum seu capitaneum gentis vel familie militum equitum vel peditum qui sunt vel erunt de cetero in servicio Regis Abenjacob apud Maruechos providimus ac duximus statuendum* (ibid., 308, note 2).

141 Masiá i de Ros, *Jaume II: Aragó, Granada i Marroc* (cf. note 113), 173–175.

142 Dufourcq, *L'Espagne catalane et le Maghrib aux XIIIe et XIVe siècles* (cf. note 42), 461–462. The *alcayt* of Tunis also had a secretary of his own in 1313: Masiá de Ros, *La corona de Aragón y los estados del norte de África* (cf. note 84), 491, doc. 186.

143 Dufourcq, "Nouveaux documents sur la politique africaine de la Couronne d'Aragon" (cf. note 138), 26; Masiá i de Ros, *Jaume II: Aragó, Granada i Marroc* (cf. note 113), 348–349.

144 On his tasks and obligations: Damien Coulon, "Négocier avec les sultans de Méditerranée orientale à la fin du moyen âge: Un domaine privilegié pour les hommes d'affaires?," in *Negociar en la Edad Media / Négocier au Moyen Âge*, ed. Maria Teresa Ferrer i Mallol, Anuario de Estudios Medievales, Anejo 61 (Barcelona: Consejo Superior de Investigaciones Científicas, 2005), 503–526, especially 524–526.

145 Masiá i de Ros, *Jaume II: Aragó, Granada i Marroc* (cf. note 113), 416 (25 April 1309), 417 (28 April 1309).

146 Ibid., 423 (3 May 1309), 424–426 (3 May 1309).

147 *E estas cosas face menester que sean secretas, que sino seria grant dayno a periglo al senyor rey*—Masiá i de Ros, *Jaume II: Aragó, Granada i Marroc* (cf. note 113), 411 (3 January 1309).

148 Ibid., 404 (2 August 1308), 409 (s.d.), 412 (16 January 1309), 424–427 (3 May 1309).

149 Mas Latrie, *Traités de paix et de commerce et documents* (cf. note 51), 2:297–300, doc. 10; Masiá i de Ros, *Jaume II: Aragó, Granada i Marroc* (cf. note 113), 424–427.

150 Dufourcq, *L'Espagne catalane et le Maghrib aux XIIIe et XIVe siècles* (cf. note 42), 358, 367, 393.

151 Masiá i de Ros, *Jaume II: Aragó, Granada i Marroc* (cf. note 113), 169; Dufourcq, "Nouveaux documents sur la politique africaine de la Couronne d'Aragon" (cf. note 138), 308, 312, 319.

152 *En R. Torro parent de Bernat Segui*—Masiá i de Ros, *Jaume II: Aragó, Granada i Marroc* (cf. note 113), 426.

153 Giménez Soler, "Caballeros españoles en Africa" (cf. note 28), 308–317; Dufourcq, "Nouveaux documents sur la politique africaine de la Couronne d'Aragon" (cf. note 138), 320; Thoden, *Abū 'l-Ḥasan 'Alī: Merinidenpolitik* (cf. note 76), 189; Dufourcq, *L'Espagne catalane et le Maghrib aux XIIIe et XIVe siècles* (cf. note 42), 380, 385, 391–393, 462 ("C'est toute une tribu"). Cf. the mercenary Gonzalo, who replaced his brother Sancho as captain of the Christian guard in Marrakesh in 1234: García Sanjuán, "Mercenarios cristianos al servicio" (cf. note 11), 184-185.

154 Cf. note 116.

155 Dufourcq, "Documents inédits sur la politique ifrikiyenne de la Couronne d'Aragon" (cf. note 97), 280; Roser Salicrú Lluch, "Más allá de la mediación de la palabra: Negociación con los infieles y mediación cultural en la Baja Edad Media,"

in *Negociar en la Edad Media / Négocier au Moyen Âge*, ed. Maria Teresa Ferrer i Mallol, Anuario de Estudiios Medievales, Anejo 61 (Barcelona: Consejo Superior de Investigaciones Científicas, 2005), 409–440, at 434–436; Jaspert, "Interreligiöse Diplomatie im Mittelmeerraum" (cf. note 84), 180–181. On intercultural brokers cf.—with further references: Höh, Jaspert, and Oesterle, *Cultural Brokers at Mediterranean Courts in the Middle Ages* (cf. note 3).

156 On the following, see Jaspert, "Zur Loyalität interkultureller Makler" (cf. note 107), 259–268.

157 ... *e sabets senyor que yo vinc primer a hobediencia vostra*—Masiá de Ros, *La corona de Aragón y los estados del norte de África* (cf. note 84), 491, doc. 186.

158 *Senyor comam me en vostra gracia ara per tots temps axi com de senyor natural*—Giménez Soler, "Caballeros españoles en Africa" (cf. note 28), 321, note 1. Cf. above, note 109.

159 *Encara senyor estic aparellat e trebaylar de nit e de dia aytant com pusca ne sapia en tot ço que yo enten que sia vostre servii*—Masiá i de Ros, *Jaume II: Aragó, Granada i Marroc* (cf. note 113), 348–349 (24 March 1309).

160 *E fiam en vos axi com senyor deu fiar als sos naturals*—Masiá i de Ros, *Jaume II: Aragó, Granada i Marroc* (cf. note 113), 361 (3 May 1309).

161 ... *e entenem en aquelles* [in Bernat's letters] *la bona voluntat que vos aviets en servir e procurar profit e honor a nos*—Masiá i de Ros, *Jaume II: Aragó, Granada i Marroc* (cf. note 113), 423 (3 May 1309).

162 For example: Masiá i de Ros, *Jaume II: Aragó, Granada i Marroc* (cf. note 113), 59 (11 July 1302).

163 Kurt-Ulrich Jäschke, "Mehrfachvasallität in Grenzregionen—ein Forschungsdesiderat?," in *Granice i pogranicza: Język i historiar*, ed. Stanisław Dubisz and Alicja Nagórko (Warsaw: Elipsa, 1994), 65–117; Roman Deutinger, "Seit wann gibt es Mehrfachvasallität?," *Zeitschrift der Savigny-Stiftung für Rechtsgeschichte: Germanistische Abteilung* 119 (2002): 78–105.

References

Printed Sources

Adorne, Anselme. *Itinéraire d'Anselme Adorno en Terre sainte, 1470–1471*, edited and translated by Jacques Heers and Georgette de Groër. Paris: Éditions du Centre national de la recherche scientifique, 1979.

Alarcón Santón, Maximiliano A., and Ramón García de Linares. *Los documentos árabes diplomáticos del Archivo de la Corona de Aragón: Editados y traducidos*. Madrid: Estanislao Maestre, 1940.

Arribas Palau, Mariano. "Cartas de recomendación cursadas al Sultán Abu' Sa'i'd Utmān III de Marruecos por el Rey de Aragón, Fernando I, el de Antequera." *Hespéris Tamuda* 1 (1956): 387–407.

Barrantes Maldonado, Pedro. *Ilustraciones de la Casa de Niebla*. Fuentes para la historia de Cádiz y su provincia 3. Cádiz: Univérsidad de Cádiz, Servicio de Publicaciones, 1998.

Desamparados Cabanes Pecourt, María, ed. *Crónica latina de los reyes de Castilla*. Textos medievales 11. Valencia: J. Nácher, 1964.

Domínguez Sánchez, Santiago, ed., *Documentos de Nicolás IV (1277–1280), referentes a España*. León: Publicaciones Universidad de León, 1999.

Dufourcq, Charles-Emmanuel. "Documents inédits sur la politique ifrikiyenne de la Couronne d'Aragon." *Analecta Sacra Tarraconensia* 25 (1952): 255–292.

Dufourcq, Charles-Emmanuel. "Nouveaux documents sur la politique africaine de la Couronne d'Aragon." *Analecta Sacra Tarraconensia* 26 (1953): 291–322.

El-Kartas, Roudh. *Histoire des souverains du Maghreb (Espagne et Maroc) et annales de la ville de Fès*, edited and translated by Auguste Beaumier. Paris: Duprat, 1860.
Feliú, Gaspar, Josep Maria Salrach Marés, and María Josepa Arnall Juan, eds. *Els pergamins de l'Arxiu Comtal de Barcelona de Ramon Borrell a Ramon Berenguer I.* Vol. 3, Col·lecció Diplomataris 20. Barcelona: Pagès Editors, 1999.
Ferrer i Mallol, María Teresa. "Marruecos y la Corona catalano-aragonesa: Mercenarios catalanes al servicio de Marruecos (1396–1410)." In *Homenaje al profesor Eloy Benito Ruano*, 251–272. Vol. 1. Murcia: Universidad de Murcia, 2010.
Giménez Soler, Andrés. "Documentos de Túnez, originales o traducidos del Archivo de la Corona de Aragón." *Anuari* 1909–10 (1911): 210–259.
Gual Torrella, Mariano. "Milicias cristianas en Berbería." *Boletín de la Sociedad Arquelogica Luliana* 34 (1973): 54–63.
Guinot Rodríguez, Enric ed. *Pergamins, processos i cartes reials: Documentació dispersa valenciana del segle XIII.* Fonts Històriques Valencianes 46. València: Universitat de València, 2010.
Ibn-Khaldun. *The Muqaddimah: An Introduction to History*, translated by Franz Rosenthal. Princeton: Princeton University Press, 1958.
Jaume I, Bernat Desclot, Ramon Muntaner, and Pere III. *Les quatre grans Cròniques: Revisió del text, pròlegs i notes per Ferran Soldevilla.* Barcelona: Edicions Selecta, 1983.
Johannis abbas S. Arnulfi. "Vita Iohannis Abbatis Gorziensis auctore Iohanne abbate S. Arnulfi." In *Annales, chronica et historiae aevi Carolini et Saxonici*, edited by Georg Heinrich Pertz, 335–377. Monumenta Germania Historia Scriptores 4. Hannover: Hahn, 1841.
Klüpfel, Ludwig. *Die äußere Politik Alfonsos III. von Aragonien bis zu den Verhandlungen von Gaeta*, 1285–1299. Berlin: Rothschild, 1912.
La Mantia, Giuseppe. *Codice diplomatico dei re Aragonesi di Sicilia.* Vol. 1, Anni 1282–1290. Palermo: Società siciliana per la storia patria, 1917.
Mas Latrie, Louis de, ed. *Traités de paix et de commerce et documents divers concernant les relations des chrétiens avec les Arabes de l'Afrique septentrionale au Moyen âge.* Paris: Plon, 1866. Reprinted. Burt Franklin Research and Source Works Series 63, 1. New York: Franklin, 1964.
Mas Latrie, Louis de, ed. *Traités de paix et de commerce et documents divers concernant les relations des Chrétiens avec les Arabes de l'Afrique septentrionale au moyen âge: Supplément et tables.* Paris: Baur et Détaille, 1872.
Masiá i de Ros, Ángels. *Jaume II: Aragó, Granada i Marroc: Aportació documental.* Barcelona: C.S.I.C., 1989.
Mettmann, Walter, ed. *Alfonso X. el Sabio: Cantigas de Santa Maria.* Clásicos Castalia. Madrid: Castalia, 1986–1989, transl. *The Songs of Holy Mary by Alfonso X, the Wise: A Translation of the Cantigas de Santa Maria*, translated by Kathleen Kulp-Hill. Medieval & Renaissance Texts & Studies. Tempe: Arizona Center for Medieval and Renaissance Studies, 2000.
Rosell, Cayetano, ed. *Crónicas de los reyes de Castilla: Desde don Alfonso el Sabio hasta los Católicos don Fernando y doña Isabel.* Vol. 2, Biblioteca de Autores Españoles 68. Madrid: Atlas, 1953.
Ruiz-Doménec, José Enrique. *Quan els vescomtes de Barcelona eren: Història, crònica i documents d'una família dels segles X, XI i XII.* Barcelona: Pagès editors, 2006.
Sánchez Belda, Luis, ed. *Chronica Adefonsi imperatoris.* Escuela de Estudios Medievales, Textos 14. Madrid: "Diana", 1950.
Sánchez Saus, Rafael. "Un linaje hispano-marroquí entre la leyenda y la historia: Los Farfán de los Godos." In *Actas del Congreso Internacional "El Estrecho de Gibraltar"*, edited by Eduardo Ripoll Perelló, 323–332. Vol. 2. Madrid: UNED, 1988.

Soldevila, Ferran, Jordi Bruguera, and Maria Teresa Ferrer i Mallol, eds. *Les quatre grans Cróniques*. Vol. 1, *Llibre dels feits del rei En Jaume*. Memòries de la Secció Historico-Arqueològica 73. Barcelona: Institut d'Estudis Catalans, 2007.
Soldevila, Ferran, Jordi Bruguera, and Maria Teresa Ferrer i Mallol, eds. *Les quatre grans Cróniques*. Vol. 2, *Crònica de Bernat Desclot*. Memòries de la Secció Historico-Arqueològica 80. Barcelona: Institut d'Estudis Catalans, 2008.

Studies

Abun-Nasr, Jamil M. *A History of the Maghrib in the Islamic Period*. Cambridge: Cambridge University Press, 1987.
Aivars Catlos, Brian. "Mahomet Abenadalill: A Muslim Mercenary in the Service of the Kings of Aragon (1290–1291)." In *Jews, Muslims, and Christians in and around the Crown of Aragon: Essays in Honour of Professor Elena Lourie*, edited by Harvey J. Hames, 257–302. The Medieval Mediterranean 52. Leiden: Brill, 2004.
Albarrán Iruela, Javier. "De la conversión y expulsión al mercenariado: Los cristianos en las fuentes almohades." In *La Península Ibérica en tiempos de Las Navas de Tolosa*, edited by Carlos Estepa and María Antonia Carmona Ruiz, 79–91. Monografías de la Sociedad Española de Estudios Medievales 5. Madrid: Sociedad Española de Estudios Medievales, 2014.
Alemany, José. "Milicias cristianas al servicio de los sultanes musulmanes de Almagreb." In *Homenaje a D. Francisco Codera en su jubilación del profesorado: Estudios de erudición oriental*, edited by Eduardo Saavedra, 133–169. Zaragoza: Mariano Escar, 1904.
Arbel, Benjamin. "Venetian Cyprus and the Muslim Levant, 1473–1570." In *Kypros kai oi Staurophories* [Cyprus and the Crusades], edited by Nicholas S. H. Coureas and Jonathan Riley Smith, 159–185. Nicosia: Cyprus Research Centre, 1995.
Baadj, Amar Salem. *Saladin, the Almohads and the Banu Ghaniya: The Contest for North Africa (12th and 13th Centuries)*. Studies in the History and Society of the Maghrib 7. Leiden: Brill, 2015.
Barton, Simon. "Traitors to the Faith? Christian Mercenaries in al-Andalus and the Maghreb, c. 1100–1300." In *Medieval Spain: Culture, Conflict and Coexistence; Studies in Honour of Angus MacKay*, edited by Roger Collins and Anthony Goodman, 23–45. Basingstoke: Palgrave/Macmillan, 2002.
Batlle Gallart, Carmen. "Noticias sobre la milicia cristiana en el Norte de África en la segunda mitad del siglo XIII." In *Homenaje al profesor Juan Torres Fontes*, edited by Universidad Murcia, 127–137. Vol. 1. Murcia: Universidad de Murcia, 1987.
Bennison, Amira K. "Drums, Banners and Baraka: Symbols of Authority during the First Century of Marinid Rule, 1250–1350." In *The Articulation of Power in Medieval Iberia and the Maghrib*, edited by Amira K. Bennison, 195–216. Oxford: Oxford University Press, 2014.
Binšarīfa, Muḥammad. "Maṣādir maġribiyya šafawiyya fī masālik al-abṣār li-l-ʿUmarī [Oral Maghribian/Moroccan Sources in the (book entitled) 'Paths of Vision/Deeper Insight' by al-ʿUmarī]." *Maǧallat Maǧmaʿ al-Luġat al-ʿArabiyya* 82 (1998): 232–255.
Brunschvig, Robert. *La Berbérie orientale sous les Ḥafsides: Des origines a la fin du XVe siècle*. 2 vols. Paris: Librairie d'Amérique et d'Orient Adrien-Maisonneuve, 1941–1947.
Burns, Robert Ignatius. "Renegades, Adventurers, and Sharp Businessmen: The Thirteenth Century Spaniard in the Cause of Islam." *The Catholic Historical Review* 58 (1972): 341–366.
Cabezuelo Pliego, José Vicente. "Cristiano de Alá, renegado de Cristo: El caso de Abdalla, fill d'en Domingo Vallés, un valenciano al servicio del islam." *Sharq al-Andalus* 13 (1996): 27–46.

Catalán, Diego. *El Cid en la historia y sus inventores*. Madrid: Fundación Ramón Menéndez Pidal, 2002.

Cenival, Pierre de. "L'église chrétienne de Marrakesch au XIIIe siècle." *Hespéris* 7 (1927): 69–83.

Clément, François. "Reverter et son fils, deux officiers catalans au service des sultans de Marrakech." *Medieval Encounters* 9 (2003): 79–107.

Constable, Olivia Remie. *Housing the Stranger in the Mediterranean World: Lodging, Trade and Travel in Late Antiquity and the Middle Ages*. New York: Cambridge University Press, 2003.

Constable, Olivia Remie. "Ringing Bells in Hafs'id Tunis: Religious Concessions to Christian *Fondacos* in the Later Thirteenth Century." In *Histories of the Middle East: Studies in Middle Eastern Society, Economy and Law in Honor of A. L. Udovitch*, edited by Roxani Eleni Margariti, Adam Sabra, and Petra M. Sijpesteijn, 53–72. Islamic History and Civilization 79. Leiden: Brill, 2011.

Coulon, Damien. "Négocier avec les sultans de Méditerranée orientale à la fin du moyen âge: Un domaine privilegié pour les hommes d'affaires?" In *Negociar en la Edad Media/ Négocier au Moyen Âge*, edited by Maria Teresa Ferrer i Mallol, 503–526. Anuario de Estudios Medievales, Anejo 61. Barcelona: Consejo Superior de Investigaciones Científicas, 2005.

de las Mercedes Costa Paretas, Maria. "El noble Jaume d'Aragó, fin bastard de Jaume II." *Estudis d'història medieval* 1 (1969): 39–60.

Deimann, Wiebke. *Christen, Juden und Muslime im mittelalterlichen Sevilla: Religiöse Minderheiten unter muslimischer und christlicher Dominanz (12. bis 14. Jahrhundert)*. Berlin: Lit-Verlag, 2012.

Deutinger, Roman. "Seit wann gibt es Mehrfachvasallität?" *Zeitschrift der Savigny-Stiftung für Rechtsgeschichte: Germanistische Abteilung* 119 (2002): 78–105.

Drocourt, Nicolas. "Al-Andalus, l'Occident chrétien et Byzance: Liens et réseaux de personnes autour des évêques Recemundo et Liutprand de Crémone; Quelques hypothèses." In *Le Maghreb, al-Andalus et la méditerranée occidentale (VIIIe–XIIIe siècles)*, edited by Philippe Sénac, 57–80. Zaragoza: Pórtico Librerías, 2007.

Drocourt, Nicolas. "Christian-Muslim Diplomatic Relations: An Overview of the Main Sources and Themes of Encounter (600–1000)." In *Christian Muslim Relations: A Bibliographical History*. Vol. 1, *(600–900)*, edited by David Thomas, Barbara Roggema, and Juan Pedro Monferrer-Sala, 29–72. History of Christian-Muslim Relations 11. Leiden: Brill, 2009.

Dufourcq, Charles Emmanuel. *L'Espagne catalane et le Maghrib aux XIIIe et XIVe siècles: De la bataille de Las Navas de Tolosa (1212) à l'avènement du sultan mérinide Abou-l-Hasan (1331)*. Paris: Presses Universitaires de France, 1966.

Echevarría, Ana. *Knights in the Frontier: The Moorish Guard of the Kings of Castile (1410–1467)*. Leiden: Brill, 2009.

Escher, Anton, and Eugen Wirth. *Die Medina von Fes: Geographische Beiträge zu Persistenz und Dynamik, Verfall und Erneuerung einer traditionellen islamischen Stadt in handlungstheoretischer Sicht*. Erlanger geographische Arbeiten 53. Erlangen: Palm Enke, 1992.

Fancy, Hussein. "The Last Almohads: Universal Sovereignty between North Africa and the Crown of Aragon." In *Spanning the Strait: Studies in Unity in the Western Mediterranean*, edited by Yuen-Gen Liang, Abigail Krasner Balbale, Andrew Devereux, and Camilo Gómez-Rivas, 102–136. Medieval Encounters 19, 1–2. Leiden: Brill, 2013.

Fancy, Hussein. *The Mercenary Mediterranean: Sovereignty, Religion, and Violence in the Medieval Crown of Aragon*. Chicago: University of Chicago Press, 2016.

Fletcher, Richard A. *The Quest for El Cid*. London: Hutchinson, 1989.

Forey, Alan. *The Fall of the Templars in the Crown of Aragon*. Aldershot: Ashgate, 2001.

Fowler, Kenneth Alan. *Medieval Mercenaries.* Vol. 1, The Great Companies. Oxford: Blackwell, 2001.

France, John, ed. *Mercenaries and Paid Men: The Mercenary Identity in the Middle Ages.* History of Warfare 47. Leiden: Brill, 2008.

Frank, István. "Reverter, vicomte de Barcelone (vers 1130–1145)." *Boletín de la Real Academia de Buenas Letras de Barcelona* 26 (1954/1956): 195–204.

Fromherz, Allen James. *The Near West: Medieval North Africa, Latin Europe and the Mediterranean in the Second Axial Age.* Edinburgh: Edinburgh University Press, 2016.

Garcia Fitz, Francisco, and Joao Gouveia Monteiro, eds. *War in the Iberian Peninsula, 700–1600.* New York: Routledge, 2018.

García Sanjuán, Alejandro. "Mercenarios cristianos al servicio de los musulmanes en el norte de África durante el siglo XIII." In *La Península Ibérica entre el Mediterráneo y el Atlántico: Siglos XIII–XV,* edited by Manuel González Jiménez and Isabel Montes Romero-Camacho, 435–448. Cádiz: Diputación de Cádiz Servicio de Publicaciones, 2006. Reprinted in García Sanjuán, Alejandro, ed. *Coexistencia y conflictos: Minorías religiosas en la Península ibérica durante la Edad Media,* 171–192. Granada: Universidad de Granada, 2015.

Gazulla, Faustino. *Jaime I de Aragón y los estados musulmanes.* Barcelona: Real Academia de Buenas Letras, 1919.

Giménez Soler, Andrés. "Caballeros españoles en Africa y africanos en España." *Revue hispanique* 12 (1905): 299–372.

Giménez Soler, Andrés. "Caballeros españoles en Africa y africanos en España II." *Revue hispanique* 16 (1907): 56–69.

Gubert, Serge. "Pratiques diplomatiques mérinides (XIIIe–XVe siècle)." In *Les relations diplomatiques entre le monde musulman et l'Occident latin (XIIe–XVIe siècle),* edited by Denise Aigle and Pascal Buresi, 435–468. Oriente moderno, n.s., 88. Rome: Istituto per l'Oriente C. A. Nallino, 2008.

Henriet, Patrick. "Xénophobie et intégration à Léon au XIIIe siècle: Le discours de Lucas de Túy sur les étrangers." In *L'étranger au Moyen Âge: XXXe Congrès de la SHMES (Göttingen, juin 1999),* 37–58. Paris: Publications de la Sorbonne, 2000.

Henriet, Patrick. "'Ad regem Cordube militandi gratia perrexit': Remarques sur la présence militaire chrétienne en al-Andalus (Xe–XIIIe siècle)." In *Regards croisés sur la guerre sainte: Guerre, idéologie et religion dans l'espace méditerranéen latin (XIe–XIIIe siècle),* edited by Daniel Baloup and Philippe Josserand, 359–380. Toulouse: Presses universitaires du Midi, 2006.

Höh, Marc von der, Nikolas Jaspert, and Jenny Rahel Oesterle. "Courts, Brokers and Brokerage in the Medieval Mediterranean." In *Cultural Brokers at Mediterranean Courts in the Middle Ages,* edited by Marc von der Höh, Nikolas Jaspert, and Jenny Rahel Oesterle, 10–31. Paderborn: Fink-Schöningh, 2013.

Hopkins, J. F. P. *Medieval Muslim Government in Barbary: Until the Sixth Century of the Hijra.* London: Luzac and Company, 1958.

Hunt, Janin, and Ursula Carlson. *Mercenaries in Medieval and Renaissance Europe.* Jefferson: McFarland, 2013.

Jadla, Brahim. "L'Ifriqiyah hafside et la chrétienté: Le temps des ruptures." In *Les relations diplomatiques entre le monde musulman et l'Occident latin (XIIe–XVIe siècle),* edited by Denise Aigle and Pascal Buresi, 311–321. Oriente moderno, n.s., 88. Rome: Istituto per l'Oriente C. A. Nallino, 2008.

Jäschke, Kurt-Ulrich. "Mehrfachvasallität in Grenzregionen—ein Forschungsdesiderat?" In *Granice i pogranicza: Język i historiar,* edited by Stanisław Dubisz and Alicja Nagórko, 65–117. Warsaw: Elipsa, 1994.

Jaspert, Nikolas. "Interreligiöse Diplomatie im Mittelmeerraum: Die Krone Aragón und die islamische Welt im 13. und 14. Jahrhundert." In *Aus der Frühzeit europäischer Diplomatie: Zum geistlichen und weltlichen Gesandtschaftswesen vom 12. bis zum 15. Jahrhundert*, edited by Claudia Märtl and Claudia Zey, 151–190. Zürich: Chronos, 2008.

Jaspert, Nikolas. "Zur Loyalität interkultureller Makler im Mittelmeerraum: Christliche Söldnerführer (alcayts) im Dienste muslimischer Sultane." In *Loyalty in the Middle Ages: Ideal and Practice of a Cross-Social Value*, edited by Jörg Sonntag and Coralie Zermatten, 235–274. Turnhout: Brepols, 2016.

Jaspert, Nikolas. "Konversion zum Islam im spätmittelalterlichen Mittelmeerraum und die Entscheidungen des Anselm Turmeda alias 'Abdallāh at-Tarǧumān'." In *Der Sultan und der Heilige: Islamisch-christliche Perspektiven auf die Begegnung des hl. Franziskus mit Sultan al-Kamil (1219–2019)*, edited by Amir Dziri, Angelica Hilsebein, Mouhanad Khorchide, and Bernd Schmies, 367–396. Münster: Aschendorff Verlag, 2021.

Jaspert, Nikolas, and Sebastian Kolditz. "Christlich-muslimische Außenbeziehungen im Mittelmeerraum: Zur räumlichen und religiösen Dimension mittelalterlicher Diplomatie." *Zeitschrift für Historische Forschung* 41 (2014): 1–88.

Kaplony, Andreas. *Konstantinopel und Damaskus: Gesandtschaften und Verträge zwischen Kaisern und Kalifen 639–750.* Berlin: Schwarz, 1996.

Kosto, Adam J. *Making Agreements in Medieval Catalonia: Power, Order, and the Written Word, 1000–1200.* Cambridge: Cambridge University Press, 2001.

Kosto, Adam J. *Hostages in the Middle Ages.* Oxford: Oxford University Press, 2012.

Ladero Quesada, Miguel Ángel. "Una biografia caballeresca del siglo XV. 'La Coronica del yllustre y muy magnifico cauallero don Alonso Perez de Guzman el Bueno.'" *En la España medieval* 22 (1999): 247–283.

Lagardère, Vincent. *Les Almoravides: Le djihâd andalou.* Paris: Éditions l'Harmattan, 1999.

Lapiedra Gutiérrez, Eva. "Christian Participation in Almohad Armies and Personal Guards." *Journal of Medieval Iberian Studies* 2 (2010): 235–250.

Lévi-Provençal, Evariste. *La conquête et l'émirat hispano-umaiyade (710–912).* Paris: Maisonneuve, 1950.

López de Coca y Castañer, José Enrique. "Caballeros Moriscos al servicio de Juan II y Enrique IV, reyes de Castilla." *Meridies* 3 (1996): 119–136.

Lower, Michael. "Ibn al-Lihyani: Sultan of Tunis and Would-Be Christian Convert (1311–18)." *Mediterranean Historical Review* 24 (2009): 17–27.

Lower, Michael. "The Papacy and Christian Mercenaries of Thirteenth-Century North Africa." *Speculum* 89 (2014): 601–631.

Maillard, Clara. *Les papes et le Maghreb aux XIIIème et XIVème siècles: Étude des lettres pontificales de 1199 à 1419.* Turnhout: Brepols, 2014.

Maillard, Clara. "Protection des chrétiens en terre d'Islam et discussion entre papes et souverains musulmans: Le cas singulier des mercenaires du Maroc." In *Religious Minorities in Christian, Jewish and Muslim Law (5th–15th Centuries)*, edited by Nora Berend, Youna Hameau-Masset, and John Tolan, 317–333. Turnhout: Brepols, 2017.

Maíllo Salgado, Felipe. "Precisiones para la historia de un grupo étnico-religioso: Los farfanes." *Al-Qantara* 4 (1983): 265–282.

Marín, Manuela, Julio Samsó, and María Isabel Fierro, eds. *The Formation of al-Andalus.* Vol. 1, History and Society. Aldershot: Ashgate, 1998.

Masiá de Ros, Ángeles. *La corona de Aragón y los estados del norte de África: Política de Jaime II y Alfonso IV en Egipto, Ifriquía y Tremecén*, 487–493. Barcelona: Instituto español de estudios mediterráneos, 1951.

Menéndez Pidal, Ramón. *La España del Cid.* Vol. 2. Madrid: Ed. Plutarco, 1929.

Montaner Frutos, Alberto. "Dichos y hechos: Tácticas de la obediencia en el Cantar de mio Cid." *Cahiers d'études hispaniques medievales* 34 (2011): 29–39.

Montoya, Jesús. "El frustrado cerco de Marrakech (1261–1262)." *Cuadernos de estudios medievales* 8/9 (1980/1981): 183–192.

Morillo, Stephen. "Mercenaries, Mamluks and Militia: Towards a Crosscultural Typology of Military Service." In *Mercenaries and Paid Men: The Mercenary Identity in the Middle Ages*, edited by John France, 243–260. Leiden: Brill, 2008.

Muldoon, James. *Popes, Lawyers, and Infidels: The Church and the Non-Christian World 1250–1550*. Liverpool: Liverpool University Press, 1979.

Ochoa Brun, Miguel Ángel. *Historia de la diplomacia española*. Vol. 3. Madrid: Ministerio de Asuntos Exteriores, 1991.

Peña Pérez, Javier. *El Cid: Historia, leyenda y mito*. Burgos: Dossoles, 2000.

Richard, Jean. "An Account of the Battle of Hattin Referring to the Frankish Mercenaries in Oriental Moslem States." *Speculum* 27 (1952): 168–177.

Rothman, Ella Natalie. *Brokering Empire: Trans-Imperial Subjects between Venice and Istanbul*. Ithaca: Cornell University Press, 2012.

Ruiz-Domènec, José Enrique. "Las cartas de Reverter, vizconde de Barcelona." *Boletín de la Real Academia de Buenas Letras de Barcelona* 39 (1983/1984): 93–118.

Ruiz-Domènec, José Enrique. *Atardeceres rojos: Cuatro vidas entre el islam y la cristianidad en la época de las cruzadas*. Barcelona: Ariel, 2007.

Safran, William. "Diasporas in Modern Societies: Myths of Homeland and Return." *Diaspora: A Journal of Transnational Studies* 1 (1991): 83–99.

Salicrú Lluch, Roser. "En busca de una liberación alternativa: Fugas y apostasía en la Corona de Aragón bajomedieval." In *La liberazione dei "captivi" tra Cristianità e Islam: Oltre la crociata e il Gihad; Tolleranza e servizio umanitario*, edited by Giulio Cipollone, 703–713. Collectanea Archivi Vaticani 46. Vatican City: ASV, 2000.

Salicrú Lluch, Roser. "Mercenaires castillans au Maroc au début du XVe siècle." In *Migrations et diasporas méditerranéennes (Xe–XVIe siècles)*, edited by Michael Ballard and Alain Ducellier, 417–434. Paris: Éditions de la Sorbonne, 2002.

Salicrú Lluch, Roser. "Más allá de la mediación de la palabra: Negociación con los infieles y mediación cultural en la Baja Edad Media." In *Negociar en la Edad Media/Négocier au Moyen Âge*, edited by Maria Teresa Ferrer i Mallol, 409–440. Anuario de Estudiios Medievales, Anejo 61. Barcelona: Consejo Superior de Investigaciones Científicas, 2005.

Salicrú Lluch, Roser. "Entre la praxis y el estereotipo: Vivencias y percepciones de lo islámico-ibérico en fuentes archivísticas y narrativas bajomedievales." In *Ritus infidelium: Miradas interconfesionales sobre las prácticas religiosas en la edad media*, edited by José Martínez Gázquez and John Victor Tolan, 99–111. Collection de la Casa de Velázquez 138. Madrid: Casa de Velazquez, 2013.

Sans i Travé, Josep Maria. *La fi dels templers catalans*. Col·lecció els ordes militars 10. Lleida: Pagès, 2008.

Savvides, Alexis G. C. "Late Byzantine and Western Historiographers on Turkish Mercenaries in Greek and Latin Armies: The Turcoples/Tourkopouloi." In *The Making of Byzantine History: Studies Dedicated to Donald M. Nicol*, edited by Roderick Beaton, 122–136. Publications of the Centre for Hellenic Studies 1. Aldershot: Ashgate, 1993.

Thoden, Rudolf. *Abū 'l-Ḥasan 'Alī: Merinidenpolitik zwischen Nordafrika und Spanien in den Jahren 710–752 H./1310–1351*. Freiburg: Schwarz, 1973.

Tweed, Thomas A. *Crossing and Dwelling: A Theory of Religion*. Cambridge: Harvard University Press, 2006.

Valdés Fernández, Fernando. "Die Gesandtschaft des Johannes von Gorze nach Cordoba." In *Otto der Grosse, Magdeburg und Europa*. Vol. 1, *Essays*, edited by Ministerium für Wirtschaft und Arbeit des Landes Sachsen-Anhalt; Kulturhistorisches Museum Magdeburg. Red.: Claus-Peter Hasse, 525–536. Mainz: Zabern, 2001.

Valérian, Dominique. "Le fondouk, instrument du contrôle sultanien sur les marchands étrangers dans les ports musulmans (XIIe–XVe siècle)." In *La mobilité des personnes en Méditerranée de l'Antiquité à l'époque moderne: Procédures de contrôle et documents d'identifications*, edited by Claudia Moatti, 677–698. Collection de l'École Française de Rome 341. Rome: École Française de Rome, 2004.

Valérian, Dominique. "Les agents de la diplomatie des souverains maghrébins avec le monde chrétien (XIIe–XVe siècles)." *Anuario de estudios medievales* 38 (2008): 886–900.

Vela Aulesa, Carles. "L'Andalus en la política de Barcelona i la Corona d'Aragó (segle XI–1213)." In: *Tractats i negociacions diplomàtiques amd els regnes peninsulars i l'Ándalus (segle XI–1213)*, edited by Maria Teresa Ferrer i Mallol and Manuel Riu i Riu, 117–180. Memòries de la Secció Històrico-Arqueològica 106. Barcelona: Institut d'Estudis Catalans, 2018.

Whalen, Brett. "Corresponding with Infidels: Rome, the Almohads, and the Christians of Thirteenth-Century Morocco." *Journal of Medieval and Early Modern Studies* 43 (2011): 487–513.

Wirth, Eugen. "Stadtplanung und Stadtgestaltung im islamischen Maghreb 1: Fès Djedid als 'ville royale' der Meriniden (1276 n. Chr.)." *Madrider Mitteilungen* 32 (1991): 213–232.

9

PROFESSIONAL TURKS OR MILITARY DIASPORA?

The Mamluks and Dynamics of Ethnicity in Late Medieval Egypt and Syria

Julien Loiseau

Military slavery has long been a privileged way to ensure strength and to build loyalty in Islamic polities. Beginning in the 870s, the wealthiest princes used to surround themselves with Praetorian guards of former slaves ("mamluks") selected for their military skills within "martial races" living at the fringes of the Islamicate world. In 1250, one of the largest regiments of mamluks of Turkic background ousted the heir of their former master and seized power in Egypt and Syria. Their history is that of an allochtonous Turkic-like military élite, recruited through slavery and manumission. During its almost three-century-long rule (1250–1517), the Mamluk military had to face dynamics of ethnicity that either buttressed or challenged its collective identity. These "ethnic trends" were linked to the patterns of slave trade that supplied the sultanate with young boys and girls, and to global migration phenomena. They also might have been brought about by political decision and by the rulers' propensity to favour their own people. This chapter therefore aims to identify military diasporic groups with para- (or imagined) ethnic background and respective ethnic self-awareness, that acted as distinct forces within the Mamluk military in late medieval Egypt and Syria.

Introduction

For almost three centuries (1250–1517), Egypt and Syria formed the core of the mightiest polity in the Middle East—the Mamluks—which relied on an allochthonous military élite for its strength.[1] Mamluks, i.e. military slaves, were professional soldiers, theoretically born outside the Islamicate world, that is in the "abode of war" (*dār al-ḥarb*), enslaved in childhood and trained to serve their master who manumitted them upon completing their military training and entering adulthood.[2] The institution of military slavery dates to late ninth-century Iraq,

DOI: 10.4324/9781003245568-10

where the Abbasid caliphs tried to free themselves from their entourage by relying on forces of undivided allegiance directly depending on their person.[3] In 1250, the last Ayyubid sultan of Egypt was murdered by the mamluk officers of his father, who provisionally placed the widow of their former master on the throne. These mamluks soon designated one of their own to act as sultan: And thus a new political regime was born.

In the fourteenth- and fifteenth-century Egypt and Syria, the role played by (mainly autochthonous) civilian élites in the management of the state, the integration of non-mamluk individuals into the military aristocracy, as well as the room made in the army for freeborn soldiers of various ethnic identities, has been well emphasized by recent scholarship.[4] Although the role of the Mamluks in the state has been recently reconsidered, historians still commonly refer to the polity born in 1250 as the Mamluk sultanate of Egypt and Syria, because of the Mamluks' deep imprints on the state and the country. Interestingly enough, however, the medieval scholars who recorded in Arabic the good and bad fortunes of their patrons and lords used to designate the regime as the *Dawlat at-Turk*, *Dawlat at-Atrāk*, or *Dawla at-turkiyya*, namely the "Dynasty (or the Reign) of the Turks".[5]

To what extent were the Mamluks "Turks" or "Turkic"? What did these categories mean to contemporaries? We know that Turkic (initially Qipchaq Turkic and later on, in the fifteenth century, Anatolian Turkic) was the vernacular language of the military society in Egypt and Syria, used in military training and, later, even in some stages of religious education.[6] We also know that most of the personal names (*ism*) borne by the Mamluks were of Turkic origin, with very few in Arabic, Persian, Mongol, or Circassian.[7] We also know, however, that languages can be learnt and that, as former slaves, Mamluks often owed their personal name to their first master; in most cases, the slave trader who initially bought them. Neither language nor (to a lesser extent) names can conclusively assert the actual origin of individuals.[8] To be "Turkic" in the medieval Islamicate world meant to be part of one of the main races (*jins*, pl. *ajnās*) into which humankind was divided. Turkic-speaking peoples were usually considered the offspring of Noah's son Japheth. Their role in human (and moreover in Islamic) history was to be compared to that of the Arabs, the former coming from the cold deserts north of the civilized world, the latter from the hot areas south.[9] The harshness of their homelands was used by contemporaries to explain the physical and moral qualities that made the "Turks" a martial race (in the same sense as Afghans did in British-ruled India), the most redoubtable in human history since the deeds of the Arabs.

Islamic history up to the nineteenth century and the advent of nationalisms provides countless pieces of evidence of the prominent role played by such Turkic military diasporas, in the loosest sense of the term, scattered from Egypt to India. Élite regiments of heavy cavalry, bodyguards of servile origin, or irregular armed bands, speaking one Turkic dialect and/or viewed as being of allochthonous Turkic stock by local populations, were involved in the military affairs and, hence, in the politics of most Islamicate countries east of the

Nile valley from the eleventh century onwards, if not earlier.[10] Seen from this loose perspective, Mamluks in late medieval Egypt and Syria would represent one (although an outstanding) example of this Turkic diaspora in which former Turkic slaves succeeded in securing the throne for themselves or for their offspring for almost three centuries. There is no doubt that the leading role of allochthonous Turkic-speaking militaries in the global history of the Islamicate world remains an important issue on the research agenda. Yet can we still speak of a diaspora on such a scale? Between the Ghaznavids in eleventh-century Iran and Afghanistan and their Seljuk foes, or between the Mamluks in fifteenth-century Egypt and Syria and the Aq Qoyunlu (or "White Sheep") Turkmen confederation they fought in Eastern Anatolia, there was neither a shared homeland nor language, nor even the same imagined ethnic background. The only thing they shared were similar opportunities offered by the division of (military) labour and political imaginations of medieval Islam, in which soldiering was usually to be devolved to warlike allochthonous groups.[11]

The history of the Mamluks in late medieval Egypt and Syria is that of an allochthonous Turkic military élite, recruited (and, hence, at least partly disciplined) through slavery and manumission. During its almost three-century-long rule, however, the "dynasty of the Turks" had to face dynamics of ethnicity that either buttressed or challenged its collective identity. These "ethnic trends" were linked to the patterns of slave trade that supplied the sultanate and its main households with young boys and girls, and/or to global migration phenomena. In some cases, they might reflect the rulers' propensity to favour their own kin or people. It might be therefore possible to identify military diasporic groups with para- (or imagined) ethnic backgrounds and respective ethnic self-awareness, acting as distinct forces within the "Dynasty of the Turks" in late medieval Egypt and Syria. This is the aim of this chapter.[12]

Mongol Conquests, the Qipchaq Diaspora, and the "Race of Mamluks"

Genghis Khān's armies reached the Islamicate world in successive waves, beginning with Central Asia in 1219. They prompted massive population displacements and drove various vanquished military groups to the south and west. Khwarezmian troops fleeing the collapse of the Khwarezm-shah's empire in Central Asia and Iran, finally destroyed in 1231, played an important role in the wars that weakened the Islamic polities of the Middle East in the following decades.[13] In the summer of 1244, they famously sacked Jerusalem. In October of the same year, they fought victoriously in the battle of La Forbie (Farbiyā) alongside the armies of the Ayyubid sultan of Egypt, against the forces of the Latin kingdom of Jerusalem allied to other Ayyubid rulers, and were later incorporated into the Ayyubid army of Egypt.[14] Quṭuz, a nephew of the last Khwarezm shah Jalāl al-Dīn (r. 1220–1231), followed a parallel path: A prisoner of the Mongols and sold in Damascus as a mamluk, Quṭuz entered the household of the first Mamluk sultan of Egypt al-Muʿizz Aybak (r. 1250–1257) and eventually succeeded his son as sultan of Egypt in 1259.[15]

In the summer of 1260, the Mongol Ilkhan Hülegü (r. 1256–1265), whose forces occupied Syria, sent a letter to Sultan Quṭuz asking for submission and slandering his origins by claiming that he was "of the race of mamluks (*jins al-mamālīk*) who fled before our sword into this country, who enjoyed its comforts and then killed its rulers".[16] This was a reference to the murder of the last Ayyubid sultan of Egypt by Mamluk officers a decade earlier. Quṭuz, however, achieved a major victory against the Mongols in the battle of ʿAyn Jālūt on 3 September 1260. The armies that had marched out of Egypt against the Mongols combined Mamluk heavy cavalry of servile origin with Türkmens, Arab Bedouins, and Kurds serving as light cavalry or infantry.[17] Quṭuz thus commanded a multi-ethnic army against the Mongols and their allies. In this respect, beyond the person of the sultan, was there any ethnic base in the "race of mamluks" vilified by the Ilkhan? Indeed, the use of the Arabic word *jins* (race, people) in Hülegü's letter implies that his calumny had an ethnic agenda and that in his view ethnicity was involved in the recruitment process of mamluks through slavery.

The Khwarezmian armies, largely consisted of Turkic-speaking individuals belonging to the Qipchaq group, one of the three main Turkic-speaking nomadic peoples of Eurasia, established since the eleventh century in the Caspian and Pontic steppes.[18] Sighnāq, one their strongholds in the valley of the Syr-Darya, was incorporated into the Khwarezmian empire at the beginning of the thirteenth century. The mother of the Khwarezm-shah ʿAlā' al-Dīn Muḥammad (r. 1200–1220), Terken Khātūn, was a Qipchaq princess and might have broadened the influence of her people in the Khwarezmian army.[19] Yet Qipchaq warriors, recruited and settled with families and herds, could also be found in other countries neighbouring what Arabic and Persian geographers used to call the Qipchaq steppe (*Dast i-Qipchaq*), such as the Balkans under Byzantine rule or the Caucasian kingdom of Georgia, from the late eleventh century onwards. Qipchaq officers, both Georgianized and Christianized, played a prominent role in Georgian politics and even fought Islamicized Qipchaq kinsmen serving Islamic rulers in Azerbaijan. The Qipchaq diaspora expanded as far as the Delhi sultanate in India, whose Turkic-speaking rulers relied upon Turkic-speaking mamluks, i.e. military slaves, part of whom were of Qipchaq stock.[20]

Military slavery already increased in the twelfth century with the scattering of Qipchaq people over and beyond the neighbouring countries of the Qipchaq steppe. But Mongol conquests critically scaled up the phenomenon. Charles J. Halperin has convincingly argued that Qipchaq people were not only considered as being part of "all those who lived in felt tents", whose submission to the Genghiskhanid empire was a divine requirement. Mongol conquerors saw them collectively as their "slaves" and "cattle-herders", as evidenced in their diplomatic correspondence with Kievan and Hungarian rulers.[21] Mongol conquests, following the defeat of a Rus' and Qipchaq coalition at the battle of the Kalka

River in 1223, drove away Qipchaq people towards the kingdom of Hungary, the Bulgarian empire, or the Byzantine Empire of Nicaea for which they fought against the Latins. Part of the vanquished Qipchaq forces were also incorporated in Genghis Khān's armies. Some of them served at the imperial court of Karakorum and during the Mongol conquest of China. The Yüan dynasty had several Qipchaq officers and even created a Qipchaq guard regiment in 1286.[22]

Yet the definitive conquest of the Qipchaq steppe by the Genghiskhanids half a century earlier already had major consequences in the history of the Middle East. Countless Qipchaq captives, be they vanquished warriors, women or children, were sold in the late 1230s in the slave markets of various regions, such as Ayyubid territories of south-eastern Anatolia, Syria, and Egypt. Ayyubid sultans had a long tradition of personal bodyguards recruited among Turkic-speaking mamluks. The length and cost of their recruitment and training usually limited the actual number of fighting-age mamluks to a few dozen, not taking into account young slave soldiers still in training. In the late 1230s, however, mamluks were available on slave markets in such quantities that Ayyubid sultans were able to have at their disposal regiments of several hundred élite fighters, such as the ʿAzīziyya regiment of al-ʿAzīz Muḥammad in Aleppo or the Nāṣiriyya Mamluks of his son and successor al-Nāṣir Yūsuf only a few years later.[23] This change of scale was nowhere more apparent than in Egypt where the Baḥriyya Mamluks (whose barracks were close to the river Nile, *baḥr* in Arabic) of al-Ṣāliḥ Ayyūb (r. 1240–1249) numbered about one thousand and represented 10% of the whole Egyptian army.[24] The Baḥriyya regiment, from which most of the officers of the Mamluk regime and its first sultans up to al-Manṣūr Qalāwūn (r. 1279–1290), a former Qipchaq slave boy, originated, played a prominent role in the genesis of the Mamluk sultanate.[25] Hence the "Dynasty of the Turks", born in the turmoil of the Mongol conquests, was part of the global military Qipchaq diaspora scattering at that time from Egypt to China.

The large numbers of Qipchaq captives in the slave markets was not the only reason for their recruitment as mamluks. Since the recruitment of Qipchaq slaves for the armies of the Delhi sultanate in the twelfth century, they were renowned for their warlike virtues and their outstanding agility as horsemen and mounted archers. Their swords and bows, saddles and stirrups, inspired others, e.g. the equipment of Rus' horsemen.[26] For Islamic rulers such as the Ayyubid sultans, Qipchaq captives also had the merit of being heathen. The non-Muslim origin of mamluks (or any other slaves) was not only a technical matter required by the Islamic law. It was indeed prohibited to reduce Muslims or non-Muslim "protected" people (*dhimmī*) living in the "House of Islam" to slavery, apart from captured rebels. Hence, slaves had to be caught or bought from beyond the lands of Islam.[27] However, slavery had never been an end but only a means in the recruitment of mamluks. The initial heathen identity of the young captives somehow guaranteed the alleged savagery and brutality that would make them outstanding warriors. In a way, this was theorized by Ibn Khaldūn (1332–1406) in his famous contrast between "Bedouins" (*ahl al-badw*) and "sedentary people" (*ahl al-amṣār*). The latter were powerless to defend themselves and willing to submit to the former or to entrust their defence to them; over time, however, "Bedouin" offspring born among sedentary people lost the martial virtues of their fathers before being overthrown by "Bedouin" newcomers.[28] In such a view, the best way for a dynasty to retain power and to face its foes was to have a continuous access to "Bedouin" reservoirs, i.e. freeborn warriors or traded captives.

The competitive edge of mamluks over other sources of military manpower came from their training and their personal loyalty. Mamluks' loyalty was, at least allegedly, fostered through integration into their master's household, the constant providing of food, home, and protection to young captives deprived of their family, and the promise of manumission. During their several years-long training (up to seven or eight years for the youngest slaves), mamluks became familiar with their new social and cultural environment, acquired rudiments of the Arabic language and Islamic religious education, were formally converted to their new faith, and continued to improve their horsemanship and weapons handling (bow, spear, sword, and mace).[29] At the time of their manumission, after completing their training, they knew whom and what they were supposed to fight for, without having lost the martial virtues of their "Bedouin" (to quote Ibn Khaldūn's argument) origins. Young Qipchaq captives of pastoralist and shamanist background therefore fitted perfectly the requirements of military slaves.

Hence, one may wonder if the fighting qualities of the mamluks were a result of their ethnic identity and a legacy of the traditions of horsemanship borne on the Qipchaq steppe, or the consequence of their élite training in Cairene barracks and training grounds. It is most likely that, in the eyes of their masters as well as in their self-representation, it was a little bit of both. The redoubtable Mamluk heavy cavalry, which succeeded in repelling Mongol armies at ʿAyn Jālūt in 1260, might have drawn strength from both its ethnic cohesiveness and the *esprit de corps* (*ʿasabiyya* in Ibn Khaldūn's wording) that united members of the same household, as well as from the specific training that brought their supposed martial virtues out. The "race of mamluks" vilified by the Ilkhan was born from the crossing of a remote ethnic origin and of a newly built military identity.

Mongol Diasporas in Early Mamluk Egypt and Syria

An excursus into the history of another diasporic group, that of Mongol immigrants in the Mamluk sultanate of Egypt and Syria, may help to better comprehend the potential diasporic dimension of the Mamluk military. As soon as the early 1260s, numerous people fleeing the internal war raging in the Caucasus between the Ilkhans and the Golden Horde sought refuge in Egypt and Syria.[30] Two hundred Mongol horsemen arrived in Damascus in 1262 and five times as many the year after, without taking account of women and children. There is nothing surprising in the destination of these immigrants (*wāfidiyya*) as Arabic chroniclers named them.[31] A few months before the first arrival, the Mamluk Sultan al-Ẓāhir Baybars (r. 1260–1277) had already engaged in diplomatic contact with Berke Khān, the sovereign of the Golden Horde, the first Mongol leader to convert to Islam.[32] The khan was seeking allies in his war against Hülegü, the leader of the Ilkhanate. As for the Mamluk sultan, a former Qipchaq slave, his main concern was to keep open the routes of the slave trade from the Qipchaq steppe, at that time under Berke's control, bringing military slaves to Egypt and Syria. Mongol refugees from the Golden Horde and later on from Ilkhanid Anatolia continued sporadically to arrive in the sultanate until 1277 and the last campaign of Sultan Baybars against the Ilkhans. Two decades later, in 1296, the Mamluk sultanate faced an important inflow of Mongol immigrants on its Eastern border in the Euphrates valley. Numbering ten-thousand (or according other reports eighteen-thousand) horsemen, these heathen Mongols of Oirat origin were most likely prompted to flee the Ilkhanid territories of Iran and Iraq by the advent of Ghāzān Khān (r. 1295–1304) who had converted to Islam and pursued a policy of Islamization of the Ilkhanate. Arrivals of Mongol immigrants in the Mamluk sultanate still occurred at the beginning of the fourteenth century, the last of importance being the arrival of two hundred horsemen in 1305.[33]

What was the actual military role of these diasporic groups and what does their arrival say about dynamics of ethnicity within the Mamluk military? The Mongol immigrants in Egypt and Syria of the 1260s were not been called-up to back the Mamluk army in the khan's name. Rather, these refugees were warring horsemen led by high-level officers of the Mongol armies, namely ancient *tümen* cavalry commanders who previously had commanded corps of ten regiments of one thousand horsemen each. Despite initial mistrust, they were allowed to settle in Cairo and to enter the service of Mamluk emirs while their officers were incorporated into the Mamluk military hierarchy. Sultan Baybars took care to not let them reach the highest ranks in the Mamluk army or assume the highest offices at the Mamluk court: Most of them served as commanders in the Ḥalqa regiments which had become at that time of secondary importance. It was not until al-Manṣūr Qalāwūn's reign (r. 1279–1290) that one of them, Sayf al-Dīn Noghāy, had access to the top rank of Emir of the Hundred.[34] There is nothing about ethnic segregation or rivalry between Mongols and Qipchaqs in this prudent policy. At the same time, several Mamluks of Mongol origin, whether they were captured on the battlefield or bought as slaves, reached the highest ranks of the Mamluk army just as Qipchaqs did. One of these Mamluk emirs of Mongol origin, Kitbughā, who had been captured during the battle of Homs in 1260, even ascended the throne in 1294 under the reigning name of al-Malik al-ʿĀdil (the Just king). The key difference between Mamluks of Mongol background and Mongol immigrants is not a question of personal prestige. Sultan Baybars himself, as several of his officers, did seek matrimonial alliances with the newcomers and married the daughters of the highest-ranked Mongol refugees. The latter, however, arrived in Egypt and Syria as freeborn men, while the former passed through the Mamluk military career path which was only open, at that time, to young captives and slaves. The loyalty expected of freeborn immigrants could not compare to that of former slaves.

The massive arrival of Oirat Mongol horsemen in 1296, under the reign of the Mamluk sultan of Oirat origin, al-ʿĀdil Kitbughā (r. 1294–1296), is a different case. The ethnic affinity between the sultan and the newcomers did not prevent Kitbughā from adopting a cautious attitude towards them, welcoming only Oirat commanders and their retinue in Cairo while forcing the remainders to settle in depopulated areas of Palestine (around Atlit) and Lebanon (in the Bekaa).[35] In the eyes of the highest Mamluk officers, however, Sultan Kitbughā was somehow perceived as associated with the immigrants.[36] Therefore, Kitbughā's overthrowing in 1296 was followed by a purge of Oirat emirs from the army. Rank and file Oirat soldiers kept nurturing a feeling of ethnic solidarity with Kitbughā; they rose up in 1299, when the Mamluk army marched against the Ilkhanids in Palestine, to bring him back to the throne.[37] Oirats and their offspring also kept their ethnic particularism within the Mamluk army until at least the early 1330s. They were downgraded at that time to menial tasks in the barracks of the Cairo citadel.[38] Mongol Oirats who immigrated in the 1290s to Egypt and Syria thus offer a true example of a conscious diasporic group within the Mamluk military. Yet even at the time of Kitbughā's short reign, ethnic solidarity has never been the main determining feature of solidarity among the Mamluks. Emir Salār, one of the highest officers of his reign, also was of Oirat birth, a shared origin that had no influence on his rivalry with the sultan or on his hostility towards the refugees.

Reuven Amitai has convincingly argued that, with the sole exception of the Oirat episode, Mongol ethnic solidarity (*jinsiyya*) should not be overstated as a driving force within the Mamluk military. The following instances of Mongol immigrants reaching Egypt and Syria are of a different order. In 1305, relatives of Emir Salār, including his mother and two brothers, arrived among a force of two hundred Mongol horsemen.[39] Between 1304 and 1326, Sultan al-Nāṣir Muḥammad summoned about two hundred members of his maternal parenthood (uncles, aunts, first and second cousins) from Iran: His mother, Ashlūn, was the daughter of an officer who had arrived in Egypt among the wake of the first Mongol immigrants.[40] In the first decades of the fourteenth century, ethnic affinity drove family reunification while, at the same time, the Mamluk aristocracy had begun to lose its genuine military identity.

Diaspora vs. Allochthony: The Dynamics of Mamluk Identity

The reunification with his maternal kin achieved by Sultan al-Nāṣir Muḥammad benefited from the regular diplomatic contacts between the Mamluk sultanate and the Ilkhanate, and was boosted by the conclusion of a peace agreement in 1323. Yet it mainly relied on the fact that princess Ashlūn was the freeborn daughter of a Mongol high officer who willingly settled in Egypt half a century before. Unlike young slaves torn away from their family, the sultan had a mother and a father. It is no coincidence that mamluks bore the same *nasab* (the genealogical component of the Arabic name). Whatever the name of their actual father was, they were always called "Ibn ʿAbd Allāh", "Son of God's servant", which was both a reminder of their conversion to Islam (ʿAbd Allāh was the name of Prophet Muḥammad's father) and a mark of their enslavement since their actual filiation had been erased and left room for a fictive genealogy. Slavery was a social death leading to the oblivion of family ties.[41]

Throughout the fourteenth century, however, Arabic chroniclers recorded family reunions in Egypt and Syria of relatives torn away by slavery as strange and noticeable facts. In the 1330s, one of the sultan's former slaves and favourites, Emir Yalbughā, brought his father and two brothers from the "land of the Mongols". Three decades later, Emir Qarābughā owed his career in the sultanate to his son and daughter, respectively, a mamluk and a concubine of Sultan al-Ashraf Shaʿbān (r. 1363–1377).[42] The most famous instance is the arrival in 1381 of the father, sisters, and numerous relatives of Emir Barqūq, at that time the country's strong man who would become sultan soon after. None other than the slave merchant, who initially had imported Barqūq himself two decades earlier, brought them to Egypt.[43] Slave traders not only used to keep close links with the young captives they sold, especially when the latter made a successful career, but they also maintained relationships with their country of birth thanks to networks of intermediaries.

However enlightening the explanation above might be to better understand the slave trade and its organization, it should not gloss over a fact that was crucial for the Mamluk military until the late fourteenth century: Most of the former slaves that formed the core of its àlite regiments had been definitively deprived of their natural family and exiled without hope of return. The scarcity of resources that prevailed in the areas from which most military slaves came sometimes led families to sell their own children. In such circumstances, there was no way back home.[44] Military slaves actually had a country of birth but they had no homeland other than the place where they were born to a new social identity: In the case of the Mamluks, Egypt and Syria.

In the Mamluk sultanate, young apprentice slave soldiers (*kuttābiyya*) were theoretically separated from the civil host society for the duration of their training. They were, however, already engaged in the processes of acculturation, social inclusion, and integration. Barracks in the courtyard of the citadel, in Cairo as well as in the main cities, kept them apart from the urban population. This separation was only interrupted by visits of their instructors and outside trips for training, as well as public bath and mosque attendance when the military base did not harbour such facilities.[45] Yet such a seclusion only applied to the sultan's mamluks and to a lesser extent to those of his main viceroys in Aleppo and Damascus. As for young slave soldiers belonging to Mamluk officers, they grew up in their master's house along with his freeborn sons and in some cases, the offspring of his civil servants. Hence, often-lifelong attachment, friendship, and loyalty were formed, not only between young slaves of the same master and age group who considered themselves as brothers (*khushdāsh*) but also between mamluks and their freeborn comrades.[46]

The seclusion of the Mamluk military was never hermetic, however. Cairo's citadel was designed as a city within a city, with thousands of permanent inhabitants, within but separate from the city of Cairo proper. The sultans and their main officers dwelt there with their mamluks, harems, and retinues up to the middle of the fourteenth century. Yet separation was shattered as newly manumitted mamluks increasingly settled outside the citadel in individual houses. Neighbourhoods mainly inhabited by the Mamluk military did expand around urban citadels: In Aleppo and Damascus as well as in Cairo, these areas were called "Beneath the Citadel" (Taḥt al-Qalʿa).[47] These districts, however, were by no means populated only by mamluks. The large houses Mamluk officers used to build or rent in these privileged districts attracted an important population of craftsmen and suppliers, not to mention the numerous servants of the various required facilities, such as mill, press, public bath, and mosque. Districts "Beneath the Citadel" were also renowned for the specialized markets where mamluks bought their horses, weapons, and military devices.

The Mamluk military was thus never hermetically secluded from the surrounding Egyptian and Syrian society, even at the time of the Baḥriyya regiment, the nucleus of the "Dynasty of the Turks" in the 1240s. If the regiment itself initially was garrisoned in a recently built fortress on al-Rawḍa Island, apart from the city, its officers' main residences were established in al-Qāhira, the central district of Cairo. Seclusion only affected young apprentice slave soldiers, in a way that never was hermetic enough in the eyes of their masters as well as in the eyes of the regime's critics. When the Cairene scholar al-Maqrīzī (1364–1442) blamed Sultan al-Ẓāhir Barqūq (r. 1382–1399) for having let his mamluks leave the citadel after their manumission, settle in the city and marry freeborn women, he was playing on the myth of a pure and virtuous Mamluk military that most likely never existed outside his nostalgic image of the past.[48]

The issues of sexual relationships and marriage are crucial for our understanding of the diasporic dimensions of Mamluk military society.[49] It is known that until at least the 1330s, slave merchants whose trade connections with the Mamluk sultanate had been buttressed by the alliance with the Golden Horde, imported slave girls (*jāriya*) as well as slave boys (*mamlūk*) from the Qipchaq steppe. Embassies from the Golden Horde, which brought slaves to be sold in Egypt, also evidence this trade in both slave boys and girls. In 1304, despite considerable loss of life during the sea journey, 400 mamluks and 200 slave girls from the Golden Horde territories arrived in Cairo with the envoys of Toqta Khān.[50] The Genoese, whose ships played a prominent role in the slave trade, were fully aware of it: Between 1304 and 1316, the city of Genoa officially prohibited the supply of military materiel to the Mamluk sultanate, including that of "male and female mamluks" (*mumulicos sive mumulichas*).[51] Unlike mamluks, the purchase of which was to some extent limited and controlled by the sultan and his officers, female slaves from the same ethnic background were bought also by masters who did not belong to the Mamluk military. The highest number of slaves, however, be they male or female, were intended for the largest and richest households of the sultanate, that of the sultan and his emirs. Hence, in Mamluk households, where most of slave soldiers were of Qipchaq origin until the last decades of the fourteenth century, young men and women sharing the same ethnic background cohabited.[52]

Islamic law allowed sexual relations in two different contexts: Between husband and wife (up to four wives at the same time), as well as between the master and his female slaves with no limitation. Mamluk sultans and emirs thus practised legal polygamy along with the purchase of numerous female slaves, before moral and economic considerations changed the situation in the fifteenth century.[53] There was, however, no ethnic homogeneity among the numerous members of Mamluk harems. Mamluks often married the widow or daughter of their former master, along with one of their favourite concubines and/or freeborn women from Syrian or Egyptian families. As for female slaves, there is no clear evidence of ethnic preference in Mamluk harems even if Qipchaq women might have constituted the largest group. Indeed, male and female slaves were not valued for the same things. At the time of Qipchaq ethnic hegemony, the Mamluk military much appreciated (free or enslaved) Mongol women, as well as young boys of the same background.[54] Beside canons of beauty and the value given at one time to a specific people (*jins*), marriage was first and foremost a question of alliance. This was true not only for senior officers but also for their manumitted slaves. Indeed, the former masters had to consent to their marriage. Chroniclers recorded severe incidents following the unauthorized marriage of manumitted mamluks: The most extreme instance is the castration of the wrongdoer by order of his former master in Damascus in 1349.[55] More often than not, Mamluk sultans and emirs exercised their rights over their former slaves by marrying one of their favourite mamluks to their own daughter.[56] Ethnic homogeneity was relatively strong in Mamluk households, as slaves of both sexes were available and because of strategic alliances. Mamluks of Qipchaq background did not marry or sexually prefer Qipchaq female slaves because they were seeking to build a Qipchaq diaspora in Egypt and Syria but because the latter were more readily available than other slaves; patterns of the slave trade played a more prominent role in this matter than ethnic preference.

Circassians, Ethnic Affinity, and the Reformation of the Mamluk Military System

Things began to change with regard to ethnic preference during the last two decades of the fourteenth century.[57] The coming to Egypt in 1381 of numerous relatives of Emir Barqūq from the "lands of the Circassians" (*Bilād al-Jarākisa*), i.e. the mountains of the north-western Caucasus, alone would not necessarily mark a profound change from previous practice. Aside from the Caucasian origin of the people brought by Barqūq's former slave trader, their arrival is in many ways similar to previous family reunions within Mamluk military society. During his reign, however, Sultan Barqūq (r. 1382–1399) increased the trade of Circassian slave boys to his kingdom and bought his kinsmen in unprecedented numbers. Barqūq left behind upon his death, according to one of his biographers, about 6,000 horses and 5,000 camels; as for the mamluks, he had bought since 1378 when he became great amir (*amīr kabīr*) and the sultanate's real leader, they reached five thousand. They were almost as numerous as those of his great predecessor Sultan al-Manṣūr Qalāwūn (r. 1279–1290).[58] Yet a large proportion of Barqūq's mamluks were of "Circassian race" (*jarkasī al-jins*). When exactly this tendency began is difficult to establish. In spring 1389, however, when Sultan Barqūq was temporarily deposed from the throne after rebellious emirs marched on Cairo, the victorious troops looked for the "houses of the Circassian mamluks" in the streets of the capital, suggesting that at that time, some of Barqūq's kinsmen had already completed their training and had been manumitted.[59] Barqūq was back on the throne by the beginning of 1390. In summer 1398, the sultan escaped a conspiracy fomented by one of his favourite Circassian mamluks, who he had promoted to the rank of Emir of the Hundred, the highest of the military. In this context of great tension, chroniclers reported that the first wife (*khawand kubrā*) of the sultan, who was of Turkic race (*turkī al-jins*), advised her husband to not exclusively choose his mamluks among his fellow Circassians: "Make your army streaked (*ablaq*) with four races (*jins*), she adds, that is to say with Tatars, Circassians, Anatolians (*Rūm*) and Turkmens".[60] It is most likely that this advice from inside the harem was less due to the princess's own ethnic background than to the concern of the sultan's entourage regarding the hegemony achieved by Circassians in the Mamluk military. How had it come to that point?

Strictly speaking, al-Ẓāhir Barqūq was not the first Mamluk sultan to bring in slaves of Circassian background. According to fifteenth-century chroniclers, Sultan Qalāwūn (r. 1279–1290) had already purchased mamluks of Mongol and Circassian origin along with Turkic slaves.[61] Scholars were eager in the fifteenth century to find precedents to Barqūq's career and speculated about the alleged Circassian origin of Sultan al-Muẓaffar Baybars (r. 1309–1310), a former mamluk of Sultan Qalāwūn.[62] Be that as it may, it is most significant that the issue of Baybars' possible Circassian background was, to the best of my knowledge, not discussed prior to the late fourteenth century.[63] Before Barqūq's advent, there was no "Circassian question" in the Mamluk military. Very few Mamluk emirs of Circassian origin made a significant career in the fourteenth century: One of the rare examples is Emir Jūkān (d. 1381), described by his biographer as "one of the ancient Circassians", i.e. belonging to generations predating that of Barqūq's mamluks.[64] There is evidence of several emirs named "Jārkas" ("Circassian") during the fourteenth century but without any certainty that this ethnic name given to them by their slave trader proves a respective actual origin. Jārkas al-Khalīlī, for instance, a brother-in-arms (*khushdāsh*) and faithful companion of Barqūq, was most likely of Turkmen background despite what his name suggests.[65] The massive purchase of Circassian mamluks beginning in the 1380s was thus unprecedented in the Mamluk sultanate.

The shortening of Turkic slave supply, due to the increased conversion of Qipchaq people to Islam, along with the ravages of the Black Death in the territories of the Golden Horde, may have played a role in this shift. One wonders how, according to the advice of Barqūq's wife, "Tatars" would have been an alternative to the purchase of Circassian mamluks, whatever the exact meaning of this ethnic designation at that time may have been (Turkified Mongol captives or Turkic slaves from territories under Mongol rule?).[66] The "lands of the Circassians" in the Caucasian mountains, however, were also under nominal control of the Golden Horde and it is likely that they were struck as well by the plague. In the late thirteenth century, Circassian slaves were already abundantly present in Crimean slave markets: Considering the low demand in the Middle East, Genoese merchants instead exported them to the Western Mediterranean.[67] Shifts in supply and demand are not a sufficient explanation of the hegemony of Circassian mamluks one century later. Barqūq's policies were critical in this respect. Ethnic affinity most likely motivated such a decision, especially since it allowed the sultan to incorporate his own kinship within the Mamluk military. The arrival of his relatives from the Caucasus carried on at least until 1388.[68] The purchase of Circassian slaves, however, also conveys a genuine understanding of the Mamluk military driving forces. The Circassians' superficial Christianism should not conceal the fact that their rustic origin was the guarantee of their unspoiled roughness and brutality required from them to defend civilization more effectively.[69]

The massive incorporation of Circassian mamluks somehow allowed the reviving of military virtue among the Mamluk aristocracy, which had lost part of its warlike identity during the peaceful decades that followed the truce with the Ilkhans in 1323 and the rallying of most Mamluk officers behind the dynastic legitimacy of the Qalawunids.[70] Moreover, Sultan Barqūq breathed a new *esprit de corps* (*ʿasabiyya*) based on ethnic solidarity into the Mamluk military. In the terms of Ibn Khaldūn's theory, a new dynasty (*dawla*) was re-founded on the ruins of the old Mamluk power, three generations (about 120 years) after its establishment by sultan Baybars (r. 1260–1277). Hence it is intriguing that Ibn Khaldūn, who arrived in Egypt ten days after Sultan Barqūq's rise to power, served him as a chief judge, and lived in the Mamluk sultanate until his death in 1406, never highlighted this ethnic re-foundation of the Mamluk military, while describing in his writings the decline of the "Dynasty of the Turks" after three generations.[71] On two occasions, Tamerlane (r. 1375–1405), the great conqueror of the time, who so deeply impressed Ibn Khaldūn, marched on the border of the Mamluk territories but decided to withdraw, when facing the move of Sultan Barqūq's (mainly Circassian) troops. There is probably no better confirmation to the latter's successful reforms.

The Circassians: Professional Turks, Mamluk Nobility

The outcome is well known. One year after the death of Sultan Barqūq, Tamerlane moved towards Syria, where he assaulted and plundered Aleppo and Damascus in winter 1400–1401. The latter, initially defended by Sultan al-Nāṣir

Faraj (r. 1399–1412), the 14-year-old son of Barqūq, had been abandoned when leading Mamluk officers heard about an attempted coup in Cairo. This failure of the Mamluk military plunged the sultanate into a ten-year-long crisis, leading to the secession of Syria's and Upper Egypt's provinces, factional fighting, and finally the overthrow and murder of Sultan Faraj in 1412. This turmoil, which resulted in the seizing of power by Sultan al-Mu'ayyad Shaykh (r. 1412–1421), a former Circassian mamluk of Barqūq, reshuffled the cards among the Mamluk military and changed the genuine nature of Circassian mamluk identity.

Evidence of this change can be found in the political image of the new sultan. Unlike his former master and some of his successors, who also came from among Barqūq's Circassian mamluks, Sultan Shaykh earned a reputation of equanimity by not favouring individuals on the sole basis of their Circassian origin.[72] His neutrality regarding ethnic affinity contrasted with Barqūq's "love" for his fellow Circassians as well as with Faraj's "hate" towards the Circassian emirs of his father. Yet what was described in narrative sources of that time as personal inclination or feelings has to be understood in terms of political choices. Meanwhile, however, a literary work composed at court in the well-established tradition of eulogy, the *Sayf al-Muhannad* of Badr al-Dīn al-ʿAynī (1361–1451), emphasized the Circassian origin of Sultan Shaykh.[73] Al-ʿAynī, who claimed that his statements were grounded in personal conversation with the sultan, portrayed him not only as the son of both a Circassian mother and father, but as an offspring of the prominent lineage of the "noblest clan" (*ashraf al-baṭn*) of the Circassians. The author even gives the (likely alleged) names of the sultan's ancestors with the sole and significant exception of his father's, usually erased in Mamluk onomastics. The logical outcome of this demonstration is that Sultan Shaykh did not owe his legitimacy to the legacy of his master's household but to the nobility of his lineage superior to that of Barqūq. Meanwhile, the author passed over in silence the murder of Barqūq's son and heir by order of Emir Shaykh himself in 1412. This obvious infringement of Mamluk ethics, which usually implies transfer of loyalty to the master's sons after his death, was deliberately ignored in order to enhance an unprecedented ethnic nobility.[74]

Al-ʿAynī's eulogy was likely intended for the civilian élites of the sultanate more than to Mamluk officers. It illustrates, however, an ongoing change in the way in which Circassians saw their position in the fifteenth-century Mamluk military. Younger (and older) slaves of Circassian origin continued to be imported from the Caucasus. Free individuals also tried their hand by claiming kinship with Mamluk officers who already served in the Mamluk military. Sultan al-Ashraf Barsbāy (r. 1422–1438) made several of his relatives come to Egypt in the first years of his reign, including two blood brothers and one breastfed brother. His concern also extended to the brothers and relatives of his favourite Circassian concubine, who had become his first wife (*khawand kubrā*). Prestige of Circassian origin could be handed over to the Mamluks' freeborn descent, especially if their mother also was of Circassian birth. Hence Sultan Barsbāy's son, who briefly succeeded him in 1438, bore the *nisba* (or name of origin) of "al-Jarkasī" (the Circassian) as did his father and mother of servile background despite his free birth in Egypt.[75] To be Circassian had become a passport to the highest circles of the Mamluk military, whatever the actual fighting qualities of the individual. The Cairene chronicler Ibn Taghrī Birdī (1409–1470), son of a Mamluk officer of Anatolian origin, who became the near-official memorialist of the Mamluk military, claimed that several individuals included in his biographic dictionary were renowned for no other reason or personal virtue than their belonging to the "Circassian race".[76] There is no better illustration of the demilitarization of Circassian identity than the comment added by the copyist of Ibn Taghrī Birdī's dictionary to the margins of the manuscript:

> Question: Could the mere fact of being Circassian be considered a quality? Almighty God has innately awarded beauty and natural nobility to the noble Circassian race.[77]

The gradual conversion of Circassian identity from a token of martial virtue to an ethnic nobility in the Mamluk military did not critically affect the relationships between the Sultanate of Egypt and Syria and the "lands of the Circassians" from where most of its rulers had hailed. Yet among the dwellers of the Caucasian mountains, it probably enhanced the desire to migrate towards a land where milk and honey flowed. Some individual careers convey the impression that porosity between the two countries was higher than in the age when most of young mamluks came from the Qipchaq steppe. In the late 1440s, a still heathen Circassian, who was acknowledged as "the emir of the lands of the Circassians", arrived in Cairo. Whatever the actual meaning of his position in his country of origin, his personal prestige was strong enough in Egypt to convince Sultan al-Ẓāhir Jaqmaq, himself a former Circassian mamluk, to marry the emir's daughter. Later on, in the early sixteenth century, a Mamluk officer who ended his career serving the new Ottoman rulers in Syria, was known by the nickname "Ibn Sulṭān Jarkas", Son of Sultan Jarkas, whatever the actual position of his father in his homeland may have been.[78] To return home to the "lands of the Circassians", however, was almost as unthinkable as it was to travel back to the Qipchaq steppe. There is very little evidence of Circassians going back and forth between Egypt and the Caucasus. There was, however, a Circassian Mamluk officer, nicknamed "the Renegade" (al-Murtadd) because he decided to go back home after the murder of his master in 1412, before changing his mind and returning to Egypt. In 1449, Sultan Jaqmaq convinced his brother to finally settle in Cairo after he had first come but returned home a decade earlier.[79]

Moreover, despite the Caucasian origin of a large proportion of high-ranking Mamluk officers in the fifteenth century, the "Circassian language" (*lisān al-jarkasī*) was hardly ever spoken at the Mamluk court. Contemporary sources provide, to the best of my knowledge, only two instances in which the language of the Circassians, which might have been close to the modern Adyghe, was used in public audiences.[80] The extreme scarcity of personal names regarded as genuinely Circassian in fifteenth-century Mamluk onomastics is also striking. This means that even at the time of the strongest affirmation of Circassian identity amidst the Mamluk military's highest circles, when Sultan al-Ashraf Qāytbāy (r. 1468–1496) named his own daughter the "lady of the Circassians" (*Sitt al-jarākisa*), neither Circassian names nor the so-called Circassian language ever supplanted Mamluk Turkic culture and language.[81]

Genoese documentation indeed suggests that natives of the Caucasian north-western mountains usually bore Turkic names.[82] It is also likely that the Turkic dialect spoken in the Golden Horde territories was used as well as a koiné in the "lands of the Circassians", even if Barqūq's father, who had arrived in Egypt in 1381, reportedly spoke neither Arabic nor Turkic.[83] Be that as it may, it does not affect the critical observation that the common identity of the Mamluk military, even during the Circassian momentum, continued to rely on Turkic culture acquired during the young slaves' training, or possibly inherited from their homeland. The oldest translations of poetic, legal, or technical works from Persian or Arabic to Qipchaq Turkic dialect ever sponsored at the Mamluk court all postdate the advent of Sultan Barqūq in 1382. This is all the more true for military treatises composed in, or translated into, Turkic.[84] The promotion of an ethnic-based nobility within the Mamluk military, that of the Circassians, did not conflict with the uninterrupted transmission of a Turkic professional identity as part of the Mamluk curriculum. Fifteenth-century sultans and emirs, be they Circassian or not, were always seen by their subjects as "Turkic" and members of the "Dynasty of the Turks".[85]

Conclusion: Diasporic Groups vs. Military Identity

The Mamluks who ruled Egypt and Syria between 1250 and 1517 were definitively not, as a whole, a military diaspora, except if they were regarded on a global scale as part of the large Turkic military diaspora that extended from Egypt to India during the Middle Ages. In a few pivotal moments of their almost three-century long rule, however, diasporic military groups blossomed in the shadow of the "Dynasty of the Turks".

The first moment saw the large diaspora of Qipchaq Turks, scattered by the Mongol conquests, giving rise to a new regime in which the army and its élite regiments of heavy cavalry recruited in childhood through slavery played a prominent role. In 1250, a "race of mamluks" born at the crossing of ethnic homogeneity and servile curriculum seized power in Cairo. The first generation of Mamluk sultans and emirs were representative of a diasporic military group forged in the household of the last Ayyubid sultan of Egypt, al-Ṣāliḥ Ayyūb (r. 1240–1249), who was their former master and the genuine father of the "Dynasty of the Turks". The second moment in the late 1290s was brief and had no real consequence on the Mamluk military, whose social reproduction has lost its diasporic dimension at that time. The massive immigration of heathen Mongol horsemen of Oirat origin did not undermine the regime but was promptly addressed by Mamluk leading officers, who settled them in Palestine and Lebanon. This diasporic group that partly succeeded to maintain itself played only a subsidiary role while serving the "Dynasty of the Turks".

The third diasporic moment in the history of the Mamluk military, however, had a major legacy. Circassians formed a genuine diasporic group within the Mamluk military in the late fourteenth century. They were not only recruited for the warlike skills theoretically rooted in their rustic origin but also for the new *esprit de corps* their ethnic solidarity would provide to the exhausted "Dynasty of the Turks". What Sultan al-Ṣāliḥ Ayyūb did for the "race of mamluks", Sultan Barqūq (r. 1382–1399) did for the "race of Circassians": He forged their diasporic group in his household. Barqūq was not only their master; he was the genuine father of the Circassians. In the fifteenth century, however, the Circassian diasporic group gradually lost its original warlike identity and mutated into an ethnic-based nobility, which restricted the access to the highest circles of the "Dynasty of the Turks" to its kinsmen. By this point, Circassians were still a diasporic group but not a genuine military one.

No one knows what would have happened if the Ottoman armies of Selim I had not achieved an unpredictable and surprising victory over the Mamluk military in 1516–1517.[86] Yet slave markets already witnessed a shift in the late fifteenth century. A growing number of mamluks trained in the barracks of Cairo or Damascus were young captives caught on the Balkan battlefield by the Ottoman armies. Another military diasporic group starting to emerge at that time amidst the Mamluk military were the Franks (*franjī*), mainly Hungarian but also Germans.[87] Yet the Ottoman conquest nipped this potential transformation of the Mamluk military in the bud.

Notes

1 Robert Irwin, *The Middle East in the Middle Ages: The Early Mamluk Sultanate, 1250–1382* (London: Croom Helm, 1986). A note on the use of the term "mamluk": I use "mamluk" (non-capitalized) when talking about military slaves more generally and "Mamluk" (capitalized) when dealing with the Mamluk sultanate and its military élite more specifically.

2 David Ayalon, "The Mamluks: The Mainstay of Islam's Military Might," in *Slavery in the Islamic Middle East*, ed. Shaun E. Marmon (Princeton: Markus Wiener, 1999), 89–117; Julien Loiseau, *Les Mamelouks (XIIIe–XVIe siècle): Une expérience du pouvoir dans l'Islam médiéval* (Paris: Seuil, 2014).

3 Étienne de la Vaissière, *Samarcande et Samarra: Élites d'Asie centrale dans l'Empire abbasside*, Studia Iranica 35 (Paris: Association pour l'avancement des études iraniennes, 2007), 237–271.

4 Bernadette Martel-Thoumian, *Les Civils et l'administration dans l'État militaire mamlūk (IXe/XVe siècle)* (Damascus: Institut Français de Damas, 1992); Jo Van Steenbergen, *Order out of Chaos: Patronage, Conflict and Mamluk Socio-Political Culture, 1341–1382*, The Medieval Mediterranean 65 (Leiden: Brill, 2006); Mathieu Eychenne, *Liens personnels, clientélisme et réseaux de pouvoir dans le sultanat mamelouk (milieu XIIIe–fin XIVe siècle)* (Beirut: Institut Français du Proche-Orient, 2013).

5 See for instance Baybars al-Manṣūrī (m. 1325), *Zubdat al-fikra fi ta'rīkh al-hijra*, ed. Donald S. Richards, Bibliotheca Islamica 42 (Beirut/Berlin: Dār al-Nashr/al-Kitāb al-ʿarabī-Das arabische Buch, 1998), 1 (*Dhikr al-dawla al-qāhira al-turkiyya wa ibtidā'ihā*

fī l-diyār al-miṣriyya: About the subjugating Turkic dynasty and its beginnings in Egypt). For a thorough discussion of the issue, see Koby Yosef, "*Dawlat al-atrāk* or *dawlat al-mamālīk*? Ethnic Origin or Slave Origin as the Defining Characteristic of the Ruling Élite in the Mamlūk Sultanate," *Jerusalem Studies in Arabic and Islam* 39 (2012): 387–410.

6 Janos Eckmann, "The Mamluk-Kipchak Literature," *Central Asiatic Journal* 8, no. 4 (1963): 304–319; Barbara Flemming, "Literary Activities in Mamluk Halls and Barracks," in *Studies in Memory of Gastion Wiet*, ed. Myriam Rosen-Ayalon (Jerusalem: Hebrew University of Jerusalem, 1977), 249–260; Ulrich Haarmann, "Arabic in Speech, Turkish in Lineage: Mamluks and Their Sons in the Intellectual Life of Fourteenth-Century," *Journal of Semitic Studies* 33 (1988): 81–114.

7 David Ayalon, "Names, Titles and 'Nisbas' of the Mamlūks," *Israel Oriental Studies* 5 (1975): 189–232, repr. in David Ayalon, *The Mamlūk Military Society* (London: Variorum Reprints, 1979).

8 Koby Yosef has recently argued, however, that from the late thirteenth century onwards, the Turkic names given to the young mamluks in the Mamluk sultanate were chosen according to their actual origin: i.e. Turkic, Mongol, Anatolian, or Caucasian mamluks were given different Turkic names according to their ethnic background. Koby Yosef, "Mamluks of Jewish Origin in the Mamluk Sultanate," *Mamlūk Studies Review* 22 (2019): 49–95, at 56–60.

9 Ulrich Haarmann, "Ideology and History, Identity and Alterity: The Arab Image of the Turks from the ʿAbbasids to Modern Egypt," *International Journal of Middle Eastern Studies* 20 (1988): 175–196.

10 Peter B Golden, *An Introduction to the History of the Turkic Peoples: Ethnogenesis and State-Formation in Medieval and Early Modern Eurasia and the Middle East* (Wiesbaden: Harrassowitz Verlag, 1992).

11 Gabriel Martinez-Gros, *Ibn Khaldûn et les sept vies de l'Islam* (Paris: Actes Sud, 2006).

12 In this chapter, I reassess previous conclusions on ethnicity in the Mamluk military initially developed in a preliminary research on the collective identity of Circassian Mamluks. See Julien Loiseau, "Soldiers Diaspora or Cairene Nobility? The Circassians in the Mamluk Sultanate," in *Union in Separation: Diasporic Groups and Identities in the Eastern Mediterranean (1100–1800)*, ed. Georg Christ et al., Viella Historical Research 1 (Rome: Viella, 2015), 207–217.

13 Peter B. Golden, "'I Will Give People unto Thee': The Činggisid Conquests and Their Aftermath in the Turkic World," *Journal of the Royal Asiatic Society of Great Britain and Ireland* (Third Series) 10, no. 1 (2000): 21–41, at 30.

14 Donald S. Richards, "Al-Malik al-Ṣāliḥ Nadjm al-Dīn Ayyūb," in *Encyclopaedia of Islam*, 2nd ed. (Brill online).

15 Donald P. Little, "Ḳuṭuz," in *Encyclopaedia of Islam*, 2nd ed. (Brill online).

16 al-Maqrīzī, *Kitāb al-Sulūk li-maʿrifat duwal al-mulūk*, ed. Muḥammad M. Ziyāda and Saʿīd A. ʿAshūr, vol. 1/3 (Cairo: Dār al-kutub, 1939), 427, quoted by Reuven Amitai-Preiss, *Mongols and Mamluks: The Mamluk-Ilkhanid War, 1260–1281* (Cambridge: Cambridge University Press, 1995), 36.

17 Amitai-Preiss, *Mongols and Mamluks*, 36.

18 Golden, *An Introduction to the History of the Turkic Peoples*, 270–282.

19 Golden, "'I Will Give People unto Thee'," 25 and 29–30; Szilvia Kovács, "Kipchak," in *Encyclopaedia of Islam*, 2nd ed. (Brill online).

20 Charles J. Halperin, "The Kipchak Connection: The Ilkhans, the Mamluks and Ayn Jalut," *Bulletin of the School of Oriental and African Studies* 63 (2000): 229–245, at 233–235.

21 Ibid., 235–236.

22 Ibid., 240–241.

23 Anne-Marie Eddé, *La principauté ayyoubide d'Alep (579/1183–658/1260)*, Freiburger Islamstudien 21 (Stuttgart: Franz Steiner Verlag, 1999), 270–279.

24 R. Stephen Humphreys, "The Emergence of the Mamluk Army," *Studia Islamica* 45 (1977): 67–99, at 94–97. According to Humphreys, the permanent army of the Ayyubid sultan of Egypt, including the *ḥalqa* (which was at that time its élite corps of armoured cavalry) and infantry, numbered at that time about 10,000 to 12,000 men.
25 Loiseau, *Les Mamelouks*, 112–118.
26 Halperin, "The Kipchak Connection," 234.
27 Robert Brunschvig, "ʿAbd," in *Encyclopaedia of Islam*, 2nd ed. (Brill online).
28 Martinez-Gros, *Ibn Khaldûn*.
29 David Ayalon, "L'esclavage du Mamelouk," *Oriental Notes and Studies* 1 (1951): 1–64, repr. in David Ayalon, *The Mamlūk Military Society* (London: Variorum Reprints, 1979); Hassanein Rabie, "The Training of the Mamluk Faris," in *War, Technology and Society in the Middle East*, ed. Vernon J. Parry and Malcolm E. Yapp (London: Oxford University Press, 1975), 153–163.
30 David Ayalon, "The Wafidiya in the Mamluk Kingdom," *Islamic Culture* 25 (1951): 89–104, repr. in David Ayalon, *Studies on the Mamlūks of Egypt* (London: Variorum Reprints, 1977); Nobutaka Nakamachi, "The Rank and Status of Military Refugees in the Mamluk Army: A Reconsideration of the *Wāfidīyah*," *Mamlūk Studies Review* 10, no. 1 (2006): 55–81; Reuven Amitai, "Mamluks of Mongol Origin and Their Role in Early Mamluk Political Life," *Mamlūk Studies Review* 12, no. 1 (2008): 119–138.
31 Cf. on the image of the *wāfidiyya* in the eyes of Egyptian and Syrian historians Koby Yosef, "Cross-Boundary Hatred: (Changing) Attitudes towards Mongol and 'Christian' *mamlūks* in the Mamluk Sultanate," in *The Mamluk Sultanate from the Perspective of Regional and World History: Economic, Social and Cultural Development in an Era of Increasing International Interaction and Competition*, ed. Reuven Amitai and Stephan Conermann, Mamluk Studies 17 (Göttingen: V&R Unipress, Bonn University Press, 2019), 149–214, at 161–165.
32 Marie Favereau, *La Horde d'Or et le sultanat mamelouk: Naissance d'une alliance* (Cairo: IFAO, 2018).
33 Ayalon, "The Wafidiya," 99–101; Amitai, "Mamluks of Mongol Origin," 130–131.
34 Amitai, "Mamluks of Mongol Origin," 127–128.
35 Cf. on Mamluk coastal defence, Georg Christ, "The Sultans and the Sea: Mamluk Coastal Defence, Dormant Navy and Delegation of Maritime Policing (14th and Early 15th Centuries)," in *The Mamluk Sultanate from the Perspective of Regional and World History: Economic, Social and Cultural Development in an Era of Increasing International Interaction and Competition*, ed. Reuven Amitai and Stephan Conermann, Mamluk Studies 17 (Göttingen: V&R Unipress, Bonn University Press, 2019), 215–256.
36 Kitbughā's reign was even vilified by his contemporaries as the "reign of the Mongols" (*dawlat al-mughūl*): Yosef, "*Dawlat al-atrāk* or *dawlat al-mamālīk*?," 395.
37 Amitai, "Mamluks of Mongol Origin," 130–131.
38 Ayalon, "The Wafidiya," 101.
39 Amitai, "Mamluks of Mongol Origin," 132; Anne F. Broadbridge, "Sending Home for Mom and Dad: The Extended Family Impulse in Mamluk Politics," *Mamlūk Studies Review* 15 (2011): 1–18; Koby Yosef, "Mamluks and Their Relatives in the Period of the Mamluk Sultanate (1250–1517)," *Mamlūk Studies Review* 16 (2012): 55–69.
40 Amitai, "Mamluks of Mongol Origin," 129. Loiseau, *Les Mamelouks*, 170.
41 Orlando Patterson, *Slavery and Social Death: A Comparative Study* (Cambridge, MA: Harvard University Press, 1982).
42 Donald S. Richards, "Mamluk Amirs and Their Families and Households," in *The Mamluks in Egyptian Politics and Society*, ed. Thomas Philipp and Ulrich Haarmann (Cambridge: Cambridge University Press, 1998), 32–54, at 36–37; Yosef, "Mamluks and Their Relatives," 57–59.
43 Julien Loiseau, *Reconstruire la Maison du sultan: Ruine et recomposition de l'ordre urbain au Caire (1350–1450)* (Cairo: IFAO, 2010), 1:198; Loiseau, *Les Mamelouks*, 177.
44 Loiseau, *Les Mamelouks*, 33–34.

45 Ayalon, "L'esclavage du Mamelouk."
46 Mathieu Eychenne, "Le *bayt* à l'époque mamlouke: Une entité sociale à revisiter," *Annales Islamologiques* 42 (2008): 275–295.
47 Loiseau, *Les Mamelouks*, 218–225.
48 Maqrīzī, *Khiṭaṭ*, 3:693, quoted in Loiseau, *Les Mamelouks*, 283.
49 Loiseau, *Les Mamelouks*, 280–284.
50 Ibn al-Dawādārī, *Kanz al-Durar wa jāmiʿ al-ghurar: Die Chronik des Ibn ad-Dawādārī*, ed. Hans Robert Roemer et al., 9 vols. (Cairo: Deutsches Archaölogisches Institut Kairo, 1960–1994), 9:128.
51 Wilhelm Heyd, *Histoire du commerce du Levant au Moyen-Âge*, 2 vols. (Leipzig: Otto Harrassowitz, 1885–1886), 2:35.
52 Hannah Barker, *That Most Precious Merchandise: The Mediterranean Trade in Black Sea Slaves, 1260–1500* (Philadelphia: University of Pennsylvania Press, 2019).
53 Yossef Rapoport, "Women and Gender in Mamluk Society: An Overview," *Mamlūk Studies Review* 11, no. 2 (2007): 1–47.
54 Loiseau, *Les Mamelouks*, 167–170.
55 Ibn Qāḍī Shuhba, *Ta'rīkh*, ed. ʿAdnān Darwīsh, 4 vols. (Damascus: Institut Français de Damas, 1977–1994), 1:662.
56 Loiseau, *Les Mamelouks*, 73–74.
57 David Ayalon, "The Circassians in the Mamluk Kingdom," *Journal of African and Oriental Studies* 69, no. 3 (1949): 135–147; Amalia Levanoni, "Al-Maqrīzī's Account of the Transition from Turkish to Circassian Mamluk Sultanate: History in the Service of Faith," in *The Historiography of Islamic Egypt (c. 950–1800)*, ed. Hugh Kennedy, The Medieval Mediterranean 31 (Leiden: Brill, 2001), 93–105; Robert Irwin, "How Circassian Were the Circassian Mamluks?," in *The Mamluk Sultanate from the Perspective of Regional and World History*, ed. Reuven Amitai and Stephan Conermann (Göttingen: V&R Unipress, Bonn University Press, 2019), 109–122.
58 Ibn Taghrī Birdī, *Al-Nujūm al-zāhira fi mulūk Miṣr wa l-Qāhira*, 16 vols. (Cairo: Dār al-kutub al-miṣriyya, 1963–1972), 12:106–107.
59 Ibn al-Furāt, *Ta'rīkh al-duwal wa l-mulūk*, ed. Qusṭantin Zurayq and Najlā ʿIzz al-Dīn, vol. 9/1 (Beirut: Bayrūt al-Maṭbaʿa al-amrikāniya, 1938), 89–90.
60 Ibn Taghrī Birdī, *Nujūm*, 12:88.
61 Maqrīzī, *Kitāb al-sulūk*, vol. 1/3, 756. Ibn Taghrī Birdī, *Nujūm*, 7:330.
62 al-ʿAynī, *ʿIqd al-jumān fi ta'rīkh ahl al-zamān*, ed. Muḥammad Muḥammad Amīn, 5 vols. (Cairo: Dār al-kutub, 1987–2009), 5:151. Ibn Taghrī Birdī, *Nujūm*, 8:232.
63 Loiseau, *Les Mamelouks*, 186 and 341, note 187.
64 Ibn Qāḍī Shuhba, *Ta'rīkh*, 3:71.
65 Loiseau, *Les Mamelouks*, 182.
66 Cf. on the use of the name "Tatar" and the origins it labelled in the late fourteenth century Yosef, "Cross-Boundary Hatred," 173–184.
67 Michel Balard, "Remarques sur les esclaves à Gênes dans la seconde moitié du XIIIe siècle," *Mélanges d'archéologie et d'histoire publiés par l'École française de Rome* 80, no. 2 (1968): 627–680, at 644–645. Joseph Trenchs Odena, "De Alexandrinis: El comercio prohibido con los musulmanes y el Papado de Avinón durante la primera mitad del siglo XIV," *Annuario de estudios medievales* 10 (1980): 237–320.
68 Loiseau, *Les Mamelouks*, 177–178.
69 Cf. on the Circassians' christianism Vladimir Kouznetsov and Iaroslav Lebedysky, *Les Chrétiens disparus du Caucase: Histoire et archéologie du christianisme au Caucase du Nord et en Crimée* (Paris: Éditions Errance, 2009).
70 Jo Van Steenbergen, "The Mamluk Sultanate as a Military Patronage State: Household Politics and the Case of the Qalāwūnid *bayt* (1279–1392)," *Journal of the Economic and Social History of the Orient* 56 (2013): 189–217.
71 Loiseau, *Reconstruire la Maison du sultan*, 1:151–152.
72 Ibn Taghrī Birdī, *Al-Manhal al-ṣāfī wa l-mustawfī baʿd al-Wāfī*, ed. Muḥammad Muḥammad Amīn et al., 13 vols. (Cairo: Dar al-kutub, 1956–2009), 6:311.

73 Al-ʿAynī, *Al-Sayf al-muhannad fi sīrat al-Malik al-Mu'ayyad*, ed. Fāḥim Muḥammad Shalṭūṭ (Cairo: al-Hay'a al-ʿāmma li-quṣūr al-thaqāfa, 2003).
74 Loiseau, *Les Mamelouks*, 192–195.
75 Ibid., 175–176 and 190–191; Ayalon, "Names, Titles and 'Nisbas' of the Mamluks."
76 Ibn Taghrī Birdī, *Al-Manhal al-ṣāfī*, 7:17 and 12:365.
77 Ibn Taghrī Birdī, *Al-Manhal al-ṣāfī*, 7:17, note 5 (with reference to the manuscript BnF ar. 2068–2072).
78 Ibn Taghrī Birdī, *Ḥawādith al-duhūr fī madā l-ayyām wa l-shuhūr*, ed. William Popper (Berkeley: University of California Press, 1930–1942), 55; Ibn Iyās, *Badā'iʿ al-zuhūr fī waqā'iʿ al-duhūr*, ed. Muḥammad Muṣṭafā, 2nd ed., 5 vols. (Cairo: al-Hay'a al-miṣriyya al-ʿāmma li-l-kitāb, 1982–1984), 5:319.
79 Ibn Taghrī Birdī, *Ḥawādith al-duhūr*, 48 and 601.
80 Ibn Taghrī Birdī, *Nujūm*, 11:224–225 and 13:69. The same author also claims that Sultan al-Ṣāliḥ Muḥammad (r. 1421–1422) who briefly succeeded his father Sultan al-Ẓāhir Ṭaṭar (r. 1421), himself a former Circassian mamluk, "knew the Circassian language": Ibn Taghrī Birdī, *Nujūm*, 14:234.
81 On Qāytbāy's daughter, see Ibn Iyās, *Badā'iʿ al-zuhūr*, 3:288.
82 Heyd, *Histoire du commerce*, 2:394–395.
83 Ibn Taghrī Birdī, *Nujūm*, 11:183.
84 Loiseau, *Les Mamelouks*, 86 and 188–189.
85 The chronicler Ibn Iyās (1448–1524) used to introduce necrologies of Mamluk (most often Circassian) amirs of the late fifteenth century by the following sentence: "Among the Turks who died this year". See for instance Ibn Iyās, *Badā'iʿ al-zuhūr*, 3:17, 37, and 46.
86 Benjamin Lellouch and Nicolas Michel, eds., *Conquête ottomane de l'Égypte: Arrière-plan, Impact, Échos* (Leiden: Brill, 2012).
87 This Frankish aborted diaspora is still missing its modern historian. Most of the evidence regarding Mamluks of Frankish background is to be found in late fifteenth-century Latin pilgrimage relations and travelogues. On the use of these texts for Mamluk history, see Ulrich Haarmann, "The Mamluk System of Rule in the Eyes of Western Travelers," *Mamlūk Studies Review* 5 (2001): 1–24.

References

Printed Sources

al-ʿAynī. *ʿIqd al-jumān fi ta'rīkh ahl al-zamān*, edited by Muḥammad Muḥammad Amīn. 5 vols. Cairo: Dār al-kutub, 1987–2009.

al-ʿAynī. *Al-Sayf al-muhannad fi sīrat al-Malik al-Mu'ayyad*, edited by Fāḥim Muḥammad Shalṭūṭ. Cairo: al-Hay'a al-ʿāmma li-quṣūr al-thaqāfa, 2003.

Baybars al-Manṣūrī. *Zubdat al-fikra fi ta'rīkh al-hijra*, edited by Donald S. Richards. Bibliotheca Islamica 42. Beirut/Berlin: Dār al-Nashr/al- Kitāb al-ʿarabī-Das arabische Buch, 1998.

Ibn al-Dawādārī. *Kanz al-Durar wa jāmiʿ al-ghurar: Die Chronik des Ibn ad-Dawādārī*, edited by Hans Robert Roemer et al. 9 vols. Cairo: Deutsches Archaölogisches Institut Kairo, 1960–1994.

Ibn al-Furāt, *Ta'rīkh al-duwal wa l-mulūk*, edited by Qusṭantin Zurayq and Najlā ʿIzz al-Dīn. vol. 9/1. Beirut: al-Maṭbaʿa al-amrikāniya, 1938.

Ibn Iyās. *Badā'iʿ al-zuhūr fī waqā'iʿ al-duhūr*, edited by Muḥammad Muṣṭafā. 2nd ed. 5 vols. Cairo: al-Hay'a al-miṣriyya al-ʿāmma li-l-kitāb, 1982–1984.

Ibn Qāḍī Shuhba. *Ta'rīkh*, edited by ʿAdnān Darwīsh. 4 vols. Damascus: Institut Français de Damas, 1977–1994.

Ibn Taghrī Birdī. *Al-Manhal al-ṣāfī wa l-mustawfī baʿd al-Wāfī*, edited by Muḥammad Muḥammad Amīn et al. 13 vols. Cairo: Dār al-kutub, 1956–2009.

Ibn Taghrī Birdī. *Ḥawādith al-duhūr fī madā l-ayyām wa l-shuhūr*, edited by William Popper. Berkeley: University of California Press, 1930–1942.
Ibn Taghrī Birdī. *Al-Nujūm al-zāhira fi mulūk Miṣr wa l-Qāhira*. 16 vols. Cairo: Dār al-kutub al-miṣriyya, 1963–1972.
al-Maqrīzī. *Kitāb al-Sulūk li-ma'rifat duwal al-mulūk*, edited by Muḥammad M. Ziyāda and Sa'īd A. 'Ashūr. 4 vols. Cairo: Dār al-kutub, 1939–1973.
al-Maqrizi, *Al-Mawā'iẓ wa l-i'tibār fi dhikr al-khiṭaṭ wa l-āthār*, edited by Ayman Fu'ad Sayyid. 5 vols. London: al-Furqan Islamic Heritage Foundation, 2002–2004.

Studies

Amitai, Reuven. "Mamluks of Mongol Origin and Their Role in Early Mamluk Political Life." *Mamlūk Studies Review* 12, no. 1 (2008): 119–138.
Amitai-Preiss, Reuven. *Mongols and Mamluks: The Mamluk-Ilkhanid War, 1260–1281*. Cambridge: Cambridge University Press, 1995.
Ayalon, David. "The Circassians in the Mamluk Kingdom." *Journal of African and Oriental Studies* 69, no. 3 (1949): 135–147.
Ayalon, David. "L'esclavage du Mamelouk." *Oriental Notes and Studies* 1 (1951): 1–64. Reprinted in Ayalon, David. *The Mamlūk Military Society*. London: Variorum Reprints, 1979.
Ayalon, David. "The Wafidiya in the Mamluk Kingdom." *Islamic Culture* 25 (1951): 89–104. Reprinted in Ayalon, David. *Studies on the Mamlūks of Egypt*. London: Variorum Reprints, 1977.
Ayalon, David. "Names, Titles and 'Nisbas' of the Mamlūks." *Israel Oriental Studies* 5 (1975): 189–232. Reprinted in Ayalon, David. *The Mamlūk Military Society*. London: Variorum Reprints, 1979.
Ayalon, David. "The Mamluks: The Mainstay of Islam's Military Might." In *Slavery in the Islamic Middle East*, edited by Shaun E. Marmon, 89–117. Princeton: Markus Wiener, 1999.
Balard, Michel. "Remarques sur les esclaves à Gênes dans la seconde moitié du XIIIe siècle." *Mélanges d'archéologie et d'histoire publiés par l'École française de Rome* 80, no. 2 (1968): 627–680.
Barker, Hannah. *That Most Precious Merchandise: The Mediterranean Trade in Black Sea Slaves, 1260–1500*. Philadelphia: University of Pennsylvania Press, 2019.
Broadbridge, Anne F. "Sending Home for Mom and Dad: The Extended Family Impulse in Mamluk Politics." *Mamlūk Studies Review* 15 (2011): 1–18.
Brunschvig, Robert. "'Abd." In *Encyclopaedia of Islam*, 2nd ed. Brill (online), s.v.
Christ, Georg. "The Sultans and the Sea: Mamluk Coastal Defence, Dormant Navy and Delegation of Maritime Policing (14th and Early 15th Centuries)." In *The Mamluk Sultanate from the Perspective of Regional and World History: Economic, Social and Cultural Development in an Era of Increasing International Interaction and Competition*, edited by Reuven Amitai and Stephan Conermann, 215–256. Mamluk Studies 17. Göttingen: V&R Unipress, Bonn University Press, 2019.
Eckmann, Janos. "The Mamluk-Kipchak Literature." *Central Asiatic Journal* 8, no. 4 (1963): 304–319.
Eddé, Anne-Marie. *La principauté ayyoubide d'Alep (579/1183–658/1260)*. Freiburger Islamstudien 21. Stuttgart: Franz Steiner Verlag, 1999.
Eychenne, Mathieu. "Le *bayt* à l'époque mamlouke: Une entité sociale à revisiter." *Annales Islamologiques* 42 (2008): 275–295.
Eychenne, Mathieu. *Liens personnels, clientélisme et réseaux de pouvoir dans le sultanat mamelouk (milieu XIIIe–fin XIVe siècle)*. Beirut: Institut Français du Proche-Orient, 2013.

Favereau, Marie. *La Horde d'Or et le sultanat mamelouk: Naissance d'une alliance*. Cairo: IFAO, 2018.

Flemming, Barbara. "Literary Activities in Mamluk Halls and Barracks." In *Studies in Memory of Gastion Wiet*, edited by Myriam Rosen-Ayalon, 249–260. Jerusalem: Hebrew University of Jerusalem, 1977.

Golden, Peter B. *An Introduction to the History of the Turkic Peoples: Ethnogenesis and State-Formation in Medieval and Early Modern Eurasia and the Middle East*. Wiesbaden: Harrassowitz Verlag, 1992.

Golden, Peter B. "'I Will Give People unto Thee': The Činggisid Conquests and Their Aftermath in the Turkic World." *Journal of the Royal Asiatic Society of Great Britain and Ireland (Third Series)* 10, no 1 (2000): 21–41.

Haarmann, Ulrich. "Arabic in Speech, Turkish in Lineage: Mamluks and Their Sons in the Intellectual Life of Fourteenth-Century." *Journal of Semitic Studies* 33 (1988): 81–114.

Haarmann, Ulrich. "Ideology and History, Identity and Alterity: The Arab Image of the Turks from the 'Abbasids to Modern Egypt." *International Journal of Middle Eastern Studies* 20 (1988): 175–196.

Haarmann, Ulrich. "The Mamluk System of Rule in the Eyes of Western Travelers." *Mamlūk Studies Review* 5 (2001): 1–24.

Halperin, Charles J. "The Kipchak Connection: The Ilkhans, the Mamluks and Ayn Jalut." *Bulletin of the School of Oriental and African Studies* 63 (2000): 229–245.

Heyd, Wilhelm. *Histoire du commerce du Levant au Moyen-Âge*. 2 vols. Leipzig Otto Harrassowitz, 1885–1886.

Humphreys, R. Stephen. "The Emergence of the Mamluk Army." *Studia Islamica* 45 (1977): 67–99.

Irwin, Robert. *The Middle East in the Middle Ages: The Early Mamluk Sultanate, 1250–1382*. London: Croom Helm, 1986.

Irwin, Robert. "How Circassian Were the Circassian Mamluks?" In *The Mamluk Sultanate from the Perspective of Regional and World History*, edited by Reuven Amitai and Stephan Conermann, 109–122. Göttingen: V&R Unipress, Bonn University Press, 2019.

Kouznetsov, Vladimir, and Iaroslav Lebedysky. *Les Chrétiens disparus du Caucase: Histoire et archéologie du christianisme au Caucase du Nord et en Crimée*. Paris: Éditions Errance, 2009.

Kovács, Szilvia. "Kipchak." In *Encyclopaedia of Islam*, 2nd ed. Brill (online), s.v.

Lellouch, Benjamin, and Nicolas Michel, eds. *Conquête ottomane de l'Égypte: Arrière-plan, Impact, Échos*. Leiden: Brill, 2012.

Levanoni, Amalia. "Al-Maqrīzī's Account of the Transition from Turkish to Circassian Mamluk Sultanate: History in the Service of Faith." In *The Historiography of Islamic Egypt (c. 950–1800)*, edited by Hugh Kennedy, 93–105. The Medieval Mediterranean 31. Leiden: Brill, 2001.

Little, Donald P. "Ḳuṭuz." In *Encyclopaedia of Islam*, 2nd ed. Brill (online), s.v.

Loiseau, Julien. *Reconstruire la Maison du sultan: Ruine et recomposition de l'ordre urbain au Caire (1350–1450)*. 2 vols. Cairo: IFAO, 2010.

Loiseau, Julien. *Les Mamelouks (XIIIe–XVIe siècle): Une expérience du pouvoir dans l'Islam médiéval*. Paris: Seuil, 2014.

Loiseau, Julien. "Soldiers Diaspora or Cairene Nobility? The Circassians in the Mamluk Sultanate." In *Union in Separation: Diasporic Groups and Identities in the Eastern Mediterranean (1100–1800)*, edited by Georg Christ et al., 207–217. Viella Historical Research 1. Rome: Viella, 2015.

Martel-Thoumian, Bernadette. *Les Civils et l'administration dans l'État militaire mamlūk (IXe/XVe siècle)*. Damascus: Institut Français de Damas, 1992.

Martinez-Gros, Gabriel. *Ibn Khaldûn et les sept vies de l'Islam*. Paris: Actes Sud, 2006.

Nakamachi, Nobutaka. "The Rank and Status of Military Refugees in the Mamluk Army: A Reconsideration of the *Wāfidīyah*." *Mamlūk Studies Review* 10, no. 1 (2006): 55–81.

Patterson, Orlando. *Slavery and Social Death: A Comparative Study*. Cambridge, MA: Harvard University Press, 1982.

Rabie, Hassanein. "The Training of the Mamluk Faris." In *War, Technology and Society in the Middle East*, edited by Vernon J. Parry and Malcolm E. Yapp, 153–163. London: Oxford University Press, 1975.

Rapoport, Yossef. "Women and Gender in Mamluk Society: An Overview." *Mamlūk Studies Review* 11, no. 2 (2007): 1–47.

Richards, Donald S. "Al-Malik al-Ṣāliḥ Nadjm al-Dīn Ayyūb." In *Encyclopaedia of Islam*, 2nd ed. Brill (online), s.v.

Richards, Donald S. "Mamluk Amirs and Their Families and Households." In *The Mamluks in Egyptian Politics and Society*, edited by Thomas Philipp and Ulrich Haarmann, 32–54. Cambridge: Cambridge University Press, 1998.

Trenchs Odena, Josep. "De Alexandrinis: El comercio prohibido con los musulmanes y el Papado de Avinón durante la primera mitad del siglo XIV." *Annuario de estudios medievales* 10 (1980): 237–320.

de la Vaissière, Étienne. *Samarcande et Samarra: Élites d'Asie centrale dans l'Empire abbasside*. Studia Iranica 35. Paris: Association pour l'avancement des études iraniennes, 2007.

Van Steenbergen, Jo. *Order out of Chaos: Patronage, Conflict and Mamluk Socio-Political Culture, 1341–1382*. The Medieval Mediterranean 65. Leiden: Brill, 2006.

Van Steenbergen, Jo. "The Mamluk Sultanate as a Military Patronage State: Household Politics and the Case of the Qalāwūnid *bayt* (1279–1392)." *Journal of the Economic and Social History of the Orient* 56 (2013): 189–217.

Yosef, Koby. "*Dawlat al-atrāk* or *dawlat al-mamālīk*? Ethnic Origin or Slave Origin as the Defining Characteristic of the Ruling Élite in the Mamlūk Sultanate." *Jerusalem Studies in Arabic and Islam* 39 (2012): 387–410.

Yosef, Koby. "Mamluks and Their Relatives in the Period of the Mamluk Sultanate (1250–1517)." *Mamlūk Studies Review* 16 (2012): 55–69.

Yosef, Koby. "Cross-Boundary Hatred: (Changing) Attitudes towards Mongol and 'Christian' *mamlūks* in the Mamluk Sultanate." In *The Mamluk Sultanate from the Perspective of Regional and World History: Economic, Social and Cultural Development in an Era of Increasing International Interaction and Competition*, edited by Reuven Amitai and Stephan Conermann, 149–214. Mamluk Studies 17. Göttingen: V&R Unipress, Bonn University Press, 2019.

Yosef, Koby. "Mamluks of Jewish Origin in the Mamluk Sultanate." *Mamlūk Studies Review* 22 (2019): 49–95.

10

STRADIOTI

A Balkan Military Diaspora in Early Modern Europe

Nicholas C. J. Pappas

This chapter will investigate how and why the *stradioti* had such widespread and diverse experiences in the European armies of the era. It will consider the factors of the leadership and organization of the *stradioti* companies, the novel nature of their hit-and-run tactics, and the low cost of their pay compared to other mercenaries. This chapter will also address the question of the role of ethnicity, religion, and region in their employment. Is it proper to identify these troops with modern ethnic appellations when evidence exists that a process of amalgamation and assimilation existed in exile? In order to understand this process, the identity of those *stradioti* who settled in Italy, Dalmatia, and the Ionian Islands and beyond needs further investigation.

To comprehend the significance of this Balkan diaspora, one has to discuss briefly the role of soldiers from the Balkans in the growth of the Ottoman state from an emirate to an empire. The Ottomans employed forces from the Balkans in their conquest of south-eastern Europe, their consolidation of Anatolia, and their expansion into the Levant, Egypt, east central Europe and beyond. The best known of these forces is the Janissary corps, consisting of newly converted Muslims, forming the elite slave infantry corps of the Sultan.[1] Besides those who were impressed as children into the Janissary corps of the Sultan, many Balkan troops served in the Ottoman armed forces without having to change their religion. The Ottomans not only employed Anatolian feudal cavalry (*sipahiler*) in their conquests but also Christian cavalrymen who acknowledged Ottoman rule. These Christian *sipahiler* retained their landholdings and were exempt from certain taxes in return for service with Ottoman forces. Christian *sipahiler* participated in campaigns with the Ottoman armies, as well as providing security in regions that otherwise would have required garrisons of Ottoman troops. Such auxiliary cavalry troops existed as long as the Ottomans were confident of their complete allegiance. Yet in this turbulent era, their allegiance was uncertain,

DOI: 10.4324/9781003245568-11

as many Balkan Christians joined western armies in the various efforts to stem the Ottoman tide in South-eastern Europe. By the end of the sixteenth century, most of the Christian *sipahiler* had been eliminated either by their conversion to Islam or by the loss of their estates and status.[2]

Besides Christian *sipahiler*, some Christian troops, mostly from Aromun-speaking and Slavic-speaking pastoral communities, served as support forces known as *voynuklar* and *martoloslar.* The *voynuklar* were auxiliary infantry which served both as logistical and combat troops, and were exempt from certain taxes. One in ten of all pastoral Christian households had to supply one *voynuk.* The higher officers of the *voynuklar* were Muslims, while the lower-echelon officers were Christians. They were under a non-territorial sancak known as *Voynugân Sancağı.* The other important Ottoman military force in which Christians widely participated in the sixteenth and seventeenth centuries were the *martoloslar.* Like the voynuklar, they were exempt from some taxes and received some irregular pay in return for military service. Initially, they had an organization resembling the Janissaries, only with unconverted Christian lower officers and men. The *martoloslar* served at various times as infantry, cavalry, marines on the riverine flotillas on the Danube and Sava, border troops in Bosnia, and rural militia in interior provinces. The Ottomans phased out both the *voynuklar* and the *martoloslar* in the army in the seventeenth and early eighteenth centuries as the internal dysfunctions and external defeats of the Ottomans led to insurrectionary movements among the Christians, sometimes led by former *voynuklar* and *martoloslar.* As *voynuklar and martoloslar* lost their privileges and reverted *reaya* status, some went over into Habsburg lands as *Militärgrenze* and into Venetian territory as *begovci* and *Morlacchi*; others became brigands (*haiduks).* In altered forms and duties, *martoloslar* (*armatoloi*) continued to serve in the Greek lands into the nineteenth century.[3]

The Ottoman armed forces, however, did not subsume all military men of the last Christian principalities of the Balkans into their ranks. Instead, many of these found refuge and employment in Hungarian, Habsburg, and Venetian territories bordering the growing Ottoman state. Following the liquidation of the vassal Serbian Despotate of Smederevo by the Ottomans in 1459, for example, the Hungarian King Matthias Corvinus formed a new Serbian Despotate in Hungarian lands to attract Serbs to his military service. During his reign, Serbs made up a significant component of the Black Legion, the King's standing army, the Danubian river flotilla (*czajikas*), and auxiliary border troops. With the defeat of the Hungarians at Mohacs, the last remnant of a Serbian vassal state, the amorphous Serbian Despotate of Hungary, disappeared. Serbs and Vlachs continued to serve in the armed forces of western powers, most notably in the border troops of the Habsburgs and the Venetian Republic.[4]

Further to the south, mounted troops, mostly of Albanian and Greek origin, entered Venetian military service during the Republic's wars with the Ottoman Empire in the fifteenth century. They had previously served Byzantine and Albanian polities. Most of these mounted troops were light cavalry. Instead of the heavily armoured cavalry provided by large landholders (*pronoiars*), the *stratiotes*[5] were petty-holding military settlers known as *stratiôtai,* which means soldier. In his study, *The Late Byzantine Army,* Mark Bartusis claims that:

> Smallholding soldiers were the best bargain and their attachment to the land upon which they lived made them better suited to hold frontier positions than either *pronoia* soldiers or mercenaries. But, as far as we can tell, smallholding soldiers were at best light cavalry, and since they were frequently backward, clannish foreigners, were not the most reliable or disciplined troops.[6]

This description fits the status of the armed forces of the last significant territory of Byzantium, the autonomous Despotate of the Morea (ca. 1349–1460). Beginning in the mid-1300, significant numbers of Albanians migrated to the Greek lands of Thessaly, Attica, the islands of the Saronic and Argolic Gulfs, and most importantly for this study, the Peloponnesus (known in this era as the Morea). Many moved south because of increasing warfare among Albanian magnates and, ultimately, the Ottoman conquest of Albania. By the end of the fourteenth and the beginning of the fifteenth century, the pace of this Albanian immigration to the Morea quickened, especially as the last Despots of the Morea began encouraging Albanian immigration to repopulate lands and to provide light cavalry troops for defence of the peninsula from Ottoman incursions. By the end of the Byzantine Empire, Albanian settlements in the Morea were so extensive that they could field forces into the thousands.[7] Byzantine authorities employed many of these migrants as smallholding, light cavalry troops.

In their organization and tactics, these troops had already developed many of the traits that would later distinguish them from cavalry in Italy and the rest of Europe. They were armed and fought as light cavalry in a manner that developed from warfare among Byzantine, Slavic, Albanian, and Ottoman forces, characterized by swift movements and indirect attacks. They carried a spear, a long sabre, mace, and dagger, and were attired in a mixture of oriental, Byzantine, and western military garb, continuing traditions of cavalry warfare that used hit-and-run attacks, ambushes, feigned retreats, counterattacks, and other tactics little known to western armies of the time. Their frugal Byzantine employers appreciated these tactics:

> Since the number of trained, professional troops was small, it was necessary not only to supplement the size of the army but to alter tactics as well. The value of each professional soldier was raised to the point where pitched battles were avoided, military commanders preferring less hazardous tactics as skirmishing, pillaging, and treachery to accomplish what could not be done through overt force or through attrition.[8]

The *stratiōtai* of the Morea organized themselves for such combat. Like some modern light infantry units, they specialized in small-unit infiltration tactics—only on horseback. The basic combat unit of Byzantine cavalry consisted of 12 troopers, which could unite with other units into larger tactical formations.[9]

The Albanian immigrants and native Greek troops resisted Ottoman incursions into the Despotate in the years 1423, 1431, 1446, and 1452. The vacillating and capricious policies of the Despots Demetrios and Thomas Palaiologos caused both Albanians and Greeks in the Morea to revolt against these last Byzantine governors.[10] In the wake of the Ottoman onslaught of 1460, many initially found asylum and employment in the Venetian strongholds of Napoli di Romagna (Nauplion), Coron (Koroni), Malvasia (Monemvasia), Brazzo di Maina (Mani Peninsula), and Modon (Methoni) in the Peloponnesus. Others submitted to the Ottomans, who initially granted them status as Christian *sipahiler.* The Venetian authorities, especially in periods of conflict with the Ottomans, found these troops useful as outliers, or even skirmishers, i.e. buffers and outposts in the outer perimeters of their fortified towns to engage the Ottomans, including some of their former comrades, as mobile forces. The Venetians employed these *stratiōtai* because they were available and because they were experienced in fighting the Ottomans as well as relatively inexpensive to maintain. Indeed, Venetian reports indicated that many in the ranks were poor and that they needed employment so that they would not defect to the Ottoman side. As a result, the Venetians incorporated a great number of them into their armed forces in the Peloponnesus between the 1460s and the 1490s.[11]

In this study, we use the Greek terms *stratiôtês/stratiôtai* and Italian terms *stradioto/stradioti* interchangeably (*stradiotto/stradiotti* are alternative spellings in Italian). This author believes that the names *stradioto* and *stradioti* (plural) are Italian variants of the Greek *stratiôtês* or *stratiôtai* that generally meant soldier(s), but in later Byzantine times more specifically indicated cavalry trooper(s). Other authors have asserted that *stradioto/stradioti* came from the Italian root *strada* (road) and that the term *stradioto* meant a wanderer or wayfarer, thus denoting an errant cavalryman or warrior.[12] The various spellings and versions of the term in the primary sources further complicates the question of the etymology of the appellation. The few Greek sources, such as the *Andragathêmata tou Merkouriou Boua,* use *stratiôtês/stratiôtai,* the Greek word for soldier.[13] Latin sources, such as the dispatches of Jacomo Barbarigo, use the variant *stratiotos/stratiotorum* or *strathiotos/strathiotorum.*[14] The bulk of primary sources in Italian, such as Coriolano Cippico, Marino Sanudo, and Venetian state documents use *stradioto/stradioti,* adopted by this study, or *strathiotto/strathioti.*[15] French sources, such as Philip de Comines, use the variation *estradiot/estradiots.*[16] Although arguments on the side of the wayfarer theory have predominated in some circles, this etymological tie seems tenuous. The *stradioti*'s earlier Byzantine service, along with the fact that some of the older Latin sources from the early fifteenth century use a variation of the Greek *stratiotes,* tends to make this writer favour the "soldier" theory. Another factor is that some Venetian sources use the term *strazia*, indicating a larger force or corps of *stradioti.* This term has to come from the Greek word for army (*stratia*), rather than the Italian word for road (*strada*).[17] Perhaps this Italian definition developed as a way of distancing troops from their origins and implying that they were just wanderers and vagabonds. Be as it may, the term *stradioti* indicated light cavalry forces of Balkan origin, chiefly from Greece or Albania.

Unfortunately, there were too many *stradioti* and not enough lands in the Morea for them to defend. By the end of the fifteenth century, Venetian authorities transferred *stradioti* companies to the Ionian Islands of Corfu, Cephalonia, and Zante.[18] Soon afterwards, they reassigned other *stradioti* to the border in Friuli, and to the Dalmatian holdings of Sebenico (Šibenik), Spalato (Split), Zara (Zadar), Trau (Trogir), and Bocche di Cattaro (Kotor).[19] Most importantly for this study, the Venetians also deployed many companies of *stradioti* to Italy. Many *stradioti* requested permanent service, and the Venetians transferred around 2,000 men and their families in two waves, where they served in Venice's armed conflicts on the *Terraferma*. The *stradioti* entered Venetian service in Italy and became embroiled in the Ferrara war (1482–1484), where they displayed the versatility of light cavalry. They were instrumental in the Venetian victory at the battle of Argenta, where their small-unit hit-and-run tactics and their manoeuvrability, caused a sensation among Italian observers. Marino Sanudo, in his account of the Ferrara war, described them thusly:

> They have sword, lance with pennant, and mace. Very few wear cuirasses; generally, they wear cotton cloaks, sewn in a particular fashion. Their horses are large, accustomed to hardships, run like birds, always hold their heads high and surpass all others in manoeuvre of battle. Countless of these *stradioti* are found in Napoli di Romagna and other areas of Greece which are under the *signoria* and they consider their fortified towns as their true armour and lance.[20]

The new troops' arrival occurred at the crucial period in which the transalpine armies of France, and later the Holy Roman Empire and Spain threatened the Italian states in the late fifteenth and early sixteenth centuries. In an important battle of the subsequent War of King Charles (1494–1498), the conduct of the *stradioti* revealed other traits. In his account of the War of Charles VIII of France in Italy, Marino Sanudo described the *stradioti* thusly:

> The *stradioti* are Greeks and they wear broad capes and tall caps, some wear corselets; they carry lance in hand, and a mace, and hang a sword at their side; they move like birds and remain incessantly on their horses … They are accustomed to brigandage and frequently pillage the Peloponnesus. They are excellent adversaries against the Turks; they arrange their raids very well, hitting the enemy unexpectedly; they are loyal to their lords. They do not take prisoners, but rather cut off the heads of their adversaries, receiving according to their custom one ducat per head.[21]

In the encounter with the French at the important battle of Fornovo, they not only displayed their ferocity and the custom for hunting heads but also their capricious nature and lack of discipline. Although the French had committed numerous atrocities in their advance into Italy, the *stradioti*'s preference for taking heads and giving no quarter to enemy leaders shocked the French. The memoirist Philip de Commines describes the practice:

> Marchal de Gie sent to the king word that he had passed the mountains, and that having sent out a party of horse to reconnoitre the enemy, they had been charged by the Estradiots, one of them called Lebeuf being slain, the Estradiots cut off his head, put it on top of a lance, carried it to their provedditor, and demanded a ducat. These Estradiots are of the same nature as the Genetaires [Spanish light cavalry of North African roots]; they are attired like Turks both on horse and on foot, except they wear no turbans on their heads.[22]

While it seems that the Venetians tried to discourage the practice, they did not eliminate it. A greater fault of the *stradioti* at Fornovo was their venality and their penchant for looting. At a key point in the battle, the *stradioti* broke ranks and rushed to plunder Charles VIII's baggage train along with the Italian cavalry, thus turning a clear-cut victory into a costly and indecisive one.[23] Their assets must have outweighed their shortcomings; since the Venetians continued and even increased the employment of *stradioti*. In the Veneto-Turkish conflicts of the first half of the sixteenth century, the Republic lost one stronghold on mainland Greece after another. In the wake of these losses, more and more *stradioti* were resettled as refugee colonists on the Ionian Islands, in Dalmatia, and in mainland Italy.[24] One Greek author has estimated that the number of Albanian and Greek *stradioti* that settled in Venetian territories and more generally in Italy reached 4,500 men; together with their families, they numbered about 15,500. If one includes those settled in Southern Italy and Sicily, the numbers reached about 25,000.[25] In Piana degli Albanesi, south of Palermo, its inhabitants still sing a traditional song that waxes nostalgic for the Morea.[26] Indeed, the region was widely known as Piana dei Greci until the Mussolini government changed the name to Piana degli Albanesi during World War II, soon after Italy's disastrous campaign in Greece in 1941.[27]

Why did the Venetians engage these troops in the Republic's service through the sixteenth century and later? There seem to be three reasons for this hiring policy. Important factors in the Venetian preference in employing *stradioti* were the troops' unorthodox tactics and methods of fighting that could be utilized in different ways. The *stradioti's* light cavalry tactics matched those of Ottoman *sipahiler* (feudal) and *akinci* (irregular) cavalry, which made them an asset to Venice in the garrisons of its Balkan and Levantine possessions, as well as in the border region in the Friuli, where they were maintained well after the sixteenth century as *cappellatti*. In Italy and elsewhere in Western Europe, they proved to be useful in scouting, and raiding enemy forces in disarray or retreat. Against western forces, the *stradioti* impressed the Venetians and their adversaries with their highly mobile tactics, which included repeated attacks and disengagement, which enticed opposing forces to pursue. Enemy forces would lose formation and become even more vulnerable to the *stradioti* attacks. Opponents would have to deploy infantry armed with the pike, or artillery in defence against the *stradioti*.[28] According to the most important study of the Venetian army, "They may have been especially praised for raiding deep into enemy-occupied country where opportunities for loot were freest". However, some Venetian officials criticized the style and conduct of the *stradioti* commenting that they were "anti-Christian, perfidious, born thieves and potential traitors (...) so disobedient that they can do us no good". The contemporary authors accused the Balkan troops of desertion, brigandage, as well as perfidy by joining the armies of other states.[29] These accusations were calumnies rooted in a bias against Orthodox Christians as schismatics and the age-old prejudices against Greeks and easterners dating back to Roman times.

A second factor in the widespread employment of *stradioti* by the Venetian Republic was economy. The pay of the *stradioti* was lower than western mercenaries who were more costly to engage, be they Italians, Swiss, Germans, or others.[30] The *stradioti* were not mercenaries in the strictest sense, they were refugees who maintained themselves and their families in exile by their skill at arms. Wherever they were garrisoned or deployed, the troops brought their families and settled them at or near their place of duty. Yet the *stradioti* seemed to appreciate honours and privileges over pay, seeking out favours in the form of parades and titles, and the frugal Venetian government was very glad to oblige them. This can be seen by the titles that their leaders accumulated and the sentiments expressed in the poems, both in Greek and Italian, which dealt with their exploits.[31]

The third reason for Venice maintaining *stradioti* companies, despite their perceived faults, was that if it did not, other states would. Since most *stradioti* originated from regions that were part of the Republic, this could undermine its strength and integrity by foreign polities recruiting men who had served Venice. Other states also discovered the *stradioti* as military and economic assets and began to entice *stradioti* from Venetian service to their own by better pay or conditions of service.

The Kingdom of Naples, under Spanish suzerainty, also recruited *stradioti* in the late fifteenth and early sixteenth century. *Stradioti* began to enter into Neapolitan or Spanish service in the 1470s, especially the wake of a revolt in the Mani under the *stradioto* Korkodeilos Kladas, who resented the Venetians concluding a peace with the Ottomans. He continued hostilities and the Republic outlawed him. A Neapolitan ship evacuated Kladas and his rebel *stradioti* and brought them to Neapolitan territory where they joined Albanian refugees under the son of Scanderbeg, John Kastrioti. The two leaders then fomented an abortive revolt in the Himara region of Epirus. After the failure of that insurrection, most of them, together with other refugees from Himara, served the Spanish in Italy.[32] Later in 1538, long after the Venetians had abandoned Coron (Korone), the Spanish government in Naples accepted many refugees from that

Peloponnesian town and region, some of whom had earlier served the Venetians as *stradioti*. These troops now took on service with the Spanish in Naples. Spain continued to employ *stradioti* in the sixteenth and seventeenth century, chiefly in Naples and elsewhere in Italy. The most important recruiting area for Balkan troops for Naples in later times was Himara which would lead to further tensions with Venice.[33]

Since Spain, Naples, and the Holy Roman Empire came under the rule of Charles V in the first half of the sixteenth century, *stradioti* began serving the Habsburgs not only in Italy but also in Germany, the Netherlands, and their eastern frontier. Some of these troops came from Italy, while others may have entered service from Hungary with the Serbs after the Kingdom's fall in the 1520s.[34] Among those who distinguished themselves in Habsburg service and became knights of the Holy Roman Empire were the captains Giorgio Basta, Merkourios Bouas, Iakovos Diassorinos, and Iakovos Vasilikos.[35]

The French also appreciated the unique fighting skills of the *stradioti* that had opposed them and rapidly worked to draw them away from service with the Venetians and other Italian states. According to de Commines and others, France under Louis XII recruited some 2,000 *stradioti* in two years after French forces in Italy encountered them at the battle of Fornovo in 1494. The French called them *estradiots* and *argoulets*. The use of the two names has led some historians to consider that there were two separate corps of light cavalry in service to the French king.[36] However, it seems that the two terms were initially interchangeable, and only later came to indicate separate forces for no clear reason except for organization or regional origin. Some historians have identified the term *argoulets* with the Greek *argêtes* or Argive, perhaps because a significant number of troops who went over to French service originally came from Napoli di Romagna (Nauplion) on the Argive plain near the ancient Greek city of Argos.[37] The French maintained a corps of light cavalry known as *estradiots* or *argoulets* until the reign of Henry III.[38] Perhaps *estradiots* meant troops recruited from other areas.

Henry VIII of England also employed *Stradioti* in France and England, notably under the captains Thomas Bouas of Argos, Theodore Luchisi, and Antonios Stesinos, the former being the colonel and commander of *stradioti* in Henry's service.[39] It seems that the English captured some *stradioti* in France and they negotiated their service with Henry. Some 200 to 550 *stradioti* served as light cavalry on the Scottish border and later in France. It was in Scotland that a travelling Greek scholar from Corfu, Nikandros Noukios, encountered them under their leader, Thomas of Argos, whose surname was Bouas, a Hellenized version of the Albanian Bua. Later, when they redeployed to France to confront their former employers, Noukios claimed to have heard a remarkable speech given by Thomas to his troops before battle:

> My men and fellow soldiers, as you can see we are on the very edge of the world, serving a King and a nation at the Northern reaches. Moreover, we have brought nothing here from our own land except our prowess and

> bravery. Lo, courageously we stand against our adversaries, even though they exceed us in numbers, they cannot match our virtue. We are children of the Greeks and the barbarian ranks do not weaken our virtue.[40]

While this classicizing quotation does not necessarily indicate the group identity of the *stradioti* in European armies, it does raise the question of the ethnic and regional identity of these mercenaries. Some scholars have attempted to implant modern national identity upon the *stradioti* just as the above-mentioned Noukios tried to link ancient Greek identity to the *stradioti* he encountered. Using modern national identities on medieval and early modern ethnic and regional identities is problematic. A good number of authors have indicated that the *stradioti* were Albanian.[41] This is true to an extent but it has to be qualified. A Greek author studied the names of *stradioti* in the most extensive documentary collection of materials dealing with them (*Mnemeia Hellênikês Historias: Documents inédits à l'histoire de la Grèce au moyen âge,* edited by Kônstantinos Sathas) and found that some 80% of names were of Albanian origin, while the rest were of Greek origin.[42] This writer perused lists of *stradioti* in the same source, as well as the indices of the 50-plus volumes of *I Diarii di Marino Sanuto*, and found that although many of the names were indeed Albanian, a good number, particularly those of officers, were of Greek origin, such as Palaiologos, Spandounios, Laskaris, Rhallês, Comnônos, Evdaimonogiannês, Psendakis, Maniatis, Spyliôtis, Alexopoulos, Psaris, Zacharopoulos, Klirakopoulos, and Kondomitis. Others seemed to be of South Slavic origin, such as Vlastimiris and Voicha.[43] Therefore the study of names does not indicate that most of these troops came directly from Albania proper, as was argued by some authors. Fernand Braudel, for example, in his famous study of the Mediterranean in the sixteenth century describes the *stradioti's* history in the following manner:

> The story of the Albanians deserves a study in itself. Attracted by the "sword, the gold trappings, and the honours", they left their mountains chiefly in order to become soldiers. In the sixteenth century they were to be found in Cyprus, in Venice, in Mantua, in Rome, in Naples, and Sicily, and as far abroad as Madrid, where they went to present their projects and their grievances, to ask for barrels of gunpowder or years of pension, arrogant imperious, always ready for a fight. In the end, Italy gradually shut its doors to them. They moved on to the Low Countries, England, and France during the Wars of Religion, soldier adventurers followed everywhere by their wives, children, and priests.[44]

This description and others do not take into account that most of the *stradioti* did not come from Albania proper, but from the Venetian holdings in southern and central Greece, that is Malvasia (Monemvasia), Modon, Coron, Napoli di Romagna, Brazzo de Maina (Mani peninsula), and Lepanto (Naupaktos). The *stradioti* who entered Italy in the late fifteenth and early sixteenth centuries, together with their families, had been born in the Peloponnesus, their progenitors having immigrated there in the late fourteenth and early fifteenth century. They had settled in southern Greece through the encouragement of the Byzantine Despots of the Morea. The Albanians served as military colonists in the Peloponnesus in the attempt of the Despotate, an appendage of the moribund Byzantine Empire, to survive the expansion of the Ottoman Empire in the Balkans.[45] In addition, the Venetians began to settle Albanians in Napoli di Romagna (Nauplion) in the Argos region.[46]

Perhaps scholars ought to consider the *stradioti* as both Albanians and Greeks, using the hyphenated term "Greco-Albanian". While the bulk of *stradioti* were of Albanian origin from Greece, by the middle of the sixteenth century, there is evidence that many had become Hellenized or even Italianized. Many had resided in the Morea for generations and were probably bilingual. Albanian speakers still exist in the Argolid and elsewhere in Greece, but are bilingual and consider themselves Greeks. Even recent immigrants from Albania since the fall of communism have learned Greek. The literary work of Tzanes Koronaios is one of the most impressive examples of this phenomenon. It is a long epic poem in vernacular Greek on the exploits of one of the most famous of *stradioti*, Merkourios Bouas, in the armies of Venice, France, and the Holy Roman Empire. The author, Koronaios, seems to have been a *stradioto*-troubadour of Zantiote origin who was a companion of Merkourios Bouas. In his poem, which is a paean to Merkourios Bouas, Koronaios gives Bouas' mythological pedigree, which includes Achilles, Alexander the Great, and Pyrrhus. The language of the poem, the pedigree, and other allusions, give an indication of the process of Hellenization of the Albanian *stradioti*.[47]

Further proof of this process can be seen in the fact that upon their arrival in Venice in the 1480s, the *stradioti* did not join the *Scuola dei Albanesi*, which was the Albanian confraternity for immigrants from Albania (mostly from Shkoder/Scutari and the north). Founded in 1441, it served as an outlet for the ecclesiastical and charitable activities among Albanians in Venice. Instead of the *Scuola dei Albanesi*, the *stradioti* joined the Greek immigrant community, which was attempting to found a *Scuola dei Greci,* including a proper church and cemetery. Venetian authorities balked at recognizing a Greek confraternity, since the Catholic Church considered the Orthodox as schismatic non-Catholics. But the *stradioti* lent the weight of their state service to the Greek effort to persuade the authorities to relent. They played an important role in the foundation of the Greek confraternity and the later Church of Saint George of the Greeks. It is not a coincidence that the confraternity and the Greek Church in Venice were named after the patron saint of the *stradioti*. The Balkan military diaspora appreciated the right to practice their religion, the Greek rite, and evidently identified with their Greek coreligionists more than with the Roman Catholic Albanians from the north.[48]

Consequently, *stradioti* were instrumental in the founding of Greek churches, Uniate or Orthodox (or both in some cases) in Venice, Naples, and Ancona in

Italy, as well as in Venetian Pola, Trogir, Zadar, Split, and Šibenik, in Northern Dalmatia. In all of these regions, the *stradioti* and their families melted into the milieu of the church communities and eventually into society. In northern Dalmatia, there was, as one author calls it in German, a *Kirchensymbiose*; a slow acculturation of Greco-Albanian (*stradioti*) and South Slav elements in the Orthodox Church communities in predominantly Catholic Dalmatia until most of the old *stradioti* families eventually identified themselves as Serbs by the nineteenth century. Similar processes may have occurred in the Greek Church communities in Italy as well. The *stradioti* first integrated into the Greek Church community and then assimilated into the general society of the Italian towns.[49] Since many served under Greek commanders and served together with Greek *stradioti*, the process among the Albanian *stradioti* was similar. Another factor in this assimilative process was that the *stradioti* and their families had active involvement and affiliation with the Greek Orthodox or Uniate Church communities in Naples, Venice, and elsewhere. Hellenization thus occurred because of common military service and church affiliation. These generations of soldiers had gone through a double emigration, from Albania to the Morea, and from there to the Ionian Islands, Dalmatia, Italy, and beyond. The *stradioti*, in effect, emigrated twice, first at the turn of the fifteenth century and then at the turn of the sixteenth.

Once *stradioti* settled in various regions, they entered other professions and gradually lost their martial profession. In some regions, like the Ionian Islands and Dalmatia, they assimilated into local Orthodox populations. In Italy, a similar assimilative process occurred; they either became part of the Greek communities in towns or *Arbëresh* (Italo-Albanian) communities in southern Italy and Sicily. Many, no doubt, in time became Italians. Elsewhere, they were like the Italian interpretation of their name *stradioti*—wayfarers.

By the end of the sixteenth century, however, the number of *stradioti* companies employed in Italian and other western armies dwindled. The creation of other light cavalry formations, borrowing from the traditions of the *stradioti*, as well from those of the Spanish *genitours* (*genitaires*) and the Hungarian *hussars*, replaced the *stradioti* in many European armies. They brought other tactics with them, such as those deriving from the steppe traditions of eastern Europe and Eurasia. These new units, made up of natives or various ethnic groups, even changed their tactics and weapons, including adopting firearms. By this point, the mention of *stradioti, cavalli greci, argoulets, estradiots, Albanese, Albains, Greci, Levantini*, etc., becomes less and less frequent in the sources.

Nevertheless, Naples continued the hiring and maintenance of *stradioti* troops until the early eighteenth century. Most of these troops were later recruited from Epirus and Southern Albania, in particular from the Greco-Albanian region of Himara. According to histories of the *Reggimento Real Macedone,* a Balkan light infantry force that served the Kingdom of the Two Sicilies between 1735 and 1820, its first commander and organizer was one Count Stratês Gkikas, who is described as a veteran *stradiotto.* This may be a further indication of *stradioti* in

Neapolitan service into the eighteenth century.[50] Similarly, Venice continued to employ *stradioti* as *cappelletti* (rural gendarmes) on the *Terraferma* well into the seventeenth century. These units consisted of former *stradioti* and others with cavalry skills. They carried out security duties along the borders and police duties in the interior, adding firearms to their panoply.[51]

In lieu of *stradioti*, Venice began to recruit new infantry companies from the seventeenth century, known by their perceived ethnicity as *Compagnie* or *Milizie Greche*, consisting of Greeks, Albanians, and some south Slavs. The common denominator of these units was that the troops were Orthodox and employed Orthodox priests as chaplains. The Venetians recruited them on the Island of Corfu from outlying mainland regions. They mustered in Venice for medical examination and initial training before being deployed to posts and garrisons.[52] Because these units came from the mainland holdings around Corfu, this suggests that many of these troops originated from the Greco-Albanian Himara area known for its warrior traditions. Later in the wake of the Morean Wars (1645–1699), the Venetians raised a regiment mostly from Himara known as the *Reggimento Cimarrioto,* probably from these *Compagnie Greche.* After 1735, Venice and Naples rivalled one another over the hiring of soldiers for the *Reggimento Cimarrioto* and the *Reggimento Real Macedone* until the Napoleonic era.[53]

On the Ionian Islands, the *stradioti* continued their service throughout the eighteenth century. These *stradioti* were descendants of refugees who had received land and privileges in exchange for cavalry service in Venice's conflicts with the Turks throughout the seventeenth century. Eventually, these units became anachronisms, their ranks virtually a hereditary caste. Some of the *stradioti* or their descendants became in time members of the Ionian nobility, while others took to farming and other pursuits. By the late seventeenth and early eighteenth centuries, Venetian authorities found it necessary to reorganize the *stradioti* companies. On Zante, for example, they reduced their numbers and privileges because of absenteeism and discipline problems in the rank and file. Nevertheless, the *stradioti* formations remained in nominal service through the eighteenth century.[54] The Corfiote *Stradioti* company existed until the end of Venetian rule and the French occupation in 1797. Some say the Ionian *stradioti* were the only ones to defend the Republic in its last stand against the French, at least symbolically.[55]

The events of 1797 also symbolize the end of a century-long process that saw the decline of mercenary free companies. The establishment of effective state bureaucracies and armed forces that could pay, supply, train, and discipline troops without the use of independent contractors from the second half of the seventeenth century brought about the end of free mercenary companies in Europe by the eighteenth century. However, outside of Western Europe, the mercenary syndrome continued, as in the Balkans and the Near East. 1797 also marked the rise of new elements at work in mercenary service, including new corps, new employers, new emigrations, and new ideologies—especially nationalism.

From that year, the Ionian Islands off the western coasts of Greece and southern Albania became a base of operations and an area of conflict in the Mediterranean in the years until 1814. In that period, the new imperial powers in the eastern Mediterranean, France, Russia, and Britain, successively occupied these Greek-populated islands, formerly Venetian possessions. Each of these powers attempted to establish a nominally independent "Septinsular Republic" under their protectorate. There were efforts by all of these powers to organize native armed forces; upwards of 5,000 troops were raised from among refugees from the mainland, including bandits (*klephtes*), former Ottoman irregulars (*armatoloi*), and clansmen from the autonomous regions of Himara, Souli, and Mani, much like the earlier Balkan refugees/mercenaries like the earlier *stradioti*.

Although these refugee-warriors were skilled in the use of weapons—flintlock firearms, sabres, and yataghans—they fought and were organized according to traditions and methods that were different and considered "obsolete" in early nineteenth-century Europe. Their tradition of arms, a warrior tradition as much as a military tradition, had its origins in the martial formations among Christians under Ottoman rule.

After 1814, these refugees/mercenaries were both in exile and unemployed. Delegations went to Russia, Naples, and perhaps elsewhere looking for work or resettlement, but to no avail. Instead, a significant number joined the Greek nationalist secret society *Philike Hetaireia* and participated in the Greek uprisings of 1821. Indeed, one can say that these warriors evolved from refugees to mercenaries to national revolutionaries with the advent of modern national sentiment.

The largest segment of the military leadership of the Greek Revolution stemmed from the Balkan warrior tradition, which had undergone great expansion and change in the decades before the revolution as a result of the Russo-Turkish and Napoleonic wars. In this pre-revolutionary period, the powers of Europe, particularly Russia and even the United States had employed Greek and other Balkan warriors.

The development of modern Balkan States saw the foundation of national armies, where there was a tension between the irregular Balkan warrior tradition of the Balkans and the regular military norms of Europe, which would continue in the nineteenth century. The contradiction between the traditional and western in Greek military affairs subsided in the early twentieth century, lasting the longest in Albania, Greece, and Montenegro, because of recrudescent tribalism, clannism, or banditry.

Notes

1 On the Janissary corps from origins through the sixteenth century, see Mesut Uyar and Edward J. Erickson, *A Military History of the Ottomans: From Osman to Atatürk* (Santa Barbara, CA: Praeger Security International, 2009), 17–22, 32. On the Ottoman draft, known as *devşirme,* see Kathryn Hain, "Devshirme is a Contested Practice," *Historia: The Alpha Rho Papers* 2 (2012): 165–176.

2 On the Christian *sipahiler,* see Bistra Cvetkova, "Novyedannye o Khristianskikh-spakhiakh na Balkanskom poluostrove v. period turetskogo gospodstva," *Vizantiiskii Vremennik* 13 (1958): 184–197; Branislav Djurdjev, "Hrisćeanispahije u Sevemoj Srbiji u XV veku," *Godišnjak istoriskog društva Bosne i Hercegovine* 4 (1952): 165–169; Halil

Inalcik, "Ottoman Methods of Conquest," *Studia Islamica* 2 (1954): 103–129; and id., "Timariotes chrétiens en Albanie au XVe siècle d'après un registre de timars ottoman," in *Mitteilungen des österreichischen Staatsarchivs* 4 (1951): 118–138, at 120–131.

3 On *voynuklar* and *martoloslar,* see Robert Anhegger "Martoloslar hakkinda," *Türkiyat Mecmuasi* 7–8 (1942): 282–320, at 283–286; Iôannês Vasdravellês, *Armatoloi kai klephtes eis tên Makedonia,* 2nd ed. (Thessalonike: Hetaireia Makedonikon Spoudon, 1970); Branislav Djurdjev, "O vojnucima sao svrtom na razvoj turskog feudalizma i na pitanije turskog agaluka," *Glasnik Zemaljskog Muzeja u Sarajevu* II (1947): 75–138; Cengiz Orhonlu, *Osmanli imperatorlugunda derbend teşkilati* (Istanbul: Istanbul Üniversitesi Edebiyat Fakültesi, 1967); Milan Vasić, *Martolosi u Jugoslovenskim zemljana pod turskom vladavinom* (Sarajevo: Akademija nauka i umjetnosti Bosne i Hercegovine, 1967).

4 On the Serbian military in the Hungarian Army, see David Nicolle, *Hungary and the Fall of Eastern Europe, 1000–1568* (London: Osprey Publishing, 1988), 37–38; Geza Perjes, *The Fall of the Medieval Kingdom of Hungary: Mohács 1526–Buda 1541,* trans. Marió D. Fenyö (Boulder, CO: Social Science Monographs; Highland Lakes, NJ: Atlantic Research and Publications, 1989), 62–63, 73–74; Wayne S. Vucinich, "The Serbian Military Tradition," in *War and Society in East Central Europe,* ed. Béla Kiraly and Gunther E. Rothenberg (New York: Brooklyn College Press, 1979), 1:298–299.

5 The terms *stratiôtês* and *pronoiar* were not synonymous. For divergent views of this Byzantine "feudalism," see Mark C. Bartusis, *The Late Byzantine Army: Arms and Society, 1204–1453* (Philadelphia, PA: University of Pennsylvania, 1997), 343–367; and Ljubomir Maksimović, *The Byzantine Provincial Administration under the Palaiologoi* (Amsterdam: Hakkert, 1988), 10–31. Bartusis has recently published a detailed monographic study on *pronoia* called *Land and Privilege in Byzantium: The Institution of* Pronoia (Cambridge: Cambridge University Press, 2012).

6 Bartusis, *The Late Byzantine Army,* 335–336. Note that I do not use the term feudatory to refer to the *pronoiars*; they did not necessarily hold their lands as fiefs in return for military service.

7 On these population movements, see Sarantos Kargakos, *Alvanoi, Arvanites, Hellênes: Meletes,* 3rd ed. (Athens: I. Sideres, 2008): 25–37; Kôstas Mpirês, *Hoi Arvanites, hoi Dôrieis tou neoterou Hellênismou: Historia ton Hellenon Arvaniton* (Athens: Melissa 1960), 273–285; and Dionysios Zakythinos, *Le despotat grec de Morée,* vol. 2 (Athens: L'hellénisme contemporain, 1953), 31–32. For a fascinating speculative comparison of the Albanian migrations with that of the Dorians at the end of the Bronze age, see Nicholas G. L. Hammond, *Migrations and Invasions in Greece and Adjacent Areas* (Park Ridge, NJ: Noyes Press, 1976), 52–53.

8 Bartusis, *The Late Byzantine Army,* 343–367. Indeed, the Byzantines had military manuals on skirmishing, such as the *Peri Paradromês* (On Skirmishing) attributed to Emperor Nicephoros Phokas from the tenth century. The text with translation can be found in George T. Dennis, *Three Byzantine Military Treatises* (Washington DC: Dumbarton Oaks, 1985), 143–241.

9 "Campaign Organisation and Tactics, i.e. *Anonymou Bibliou Taktikou* (Anonymous Book on Tactics)," in *Three Byzantine Military Treatises,* ed. George T. Dennis (Washington DC: Dumbarton Oaks, 1985), 274–275.

10 Zakythinos, *Le Despotat grec de Morée,* vol. 1, *Histoire politique* (London: Variorum, 1975), 265–274, 285–290.

11 In the second volume of his monumental *The Papacy in the Levant,* Kenneth Setton described the use of *stradioti* in the Veneto-Ottoman conflict. Many of the sources on their initial Venetian employment can be found in *Mnemeia* Kônstantinos N. Sathas, ed. *Hellenikes Historias: Documents inédits relatifs à l'histoire de la Grèce au moyen âge,* vol. 1 (Paris: Maissonneuve, 1880). The outstanding studies of Venetian holdings in Nauplia by Diana Wright, especially her doctoral dissertation, "Bartolomeo Minio: Venetian Administration in 15th Century Nauplion" (PhD diss., Catholic University of America, 1999) and her co-edited volume, Diana Wright and John

Melville-Jones, eds., *The Greek Correspondence of Bartolomeo Minio*, vol. 1, *Dispacci from Nauplion, 1479–1483* (Padua: Unipress, 2008) have corrected many of the shortcomings of Sathas' pioneering collections and works.

12 *Mnemeia Hellenikes Historias*, 9 vols. (Paris: Maissonneuve, 1880–1890), 4:liv–lvi.

13 Tzanes Koronaios, "Andragathêmata tou Merkouriou Boua," in *Hellênika Anekdota*, ed. Kônstantinos Sathas (Athens: Karabias, 1873), 1:1–153.

14 Jacomo Barbarigo, "Dispacci della Guerra di Peloponneso," in *Mnemeia Hellenikes Historias: Documents inédits relatifs à l'histoire de la Grèce au moyen âge*, ed. Kônstantinos N. Sathas, 9 vols. (Paris: Maissonneuve, 1885), 6:1–92.

15 See materials published in *Mnemeia Hellenikes Historias*, vols. 1, 4, 6–9; *Commissiones et Relationes Venetae*, vol. 5, 7, *Annorum 1591–1600, 1621–1671*, in *Monumenta Spectantia Historiam Slavorum Meridionalium*, ed. Grga Novak, vol. 48, 50 (Zagreb: Jugoslovenska Akademija Znanostii Umjetnosti, 1966, 1972); and *Secrets d'état de Venise: Documents extraits notices et études, servant à éclaircir les rapports de la seigneurie avec les grecs, les slaves et la Porte ottomane à la fin du XVe et au XVIe siècle*, ed. Vladimir Ivanovič Lamanskij (St. Petersburgh: L'acad. imper. des sciences, 1884).

16 Philippe de Commynes, *Mémoires de Messire Philippe de Comines, Seigneur d'Argenton, où l'on trouve l'Histoire des Rois de France Louis XI. & Charles VIII.*, 4 vols. (London: Rollin, 1747), 2:27–28. The French also employed light cavalry known as *argoulets*, which may derive from the toponym Argos (Argive) in the Peloponnesus. The Spanish variant was *estradiotes*.

17 See for an example its use in *Monumenta spectantia historiam slavorum meridionalium*, ed. Simeon Ljubić (Zagreb: Officina Societatis Typographicae, 1877), 8:231.

18 Mpirês, *Hoi Arvanites*, 156–162; Michael E. Mallet and John R. Hale, *The Military Organization of a Renaissance State: Venice c. 1400 to 1617* (Cambridge: Cambridge University Press, 1984), 447, 450. For more details on the *stradioti* on the Ionian Islands, see the excellent study by Katerina Korrê, "Misthophorikê hypêresia, engeios, idioktêsia, klêronomikê diadochê: Hoi stradiotikês ktêseis tou Ioniou (16os Aiônas)," in *Praktika tou Panioniou Synedriou: Paxoi, 26–30 Maiou 2010*, ed. Alike Nikephorou (Paxos: Etaireia Paxinôn Meletôn, 2014), 1:197–226. Korrê is a faculty member of the University of Crete and a scholar at the Byzantine and Post-Byzantine center in Venice.She has an impressive corpus of work. Earlier works on the *stradioti* on the Ionian Islands include: Ermannos Lunzi, *Peri tis politikes katastaseos tis Heptanesou epi Eneton* (Athens: Philadelphis, 1856; repr. 1969), 264–265; Apostolos Vakalopoulos, *Historia tou Neou Hellenismou* (Thessalonike: Sphakianakes, 1968), 3:79–88; Laurentios Vrokinês, "He peri ta mesa tou IST' aiônos en Kerkyrai apoikêsis tôn Naupleiôn kai Monemvaseiôn," in *Erga*, ed. Kostas Daphnês, vol. 16–17 (Corfu: Kerkyraika Chronika, 1972); Leonidas Zoes, "Hellenikos lochos en Zakynthoi kata tous chronous tes douleias," Ho Hellenismos 14 (1911): 367–371.

19 Kônstantinos N. Sathas, ed. *Mnemeia Hellenikes Historias: Documents inédits relatifs à l'histoire de la Grèce au moyen âge*, vol. 8 (Paris: Maisonneuve, 1888); Mallet and Hale, *The Military Organization*, 173.

20 Marino Sanudo, *Commentarii della guerra di Ferrara tra gli Viniziani ed il duca Ercole d'Este nel 1482, per la prima volta pubblicati* (Venice: Picotti, 1829), 115.

21 Marino Sanudo, *La spedizione di Carlo VIII in Italia*, ed. Rinaldo Fulin (Venice: Tip. del commercio di M. Visentini, 1883), 313–314.

22 De Commynes, *Mémoires*, 2:27–28; and id., *The Memoirs of Philip de Commines, Lord of Argenton: Containing the Histories of Louis XI and Charles VIII, Kings of France, and of Charles the Bold, Duke of Burgundy; to Which Is Added, The Scandalous Chronicle, or Secret History of Louis XI. by Jean de Troyes*, ed. and trans. Andrew R. Scoble, 2 vols. (London: Henry G. Bohn, 1856), 2:200–201. Headhunting was a practice in the Balkans until at least the nineteenth century and is still used to humiliate and dehumanize an enemy by ISIS and other groups in the Middle East. One reads about towers of heads raised at battle sites by the Ottomans, however, this writer believes that it predates and post-dates the Ottomans. See Mary Edith Durham, "Head-Hunting in the

Balkans," *Man* 23 (February 1923): 19–21. This writer has also seen evidence of ritual decapitation during the Greek Civil War (1946–1949).

23 On the *stradioti* debacle at Fornovo, see John S. C. Bridge, *A History of France from the Death of Louis XI*, 5 vols. (Oxford: Clarendon Press, 1924), 2:263.

24 William Miller, *The Latins in the Levant: A History of Frankish Greece (1204–1566)* (New York: Barnes and Noble, 1937), 489–511.

25 Mpirês, *Hoi Arvanites*, 172; see below for Southern Italian and other non-Venetian *stradioti* troops.

26 *Balada shqiptare*, ed. Vladimir Zoto (Tirana: Dasara 2006), 311–312. An English translation and an explanation is presented by the noted Albanologist, Robert Elsie, as "Oh, my Beautiful Morea," in his Albanian literature website, found at http://www.albanianliterature.net/oralverse/verse_02.html.

27 J. Fracchia, "Hora: Social Conflicts and Collective Memories in Piana Degli Albanesi," *Past and Present*, no. 209 (November 2010): 181, note. Most Italians and Albanians believe the name change to correct the confusion of the Greek rite with-Greek ethnicity. I believe that the explanation of the confounding of *Greci* and *Albanese* is a combination of the identity with Greek rite and the fact that many of the Italo-Albanians in southern Italy came from the Peloponnesus.

28 Frederick L. Taylor, *The Art of War in Italy, 1494–1529* (Cambridge: Cambridge University Press, 1921), 72–73.

29 Mallet and Hale, *The Military Organization*, 376. They also supposedly took double pay by receiving pay in Italy and having relatives collect it in Greece or Albania. Eighteenth-century Venetian officials claimed the same for troops of the Venetian *Reggimento Cimarrioto*, who collected pay from both the Venetian regiment and the Neapolitan *Reggimento Real Macedone* while on leave in their homeland in Himara. Whether this is true or a trope used by bureaucrats to explain shortfalls is open to debate. Fabio Mutinelli, *Memorie storiche degli ultimi cinquant' anni della Repubblica Veneta, tratte da scritti e monumenti contemporanei* (Venice: G. Grimaldo, 1854), 159–160.

30 Mallet and Hale, *The Military Organization*, 375–380. See pages 447–447, 451 on pay scales of *stradioti*.

31 Ibid., 376–377; Manoli Blessi, "Balzeletta," in *Mnemeia Hellenikes Historias*, 8: 461–465; Blessi, "Manoli Blessi sopra la presa de Margaritin con un dialogo di un Greco et di un Fachino," in *Mnemeia Hellenikes Historias*, 8:466–470, Blessi, "Il vero successo della presa di Nicosia in Cipro di Manoli Blessi Strathiotto," in *Mnemeia Hellenikes Historias: Documents inédits relatifs à l'histoire de la Grèce au moyen âge*, ed. Kônstantinos N. Sathas, 9 vols. (Paris: Maissonneuve, 1890), 9:262–280.

32 Panagiotes Aravantinos, *Chronographiatês Epeirou ton tê homorôn hellênikôn kai Illyrikôn chorôn diatrechousa kata seiran ta enautai symbanta apo tou sôtêriou etous mechri tou 1854* (Athens: S. K. Vlastou, 1856), 1:191.

33 Ioannis K. Hassiotis, "La comunità greca di Napoli e i moti insurrezionali nella penisola Balcanica meridionale durante la seconda metà del XVI secolo," *Balkan Studies* 10 (1969): 279–288; id., "Me to Hispanous kai tous Apsbourgous," *Hellenes Misthophoroi ston Kosmo-Kathemerine Aphieroma 2–32* (January 2003): 12–14; Vincenzo Giura, "La Comunita Greca di Napoli (1534–1861)," in *Storie de Minoranze: Ebrei, Greci, Albanesi nel Regno di Napoli* (Naples: Edizioni scientifiche italiane, 1984), 119–156; Attanasio Lehasca, *Cenno storico dei servigi militari prestati nel Regno delle Due Sicilie dai Greci, Epiroti, Albanesi e Macedoni in epoche diverse* (Corfu, 1843), 3–15.

34 Two military men, *stradioti* by the name of Palaiologos, served in Hungarian forces, probably along with the Serbs in Hungarian service, ending their lives in 1528 and 1529, a few years after the Hungarian disaster at Mohacs. Georgios Ploumides, "Nikolaos (†1528) kai Iôannês (†1529) Palaiologoi Stradioti," *Dôdônê* 1 (1995): 229–234.

35 *Mnêmeia Hellênikês Historias*, 9:xiv–xxviii; Hassiotes, "Me to Hispanous kai tous Apsbourgous," 12–14. Basta became a Habsburg commander and a brutal supporter of the Catholic cause in Transylvania in the wake of the Counterreformation. He also wrote one of the first treatises on light cavalry warfare in Western Europe (Giorgio Basta,

Governo della Cavalleria Leggiera [Venice: Gionti, 1612]). Conversely, both Diasorrinos and Vasilikos corresponded with the Lutheran reformer, Melanchthon. Vasilikos became *hospodars* of Moldavia for a brief time, where he tried to introduce Protestantism. Émile Legrand, ed., *Deux vies de Jacques Basilicos: Seigneur de Samos, Marquis de Paros, Comte Palatin et Prince de Moldavie, l'une par Jean Sommer, l'autre par A.-M. Graziani, suivies de pièces rares et inédites* (Paris: Maisonneuve, 1889) and Epameinondas I. Stamatidiades, *Bios Iakobou Basilikou, Despotou Samou, Markettíou Parou, Kometos Palatinou kai Egemonos tes Moldavias* (Samos: Egemonikon Typographeion, 1894).

36 de Commynes, *Mémoires de Messire Philippe de Comines*, 2:27–28.

37 Kônstantinos N. Sathas, *Hellênikoi Stratiôtai en te Dysei kai anagennesis tes Hellenikes taktikes,* Neoellenike historia (Athens: Periodiko Hestia, 1885, repr. Ekdoseis Philomythos, 1993), 11–14.

38 Gabriel Daniel in his *Histoire de la milice françoise et des changements qui s'y font faits depuis l'établissement de la monarchie dans les Gaules jusqu'à la fin du regne de Louis le Grand*, 2 vols. (Paris 1721), 2:168, divides the *stradioti* in the sixteenth century French army into two separate corps of *argoulets* and *estradiots*. See also Eugène Fieffé, *Histoire des troupes étrangères au service de France depuis leur origine jusqu'à nos jours* (Paris: Librairie militaire, 1854), 1:66–71.

39 Gilbert J. Millar, "The Albanians: Sixteenth-Century Mercenaries," *History Today* 26 (1976): 468–472, at 470, 472; id., *Tudor Mercenaries and Auxiliaries 1485–1547* (Charlottesville: University Press of Virginia, 1980), 44, 48, 69, 73, 133, 146, 148–149, 151, 161, 164–165; Vakalopoulos, *Historia tou Neou Hellênismou* (Thessalonike, 1968), 3:191.

40 Andreas Moustoxydês, "Nikandros Noukios," *Pandôra* 7, no. 154 (1856): 222; see also P. Kallinikos. "O Nikandros Noukios ste Bretannia," *Deltion Anagnôstikes Hetairieas Kerkyras* 7 (1970): 35–42.

41 See, for example: Franz Babinger, "Albanische Stradioten im Dienste Venedigs im ausgehenden Mittelalter," *Studia Albanica* 1 (1964): 95–105; Fernand Braudel, *The Mediterranean and the Mediterranean World in the Age of Philip II*, trans. Sian Reynolds, 2 vols. (New York: Harper and Row, 1975) 1:48–49; Millar, "The Albanians," 468–472; and Paolo Petta, *Stradioti: soldati albanesi in Italia, sec. XV–XIX (*Lecce: Argo, 1996), passim.

42 Mpirês, 191–192.

43 *Mnêmeia Hellênikês Historias*, vols. 1, 4, 6–9.

44 Braudel, *The Mediterranean and the Mediterranean World*, 1:48–49.

45 Nicholas Cheetham, *Medieval Greece* (New Haven, CT: Yale University Press, 1981), 195–207; Mallet and Hale, *The Military Organization*, 47, 50; Dionysios A. Zakythinos, *Le Despotat grec de Morée*, vol. 2: *Vie et institutions* (London: Variorum, 1975), 31–37, 135–145.

46 Peter W. Topping, "Albanian Settlements in Medieval Greece: Some Venetian Testimonies," in *Charanis Studies: Essays in Honor of Peter Charanis*, ed. Angeliki E. Laiou-Thomadakis (New Brunswick, NJ: Rutgers University Press, 1980), 261–272.

47 Koronaios, "Andragathêmata tou Merkouriou Boua," 1:1–153. On the influence of *stradioti* exploits and style upon Greek and Italian culture, see Stathis Birtachas, "La memoria degli stradioti nella letteratura italiana del tardo Rinascimento," in *Tempo, spazio e memoria nella letteratura italiana: Omaggio ad Antonio Tabucchi / Χρόνος, τόπος και μνήμη στην ιταλική λογοτεχνία: Τιμή στον Antonio Tabucchi*, ed. Zosi Zografidou (Rome: Aracne; Thessaloniki: University Studio Press, 2012), 123–141.

48 Giorgio Fedalto, "Le minoranze straniere a Venezia tra politica e legislazione," in *Venezia: Centro di mediazione tra Oriente e Occidente (secoli XV–XVII); Aspetti e problemi, Atti del II Convegno internazionale di storia della civiltà veneziana promosso e organizzato dalla Fondazione Giorgio Cini, dal Centro Tedesco di Studi Veneziani, dall'Istituto Ellenico di Studi Bizantini e Post-Bizantini (Venezia, 3–6 ottobre 1973)*, ed. Hans-Georg Beck,

Manoussos I. Manoussakas, and Agostino Pertusi, 2 vols. (Florence: Olschki, 1977), 1:143–162; Iôannês Veloudos, *Hellênôn Orthodoxôn apoikiaen Venetia: Historikôn hypomnêma*, 2nd ed. (Venice: Chrestos Triantapyllou 1893), 16–27.

49 Veloudos, *Hellênôn Orthodoxôn apoikiaen Venetia*, 16–27; Giura, "La Comunità Greca di Napoli (1534–1861)," 121–127; Dušan Lj. Kašić, "Die griechisch-serbische Kirchensymbiose in Norddalmatien vom XV. bis zum XIX. Jahrhundert," *Balkan Studies* 15 (1974): 21–48.

50 Nicholas C. Pappas, "Balkan Foreign Legions in Eighteenth Century Italy: Reggimento Real Macedone and Its Successors," in *Nation and Ideology: Essays in Honor of Wayne S. Vucinich*, ed. Ivo Banac, John C. Ackerman, and Roman Szporluk (Boulder, CO: Columbia University Press, 1981), 35–39.

51 Stathis Birtachas, "Stradioti, Cappelletti, Compagnie or Milizie Greche: 'Greek' Mounted and Foot Mercenary Companies in the Venetian State (Fifteenth to Eighteenth Centuries)," in *A Military History of the Mediterranean Sea Aspects of War, Diplomacy, and Military Elites*, ed. Georgios Theodokis, and Aysel Yildiz (Leiden: Brill, 2018), 325–346, at 334–335.

52 Birtachas, "Stradioti, Cappelletti, Compagnie or Milizie Greche," 335–338.

53 Pappas, "Balkan Foreign Legions in Eighteenth Century Italy," 38–39.

54 On the *stradioti* in the Ionian Islands, see Lunzi, *Peri tis politikes katastaseos tis Heptanesou epi Eneton*, 264–265; Laurentios Vrokines, *He peri ta mesa tou 16 aionos en Kerkyrai apoikesis ton Naupleion kai Monemvaseion* (Corfu, 1905); and Zoes, "Hellenikos lochos en Zakynthoi kata tous chronous tes douleias," 367–371.

55 Nicholas C. Pappas, *Greeks in Russian Military Service in the Late Eighteenth and Early Nineteenth Centuries* (Thessaloniki: Institute for Balkan Studies, 1991), 57. It is unsure if they mustered as a cavalry force or another arm of the military. The above monograph deals with the last generation of Greco-Albanian refugees/mercenaries, who served in the armies of Russia, France, and Great Britain and later provided many leaders in the Greek Revolution of 1821–1830.

References

Printed Sources

Aravantinos, Panagiotes. *Chronographiatês Epeirou ton tê homorôn hellênikôn kai Illyrikôn chorôn diatrechousa kata seiran ta enautai symbanta apo tou sôtêriou etous mechri tou 1854*. Vol. 1. Athens: S. K. Vlastou, 1856.

Barbarigo, Jacomo. "Dispaccidella Guerra di Peloponneso." In *Mnemeia Hellenikes Historias: Documents inédits relatifs à l'histoire de la Grèce au moyen âge*, edited by Kônstantinos N. Sathas, 1–92. Vol. 6. Paris: Maissonneuve, 1885.

Basta, Giorgio. *Governo della Cavalleria Leggiera*. Venice: Gionti, 1612.

Blessi, Manoli (Antonio Molin detto il Burchiella). "Balzeletta." In *Mnemeia Hellenikes Historias: Documents inédits relatifs à l'histoire de la Grèce au moyen âge*, edited by Kônstantinos N. Sathas, 461–465. Vol. 8. Paris: Maissonneuve, 1888.

Blessi, Manoli (Antonio Molin detto il Burchiella). "Manoli Blessi sopra la presa de Margaritin con un dialogo di un Greco et di un Fachino." In *Mnemeia Hellenikes Historias: Documents inédits relatifs à l'histoire de la Grèce au moyen âge*, edited by Kônstantinos N. Sathas, 466–470. Vol. 8. Paris: Maisonneuve, 1888.

Blessi, Manoli (Antonio Molin detto il Burchiella). "Il vero successo della presa di Nicosia in Cipro di Manoli Blessi Strathiotto." In *Mnemeia Hellenikes Historias: Documents inédits relatifs à l'histoire de la Grèce au moyen âge*, edited by Kônstantinos N. Sathas, 262–280. Vol. 9. Paris: Maissonneuve, 1890.

"Campaign Organisation and Tactics" (i.e. Anonymou Bibliou Taktikou [Anonymous Book on Tactics])." In *Three Byzantine Military Treatises*, edited by George T. Dennis, 241–335. Washington, DC: Dumbarton Oaks, 1985.

Commissiones et Relationes Venetae. Vol. 5, 7, *Annorum 1591–1600, 1621–1671*, in *Monumenta Spectantia Historiam Slavorum Meridionalium*, edited by Grga Novak. Vol. 48, 50. Zagreb: Jugoslovenska Akademija Znanostii Umjetnosti, 1966, 1972.

Commynes, Philippe de. *Mémoires de Messire Philippe de Comines, Seigneur d'Argenton, où l'on trouve l'Histoire des Rois de France Louis XI. & Charles VIII.* Vol. 2. London: Rollin, 1747.

Commynes, Philippe de. *The Memoirs of Philip de Commines, Lord of Argenton: Containing the Histories of Louis XI and Charles VIII, Kings of France, and of Charles the Bold, Duke of Burgundy; to Which Is Added, The Scandalous Chronicle, or Secret History of Louis XI. by Jean de Troyes*, edited and translated by Andrew R. Scoble. Vol. 2. London: Henry G. Bohn, 1856.

Daniel, Gabriel. *Histoire de la milice française et des changements qui s'y font faits depuis l'établissement de la monarchie dans les Gaules jusqu'à la fin du règne de Louis le Grand.* Vol. 2. Paris: Denis Mariette, 1721.

Elsie, Robert. "Oh, My Beautiful Morea," in the webpage on Albanian Literature. http://www.albanianliterature.net/oralverse/verse_02.html.

Koronaios, Tzanes. "Andragathêmata tou Merkouriou Boua." In *Hellênika anekdota perisynachthenta kai ekdidomena kat' enkrisin tēs bulēs ethnikē dapanē*, edited by Kônstantinos Sathas (Vol. 1), 1–153. Athens: Typois tou Phôtos, 1867.

Lamanskij, Vladimir Ivanovič, ed. *Secrets d'état de Venise: Documents extraits notices et études, servant a éclaircir les rapports de la seigneurie avec les grecs les slaves et la Porte ottomane à la fin du XVe et au XVIe siècle.* St. Petersburgh: L'acad. imper. des sciences, 1884.

Legrand, Émile, ed. *Deux vies de Jacques Basilicos: Seigneur de Samos, Marquis de Paros, Comte Palatin et Prince de Moldavie, l'une par Jean Sommer, l'autre par A.-M. Graziani, suivies de pièces rares et inédites.* Paris: Maisonneuve, 1889.

Lehasca, Attanasio. *Cenno storico dei servigi militari prestati nel Regno delle Due Sicilie dai Greci, Epiroti, Albanesi e Macedoni in epoche diverse.* Corfu, 1843.

Ljubić, Simeon, ed. *Monumenta spectantia historiam slavorum meridionalium.* Vol. 8. Zagreb: Officina Societatis Typographicae, 1877.

Lunzi, Ermannos. *Peri tis politikes katastaseos tis Heptanesou epi Eneton.* Athens: Philadelphis, 1856; reprinted 1969.

Moustoxydês, Andreas. "Nikandros Noukios." *Pandôra* 7, no. 154 (1856): 222.

Mutinelli, Fabio. *Memorie storiche degli ultimi cinquant' anni della Repubblica Veneta, tratte da scritti e monumenti contemporanei.* Venice: G. Grimaldo, 1854.

Sanudo, Marino. *Commentarii della guerra di Ferrara tra gli Viniziani ed il duca Ercole d'Este nel 1482, per la prima volta pubblicati.* Venice: Picotti, 1829.

Sanudo, Marino. *La spedizione di Carlo VIII in Italia*, edited by Rinaldo Fulin. Venice: Tip. del commercio di M. Visentini, 1883.

Sathas, Kônstantinos N., ed. *Mnemeia Hellenikes Historias: Documents inédits relatifs à l'histoire de la Grèce au moyen âge.* Vol. 1. Paris: Maissonneuve, 1880.

Sathas, Kônstantinos N., ed. *Mnemeia Hellenikes Historias: Documents inédits relatifs à l'histoire de la Grèce au moyen âge.* Vol. 4. Paris: Maissonneuve, 1883.

Sathas, Kônstantinos N., ed. *Mnemeia Hellenikes Historias: Documents inédits relatifs à l'histoire de la Grèce au moyen âge.* Vol. 7. Paris: Maissonneuve, 1888.

Stamatidiades, Epaminondas I. *Bios Iakobou Basilikou, Despotou Samou, Markettíou Parou, Kometos Palatinou kai Egemonos tes Moldavias.* Samos: Egemonikon Typographeion, 1894.

Veloudos, Iôannês. *Hellênôn Orthodoxôn apoikiaen Venetia: Historikôn hypomnêma*, 2nd ed. Venice: Phoinix, 1893, reprinted ca. 1982.

Vrokinês, Laurentios S. *He peri ta mesa tou 16. aiônos en Kerkyrai apoikêsis tôn Naupleiôn kai Monemvaseiôn*. Corfu: Typografeion A. Lantza, 1905, also: *Erga*, edited by Kostas Daphnês. Vol. 16–17. Corfu: Kerkyraika Chronika, 1972.

Wright, Diana, and John Melville-Jones, eds. *The Greek Correspondence of Bartolomeo Minio*. Vol. 1: *Dispacci from Nauplion, 1479–1483*. Padua: Unipress, 2008.

Zoto, Vladimir, ed. *Balada shqiptare*. Tirana: Dasara, 2006.

Studies

Anhegger, Robert. "Martoloslar hakkinda." *Türkiyat Mecmuasi* 7–8 (1942): 282–320.

Babinger, Franz. "Albanische Stradioten im Dienste Venedigs im ausgehenden Mittelalter." *Studia Albanica* 1 (1964): 95–105.

Bartusis, Mark C. *The Late Byzantine Army: Arms and Society, 1204–1453*. Philadelphia, PA: University of Pennsylvania Press, 1997.

Bartusis, Mark C. *Land and Privilege in Byzantium: The Institution of Pronoia*. Cambridge: Cambridge University Press, 2012.

Birtachas, Stathis. "La memoria degli stradioti nella letteratura italiana del tardo Rinascimento." In *Tempo, spazio e memoria nella letteratura italiana: Omaggio ad Antonio Tabucchi/Χρόνος, τόπος και μνήμη στην ιταλική λογοτεχνία: Τιμή στον Antonio Tabucchi*, edited by Zosi Zografidou, 123–141. Rome: Aracne; Thessaloniki: University Studio Press, 2012.

Birtachas, Stathis. "Stradioti, Cappelletti, Compagnie or Milizie Greche: 'Greek' Mounted and Foot Mercenary Companies in the Venetian State (Fifteenth to Eighteenth Centuries)." In *A Military History of the Mediterranean Sea Aspects of War, Diplomacy, and Military Elites*, edited by Georgios Theodokis and Aysel Yildiz, 325–346. Leiden: Brill, 2018.

Braudel, Fernand. *The Mediterranean and the Mediterranean World in the Age of Philip II*, translated by Sian Reynolds. Vol. 1. New York: Harper and Row, 1975.

Bridge, John S. C. *A History of France from the Death of Louis XI*. Vol. 2. Oxford: Clarendon Press, 1924.

Cheetham, Nicholas: *Medieval Greece*. New Haven, CT: Yale University Press, 1981.

Cvetkova, Bistra. "Novyedannye o Khristianskikh-spakhiakh na Balkanskom poluostrove v. period turetskogo gospodstva." *Vizantiiskii Vremennik* 13 (1958): 184–197.

Djurdjev, Branislav. "O vojnucima sao svrtom na razvoj turskog feudalizma i na pitanije turskog agaluka." *Glasnik Zemaljskog Muzeja u Sarajevu* 2 (1947): 75–138.

Djurdjev, Branislav. "Hriscéani spahije u Sevemoj Srbiji u XV veku." *Godišnjak istoriskog društva Bosne i Hercegovine* 4 (1952): 165–169.

Durham, Mary Edith. "Head-Hunting in the Balkans." *Man* 23 (February 1923): 19–21.

Fedalto, Giorgio. "Le minoranze straniere a Venezia tra politica e legislazione." In *Venezia: Centro di mediazione tra Oriente e Occidente (secoli XV–XVII); Aspetti e problemi, Atti del II Convegno internazionale di storia della civiltà veneziana promosso e organizzato dalla Fondazione Giorgio Cini, dal Centro Tedesco di Studi Veneziani, dall'Istituto Ellenico di Studi Bizantini e Post-Bizantini (Venezia, 3–6 ottobre 1973)*, edited by Hans-Georg Beck, Manoussos I. Manoussakas, and Agostino Pertusi, 143–162. Vol. 1. Florence: Olschki, 1977.

Fieffé, Eugène. *Histoire des troupes étrangères au service de France depuis leur origine jusqu'à nos jours*. 3 Vols. Paris: Librairie militaire, 1854.

Fracchia, Joseph, "Hora: Social Conflicts and Collective Memories in Piana Degli Albanesi." *Past and Present*, no. 209 (November 2010): 181–218.

Giura, Vincenzo. "La Comunita Greca di Napoli (1534–1861)." *Clio* 4 (1982): 119–156, also *Storie di minoranze: Ebrei, greci, albanesi nel regno di Napoli*. Naples: Edizioni scientifiche italiane, 1984.

Hain, Kathryn. "Devshirme Is a Contested Practice." *Historia: The Alpha Rho Papers* 2 (2012): 165–176.

Hammond, Nicholas G. L. *Migrations and Invasions in Greece and Adjacent Areas*. Park Ridge, NJ: Noyes Press, 1976.

Hassiotis, Ioannis K. "La comunità greca di Napoli e i moti insurrezionali nella penisola Balcanica meridionale durante la seconda metà del XVI secolo." *Balkan Studies* 10 (1969): 279–288.

Hassiotis, Ioannis K. "Me to Hispanous kai tous Apsbourgous." *Hellenes Misthophoroi ston Kosmo. Kathemerine Aphieroma 2–32* (Athens, 26 January 2003): 12–14.

Inalcik, Halil. "Timariotes chrétiens en Albanie au XVe siècle d'après un registre de timars ottoman." *Mitteilungen des österreichischen Staatsarchivs* 4 (1951): 118–138.

Inalcik, Halil. "Ottoman Methods of Conquest." *Studia Islamica* 2 (1954): 103–129.

Kallinikos, P. "O Nikandros Noukios ste Bretannia." *Deltion Anagnôstikes Hetairieas Kerkyras* 7 (1970): 35–42.

Kargakos, Sarantos I. *Alvanoi, Arvanites, Hellênes: Meletes*, 3rd ed. Athens: I. Sideres, 2008.

Kašić, Dušan Lj. "Die griechisch-serbische Kirchensymbiose in Norddalmatien vom XV. bis zum XIX. Jahrhundert." *Balkan Studies* 15 (1974): 21–48.

Korrê, Katerina. "Misthophorikê hypêresía, éngeios, idioktêsía, klêronomikê diadochê: Hoi stradioti-kês ktêseis tou Ioníou (16os aiônas)." In *Praktika tou Panioniou Synedriou: Paxoi, 26–30 Maiou 2010*, edited by Alike Nikephorou, 197–226. Vol. 1. Paxos: Etaireia Paxinôn Meletôn, 2014.

Maksimović, Ljubomir. *The Byzantine Provincial Administration under the Palaiologoi*. Amsterdam: Hakkert, 1988.

Mallet, Michael E., and John R. Hale. *The Military Organization of a Renaissance State: Venice c. 1400 to 1617*. Cambridge: Cambridge University Press, 1984.

Millar, Gilbert J. "The Albanians: Sixteenth-Century Mercenaries." *History Today* 26 (1976): 468–472.

Millar, Gilbert J. *Tudor Mercenaries and Auxiliaries 1485–1547*. Charlottesville: University Press of Virginia, 1980.

Miller, William. *The Latins in the Levant: A History of Frankish Greece (1204–1566)*. New York: Barnes and Noble, 1937.

Mpirês, Kôstas. *Hoi Arvanites, hoi Dôrieis tou neoterou Hellênismou: Historia ton ellênôn Arvanitôn*. Athens: Melissa, 1960.

Nicolle, David. *Hungary and the Fall of Eastern Europe, 1000–1568*. London: Osprey, 1988.

Orhonlu, Cengiz. *Osmanli imperatorlugunda derbend teşkilati*. Istanbul: Istanbul Üniversitesi Edebiyat Fakültesi, 1967.

Pappas, Nicholas C. "Balkan Foreign Legions in Eighteenth Century Italy: Reggimento Real Macedone and Its Successors." In *Nation and Ideology: Essays in Honor of Wayne S. Vucinich*, edited by Ivo Banac, John C. Ackerman, and Roman Szporluk, 35–39. Boulder, CO: East European Quarterly xi; New York: Columbia University Press, 1981.

Pappas, Nicholas C. *Greeks in Russian Military Service in the Late Eighteenth and Early Nineteenth Centuries*. Thessaloniki: Institute for Balkan Studies, 1991.

Perjes, Géza. *The Fall of the Medieval Kingdom of Hungary: Mohács 1526–Buda 1541*, translated by Marió D. Fenyö. Boulder, CO: Social Science Monographs; Highland Lakes, NJ: Atlantic Research and Publications, 1989.

Petta, Paolo. *Stradioti: Soldati albanesi in Italia, sec. XV–XIX*. Lecce: Argo, 1996.

Ploumides, Georgios. "Nikolaos (†1528) kai Iôannês (†1529) Palaiologoi Stradioti." *Dôdônê* 24, no. 1 (1995): 227–234.

Sathas, Kônstantinos N. *Hellênikoi Stratiôtai en te Dysei kai anagennesis tes Hellenikes taktikes*, Athens: Periodiko Hestia, 1885. Reprint: Neoellenike historia. Ekdoseis Philomythos, 1993.

Setton, Kenneth M. *The Papacy in the Levant (1204–1571)*. Vol. 2, *The Fifteenth Century*. Philadelphia, PA: Philadelphia American Philosophical Society 1997.

Taylor, Frederick L. *The Art of War in Italy, 1494–1529*. Cambridge: Cambridge University Press, 1921.

Topping, Peter W. "Albanian Settlements in Medieval Greece: Some Venetian Testimonies." In *Charanis Studies: Essays in Honor of Peter Charanis*, edited by Angeliki E. Laiou-Thomadakis, 261–272. New Brunswick, NJ: Rutgers University Press, 1980.

Uyar, Mesut, and Edward J. Erickson. *A Military History of the Ottomans: From Osman to Atatürk*. Santa Barbara, CA: Praeger Security International, 2009.

Vakalopoulos, Apostolos E. *Historia tou Neou Hellenismou*. Vol. 3. Tourkokratia, 1453–1669: hē agōnes gia tēn pistē kai tēn eleutheria. Thessalonike: Sphakianakes, 1968.

Vasdravellês, Iôannês. *Armatoloi kai klephtes eis tên Makedonia*, 2nd ed. Thessalonike: Hetaireia Makedonikon Spoudon, 1970.

Vasić, Milan. *Martolosi u Jugoslovenskim zemljama pod turskom vladavinom: Djela 29/ Odjeljenje istorijsko-filoloshkih nauka 17*. Sarajevo: Akademija nauka i umjetnosti Bosne i Hercegovine, 1967.

Vucinich, Wayne S. "The Serbian Military Tradition." In *War and Society in East Central Europe*, edited by Béla Kiraly and Gunther E. Rothenberg, 285–324. Vol. 1. New York: Brooklyn College Press, 1979.

Wright, Diana. "Bartolomeo Minio: Venetian Administration in 15th Century Nauplion." PhD diss., Catholic University of America, 1999.

Zakythinos, Dionysios A. *Le despotat grec de Morée*. 2 Vols. (Vol. 1 *Histoire politique*, Vol. 2, *Vie et institutions*). Athens: L'hellénisme contemporain, 1953. Reprint London: Variorum, 1975.

Zoes, Leonidas. "Hellenikos lochos en Zakynthoi kata tous chronous tes douleias." *Ho Hellenismos* 14 (1911): 367–371.

11

MILITARY AUXILIARIES IN TWELFTH- AND THIRTEENTH-CENTURY HUNGARY

Nomads vs. Crusader Knights

László Veszprémy

Auxiliary light and heavy cavalry played a much more significant role in the medieval Kingdom of Hungary, than in other Central European kingdoms like Bohemia or Poland.[1] Hungary was one the largest kingdoms in Europe (roughly 300,000 square kilometres), but rather sparsely populated with large uninhabited marches along its borders. In the thirteenth century, for example, only around two thirds of the kingdom was populated, with an estimated population of 1.5–2 million.

On the eastern border of Hungary nomadic people occasionally appeared either as enemies or potential settlers, and the Hungarian rulers regularly recruited these newcomers as auxiliaries and/or royal bodyguards. After their temporary or final settlement they depended in every respect on the king, and their unconditional obedience to the royal court was of great political benefit, not to speak of their military skills and ability to mobilize quickly even at the price of social conflicts within society.

Being regularly called to action, these nomadic, or semi-nomadic, people had a level of combat experience that largely set them apart from peoples with a more settled way of life. Almost all of them, and especially the Pechenegs, Muslim archers, Cumans and Vlachs, utilized light cavalry tactics. Though the native Hungarians had originally used similar nomadic battle tactics, they had lost these skills when they adopted a more sedentary way of life during the eleventh and twelfth centuries. One reason for this was the emergence of the military orders of the Teutonic Knights and the Hospitallers in Hungary in the thirteenth century, with their western heavy cavalry tactics and armament. These orders made a positive impression on the royal court, whose nobles began to adopt their chivalric ideas and styles of warfare. However, at times this led to conflicts, such as with the Teutonic Knights, who were eventually expelled from Hungary because their unrestrained aspirations to independence and loyalty to

DOI: 10.4324/9781003245568-12

the papacy. In contrast, as nomadic people had no real homeland, their loyalty was not burdened and threatened by foreign relations. For example, the Székelys, a well-organized community of light cavalry warriors living off cattle breeding—whose origin is still debated in Hungarian scholarship—were dispersed widely throughout the kingdom up to the eleventh century, and only settled in Transylvania in the twelfth, having preserved their former privileges and duties (and also their special Hungarian dialect).[2]

The settlement of these peoples in regions of strategic significance and their rendering of military service was a feature of Hungarian rulership from the early Middle Ages. The military deployment of Pechenegs and Székelys in the eleventh century may be inferred from toponyms as in later periods they are alluded to in narrative and diplomatic sources. References in the Hungarian national chronicle, although difficult to date,[3] state quite clearly that auxiliary peoples formed an important part of Hungarian military organization and were involved in border defence. This was true for the Pechenegs, Székelys, and later for the Muslim archers,[4] Vlachs (Romanians),[5] and German settlers.[6] Their presence can be compared to that of the Teutonic Knights and the Hospitallers, who were also settled on the frontier,[7] although in smaller numbers. The first to be settled in the central area of the country, without direct border defence duties, were the Cumans in the middle of the thirteenth century.[8] The Pechenegs were settled at scattered points—at some 30–40 locations—through the country, so that their assimilation—also linguistically—was well advanced by the thirteenth century, while the Székelys, Cumans and Germans formed true diasporas: They had territorial privileges and to a large extent maintained their autonomy in their settlement areas until the administrative reforms of 1878.[9]

The Contradictory Message of the Narrative Sources: "Vilissimi et pessimi"

Nomadic auxiliary peoples first appear in narrative sources in the twelfth and thirteenth centuries. Given the military function of their light cavalry units, we might expect appreciative reports of them.[10] They regularly did what was expected of such units anywhere: Using composite bows, javelins, sabres and maces, and by keeping a large number of horses in reserve, they acted as the vanguard and rearguard of the army, launched surprise attacks as cavalry archers, and carried out feigned retreats and ambushes.

Surprisingly the surviving Hungarian narrative sources recount the supposed "cowardice" of these units and their flight during battle. The Hungarian national chronicle, the anonymous chronicle of c.1200 and the Illuminated Chronicle of 1358, for example, all share this negative judgement. Passages from the lost Hungarian national chronicle surviving in the Illuminated Chronicle of 1358 (chapters 153[11] and 165[12]) include disparaging remarks, with almost identical wording, about how the Pechenegs and the Székelys conducted themselves in battle. In one passage in chapter 165 a battle between King Géza II and the

Bavarian army of Henry Jasomirgott, Margrave of Austria, in 1146, is recounted in which the Pechenegs and the Székelys are described as riding ahead of the army to unsettle the enemy, but then fleeing from the battle before its end. Interestingly this can be compared to an account of the same battle by Otto of Freising, who reports that the mounted archers—presumable the same Pechenegs and Székelys—went ahead of the Hungarian army and, led by their two *ispans*, attacked, but were repulsed by the Germans. Here the author makes no disparaging remarks about the cowardice of these auxiliary units.[13]

The above mentioned Anonymous chronicler's opinion of the light cavalry auxiliary forces was no more complimentary. His remarks on the use of the bow by peoples other than the Hungarians and the ancient Hungarians (Scythians) need some further explanation. In chapter 25, he refers to the Vlachs and the Slavs as *viliores homines*. "The inhabitants of that land were the basest of the whole world, because they were Vlachs and Slavs, and because they had nothing else for arms than bows and arrows".[14] This view, however, is only superficially based on the difference in weaponry, and refers really at least as much to tactics. Indeed, this is not the only point where the author employs a kind of archaism: In the summary of the first chapter, we hear of the Scythians' unmatched expertise with the bow and arrow, but in chapter 46, he refers to the bow with contempt (*more paganismo*). By around 1200, the Hungarians, especially the elite contingents, had adopted what were, at least in the chronicler's opinion, the far superior Western arms and customs, such as the tournament. This text thus compares the modern tactics of around 1200 with older ones, suggesting that the prevailing attitude was contemptuous of light cavalry tactics.[15]

It might be thought that this criticism stemmed from a practical observation of how the combat value of the long-settled Pecheneg and Székely military forces had diminished, and how they compared unfavourable to the more structured "Western" discipline of the regular forces of the kingdom. The words of the chroniclers may allude to a deep and, more importantly, protracted crisis. Historical events, however, refute this: The second half of the twelfth century was when the Székelys with their border defence duties were relocated from western Hungary to Transylvania, and we hear of their deployment abroad in the decades following 1200: A charter of 1228, for example, reports—without any damning remarks—that the leader of the Székelys had been taken captive when marching against the Bulgarians in that year.[16]

Here we should mention the Bohemian-Hungarian clash on the River Olšava in 1116, where chapter 153 of the Hungarian national chronicle again highlights the "cowardice" of the Székelys and the Pechenegs. This chapter recounts a full and eventful story which ends in Hungarian victory. It contrasts starkly with the text of the Bohemian chronicler Cosmas (III. 42), which attributes victory in the same battle to the Bohemians.[17] Cosmas writes with words of appreciation of legions of Hungarian *hospites*, a word which may mean Western mercenaries, or rather "foreign guests" in the Hungarian usage, but more likely corresponds to the lightly armed auxiliary forces of the Hungarian army. If Cosmas was right in paying tribute to the three Hungarian vanguard columns and appreciating their mastery in crossing the river, then the tendentious account of the Hungarian national chronicle probably reflects the widespread courtly contempt for the light cavalry of the early thirteenth century, rather than attitudes at the time of the battle. Anyway, we have very few reports of Western mercenaries in the Hungarian army, and they are never called *hospites*. The mention of "Latins" in King Solomon's army in chapter 121 of the Hungarian national chronicle may fall into this category.[18] In connection with the Battle of Zemun (Zimony, Semlin,) of 1167, which ended in a Byzantine victory over the Hungarians, the Byzantine historian Niketas Choniates notes that there were many Western mercenaries, including Germans in the Hungarian army.

Returning to the disparaging comments on the light cavalry in the Hungarian national chronicle, as has been shown, they were unlikely to have been based on contemporary accounts from the eleventh or twelfth centuries. It is more plausible that they were inserted retrospectively into the text in the early thirteenth century. Paradoxically, this may have been prompted by the very rise of the light cavalry at the time when the Byzantine Empire was weakening and in crisis.[19] The Bulgarian Tsarate, revived with Vlach support, raised armies in the northern Balkans, and thus blocked the Hungarian expansion towards the Balkans and threatened the borders of the Kingdom.[20] Therefore more auxiliary troops and people were needed to defend the border regions (see also Figure 11.1) and as a result the respective light cavalry forces dominated the Hungarian military

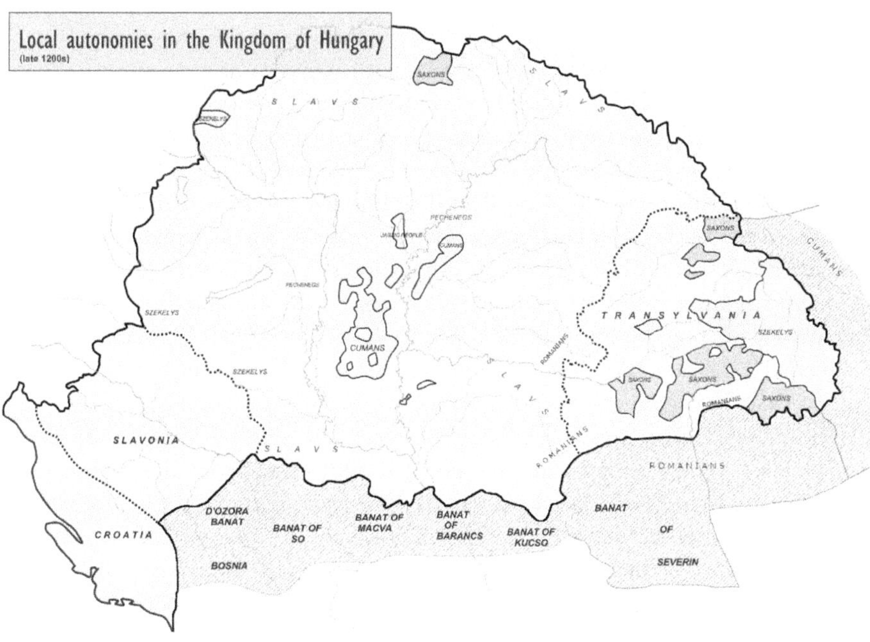

FIGURE 11.1 Local autonomies in the Kingdom of Hungary (late 1200s), Wikipedia Commons (~riley/Ceha/Mhare), accessed 21.07.2022, https://hu.m.wikipedia.org/wiki/F%C3%A1jl:Hungary_13th_cent.png.

organization, at the expense—and to the injury—of the traditional county forces, clearly leading to domestic political tensions.[21]

This argument may be reinforced by Byzantine sources. Choniates also has a contemptuous and condemnatory view of the Cumans. He describes how the Cumans quickly crossed the Danube in 1152, took enormous plunder and then quickly disappeared "as is their habit".[22] The complaint of the Byzantine author and the remarks of the Hungarian chroniclers have a common source: Court authors' antipathy to rapidly moving horsemen plundering at will. The auxiliary peoples certainly appear to have taken a broad interpretation of their right to plunder, and if they could, looted whatever they found.

Despite these objections from the chroniclers, the Byzantine and Hungarian courts had a great need for the military assistance the auxiliary peoples could provide, as evidenced by constant efforts to win them over, and the successful deployment of them together with regular troops. After the employment of the Teutonic and Hospitaller orders in Hungary ended in spectacular political failure, despite their military effectiveness (see below), recourse to the nomadic warriors became all the more inevitable. In the case of the Cumans, it is also true that the problem of their paganism and delayed Christianization generated serious domestic and foreign tensions in the thirteenth century, particularly between the Hungarians and the papacy. Nevertheless, chapter 159 of the Hungarian national chronicle indicates that King Stephen II (r. 1116–1131) had a retinue—probably a bodyguard—consisting solely of Cumans.[23] The description places the appearance of the Cumans at such an early date that Hungarian historians usually consider the reference to concern Pechenegs.[24] Given the custom of employing a foreign ethnic group as royal guards, however, the possibility should not be rejected outright.[25] Indeed, a contingent of Cumans (*Falwen*) raised by the Bohemians to assist the German King Lothair III in 1132 may have actually come from Hungary, as has been proposed by Schünemann.[26] The family relations supports this theory as King Béla II of Hungary was brother-in-law of the Bohemian ally of the German emperor, Duke Soběslav I (r. 1125–1140), who had married Béla's sister Adelhaid in 1125. According to the Byzantine historian John Kinnamos, Cumans fought alongside Russians in the Hungarian army against Byzantium during the reign of Stephen III (r. 1162–1172).[27] This information does not prove that the Cumans permanently settled in Hungary, but it does suggest that they were at least being employed for military use.

The view of Western chroniclers, who were also unflattering in their comments about lightly armed auxiliary peoples, adds complexity to our image of the Cumans. Arnold of Lübeck, in the Chronica Slavorum of 1208 also characterizes the Cuman auxiliary troops fighting beside the Hungarians in German lands as *pessimi* (worthless).[28] This is not a random remark, because he also writes in highly condemnatory terms of their cruelty in 1203, when referring to a Cuman contingent sent by the king of Bohemia. This probably originated from Hungary, most likely from the retinue of Constance, sister of King Emeric of Hungary, who had recently married the Bohemian King Ottokar I.[29] Arnold of Lübeck's remarks may prove the Hungarian kings' occasional use of Cuman military services for a time around 1200. For the Byzantines—similar to the Hungarians—, as we have seen, the Cumans were by turns enemies and paid mercenaries. The *Gesta Hungarorum* also mentions the Cumans, describing them as a people who "cause damage",[30] in full agreement with the widespread contemporary view. It also notes that the Hungarians knew all about the Cumans' military virtues, strengths and usefulness several decades before the Mongol invasion, and that a prominent motive for inviting the Teutonic Knights to Hungary, besides to pacify the area, was to counterbalance the strength of the Cumans (see below). Cuman-Hungarian relations were rather complex and contradictory: Until the 1220's different groups of Cumans were allies, mercenaries or foes almost simultaneously. From the 1220's the overwhelming majority of the Cumans living in the Hungarian frontier zone—as a result of Mongol pressure—accepted Hungarian dominance, and finally settled in the kingdom after the Mongol invasion of the 1240s.

Rarely Mentioned Auxiliary People from the Twelfth Century: The Muslims

Muslim archers first appeared in Hungary in the twelfth century and presumably remained until the Mongol invasion of 1241. Little is known about them,[31] and only a few—reasonably credible—Byzantine and Islamic sources attest to their existence.[32] Abu Hamid knew of them in the middle of the twelfth century, and indeed claimed that he had been specifically commissioned by King Géza II (r. 1141–1162) to recruit Muslim archers. Abu Hamid noted that they did not all come from the same place and were no longer fully familiar with Arabic and the Islamic faith; however they had the protection and support of the king.[33] In the middle of the twelfth century, King Géza II sent Hungarians, Pechenegs and Muslims to join a Serbian army which waged an unsuccessful campaign against the Byzantines. Archaeologists have not yet been able to locate the settlements of these archers in Hungary, but Kinnamos claims that one group settled in Syrmia (Szerémség, today in Croatia, Srijem, and in Serbia, Srem), and also states that some of them were resettled in Byzantine territory by Emperor Manuel Komnenos in 1165.[34] These may be the same people who the thirteenth-century Arab writer Yakut stated lived in 30 villages.[35] It is possible that the Seljuk Turkish population of Syrmia became subjects of the Hungarian king when the area fell under Hungarian control.[36] The German chronicler Rahewin reports that, King Géza II promised around 600 archers to the German emperor for his war against Milan in 1158, and Vincentius Pragensis, an eyewitness of the siege of Milan testifies that the same Hungarian king sent 500 "Saracens" to the emperor.[37] If the numbers are reliable, then the Hungarian Muslims must have been a substantial military force, deepening the mystery as to why we do not hear more of them until the Mongol invasion. Certainly, when they were later mentioned in the Hungarian campaign against the Bohemians in 1260, their supposed presence as Muslim soldiers provided the opportunity for contemporaries to discredit the Hungarian army.[38]

Rarely Mentioned Auxiliary People from the Thirteenth Century: The Vlachs

The Vlachs appear in Hungary in a military capacity from the thirteenth century onwards. The earliest written record of them is a royal charter dated around 1210. This indicates their presence in the forces of the *ispán* of Sibiu (Szeben, Hermannstadt), who marched to support the Bulgarian Tsar together with Saxons, Székelys and Pechenegs.[39] Success in battle is a good indicator of the military value of these auxiliary peoples. This charter evidence is supported by the *Gesta Hungarorum*, written at roughly the same time.[40] We do not know the exact course of their settlement within the kingdom, but the charter which King Andrew II granted to the Transylvanian Germans in 1224 (the so called *Andreanum*) mentions a forest of the Pechenegs and Vlachs which may be located in the area of the Făgăraş Alps.[41] This is consistent with the charter connected to the Hospitallers a few decades later, which obliged the local Vlachs to provide military assistance to the order in case of need (see below).[42]

A very important source is the Nibelungenlied, dated to around 1200. It has much to say about the diversity of warrior ethnic groups in the Kingdom of Hungary. It claims that those who were in Attila's army could have also been in the army of the Hungarian king—the Pechenegs, Vlachs and Rus.[43] Romanian historians have investigated the identity of a certain Ramunc, a prince mentioned by name as leader of the Vlachs, but it appears that he is a fictive character.[44] The charters are silent on the military performance of the Vlachs, but the Styrian Rhymed Chronicle of the early fourteenth century mentions them as among the auxiliary peoples of the Hungarian king in several places. A large number of auxiliaries fought against the Bohemians at the Battle of Kroissenbrunn in 1260, and those mentioned are Russians, Bosnians, Cumans, Székelys and Vlachs (lines 6827 and 7389).[45]

Western-Type Military Diasporas: The Teutonic and Hospitaller Knights

German settlers, here called Saxons, arrived during the large colonization movements of the twelfth and thirteenth centuries in the almost uninhabited areas of present-day northern Slovakia and the southern and eastern marches of Transylvania. The principal reason for their settlement was economic rather than military, although according to their charter of privileges issued in 1224 they had to equip 500 warriors for campaigning within the kingdom and 100 for outside the realm.

The Teutonic Knights, on the other hand, were brought in mainly for military purposes in the campaigns against the Cumans, who were pagan at the time, in the Barcaság (Burzenland, Ţara Bârsei) area around what is now the city of Braşov (Brassó, Kronstadt) in the south of Transylvania.[46] They secured the borders and furthered a process which culminated in the submission of the Cuman and Vlach populations who lived in the frontier zone called Cumania, in the territory of the future Wallachia. The Teutonic Knights also prepared the way for the expansion of the kingdom beyond the border, an ambition thwarted by the Mongol invasion of 1241–1242.

In an extraordinary exciting period between 1211 and 1225, the Teutonic Knights won the right to build "*castra lignea et urbes ligneas*" (wooden castles and cities) in defence of the south-eastern border zone, which resulted in the construction of half a dozen strong castles, some of them even with stone walls. The Knights undoubtedly met with military success, perhaps too much. Their ambitions seem to have put a strain on the secular and ecclesiastical structure of Hungary, and that, combined with the decreasing power of the Cumans after their defeat in the battle of the River Kalka in 1223 by the Mongols, led to the expulsion of the Order in 1225.[47] Their region was settled soon by Germans from the neighbouring marches of Transylvania.

The other lasting outcome of the actions of the Teutonic Knights was the establishment of Hungarian influence over Szörény (Severin), the area between the Danube and the Olt rivers. This was of key importance for the defence of the southern border, although it remained an unstable corner of the kingdom for centuries to come. The name of Buzád, the first ban of Szörény, is known only from 1233, although Attila Zsoldos has proposed that he might have attained this office in 1226.[48] An indication of the area's importance is a charter of 2 July 1247 designed to pacify Szörény by settling the Hospitallers there. Unfortunately, we know nothing about the Order's activities in the region, but according to a contemporary charter the local Vlachs were looked upon as military allies.[49] The Hospitallers were expected to send contingents not only against the pagans, but against all possible enemies of the kingdom, be they Orthodox or Roman Catholic, such as the neighbouring Bulgarians, Greeks or Austrians. Ultimately, the Hospitallers were unable to fulfil their duties, and soon gave up their border settlements and left the kingdom.

The Last Successful Settlement of Nomadic Peoples in Hungary: Cumans and Iasians

When Prince Béla (later King Béla IV) extracted tribute from the Cumans in lands beyond the Carpathians in the 1220's, he was also building on the achievement of the Teutonic Knights.[50] After 1226, the Dominicans set up a missionary diocese centred on Milkó (Milcovul) between the Olt and Siret rivers, and in 1227, the Cuman leader Borc (Boricius) was baptized by the Archbishop of Esztergom and the Hungarian king assumed the title *rex Cumanie*.[51] The Cumans could be foes and allies at the same time. In the year 1230, the Cuman Borc (also called Begovar in the sources) joined the Hungarian army with his troops as an auxiliary, while the Hungarians' enemy in Galicia was also supported by Cuman archers. These archers were led possibly by the same Kuthen (var. Kötöny), who—after being chased by the Mongols in 1239—asked as their king for his people's admittance to Hungary. In the shadow of the impending Mongol invasion they were welcome by King Béla IV, but their first stay in the country was relatively short-lived. The Cumans did not actually participate in the fights of the Hungarians against the invading Mongols. In the spring of 1241, a few weeks before the Mongol invasion, King Kuthen and his entourage were killed by the Hungarians in Pest as a consequence of conflicts between the nomadic and settled ways of life of the newcomers and the natives respectively, and King Kuthen's enraged people left the kingdom for the Balkans.[52]

Cumans were also present in the Mongol armies.[53] After the Mongol invasion, in 1246 King Béla again invited the Cumans to settle in areas of the Great Plain between the Danube and the Theiss rivers; in a region that had become almost uninhabited after the Mongol raids of 1241–1242. The Cuman tribes (a population of ca. 40,000–60,000, with ca. 2,000–4,000 warriors), subsequently settled throughout the Great Hungarian Plain, creating two regions incorporating the name Cumania (Kunság in Hungarian): Greater Cumania (Nagykunság) and Little Cumania (Kiskunság), covering a territory of some 8,000–8,500 square kilometres.

Soon after their final settlement, the Cumans became the crucial light-cavalry component of the royal armies. As a part of the Hungarian army they fought in the Babenberg wars of succession (1246–1260), and participated in ten campaigns against Austria, Moravia, Carinthia and Styria, where they terrifying the Austrian lands with their nomadic mode of war, which involved looting and taking Christian slaves. Indeed, the Austrian chroniclers reporting on the battle of Marchfeld at Dürnkrut in 1278 condemned the Cumans for their cruelty, but did not leave any doubt about their military effectiveness.[54]

The Cumans took part in all the major royal campaigns of the late thirteenth and early fourteenth century, and it is probably their leaders who are depicted on the frontispiece of the aforementioned Illuminated Chronicle. The Chronicle is an outstanding masterpiece of Angevin art in Hungary but also in Central Europe.[55] On the frontispiece King Louis the Great (r. 1442–1482) is seated in the middle,[56] flanked by western-looking heavily armoured knights standing to his right, and eastern-looking lightly armoured soldiers to his left. The meaning of this depiction is still debated, but could it possibly represent the two-sided Hungarian military system? Does it depict westernized Hungarian knights on the right, and Cuman and Székely captains, or members of the Jasian (Hung. Jász) bodyguard, on the left?[57] This would not be surprising as the Austrian chroniclers, like the author of the *Steirische Reimchronik* (Styrian Rhymed Chronicle, regularly stressed the exotic character of the Hungarian army and its multi-ethnic auxiliaries.[58] Perhaps the king of Hungary is pictured in a more sophisticated way as the lord of East and West, with his eastern "Hungarians" to the left, and his western vassals and allies to the right.

King Béla IV tried to secure Cuman loyalty by various means, including intermarriage between the Cumans and the Hungarian royal family. The son of King Béla IV, Stephen. married a Cuman princess, Elizabeth the Cuman.[59] Their son, Ladislaus IV "the Cuman" (r. 1272–1290) showed particular affinity to his mother's ancestry, abandoning Hungarian culture and dress for Cuman culture, dress, and hairstyle, which provoked protests and eventually interdicts from the papacy.[60]

The second half of the thirteenth century was the heyday of the Cumans in Hungary. During the lasting civil wars of the 1260's they supported the minor king, Stephen (minor king 1262–1270, king as Stephen V, r. 1270–1272) against his father. A unique written document survives in a Venetian archive, which is the accounts of a royal money-lender from 1264, a certain Syr Wullamus.[61] It is not by chance that 11 percent of all the royal revenue was expended on the purchase of gifts for the Cuman leaders (mostly textiles) to secure their support.[62] According to the accounts, the Cumans received disproportionately more gifts than the Hungarian dignitaries.

The most remarkable chivalric objects in Hungary survive in Cuman graves. The Cumans were Christianized, but still preserved their heathen burial practices for at least a century following their migration to Hungary in the mid-thirteenth century, placing animal bones, horse skeletons, jewellery, and swords in their graves. The most impressive find remains a gold buckle of a sword belt, with four buttons, found in Kígyóspuszta in the middle of Little Cumania (Kiskunság), in 1816.[63] On the buckle one can see a tournament scene with knights in flat-topped great helms and mail hauberks, with musicians surrounding them. We even have parallels to this belt from two other Cuman graves; all are Western and Byzantine artefacts, and not nomadic products.[64] From these objects we can glean some ideas about the material culture of the Cuman elite and warriors whose heritage disappeared without trace.[65] Most probably these objects were royal presents to Cuman warlords, such as the Western type double-edged sword with a dynastical coat of arms, found in Kunszentmárton.[66] David Nicolle was incorrect when he identified the Kígyóspuszta buckle as a genuine Hungarian product.[67] This belt may be dated to the late thirteenth century, as we have comparable tournament scenes in contemporary French and Northern Italian manuscript illuminations. The heraldic devices on the standards and shields are not identified, but they are certainly not Hungarian. On the four buttons there are short Latin orations to Saints Bartholomew, Marguerite, Jacob, and Stephen the Protomartyr, etched with fourteenth-century letters.

Between 1279 and 1290 Cuman society went through a period of crisis. The king was forced by papal legates and the Hungarian opposition to issue the first and second Cuman laws, which compelled the Cumans to abandon their heathen customs, to accept permanent settlement, to leave illegally occupied lands, and to free Christian slaves who had been captured within the kingdom (although those captured abroad could be kept). At the same time their typical style of beard and hair was outlawed. There were armed clashes between the Hungarians and Cumans in 1280 and 1282. The king tried to convince the Cumans not to leave the country, yet a small group still moved to Wallachia, later in Hungary a battle took place between Cuman rebels and the king's forces in 1282. The crises culminated in the assassination of King Ladislas IV, a Cuman himself through his mother, by disaffected Cuman rebels in 1290. The golden age of the Cumans ended in the 1290s when they lost their influence at the royal court and political decision making. Thereafter, a gradual and peaceful acculturation process started that lasted for the next century. Their military importance also diminished, by 1435 the size of their light cavalry contingent was reduced to 200 (together with the Iasians), though this had increased to 600 by the 1450s. The Cumans initially lived in felt yurts, but as time went by they gradually gave up their nomadic way of life. By the fifteenth century, the Cumans were permanently settled in Hungary, in villages whose structures corresponded to that of the native population. Even as Christians they remained bilingual for a long time, the last person speaking Cuman died in the 1740s. The Cumans came directly under the power of the king of Hungary and the title of *dominus Cumanorum* (judge of the Cumans) had passed to the count palatine, who was the highest official after the king. The Cumans had their own representatives and were exempt from the jurisdiction of county officials.

It is still an open question as to when the Iasians entered the kingdom, mentioned for the first time in 1318, with a much smaller population than the Cumans. Most scholars suggest a common arrival with the Cumans based on a shared eastern nomadic tradition. Recently a later arrival date has been suggested, that connects this military group to the entourage and bodyguard of the first Angevin king in Hungary, Charles at the beginning of the fourteenth century.[68]

Conclusion

In the twelfth and thirteenth centuries, Hungary's armed forces had a constant need for auxiliary troops that represented military diasporas within Hungary. There were many conflicts in which the same ethnic groups were fighting on both the Hungarian side and for their enemies. As with the Byzantines, who also frequently employed foreign auxiliaries, this did not usually cause many problems. The critical remarks of the chroniclers proved to be a temporary view, and even they reflected how the auxiliary peoples were becoming an increasingly significant part of Hungarian military organization in the thirteenth century, providing a permanent framework for combined heavy and light cavalry tactics. In the cases we know of, the Hungarians were able to deploy nomadic auxiliary troops in large numbers, provoking the outrage, or perhaps the envy, of contemporaries. Whereas it was the tactics of the light cavalry which caused misgivings in Hungary, it was the brutality of the troops which disturbed observers in the West.

All this happened at a time of unequalled expansiveness in Hungarian foreign policy,[69] when campaigns were undertaken almost every year. The principal targets of the campaigns were Galicia to the north-west, the Babenberg inheritance lands to the west, Cumania to the south-west, and the formation of new provinces, called *banates* in Szörény (Severin) and Macsó (Mačva), plus campaigns against Serbian and Bulgarian territories in the south. The nomadic auxiliaries were deployed on every front, even in Austria (specifically Styria), where their breach of Western norms of conduct provoked indignation.

The Hungarian court tried out and eventually rejected another option, the Western-type military diaspora represented by the Teutonic Knights and Hospitallers. The Cumans, who settled in the thirteenth century, were the people who best lived up to the expectations placed on military auxiliaries. Many charters and narrative sources tell of their military deeds inside and outside the kingdom. The military success of the light-cavalry peoples, especially the Muslims and Cumans, came at a price, blighting the reputation of Hungary as a kingdom which had adopted Western and Christian norms, and prolonging—at least in Central Europe—the horrific vision of merciless pagan "Hungarian" horsemen.

Notes

1 Hansgerd Göckenjan, *Hilfsvölker und Grenzwächter im mittelalterlichen Ungarn* (Wiesbaden: Steiner, 1971); György Györffy, *A magyarság keleti elemei* [Eastern Groups in Medieval Hungary] (Budapest: Gondolat, 1990); Attila Bárány, "Auxiliary peoples," in *Medieval Warfare and Military Technology: An Encyclopedia*, ed. Clifford J. Rogers. (Oxford: Oxford University Press, 2010), 1:98–100; Nora Berend, Przemysław Urbańczyk, and Przemysław Wiszewski, *Central Europe in the High Middle Ages: Bohemia, Hungary and Poland, c. 900–c. 1300* (Cambridge: Cambridge University Press, 2013), 252–258. In Poland, exceptionally once were mentioned "Saracens" as allies of Wladyslaw II during the siege of Poznan in 1146 who may be identified as Prussians, or rather Cumans as Gładysz suggests. Mikolaj Gładysz, *The Forgotten Crusaders: Poland and the Crusader Movement in the Twelfth and Thirteenth Centuries* (Leiden: Brill 2012), 59, 95.

2 Göckenjan, *Hilfsvölker*, 96–125. The Székely ispán (comes) was mentioned in the Hungarian sources for the first time in 1228. Attila Zsoldos, *Magyarország világi archontológiája 1000–1301* [Secular Archontology of Hungary, 1000–1301] (Budapest: MTA TTI, 2011), 239; Imre Szentpétery and Iván Borsa, eds., *Regesta regum stirpis Arpadianae critico-diplomatica* (Budapest: MTA, Akadémiai, 1923–1987), 1:277–281 (no. 926); Zsigmond Jakó, ed., *Erdélyi Okmánytár*, vol. 1: *1023–1300* (Budapest: Akadémiai, 1997), 133 (no. 37); Victor Spinei, *Moldavia in the 11th–14th Centuries* (Bucharest: Editura Academiei, 1986), 97.

3 The beginnings of Hungarian chronicle writing can be traced back to the second half of the eleventh century, but unfortunately, only manuscripts from the fourteenth century have survived, among them the lavishly decorated Illuminated Chronicle of 1358. This fourteenth-century redaction became the basis of all later chronicles, which is why we refer to it as the Hungarian national chronicle. The oldest surviving chronicle was written by an anonymous notary, also called Master P, in c. 1200, referred to in Hungarian scholarly literature as the Anonymous chronicler. We refer to his work as the *Gesta Hungarorum*.

4 Göckenjan, *Hilfsvölker*, 82–89.

5 László Makkai, "Transylvania in the Medieval Hungarian Kingdom, 896–1526," in *History of Transylvania*, ed. id. and András Mócsy (Boulder, CO: Social Science Monographs, 2001), 333–587, at 440.

6 Harald Zimmermann, *Siebenbürgen und seine Hospites Theutonici* (Vienna: Böhlau, 1996); Harald Zimmermann, *Der Deutsche Orden im Burzenland: Eine diplomatische Untersuchung* (Vienna: Böhlau, 2000).
7 Zimmermann, *Siebenbürgen*, 187–225; Zimmermann, *Der Deutsche Orden*, 112–158; Konrad Gündisch, *Generalprobe Burzenland: Neue Forschungen zur Geschichte des Deutschen Ordens in Siebenbürgen und im Banat* (Vienna: Böhlau, 2013); László Pósán, *Hungary and the Teutonic Order in the Middle Ages* (Budapest: MTA Bölcsészettudományi Kutatóközpont, 2021), 27–108; Şerban Papacostea, *Between the Crusade and the Mongol Empire: The Romanians in the 13th Century* (Cluj-Napoca: Centre for Transylvanian Studies, 1998), 40–46.
8 Ádám Pálóczi-Horváth, *Pechenegs, Cumans, Iasians: Steppe Peoples in Medieval Hungary* (Budapest: Corvina, 1989), 54–85; Nora Berend, *At the Gate of Christendom: Jews, Muslims and "Pagans" in Medieval Hungary, c. 1000–c. 1300* (Cambridge: Cambridge University Press, 2001), 140–147; recently id., "Immigrants and Locals in Medieval Hungary: 11th–13th Centuries," in *The Expansion of Central Europe in the Middle Ages*, ed. Nora Berend (Burlington, VT: Ashgate Variorum, 2013), 310–313.
9 In 1876 the district of the Székelys extended over an area of ca. 12,000 square kilometres, the district of the Cumans 8,000–8,500, that of the Isaians ca. 1,200 square kilometres, and those of the Germans in Transylvania extended over ca. 11,000 square kilometres.
10 For the advantages of the light cavalry force, see Russell Mitchell, "Light Cavalry, Heavy Cavalry, Horse Archers, Oh My! What Abstract Definitions Don't Tell Us about 1205 Adrianople," *Journal of Medieval Military History* 6 (2008): 95–118.
11 "Bisseni atque Syculi vilissimi usque ad castrum regis absque vulnere fugierunt" / "the worthless Pechenegs and Székely fled unharmed to the king's camp." Imre Szentpétery, ed., *Scriptores rerum Hungaricarum tempore ducum regumque stirpis Arpadianae gestarum* (henceforth SRH) (Budapest: Egyetemi, 1937), 1:436 (chap. 153 for the year 1116). A new English translation: János M. Bak and László Veszprémy, eds. and trans., *Chronica de gestis Hungarorum e codice picto saec. xiv. Chronicle of the Deeds of the Hungarians from the Fourteenth-Century Illuminated Codex* (Budapest, New York: CEU Press, OSzK, 2018).
12 "Bisseni vero pessimi et Siculi vilissimi omnes pariter fugierunt sicut oves a lupis, qui more solito preibant agmina Hungarorum" / "All the wretched Pechenegs and the worthless Székely, who, as usual, went before the Hungarian army, took to flight like sheep before the wolves" SRH 1:456 (chap. 165 for the year 1146).
13 Györffy, *A magyarság*, 120; Otto of Freising, *Gesta Friderici* 1,33, MGH SSrG 46:52.
14 "… et habitatores terre illius viliores homines essent tocius mundi, quia essent Blasii et Sclavi, quia alia arma non haberent, nisi arcum et sagittas" (chap. 25); Anonymous, Notary of King Béla, *The Deeds of the Hungarians, Master Roger's Epistle to the Sorrowful Lament about the Destruction of Hungary by the Tartars*, ed. János M Bak, Martyn Rady, and László Veszprémy (Budapest, New York: CEU Press, 2010), 60–61; Victor Spinei, *The Romanians and the Turkic Nomads North of the Danube Delta from the Tenth to the Mid-Thirteenth Century: East Central and Eastern Europe in the Middle Ages, 450–1450* (Leiden: Brill, 2009), 90.
15 Gyula Kristó, *A történeti irodalom Magyarországon a kezdetektől 1241-ig* [Historical Literature in Hungary until 1241] (Budapest: Argumentum, 1994), 46–48; see László Veszprémy, "More paganismo: Reflections on Pagan and Christian Past in the *Gesta Hungarorum* (GH) of the Hungarian Anonymous Notary," in *Historical Narratives and Christian Identity on a European Periphery: Early History Writing in Northern, East Central, and Eastern Europe (c. 1070–1200)*, ed. Ildar H. Garipzanov (Turnhout: Brepols, 2011), 183–201.
16 *Regesta regum*, 277–281; *Erdélyi Okmánytár* 1:133 mentioned earlier in a different context in note 2.
17 SRH 1:436–437; Cosmas, *Chronica* 3.42 (MGH SS NS, 2:214–217); Szabados offers a good overview of the battle, though he argues for an indisputable Hungarian victory.

György Szabados, "A 12. századi magyar hadtörténet forrásproblémáiból [Problems of Sources in Twelfth Century Hungarian Military History]," *Aetas* 22, no. 4. (2007): 153–160.

18 SRH 1:391 (chap. 121) can be supported by Kinnamos, who mentions Italian mercenaries in the Byzantine army in 1128. Kinnamos for the year 1128 (1.4) mentions Lombards. John *Kinnamos, Deeds of John and Manuel Comnenus,* trans. Charles M. Brand (New York: Columbia University Press, 1976), 18, for a general overview, see Marko Drašković, "Recruitment Methods in Case of Westeuropean Mercenaries in Byzantium, 1081–1185," *Istorijski Casopis* 61 (2012): 27–44.

19 Papacostea, *Between the Crusade*, 23–40.

20 For the military role of the Cumans at that time see Spinei, *The Romanians*, 144–145.

21 Györffy, *A magyarság*, 13; Szabados, "A 12. századi," 157.

22 Niketas Choniates, *O City of Byzantium: Annals*, trans. Harry J. Magoulias (Detroit: Wayne State University Press, 1984), 54; SRH 1:259. Further examples in the work of Choniates: in 1176 and 1186 Cumans were paid by the Byzantines, ibid., 100, 206. He describes their tactics of the feigned retreat in 1187 and 1205, ibid., 218, 337. After all, Kinnamos isn't hostile towards the Cumans used as mercenaries in the Byzantine army with the task of reconnaissance and border guard, e.g. in the years 1155–6, 1167 (here against the Hungarians), ibid., 112, 203. On the other hand, during the battle of Manzikert they left the field and deserted, see Alexandru Madgearu, "The Pechenegs in the Byzantine Army," in *The Steppe Lands and the World beyond Them, Studies in Honor of Victor Spinei on His 70th Birthday*, ed. Florin Curta and Bogdan-Petru Maleon (Iaşi: Editura Universitatii, 2013), 207–218. Earlier Psellos presented an extremely negative view on the Pechenegs, see *Chronographia* VII, 68, ed. Emile Renauld (Paris: Les Belles Lettres, 1928), 125–126; transl. Edgar R. A. Sewter (New Haven, CT: Yale University Press, 1953), 242; quoted by Florin Curta, "The Image and Archaeology of the Pechenegs," *Banatica* 23 (2013): 143–202, at 149.

23 SRH 1:444–445 (chap. 59).

24 Ferenc Makk, "Megjegyzések II. István történetéhez [Remarks about the history of King's Stephen II]," in *Középkori kútfőink kritikus kérdései*, ed. János Horváth and György Székely (Budapest: Akadémiai, 1974), 253–261, at 253–254. Recently they were identified with a different nomadic group called "Berendei" in the Russian Primary Chronicle, see Aleksey A. Shakhmatov, ed., *Polnoe sobranie russkikh letopisei*, vol. 2: *Ipatievskaia letopis* (Saint Petersburg: Imperatorskaia Arkheograficheskaia kommissia, 1908; repr. Moscow: Nauka, 1962; 1998), 286; Szilvia Kovács, *A kunok története a mongol hódításig* [History of the Cumans until the Mongol Invasion] (Budapest: Balassi, 2014), 189.

25 István Vásáry, *Cumans and Tatars* (Cambridge: Cambridge University Press, 2005), 11. Spinei reluctantly accepts their identity with the Cumans, cf. Victor Spinei, "The Cuman Bishopric—Genesis and Evolution," in *The Other Europe in the Middle Ages: Avars, Bulgars, Khazars, and Cumans; East Central and Eastern Europe in the Middle Ages, 450–1450*, ed. Florin Curta and Roman Kovalev (Leiden: Brill, 2008), 413–456, at 416.

26 Konrad Schünemann, "Ungarische Hilfsvölker in der Literatur des deutschen Mittelalters," *Ungarische Jahrbücher* 4 (1924): 99–115, at 105. A letter from the bishop of Augsburg to Otto of Bamberg: "Rex Christianus [Lothar, 1132] inducit super ecclesiam Christianam homines inhumanos et paganos, Boemos videlicet ac Flavos, qui vulgari nomine Valwen dicuntur, qui persecutores Christi et ecclesiae esse ac fuisse semper manifeste ab omnibus cognoscuntur" ("The Christian King sends upon the Christian church inhuman and pagan men, Bohemians and Flavians [i.e. Cumans], who are known in the vernacular as Valwen. Everybody knows that such peoples are and always have been persecutors of Christ and the church"). Philipp Jaffe, ed., *Monumenta Bambergensia: Udalrici Babenbergensis codex* (Berlin: Weidmann, 1869), 446–447 (no. 260).

27 For the year 1165, see *Kinnamos, Deeds*, 182 (5.15); Spinei, "The Cuman Bishopric," 417.

28 Schünemann, "Ungarische Hilfsvölker," 106; József Deér, "*Aachen und* die *Herrschersitze* der Arpaden," *Mitteilungen des Instituts für Österreichische Geschichtsforschung* 79 (1971): 1–56, at 56; Arnold of Lübeck, *Chronica Slavorum* (7,19): "Et contracto innumero exercitu de omni imperio, ubi aderant innumeri de Ungarorum finibus. contrahens secum auxilia pessimorum, qui Valve dicuntur" MGH SrG 14:281.

29 Schünemann, "Ungarische Hilfsvölker," 106. Arnold of Lübeck, *Chronica Slavorum* (6.5): "Nec defuit ibi illud perditissimum hominum genus, qui Valvae dicuntur, crudelitates suas et nequitias exercentes, de quibus loqui non est edificatio sed miseria" MGH SrG 14:224. The term "pessimus" may have had a meaning "cruel," but it was not emphasized in the Hungarian sources.

30 Anonymous, Notary of King Béla, 61.

31 Katarína Štulrajterová, "Convivenza, Convenienza and Conversion: Islam in Medieval Hungary (1000–1400 ca)," *Journal of Islamic Studies* 24 (2013): 175–198; Berend, *At the Gate of Christendom*, 140–141.

32 Berend, *At the Gate of Christendom*, 140; *Annales Otakariani* MGH SS 9:185.

33 Jean-Charles Ducène, ed., *De Grenade à Bagdad: La relation de voyage d'Abu Hamid al-Gharnati (1080–1168)* (Paris: L'Harmattan, 2006), 93–94; Berend, *At the Gate of Christendom*, 66–67, 238–239.

34 Kinnamos, *Deeds*, 186.

35 Berend, *At the Gate of Christendom*, 67.

36 Attila Katona-Kiss, "A sirmioni hunok: Egy muszlim katonai kötelék a XII. századi magyar királyi erőkben [The Huns of Sirmion: A Muslim Contingent on the Hungarian Royal Army of the Twelfth Century]," in *"Fons, skepsis, lex". Ünnepi tanulmányok a 70 esztendős Makk Ferenc tiszteletére*, ed. Tibor Almási, Éva Révész, and György Szabados (Szeged: JATE, 2010), 159–171.

37 Rahewini, *Gesta Friderici* (3, 26): MGH SSrG 46:198; Vincentii Pragensis, Annales: MGH SS 17:667.

38 Berend, *At the Gate of Christendom*, 242.

39 "associatis sibi Saxonibus, Olacis, Siculis et Bissenis", see note 2. Papacostea, *Between the Crusade*, 47, who dates the campaign between 1211 and 1213; Spinei, *The Romanians*, 145.

40 The military use of the Valchs is mentioned by Anna Komnene for the year 1091, see Vásáry, *Cumans*, 20. King Andreas II's charter of 1222 issued for the Teutonic Order refers to the land of the Székely and Vlachs ("per terram Siculorum et per terram Vlachorum"); Elek Benkő, *A középkori Székelyföld* (Budapest: MTA BTK, 2012), 2:44–45.

41 Vásáry, *The Cumans*, 28–29; Spinei, *The Romanians*, 77–78, 134.

42 *Regesta regum*, 257 (no. 853.); *Erdélyi Okmánytár* 1:191–192 (no. 205).

43 Schünemann, "Ungarische Hilfsvölker," 114; Francis P. Magoun Jr., "Geographical and Ethnic Names in the Nibelungenlied," *Mediaeval Studies* 7 (1945): 85–138, at 129–130; Bálint Hóman, "Magyar történeti elemek a Nibelung énekben [Hungarian Historical Fragments in the Nibelungenlied]," in id., *Történetírás és forráskritika* (Budapest: Magyar Történelmi Társulat, 1938), 293–336; Simon V. Péter, "A Nibelung ének magyar vonatkozásai [Hungarian Connections of the Nibelungenlied]," *Századok* 112 (1978): 271–325.

44 Adolf Armbruster, "Nochmals 'Herzoge Ramunc Uzer Vlachen Lant'," *Revue Roumaine d'Histoire* 12 (1973): 87–100.

45 *Ottokars Österreichische Reimchronik*, ed. Joseph Seemüller. MGH Dt. Chron. V/1: lines 7390–7400 (p. 98); lines 10960–10965 (p. 145); *Annales Ottokariani*. MGH SS 9:184, cited by Adolf Armbruster, *Der Donau-Karpatenraum in den mittel- und westeuropäischen Quellen des 10.–16. Jahrhunderts* (Vienna: Böhlau, 1990), 76–77; id., "Romanii in cronica lui Ottokar de Styria: o nova interpretare," *Studii, Revista de istorie* 25 (1972): 463–483; Berend, *At the Gate of Christendom*, 242; Tünde Radek, *Das Ungarnbild in der deutschsprachigen Historiographie des Mittelalters* (Frankfurt am Main: Peter Lang, 2008), 153–156. The Russians mentioned in the thirteenth-century sources are the

auxiliary troops sent from Galicia (Halytsch) following its conquest. These must be distinguished from the "Varangian" Russians acting as royal guards, which can be traced from the eleventh century onwards.

46 See notes 5 and 7.

47 Zsolt Hunyadi, "The Teutonic Order in Burzenland (1211–1225): New Re-Considerations," in *L'Ordine Teutonico tra Mediterraneo e Baltico: incontri e scontri tra religioni, popoli e culture*, ed. Hubert Houben and Kristjan Toomaspoeg (Galatino: Mario Congedo, 2008), 151–170.

48 Some historians argue for an early foundation of the banate of Szörény (Severin), like Toru Senga, "Béla királyfi bolgár, halicsi és osztrák hadjárataihoz, 1228–1232 [On the Bulgarian, Galician and Austrian Campaigns of Prince Béla]," *Századok* 122 (1988): 36–51, at 44; Zsoldos, *Magyarország*, 49, 291–292.

49 See note 40.

50 Prince Béla became the governor of Transylvania as early as 1226, for the background, see Senga, "Béla királyfi," 44.

51 Spinei, *The Romanians*, 155–156; Nora Berend, "The Mendicant Orders and the Conversion of Pagans in Hungary," in *Alle frontiere della cristianità, I frati mendicanti e l'evangelizzazione tra '200 e '300: Atti del XXVIII Convegno internazionale Assisi, 12–14 ottobre 2000* (Spoleto: Centro italiano di studi sull'alto Medioevo, 2001), 253–279.

52 It is relevant if the Cuman Begovar is identical with Borc, mentioned differently in the sources. Hypathian Chronicle in *Polnoe sobranie ruskikh letopisei,* col. 759–761, though still there are some doubts, see Senga, "Béla királyfi," 45; Spinei, "The Cuman Bishopric," 422–423; Szilvia Kovács, "Bortz, A Cuman Chief in the 13th Century," *Acta Orientalia Academiae Scientiarum Hungariae* 58 (2005:3): 255–266. The duke of Galicia (Halytsch) hired several times Cumans as auxiliaries against the Hungarians, e.g. in the years 1221, 1230, and 1233, see Márta Font, *Völker—Kultur—Beziehungen: zur Entstehung der Regionen in der Mitte des mittelalterlichen Europa* (Hamburg: Kovač, 2013), 151.

53 "But when the king of the Cumans ... began to roam about Hungary, since they had innumerable herds of cattle, [they] caused serious damage to pastures, farm lands, gardens, orchards, vineyards and other property of the Hungarians." Anonymous, Notary of King Béla, Master Roger's, 141 (chap. 3).

54 Andreas Kusternig, "Probleme um die Kämpfe zwischen Rudolf und Ottokar und die Schlacht bei Dürnkrut und Jedenspeigen am 26. August 1278," in *Ottokar-Forschungen,* ed. Max Weltin (Vienna: Verein für Landeskunde für Niederösterreich und Wien, 1979), 226–311, at 253; *Continuatio Vindobonensis*, MGH SS 9:709–711.

55 László Veszprémy, "Chronicon pictum," in *Encyclopedia of the Medieval Chronicle*, ed. Graeme Dunphy (Leiden: Brill, 2010), 1:391.

56 Ernő Marosi, "Das Frontispiz der Ungarischen Bilderchronik (Cod. lat. 404 der Széchényi-Nationalbibliothek in Budapest," *Wiener Jahrbuch für Kunstgeschichte* 46–47 (1993–1994): 364; David Nicolle: *Medieval Warfare Source Book: Warfare in Western Christendom* (London: Weidenfeld, 1995), 175.

57 Pálóczi, *Pechenegs, Cumans, Iasians*, 65–68.

58 Ernő Marosi, "Magyarok középkori ábrázolásai és az orientalizmus a középkori művészetben [Hungarians Depicted in Medieval Art and Orientalism in the Medieval Art]," in *Magyarok Kelet és Nyugat közt: A nemzettudat változó jelképei*, ed. Tamás Hofer (Budapest: Nemzeti Múzeum, Balassi, 1996), 77–97; Radek, *Das Ungarnbild*, 153–156.

59 According to John of Plano Carpini: "At his wedding-feast, ten of the Cumans came together and made an oath according to their custom, with their swords on a dog that had been sundered in two, that they defend the lands of the Magyars as would the king's own supporters against the Tartars and barbarian peoples", quoted by Berend, *At the Gate of Christendom*, 98–99.

60 Berend, *At the Gate of Christendom*, 173–176.

61 Balázs Nagy, "Foreign Trade of Medieval Hungary," in *The Economy of Medieval Hungary*, ed. József Laszlovszky et al. (Leiden: Brill, 2018), 473–490, at 479.

62 Pálóczi, *Pechenegs, Cumans, Iasians*, 69.
63 Zoltán Tóth, "La boucle de Kígyóspuszta," *Archaeologiai Értesítő* 71 (1943): 174–184; Iván Bertényi, "A címerek katonai felhasználása Magyarországon a XIII–XIV. Században [The Military Use of Coats of Arms in Hungary, Thirteenth–Fourteenth Centuries]," *Hadtörténelmi Közlemények* 34 (1987): 395–412.
64 Vásáry, *The Cumans*, 149–155.
65 Nomadic warriors are often depicted on the wall paintings of the St. Ladislas legend (fourteenth–fifteenth century) in Hungary. According to the legend, recorded in the Hungarian national chronicle, St. Ladislas defeats a Cuman warrior abducting a woman. The battle and the chase of the Cuman rider in order to free the abducted maiden is extremely bloody, yet this narrative scene, or rather a series of battle actions, found its way onto the walls of dozens of Hungarian churches from the early fourteenth century onwards. It is a mystery how this story developed in the courtly imaginary and was finally propagated throughout the kingdom. The figure of the "bad Cuman" recalls the memory of the Cumans as enemies of the Hungarians, but the real ethnicity of the pagan warrior is of minor importance, as the historical raiders in 1068 were probably not Cumans (Kipchak-Cumans) at all, as at that time they were far away from the Carpathian Basin, but some other nomadic people instead, perhaps the Oğuz. On the wall paintings, the Cumans wear a mixture of Western and nomadic dress and equipment. See Nicolle, *Medieval Warfare Source Book*, 176–177; András Vizkelety, "Nomádkori hagyományok, vagy udvari-lovagi toposzok? Észrevételek Szent László és a leányrabló kun epikai és képzőművészeti ábrázolásaihoz [Nomadic Traditions or Courtly-Chevaleresque Models, Remarks about the Narrative and Figurative Representations of St Ladislas and the Cuman Raider]," *Irodalomtörténeti Közlemények* 85 (1981): 253–275.
66 István Szathmáry, "Rovásjeles címer egy régi kardon [Runic Coat of Arms on an Old Sword]," *Tisicum: A Jász-Nagykun-Szolnok Megyei Múzeumok Évkönyve* 15 (2006): 99–103, at 99.
67 David Nicolle, *Arms and Armour of the Crusading Era, 1050–1350* (White Plains: Kraus, 1998), 1:544 and 2:940 (no. 1513).
68 Attila Zsoldos, "Az első évtizedek: Jászok a középkori Magyarországon [The First Decades: The Iasians in Medieval Hungary]," *Rubicon* 28, no. 6 (2017): 28–31.
69 Attila Bárány, "Attempts for Expansion: Hungary, 1000–1500," in *The Expansion of Central Europe in the Middle Ages*, ed. Nora Berend (Farnham: Ashgate, 2012), 330–380; Spinei, *The Romanians*, 133.

References

Anonymous, Notary of King Béla. *The Deeds of the Hungarians, Master Roger's Epistle to the Sorrowful Lament about the Destruction of Hungary by the Tartars*, edited and translated by János M. Bak, Martyn Rady, and László Veszprémy. Budapest, New York: CEU Press, 2010.

Armbruster, Adolf. "Romanii in cronica lui Ottokar de Styria: o noua interpretare." *Studii, Revista de istorie* 25 (1972): 463–483.

Armbruster, Adolf. "Nochmals 'Herzoge Ramunc Uzer Vlachen Lant'." *Revue Roumaine d'Histoire* 12 (1973): 87–100. Reprinted in id., *Auf den Spuren der eigenen Identität: Ausgewählte Beiträge zur Geschichte und Kultur Rumäniens*, 75–93. Bucharest: Enciclopedică, 1991.

Armbruster, Adolf. *Der Donau-Karpatenraum in den mittel- und westeuropäischen Quellen des 10.–16. Jahrhunderts*. Vienna: Böhlau, 1990.

Arnold of Lübeck. *Chronica Slavorum*, edited by Georg Pertz. Monumenta Germaniae Historica. Scriptores rerum Germanicarum, vol. 14. Hannover: Hahn, 1868.

Bak, János M., and László Veszprémy, eds. and trans. *Chronica de gestis Hungarorum e codice picto saec. xiv. Chronicle of the Deeds of the Hungarians from the Fourteenth-Century Illuminated Codex*. Budapest, New York: CEU Press, OSzK, 2018.

Bárány, Attila. "Auxiliary Peoples." In *Medieval Warfare and Military Technology: An Encyclopedia*, edited by Clifford J. Rogers, 98–100. Vol. 1. Oxford: Oxford University Press, 2010.

Bárány, Attila. "Attempts for Expansion: Hungary, 1000–1500." In *The Expansion of Central Europe in the Middle Ages*, edited by Nora Berend, 330–380. Farnham: Ashgate, 2012.

Benkő, Elek. *A középkori Székelyföld* [The Medieval Székelys]. 2 vols. Budapest: MTA BTK, 2012.

Berend, Nora. *At the Gate of Christendom: Jews, Muslims and "Pagans" in Medieval Hungary, c. 1000–c. 1300*. Cambridge: Cambridge University Press, 2001.

Berend, Nora. "The Mendicant Orders and the Conversion of Pagans in Hungary." In *Alle frontiere della cristianità, I frati mendicanti e l'evangelizzazione tra '200 e '300: Atti del XXVIII Convegno internazionale Assisi, 12–14 ottobre 2000*, 253–279. Spoleto: Centro italiano di studi sull'alto Medioevo, 2001.

Berend, Nora. "Immigrants and Locals in Medieval Hungary: 11th–13th Centuries." In *The Expansion of Central Europe in the Middle Ages*, edited by Nora Berend, 310–313. Burlington, VT: Ashgate Variorum, 2013.

Berend, Nora, Przemysław Urbańczyk, and Przemysław Wiszewski. *Central Europe in the High Middle Ages: Bohemia, Hungary and Poland, c. 900–c. 1300*. Cambridge: Cambridge University Press, 2013.

Bertényi, Iván. "A címerek katonai felhasználása Magyarországon a XIII–XIV. században [The Military Use of Coats of Arms in Hungary, Thirteenth–Fourteenth Centuries]." *Hadtörténelmi Közlemények* 34 (1987): 395–412. Reprinted in id., *A címertan reneszánsza*. Budapest: Argumentum, 2010.

Bretholz, Bertold, ed. *Cosmae Pragensis Chronica Boemorum: Die Chronik der Böhmen des Cosmas von Prag*. Monumenta Germaniae Historica. Scriptores rerum Germanicarum. n.s., vol. 2. Berlin: Weidmann, 1923.

Choniates, Niketas. *O City of Byzantium: Annals*, translated by Harry J. Magoulias. Detroit: Wayne State University Press, 1984.

Curta, Florin. "The Image and Archaeology of the Pechenegs." *Banatica* 23 (2013): 143–202.

Deér, József. "Aachen und die Herrschersitze der Arpaden." *Mitteilungen des Instituts für Österreichische Geschichtsforschung* 79 (1971): 1–56.

Drašković, Marko. "Методи врбовања западноевропских најамника у Византији (1081–1185) [Recruitment Methods in Case of Westeuropean Mercenaries in Byzantium, 1081–1185]." *Istorijski Casopis* 61 (2012): 27–44.

Ducène, Jean-Charles, ed. *De Grenade à Bagdad: La relation de voyage d'Abu Hamid al-Gharnati (1080–1168)*. Paris: L'Harmattan, 2006.

Font, Márta. *Völker—Kultur—Beziehungen: Zur Entstehung der Regionen in der Mitte des mittelalterlichen Europa*. Hamburg: Kovač, 2013.

Gładysz, Mikolaj. *The Forgotten Crusaders: Poland and the Crusader Movement in the Twelfth and Thirteenth Centuries*. Leiden: Brill, 2012.

Göckenjan, Hansgerd. *Hilfsvölker und Grenzwächter im mittelalterlichen Ungarn*. Wiesbaden: Steiner, 1971.

Györffy, György. *A magyarság keleti elemei* [Eastern Groups in Medieval Hungary]. Budapest: Gondolat, 1990.

Hóman, Bálint. "Magyar történeti elemek a Nibelung énekben [Hungarian Historical Fragments in the Nibelungenlied]." In id., *Történetírás és forráskritika*, 293–336. Budapest: Magyar Történelmi Társulat, 1938.

Hunyadi, Zsolt. "The Teutonic Order in Burzenland (1211–1225): New Re-Considerations." In *L'Ordine Teutonico tra Mediterraneo e Baltico: incontri e scontri tra religioni, popoli e culture*, edited by Hubert Houben and Kristjan Toomaspoeg, 151–170. Galatino: Mario Congedo, 2008.

Jaffe, Philipp, ed. *Monumenta Bambergensia: Udalrici Babenbergensis codex*. Berlin: Weidmann, 1869.

Jakó, Zsigmond, ed. *Erdélyi Okmánytár*. Vol. 1: 1023–1300. Budapest: Akadémiai, 1997.

Katona-Kiss, Attila. "A sirmioni hunok: Egy muszlim katonai kötelék a XII. századi magyar királyi erőkben [The Huns of Sirmion: A Muslim Contingent on the Hungarian Royal Army of the Twelfth Century]." In *"Fons, skepsis, lex": Ünnepi tanulmányok a 70 esztendős Makk Ferenc tiszteletére*, edited by Tibor Almási, Éva Révész, and György Szabados, 159–171. Szeged: JATE, 2010.

Kinnamos, John. *Deeds of John and Manuel Comnenus*, translated by Charles M. Brand. New York: Columbia University Press, 1976.

Köpke, Rudolf, ed. *Annales Otakariani*. Monumenta Germaniae Historica. Scriptores in folio, vol. 9, 181–194. Hannover: Hahn, 1851.

Kovács, Szilvia. "Bortz, A Cuman Chief in the 13th Century," *Acta Orientalia Academiae Scientiarum Hungariae* 58, no. 3 (2005): 255–266.

Kovács, Szilvia. *A kunok története a mongol hódításig* [History of the Cumans until the Mongol Invasion]. Budapest: Balassi, 2014.

Kristó, Gyula. *A történeti irodalom Magyarországon a kezdetektől 1241-ig* [Historical Literature in Hungary until 1241]. Budapest: Argumentum, 1994.

Kusternig, Andreas. "Probleme um die Kämpfe zwischen Rudolf und Ottokar und die Schlacht bei Dürnkrut und Jedenspeigen am 26. August 1278." In *Ottokar-Forschungen*, edited by Max Weltin, 226–311. Vienna: Verein für Landeskunde für Niederösterreich und Wien, 1979.

Madgearu, Alexandru. "The Pechenegs in the Byzantine Army." In *The Steppe Lands and the World beyond Them, Studies in Honor of Victor Spinei on His 70th Birthday*, edited by Florin Curta and Bogdan-Petru Maleon, 207–218. Iaşi: Editura Universitatii, 2013.

Magoun, Francis P. Jr. "Geographical and Ethnic Names in the Nibelungenlied." *Mediaeval Studies* 7 (1945): 85–138.

Makk, Ferenc. "Megjegyzések II. István történetéhez [Remarks about the History of King's Stephen II]." In *Középkori kútfőink kritikus kérdései*, edited by János Horváth and György Székely, 253–261. Budapest: Akadémiai, 1974.

Makkai, László. "Transylvania in the Medieval Hungarian Kingdom, 896–1526." In *History of Transylvania*, edited by id. and András Mócsy, 333–587. Vol. 1. Boulder, CO: Social Science Monographs, 2001.

Marosi, Ernő. "Das Frontispiz der Ungarischen Bilderchronik (Cod. lat. 404 der Széchényi-Nationalbibliothek in Budapest)." *Wiener Jahrbuch für Kunstgeschichte* 46–47 (1993–1994): 364.

Marosi, Ernő. "Magyarok középkori ábrázolásai és az orientalizmus a középkori művészetben [Hungarians Depicted in Medieval Art and Orientalism in the Medieval Art]." In *Magyarok Kelet és Nyugat közt. A nemzettudat változó jelképei*, edited by Tamás Hofer, 77–97. Budapest: Nemzeti Múzeum, Balassi, 1996.

Mitchell, Russell. "Light Cavalry, Heavy Cavalry, Horse Archers, Oh My! What Abstract Definitions Don't Tell Us about 1205 Adrianople." *Journal of Medieval Military History* 6 (2008): 95–118.

Nagy, Balázs. "Foreign Trade of Medieval Hungary." In *The Economy of Medieval Hungary*, edited by József Laszlovszky, Balázs Nagy, Péter Szabó, and András Vadas, 473–490. Leiden: Brill, 2018.

Nicolle, David. *Medieval Warfare Source Book: Warfare in Western Christendom*. London: Weidenfeld, 1995.

Nicolle, David. *Arms and Armour of the Crusading Era, 1050–1350*. 2 vols. White Plains: Kraus, 1998.

Pálóczi-Horváth, Ádám. *Pechenegs, Cumans, Iasians: Steppe Peoples in Medieval Hungary*. Budapest: Corvina, 1989.

Pósán, László. *Hungary and the Teutonic Order in the Middle Ages*. Budapest: MTA Bölcsészettudományi Kutatóközpont, 2021.

Psellos. *Chronographia*, edited by Emile Renauld. Paris: Les Belles Lettres, 1928. Translated by Edgar R. A. Sewter. New Haven, CT: Yale University Press, 1953.

Radek, Tünde. *Das Ungarnbild in der deutschsprachigen Historiographie des Mittelalters*. Frankfurt am Main: Peter Lang, 2008.

Schünemann, Konrad. "Ungarische Hilfsvölker in der Literatur des deutschen Mittelalters." *Ungarische Jahrbücher* 4 (1924): 99–115.

Seemüller, Joseph, ed. *Ottokars Österreichische Reimchronik*. Monumenta Germaniae Historica, Deutsche Chroniken, vol. 5. Hannover: Hahn, 1890–1893.

Senga, Toru. "Béla királyfi bolgár, halicsi és osztrák hadjárataihoz, 1228–1232 [On the Bulgarian, Galician and Austrian Campaigns of Prince Béla]." *Századok* 122 (1988): 36–51.

Shakhmatov, Aleksey A., ed. *Polnoe sobranie russkikh letopisei*. Vol. 2, *Ipatievskaia letopis*. Saint Petersburg: Imperatorskaia Arkheograficheskaia kommissia, 1908. Reprinted Moscow: Nauka, 1962; 1998.

Simon, V. Péter. "A Nibelung ének magyar vonatkozásai [Hungarian Connections of the Nibelungenlied]." *Századok* 112 (1978): 271–325.

Spinei, Victor. *Moldavia in the 11th–14th centuries*. Bucharest: Editura Academiei, 1986.

Spinei, Victor. "The Cuman Bishopric—Genesis and Evolution." In *The Other Europe in the Middle Ages: Avars, Bulgars, Khazars, and Cumans; East Central and Eastern Europe in the Middle Ages, 450–1450*, edited by Florin Curta and Roman Kovalev, 413–456. Leiden: Brill, 2008.

Spinei, Victor. *The Romanians and the Turkic Nomads North of the Danube Delta from the Tenth to the Mid-Thirteenth Century: East Central and Eastern Europe in the Middle Ages, 450–1450*. Leiden: Brill, 2009.

Štulrajterová, Katarína. "Convivenza, Convenienza and Conversion: Islam in Medieval Hungary (1000–1400 ca)." *Journal of Islamic Studies* 24 (2013): 175–198.

Szabados, György. "A 12. századi magyar hadtörténet forrásproblémáiból [Problems of Sources in Twelfth Century Hungarian Military History]." *Aetas* 22, no. 4. (2007): 153–160.

Szathmáry, István. "Rovásjeles címer egy régi kardon [Runic Coat of Arms on an Old Sword]." *Tisicum: A Jász-Nagykun-Szolnok Megyei Múzeumok Évkönyve* 15 (2006): 99–103.

Szentpétery, Imre, ed. *Scriptores rerum Hungaricarum tempore ducum regumque stirpis Arpadianae gestarum*. 2 vols. Budapest: Egyetemi, 1937–1938.

Szentpétery, Imre, and Iván Borsa, eds. *Regesta regum stirpis Arpadianae critico-diplomatica*. 2 vols. Budapest: MTA, Akadémiai, 1923–1987.

Tóth, Zoltán. "La boucle de Kígyóspuszta." *Archaeologiai Értesítő* 71 (1943): 174–184.

Vásáry, István. *Cumans and Tatars*. Cambridge: Cambridge University Press, 2005.

Veszprémy, László. "Chronicon pictum." In *Encyclopedia of the Medieval Chronicle*, edited by Graeme Dunphy, 391. Vol. 1. Leiden: Brill, 2010.

Veszprémy, László. "'More paganismo': Reflections on Pagan and Christian Past in the Gesta Hungarorum (GH) of the Hungarian Anonymous Notary." In *Historical Narratives and Christian Identity on a European Periphery: Early History Writing in Northern, East Central, and Eastern Europe (c. 1070–1200)*, edited by Ildar H. Garipzanov, 183–201. Turnhout: Brepols, 2011.

Vizkelety, András. "Nomádkori hagyományok, vagy udvari-lovagi toposzok? Észrevételek Szent László és a leányrabló kun epikai és képzőművészeti ábrázolásaihoz [Nomadic Traditions or Courtly-Chevaleresque Models, Remarks about the Narrative and Figurative Representations of St Ladislas and the Cuman Raider]." *Irodalomtörténeti Közlemények* 85 (1981): 253–275.

Waitz, Georg, and Bernhard von Simson, eds. *Ottonis et Rahewini Gesta Friderici I. imperatoris*. Monumenta Germaniae Historica. Scriptores rerum Germanicarum, vol. 46. Hannover: Hahn, 1912.

Wattenbach, Wilhelm. ed. *Continuatio Vindobonensis*. Monumenta Germaniae Historica XI. Scriptores in folio 9, 698–722. Hannover: Hahn, 1851.

Wattenbach, Wilhelm, ed. *Vincentii Pragensis Annales*. Monumenta Germaniae Historica. Scriptores rerum Germanicarum, vol. 17, 658–686. Hannover: Hahn, 1861.

Zimmermann, Harald. *Siebenbürgen und seine Hospites Theutonici*. Vienna: Böhlau, 1996.

Zimmermann, Harald. *Der Deutsche Orden im Burzenland: Eine diplomatische Untersuchung*. Vienna: Böhlau, 2000.

Zsoldos, Attila. *Magyarország világi archontológiája 1000–1301 [Secular Archontology of Hungary, 1000–1301]*. Budapest: MTA TTI, 2011.

Zsoldos, Attila. "Az első évtizedek: Jászok a középkori Magyarországon [The First Decades: The Iasians in Medieval Hungary]." *Rubicon* 28, no. 6 (2017): 28–31.

12

MEDIEVAL QUEENS AND THE DIASPORA OF ESCORT, CONQUEST, THE CRUSADES, AND MILITARY ORDERS

Christopher Mielke

Women in the Middle Ages led battles, raised armies, held captives, and defended castles; in short, their gender did not preclude them from taking part in some military duties.[1] Yet this study is not about medieval women warriors, rather, it is about the role of a specific class of women in the movement of soldiers and military elites across borders. János Bak identifies one of the functions of medieval Hungarian queens as active agents who brought fellow countrymen with them to kingdoms upon their marriage.[2] In addition to clerks, confessors, artists, architects, ladies, and even minstrels, incoming queens could bring lords, knights, esquires, valets, grooms, and pages, some in a military capacity and others simply as attendants.[3] The foreign retinue of a queen is a frequent trope, especially when it was felt the queen had acted suspiciously, yet systematic research on the subject has remained elusive thus far.

There are several difficulties in delineating the militarized aspects in the household of a medieval queen, chief among them terminology. On the field, the basic military unit in the later medieval West was the man-at-arms. In France and Burgundy, they could be *chevaliers bannerets* (nobles who could have a banner), *chevaliers bacheliers* (also known as *bas chevaliers*—members of the aristocracy who could have a forked pennon) and *écuyers* (squires) from the minor nobility or burgher class who could afford martial equipment themselves.[4] There was a similar hierarchy at the royal court; at the top of the queen's household were her chief administrators, such as the treasurer, steward, marshal, chaplains, and clerks. The most prominent member of the queen's staff with some martial capacity was the knight. The knights were on a par with the ladies attending the queen. Below that were various different types of servants, usually under the marshal or the steward. One particular type, the *scutifer* (i.e. esquire or groom), may have originally had a martial aspect as a shield-bearer, but by the High Middle Ages, this class of servant was usually used interchangeably with valets, having various

DOI: 10.4324/9781003245568-13

different tasks arranging the household.[5] The wardrobe account of Eleanor of Castile, the wife of Edward I of England (r. 1272–1307), taken before her death in 1290 shows that she had a total of 12 knights in her service out of a known staff of 148; many of these knights had names indicating a French origin of some kind.[6] There were also men called "outriders" who escorted the queens for safety. In addition to the twelve knights in the service of Eleanor of Castile, a total of ten outriders are mentioned: Three safeguarding the queen, two for her ladies, two for the carts of the queen's wardrobe and ladies' chamber, and three for the carts of the queen's robes, the pantry, the butlery, and the kitchen.[7] Where wardrobe or household accounts do not survive, the men accompanying the queen are only identifiable if they appear in charters. This is why in the 300 years of the Árpádian dynasty in Hungary (1000–1301), the names of only five of the queens' Masters of the Horse (i.e. equerries) are known; these cases only occurred from 1257 to 1290.[8]

Groups of knights and soldiers associated with medieval queens fit the definition of a military diaspora in various ways. Their identity was shaped by transcultural interactions in local contexts that happened to be heavily dependent on the person of the queen.[9] This study explores three ways that queens functioned as agents of diaspora in the military classes of medieval Europe from the eleventh to the sixteenth centuries. The first approach will examine the personal attendants of queens with military functions who settled in their new country; the second part focuses on the role of dynastic marriages in bringing over armies as part of a plan of immediate conquest; finally, the role of queens in supporting crusading ventures and institutions like the military orders (e.g. the Hospitallers and Templars) will be elucidated.

Marriage and the Queen's Military Escort

The soldiers who accompanied the queen to her new homeland were usually those in her personal entourage. A good example of this is Felicia of Sicily (d. 1102?), who was accompanied by 300 Sicilian nobles when she arrived in Hungary in May of 1097 to marry King Coloman "the Book-Lover" (r. 1095–1116). The Hungarian noble family of Rátót can trace their origins from a knight named Ratold, who had been one of the 300 countrymen of Felicia.[10] According to John Tuzson, the knights who came over with Felicia became so powerful and avaricious that they quickly became the ones wielding power during Coloman's reign and that of his son Stephen II (r. 1116–1131). One of their most heinous acts was to blind Coloman's brother and nephew, Álmos and Béla, to prevent either from inheriting the throne. However, Stephen II nominated his blind cousin Béla as his successor before his death in 1131. Béla II (r. 1131–1141) and his queen, Helen of Serbia (d. 1146?), called a council at the town of Arad, where Helen ordered the execution of 68 nobles who had been complicit in blinding her husband when he was a child.[11] Thus one queen imported a martial diasporic community while another eliminated them.

Although this is a rather dramatic example, it nonetheless illustrates the common practice in Hungary of granting lands or titles to foreign military members of a queen's retinue. Many prominent Hungarian families tried to trace their lineage back through this practice. A thirteenth-century chronicler ascribes the origin of the Hermány clan to two knights in the retinue of Gisela of Bavaria, the first queen of Hungary. In reality, they only arrived in Hungary half a century after Queen Gisela.[12] Two knights from the entourage of Judith of Swabia (d. 1090s?), wife of King Salamon of Hungary (r. 1063–1074), appear to be the progenitors of the Gutkeled family in Hungary.[13] A French knight accompanying Margaret (d. 1197), sister of Philip II Augustus of France and later second wife of Béla III of Hungary (r. 1173–1196), became the ancestor of the Kukenus-Renold family in Hungary.[14] Furthermore, the Martinsburg family is descended from the knights Simon and Martin, who came from Aragon with the princess Constance (d. 1222), the wife of Emeric of Hungary (r. 1196–1204).[15] Not all who came with the queens stayed, however—a knight named William who came either with Gertrude of Andechs-Meran or Yolanda of Courtenay sold the land he was given shortly after the latter's marriage in 1215.[16]

We are fortunate that the origin of some of these noble families is mentioned in the work of the chronicler Simon of Keza—he even lists the queens Gisela and Constance by name, providing some rare insight into their entourage. In Hungary, it seems that there was some concern about foreigners entering the country because these persons are specifically mentioned in the Golden Bull of 1222 and the legal ordinance of Andrew II (r. 1205–1235) from 1231. The latter states that foreigners shall not be elevated to any title without the consent of the kingdom, and the later edition specifies that this was because such persons "take away the riches of the realm", but they were not subject to such scrutiny if they wished to become residents.[17]

The phenomenon under consideration is, of course, not restricted to Hungary. Take, for example, Yolanda can be omitted, the daughter of the aforementioned Andrew II of Hungary. She married King James I of Aragon (r. 1213–1276) in 1235 and was the queen of Aragon until her death in 1251. The names of several of her retinue are known, such as the queen's men J. de Hungria and Benedictus, as well as her doorkeeper or usher, Adam. More importantly for our purpose is the figure of *comes* Dénes/Dionysius from Szepes/Spis, a member of the queen's retinue who led a small army to al-Andalus for James I. He was rewarded for his services with two buildings near the palace of the bishop of Valencia. He was thus to stay, and he was not the only one; most servants and tradesmen who accompanied Queen Yolanda seemed to have stayed with her in Barcelona. After the queen's death in 1251, there are still records of people whose surnames indicate a Hungarian origin well until the end of the thirteenth century. The last one is Elizabeth, the daughter of the aforementioned *comes* Dénes, who died in 1294.[18] What all of these examples show is that queens were agents of bringing in members of a military elite from their homeland to their new country, but that these diasporic communities were (1) relatively small and (2) integrated within a generation or two of their new home country.

In a case from England, Queen Eleanor of Provence (d. 1291) sent large quantities of gifts (particularly rings) to members of the Flemish and northern French military elite to win their favour in 1259 and 1260. She gave similar gifts in 1263 to the (presumably foreign)[19] garrison at Windsor and raised an entirely foreign army in the crisis of 1264 to support her husband.[20] That she showed such favour to foreign knights earned her a reputation not only as a spendthrift but also as a patron of "greedy" foreigners. The fear and hostility to foreign knights in the service of the royal family could be quite palpable at times. Helene Kottanner, an attendant of Elizabeth of Luxemburg (d. 1444), wife of Albert II of Austria, Bohemia, and Hungary (r. 1437–1439), mentions how villagers in western Hungary were so terrified of the 2,000 German and Bohemian knights accompanying the widowed queen that they fled to the forests in the mountains upon news of their imminent arrival.[21] In both cases, the queens were heavily involved with foreign militaries during a time of intense civil war. For them to maintain their power, it was necessary to bring in foreign knights, a phenomenon which was not without consequences.

This tradition of bringing over military men with queens continued in the fourteenth century, as seen in another case from Hungary. The massive establishment of the Clarisses cloister in Óbuda was founded by Elizabeth of Poland (d. 1380), wife of Charles Robert of Hungary (r. 1308–1342). The queen wanted it to be a sepulchral chapel, and many of her close associates not only founded altars but chose to be buried there. In one case, the queen's equerry, a Polish knight named Mroczko, not only donated an altar dedicated to the Holy Trinity to the nunnery but he was eventually buried in the sepulchral chapel along with the queen and a few of the queen's associates.[22] Later in life, when Elizabeth became regent of Poland from 1370 to 1375, she brought a massive retinue of Hungarian soldiers with her, which contributed to triggering unrest and resistance to her rule. This occurred not only when she appointed Hungarian knights to the position of voivode of Kalisz, but also towards the end of her regency, when the *starosta*[23] of Kraków was killed by a Hungarian and 160 "Hungarians" were massacred in retaliation. After this, Elizabeth closed the castle gates and resigned her post as regent.[24]

The account books of Henry IV of England (r. 1399–1413) are remarkably detailed concerning the marriage of his daughter Philippa (d. 1430) to Eric of Pomerania, the King of Denmark, Sweden, and Norway (d. 1459). There were a total of 204 people who attended her on her journey from England to Denmark, 41 of them from Denmark. Nine English knights accompanied her, a Danish knight escorted her as an ambassador, another Danish knight, and 20 Danish esquires made up the party, and her English sergeants-at-arms were included in a staff of 30 lesser attendants (including shield-bearers, i.e. *scutifers*). These knights received more cloth for their livery than the queen's personal attendants, but not as much as the official representatives accompanying Philippa.[25] We find a similarly large (and mostly military) retinue in 1449, when Mary of Guelders (d. 1463), the bride of James II of Scotland (r. 1437–1460), was conducted by sea and escorted by Lord Henric of Veere, the Admiral of Holland who was guarded by 300 men in a total of 12 ships.[26]

Yet interestingly enough, for the most part, escorts of royal brides in the later Middle Ages show a different picture with regard to the relationship between queens and the soldiers they brought with them. While royal brides nonetheless were expected to bring a rich retinue to their new lands, the military nature of their attendants seems to have been much less emphasized. Philippa of Lancaster (d. 1415), the wife of John I of Portugal (r. 1385–1433), had 40 ladies and 76 officials and servants in her retinue; her chancellor, confessor, cook, tailor, and several clerics were all known to be English, but there is no mention of any English soldiers among her staff.[27] There are 50 known names of people in the service of Mary of Austria (d. 1558), the wife of Louis II of Bohemia and Hungary (r. 1516–1526), and practically no information on knights in her service. The majority of the fifty known servants are from Austrian, German, and Flemish families, and there seems to have been little effort on the part of the queen in having established Bohemian or Hungarian nobility at her court. The only exception to disfavouring locals seems to have been the employment of castellans from the Hungarian burgher class at some of her castles.[28] When Henry VIII (r. 1509–1547) married Anne of Cleves (d. 1557) in 1540, 263 people made up her bridal retinue. Eighty-eight of this number were Germans who planned to stay in England after the marriage—40 of their professions are unspecified, and there is no specific mention of any soldiers in the retinue.[29] The accounts for the marriage of Madeleine of Valois (d. 1537) with the Scottish king James V (r. 1513–1542) in 1536 show a similar decline in the importance of soldiers brought over from the queen's homeland, a tendency with which we will also be confronting in the next section.[30] This development is largely related to the growing importance of dowries in the form of cash payments as opposed to land[31]; this allowed the queen's husband to purchase knights and soldiers without being dependent on the intermediary role of his wife.

Dynastic Marriage, Soldiers, and Conquest

Two royal marriages show how queens could bring foreign soldiers to a new land as part of their dowry. When Frederick I of Sicily (r. 1198–1250, also Holy Roman Emperor Frederick II) came of age, he married Constance of Aragon (d. 1222), the widowed queen of Hungary. Constance arrived in Palermo in August of 1209, accompanied by her brother, Count Alfonso of Provence, and 500 Catalan and Provençal knights in the hope of re-asserting Frederick's authority over the southern Italian mainland. Unfortunately, soon after their arrival, the majority of the knights, including the queen's brother, were struck down by plague. The few demoralized survivors returned home and an embarrassed Frederick had to postpone his mainland expedition.[32] Had the knights in Constance's entourage survived, some might have stayed in Sicily after the fighting had ended.

The case of Philippa of Hainault (d. 1369) represents a more successful case for a dynastic marriage that enabled the husband's family to carry out a military coup. Philippa and the future Edward III of England (r. 1327–1377) were betrothed in 1326, in a move that enabled Edward's mother Isabella of France (d. 1358) to raise an army to combat her husband Edward II (r. 1307–1327) and his favourites, the Despensers. In lieu of a dowry, Philippa's father, William of Hainault agreed to provide troops for the planned invasion of England. Philippa's uncle, Jean de Beaumont, managed to raise a force of roughly 700 mercenaries from Hainault and the German lands, with further support coming from English exiles. Taken together, the army raised against Edward II on the continent totalled roughly 1,000–1,500 soldiers.[33] It is uncertain whether or not some of the mercenaries chose to stay in England, but some members of Philippa's retinue clearly did; one monk in the Westminster chronicle of John of Reading blames the Black Death on the outlandish clothes that the queen's countrymen made fashionable in England in the years after their arrival.[34] There is a similar martial motive behind the marriage of Philippa's granddaughter, Philippa of Lancaster, to John I of Portugal in 1387. The marriage took place because the future queen's father wanted to assert his claims to the kingdom of Castile, and John I was trying to secure his recent foothold in Portugal. The importance of this arrangement is demonstrated by the fact that the princess was not provided with a dowry.[35] Around 8,000 soldiers from the British Isles took part in the campaigns in Portugal in the 1380s. Although it can be assumed that some of these soldiers settled in Portugal, there is only one piece of direct evidence. A document from 1394 mentions a resident of Winchester and his Portuguese wife, indicating that after being married in Portugal, the pair eventually settled in England.[36] For many of these "rank and file" soldiers, finding such proof of their connection to a diasporic community through a queen's marriage is often a happy accident based on what survives in the written record.

In the eleventh century, many diplomatic marriages in Central-Eastern Europe were arranged with the prospect of military support. Béla I (r. 1060–1063), and Salamon (r. 1063–1074) of Hungary all came to the throne with military aid directly from their father and brother in law respectively; it is possible that some of these troops were promised or rewarded lands in Hungary by the new kings.[37] An artefact known as the "Sword of Attila" was part of a diplomatic exchange that Salamon's mother, Anastasia of Kiev (d. 1096?), gave as a gift to Otto of Nordheim, duke of Bavaria after he and the German knights in his service helped her son regain the throne in 1063.[38] According to the chronicler Jan Długosz, Casimir I "the Restorer" of Poland (r. 1040–1058) received foreign soldiers from not one but two royal women on his way to the throne. When setting out in 1041, his mother Richeza of Lorraine (d. 1063) gave him gold, silver, and jewels from her treasury as well as an escort of German knights "worthy of a king returning to his country".[39] Furthermore, after his marriage to the Kievan princess Maria (d. 1087), her brother Yaroslav I "the Wise" (r. 1019–1054) sent forth troops to Casimir to "deal with his neighbours and restore his kingdom to its former state".[40] These examples show us that in certain instances, military support was expected as part of a marital alliance.

On the other hand, when soldiers were promised as part of a marriage that did not arrive, it could spell trouble for the young bride in question. Ingeborg of Denmark (d. 1236) was famously repudiated by her spouse, Philip II Augustus of France (r. 1180–1223), shortly after her marriage and coronation. For centuries, speculation for this sudden change in behaviour ranged from witchcraft to temporary impotence, halitosis, sweating sickness, and the question of the bride's virginity.[41] One theory places blame on Ingeborg's brother Canute VI (r. 1182–1202), who had failed to attack the English coast in 1193 as had allegedly been pre-arranged with Philip.[42] If this was the real reason behind Philip's strange behaviour, it doomed Ingeborg to 20 years of imprisonment, humiliation, and a constant insistence on her legitimacy as queen of France.[43]

Military needs were still in the background of dynastic marriages in the later medieval and early modern periods, but rather than sending soldiers, it became a common practice that the bride's family provided a dowry in the form of a cash payment. Troubled by the Turks who had made recent inroads into Croatia and eager to gain a greater influence in Italy, Maximilian I (d. 1519) decided to marry Bianca Sforza (d. 1510) as soon as the duke of Milan agreed to a sizable dowry of 400,000 golden ducats.[44] Regrettably, the empress' gigantic dowry was spent very quickly and Bianca gained a reputation for being fiscally irresponsible to the extent that she had to pawn her linen and undergarments to meet the monthly expenses when she was a guest of the city of Worms in 1497.[45] In the High Middle Ages, royal marriages contrived in times of war could involve the queen as an agent of transferring military men. Yet there were many risks posed with such a move; in later periods, dynastic marriages lost their significance for moving or obtaining knights or soldiers from the new queen's homeland. Ready cash would have been more desirable than military men of different languages, customs, and faiths.

Queens, the Crusades, and Military Orders

While women were never prohibited from going on crusade as a general rule, Pope Innocent III supported the idea that they would receive the same spiritual benefits if they helped fund the enterprise rather than going on crusade themselves.[46] Nonetheless, many women were swept up in the crusading fervour of the era. Queen Bodil (d. 1103), wife of Eric I of Denmark (r. 1095–1103),[47] Ida of Cham (d. 1101), wife of Leopold II of Austria (r. 1075–1095),[48] and Margaret of France, the aforementioned widow of Béla III of Hungary,[49] all died in the Holy Land during the First and Third Crusades. Margaret of Provence (d. 1295), and her husband, St. Louis IX of France (r. 1226–1270), departed for the Holy Land in 1250. Margaret defended the city of Damietta and secured the ransom of her husband and the French army, which had been defeated en route to Cairo.[50] The last of the major western queens to journey to the Holy Land seems to have been Eleanor of Castile (d. 1290), wife of Edward I of England.[51]

Most queens, however, were unable to go to the Holy Land themselves. Instead, they found ways to support the mission from a distance by involving themselves with the active recruitment or sponsoring of Latin soldiers in the East. Eleanor of Aquitaine (d. 1204) joined her first husband, Louis VII of France (r. 1137–1180), who led the Second Crusade with Conrad III, King of Germany (r. 1138–1152). Her determination is shown from the popular (though apocryphal) image of her rallying troops in France by riding around with other ladies like Amazons and shaming men did not want to join the enterprise. At Antioch, there was a very real danger that Eleanor and her forces would splinter from Louis in order to support her uncle, Raymond of Antioch (r. 1136–1149). As a consequence, Louis' advisors forcibly removed her when the king left the city and thus deprived her of an opportunity to pursue her own military goals.[52] While Melisende, Queen of Jerusalem (r. 1131–1153), did not lead armies herself, she appointed an occidental kinsman of hers as the constable in charge of her army.[53]

Royal women were also active supporters of military orders such as the Templars and Hospitallers. Early on, several Iberian consorts made significant donations to the Hospitaller Order between 1114 and 1119, which allowed the Order to expand.[54] Melisende and her husband, Fulk of Anjou (r. 1131–1143), jointly donated the castle of Bethgibelin to the Knights Hospitaller in 1136. The donation of this strategic fort to the order had a profound impact on the Hospitallers, and modern scholarship has regarded the queen's gift as a catalyst that helped to militarize the charitable order.[55] Indeed, by the second half of the twelfth century, the role of the Hospitallers became explicitly militaristic. Some royal women who made such donations were even granted a status akin to a lay member. Examples are provided by Constance (d. 1176), daughter of Louis VI of France (r. 1108–1137), whose donation of land in Jerusalem to the Hospitallers brought her the status of a *consoror* with special provisions for her burial at a Hospitaller church and by a donation of Balian of Ibelin (d. 1193) and his wife Maria Komnena (d. 1217, formerly the Queen of Jerusalem) to the Order which identifies them as *confratres*.[56]

This interest of queens in supporting the crusading military diasporas, such as the military orders, is similar in the case of the Baltic and Iberian realms. Agatha (d. 1248), the Russian wife of the Polish prince Konrad of Masovia (d. 1247), played a critical role in inviting the Teutonic Order to the Baltic. Upon their arrival, there were tensions with Prussia, and since Konrad was away, Agatha asked the Teutonic Order to organize an army of Poles to confront the Prussians. After the battle was over, she rounded up the survivors and made sure that their wounds were treated. Once the members of the order had recovered, Konrad, along with his wife and their three sons, granted the Teutonic Order the land of Chełmno (Kulm) and Lubawa (Löbau) in perpetuity.[57] The Teutonic Order's presence in the Baltic would last until the early fifteenth century when the Polish and Lithuanian rulers had defeated them at the Battle of Tannenberg/Grunwald/Žalgiris in 1410.[58] In the will of Magnus IV Erikson of Sweden (r. 1319–1364) from 1346, he and his wife Blanche of Namur (d. 1363) pledged to send

100 knights to "fight against the enemies of God" if Magnus died before fulfilling his oath to go on crusade himself. It is unclear whether the royal couple referred to Muslims, pagan Lithuanians or Orthodox Russians as the enemies of God.[59] In 1400, Margaret I, Queen Regnant of Denmark, Norway, and Sweden, (r. 1387–1412) wrote to Pope Boniface IX (r. 1389–1404) describing how the borders of her kingdom were under attack by pagan and Christian enemies. It is unclear whether she referred to the Russians in Finland or the Teutonic Knights occupying Gotland. The Pope quickly acceded to her request and instructed the archbishops of all three kingdoms to preach the crusade with all the usual spiritual provisions.[60]

In Iberia, royal women had a front row seat to the action of the crusades: While the chroniclers Zurara and Fernão Lopes portray the Portuguese queen Philippa of Lancaster (d. 1415) as an archetypal ideal queen who avoided political meddling, she appears as a key figure in the Portuguese conquest of Ceuta in 1415, by refusing a bribe from the queen of Granada, urging her sons to participate in the crusades, and even handing them over swords for use in combat on her deathbed.[61] Ceuta was held by the Portuguese and ended up being the gateway for the European exploration of Africa in the fifteenth century.[62] The support royal women had for crusading continued to live on in Philippa's daughter Isabella (d. 1471), the third wife of Philip the Good of Burgundy (r. 1419–1467). A fierce advocate for the crusades, she corresponded with the preacher St. John of Capistrano, who led troops against the Ottomans.[63] After the fall of Constantinople, Philip ordered the construction of three ships for the purpose of re-conquering the city in 1455; Isabella focused on the construction of a *grand nave*, manned in part by a Portuguese crew. While neither this crusade nor another one called in 1464 was successful, the Burgundian operations nonetheless show how the crusade ideology continued well into the fifteenth century.[64]

The support that women had for the military orders led to an interest in founding Hospitaller and Templar foundations in their home countries, not just in the Holy Land. When Euphrosyne of Kiev (d. 1193), wife of Géza II of Hungary (r. 1141–1162), co-founded the first Hospitaller church in Székesfehérvár with Archbishop Martyrius of Esztergom, its purpose seems to have been more charity related than expressly military.[65] A later Hungarian donation was most likely prompted by the growing military aspects of the Templars. While Yolanda of Courtenay (d. 1233), the second wife of the Hungarian king Andrew II (r. 1205–1235), did not accompany her husband on the Fifth Crusade, the king donated land in Croatia to the Templar Order after his return in 1219. The justification for this donation is gratitude for the aid the Templars granted to the king during the Crusade and to the queen while she was managing day-to-day affairs in her husband's absence.[66] Another donation dates from 1224 when the long-suffering Ingeborg of Denmark (d. 1237) founded Saint-Jean-en-Île, a priory and church of the Hospitaller Order. This was not only the seat of her retirement and eventual place of burial but also the most important Hospitaller foundation in France until 1315, which shows the queen's active interest in affairs in the Holy Land.[67] Sancha of Castile, Queen of Aragon (d. 1208), even founded a branch of the Hospitaller Order for women at Sigena after an earlier attempt at Grisén failed. While it was earlier thought that the Hospitaller sisters were composed of women who fled Jerusalem after the Battle of Hattin, recent scholarship has disproved this view.[68]

Conclusions

In the Middle Ages, it was expected that a queen would employ a variety of knights and escorts to ensure her security. The numbers varied from year to year, but it is clear that until the fifteenth century, there was a suitable need for a large, impressive trousseau of foreign knights to accompany the queen. As the court structure became more hierarchical, it seems that there was less need for the services of foreign knights; by the Renaissance, dowries given in cash or supplies became the usual practice, further reducing the military aspect of the queen's entourage. By the end of the fifteenth century, the queen was (more or less) welcome to bring ladies and chaplains from her home country, but the importance of having men with a military background in her retinue declined.

That soldiers could accompany the queen in lieu of a dowry in order to support the political goals of the queen's husband demonstrates the pivotal role she played in terms of dynastic marriage policies. Such operations could cause problems since the soldiers could die soon after their arrival, and it is thus no wonder that by the later Middle Ages, a dowry in the form of a cash payment seems to have been preferred to a large body of soldiers from the queen's homeland.

There is still much evidence indicating that the foreign knights in the service of the queen lived quite well, integrated into their host country, and, in some cases, stayed for several generations. For these knights, the person of the queen was of key importance because the "community" of foreigners at a particular court had its primary ally in her. Combined with the pressure to assimilate, this is perhaps why the retinue of a queen can only be conceived as a diasporic community within one or two generations after her arrival at the new court.

The main impetus for the earlier crusades came from the kingdoms of Western Europe and the papacy. Yet as time went by, the number of supporters of the Crusades and military orders increased and included authorities in the Holy Land, Iberia, Scandinavia, and Poland. Personal piety could be expressed by a particular queen through her support of a diaspora of Western knights in the Holy Land or on the fringes of European Christianity. It is in these countries, which shared borders with non-Christian neighbours where the crusading ideas lasted for the longest time and where this ideology was still supported by queens and royal women long after it had faded elsewhere. Regarding the relationship between queens and military orders, there is, of course, much more systematic work to be done in the future. However, already now it is quite obvious that the majority of royal women chose to support the Hospitaller Order for both charitable and martial reasons. The reason for this preference could lie in the nature of the Templar Order, which principally prohibited such interferences from women, although deviations must be expected.[69]

Medieval queens were women who could grow to be incredibly powerful, yet they were fundamentally defined by their identity as a woman and by (in most cases) their status as a foreigner. As the king's companion, the queen played a complementary role in medieval society, particularly in her actions as intercessor and supporter of the church.[70] Rather than understanding these powerful women in a vacuum, understanding their role as part of the king allows all sorts of possibilities to explore what part they played in military ventures and as central figures in certain ethnic groups abroad. While queens were neither expected nor encouraged to play an active role in fostering, creating, and reinforcing different kinds of military diasporas, the evidence presented here suggests that they did.

Notes

1 Colleen Slater, "'So Hard Was It to Release Princes whom Fortuna Had Put in Her Chains': Queens and Female Rulers as Hostage- and Captive-Takers and Holders," in *Medieval Feminist Forum: A Journal of Gender and Sexuality* 45 (2009): 12–40, at 16–24; Annie Renoux, "Elite Women, Palaces, and Castles in Northern France (ca. 850–1100)," in *Reassessing the Roles of Women as "Makers" of Medieval Art and Architecture*, ed. Therese Martin (Leiden: Brill, 2012), 2:739–782, at 741, 753–754; Elisabeth Van Houts, "Queens in the Anglo-Norman/Angevin Realm 1066–1216," in *Mächtige Frauen? Königinnen und Fürstinnen im europäischen Mittelalter (11–14. Jahrhundert)*, ed. Claudia Zey, Vorträge und Forschungen 81 (Ostfildern: Jan Thorbecke Verlag, 2015), 199–224, at 203–207, 216.

2 János M. Bak, "Roles and Functions of Queens in Árpádian and Angevin Hungary (1000–1386 A.D.)," in *Medieval Queenship*, ed. John Carmi *Parsons* (New York: St. Martin's Press, 1993), 13–24, at 16–17.

3 W. Paley Baildon, "The Trousseaux of Princess Philippa, Wife of Eric, King of Denmark, Norway, and Sweden," *Archaeologia* 67 (1916): 163–188, at 170–172; Theresa Earenfight, *Queenship in Medieval Europe* (New York: Palgrave Macmillan, 2013), 14.

4 Nicholas Michael and G. A. Embleton, *Armies of Medieval Burgundy, 1364–1477* (London: Osprey, 1983), 6.

5 John Carmi Parsons, *The Court and Household of Eleanor of Castile in 1290* (Toronto: Pontifical Institute of Mediaeval Studies, 1977), 28–30; Baildon, "The Trousseaux of Princess Philippa," 170.

6 It is quite possible that there originally would have been anywhere up to 200 people in her employment, though the names of only 148 are known. There are also 17 names known of men listed as *scutifer* or *vallettus*. This is similar to the military household of kings that also included foreign elements; for instance, the military household of Edward I had not only English members but also men from Gascony, Savoy, Wales, and Scotland. Parsons, *The Court and Household of Eleanor of Castile in 1290*, 28, 154–160; J. O. Prestwich, "The Military Household of the Norman Kings," *The English Historical Review* 96, no. 378 (January 1981): 1–35, at 4.

7 Parsons, *The Court and Household of Eleanor of Castile in 1290*, 159.

8 Attila Zsoldos, *The Árpáds and Their Wives: Queenship in Early Medieval Hungary 1000–1301* (Rome: Viella, 2019), 202.

9 Georg Christ, "Diasporas and Diasporic Communities in the Eastern Mediterranean: An Analytical Framework," in *Union in Separation: Diasporas and Diasporic Groups in the Wider Mediterranean (1100–1800)*, ed. Georg Christ et al. (Rome: Viella, 2015), 19–40, at 28, 31–39.

10 Five thousand Hungarian knights escorted the queen from the Croatian coast to Székesfehérvár, where the royal couple was wed. Simon of Keza, Jenő Szűcs, and László Veszprémy, eds. *Gesta Hungarorum: The Deeds of the Hungarians* (Budapest: Central European University Press, 1999), 166–167; Zoltán J. Kosztolnyik, *From Coloman the Learned to Béla III (1095–1196)*, Eastern European Monographs 220 (New York: Columbia University Press, 1987), 30; Bak, "Roles and Functions of Queens in Árpádian and Angevin Hungary (1000–1386 A.D.)," 16; Zsoldos, *The Árpáds and Their Wives*, 180.

11 John Tuzson, *István II (1116–1131): A Chapter in Medieval Hungarian History* (Boulder, CO: East Europe Monographs, 2002), 143–145; Christopher Mielke, *The Archaeology and Material Culture of Queenship in Medieval Hungary, 1000–1395* (Cham: Palgrave Macmillan, 2021), 80–84.

12 Simon of Keza, *Gesta Hungarorum*, 168–169; Bak, "Roles and Functions of Queens in Árpádian and Angevin Hungary (1000–1386 A.D.)," 16.

13 Bak, "Roles and Functions of Queens in Árpádian and Angevin Hungary (1000–1386 A.D.)," 16.

14 Another knight accompanying Queen Margaret was named Smaragdus Aynard. Ibid., 16–17.

15 Simon of Keza, *Gesta Hungarorum*, 168–173; Bak, "Roles and Functions of Queens in Árpádian and Angevin Hungary (1000–1386 A.D.)," 16–17.

16 Another of Yolanda's knights sold his land to accompany Yolanda's daughter when she married James I of Aragon in 1235. Zsoldos, *The Árpáds and Their Wives*,123–124.

17 "XI. Hospites nobiles ad regnum venientes, nisi incole esse velint, ad dignitates non promoveantur: per tales enim divicie regni extrahuntur." János M. Bak, György Bónis, James Ross Sweeney, trans. and ed. *The Laws of the Medieval Kingdom of Hungary* (Bakersfield: Charles Schlacks, Jr., 1989), 1:40. An earlier, less explicit version can be found on page 35.

18 Szabolcs de Vajay, "*Dominae reginae milites*: Árpád-házi Jolánta magyarjainak meghonosodása Valencia visszavétele idején," [*Dominae reginae milites*: The Settlement of the Hungarians of Yolanda of the Árpád House and the Re-Conquest of Valencia] in *Királylányok messzi földről: Magyarország és Katalónia a középkorban* [Princesses from Afar: Hungary and Catalonia in the Middle Ages], ed. Ramon Sarobe and Csaba Toth (Budapest: Hungarian National Museum, 2009), 243–257, at 252–255.

19 Henry III and Eleanor were very unpopular at this time and many of the English and Anglo-Norman military leaders were siding with Simon of Montfort. The garrison was thus likely to be composed of hangers-on from southern France, who came to Windsor through Eleanor, cf. Margaret Howell, *Eleanor of Provence: Queenship in Thirteenth-Century England* (Oxford: Blackwell, 1998).

20 The knights she recruited came from the major regions in France, as well as Flanders, Germany, Burgundy, and Spain. Howell, *Eleanor of Provence*, 168–170, 182–183, 198, 212–221.

21 The fears were amplified by the succession crisis following the death of Albert; this anecdote relates from 1440, after the birth of Elizabeth's son Ladislas the Posthumous. Maya Bijvoet Williamson, *The Memoirs of Helene Kottanner (1439–1440)* (Cambridge: D. S. Brewer, 1998), 47.

22 Eva Sniezynska-Stolot, "Queen Elizabeth as Patron or Architecture," *Acta Historiae Artium* 20 (1974): 13–36, at 17–18.

23 North Slavic term (lit. "senior") for a royal administrator similar to a seneschal or sheriff.

24 Długosz lists this incident occurring in December 1376, and his portrayal of Elizabeth as regent is worth questioning as it adheres to many of the misogynist stereotypes of medieval chroniclers. Jan Długosz, *The Annals of Jan Długosz*, ed. and trans. Maurice Michael (Chichester: I M Publications, 1997), 326–331.

25 The Danish knight acting as ambassador received two cloths of gold of Cyprus for his livery while the other knights and esquires (both Danish and English) received scarlet

and green fabric for their livery. Baildon, "The Trousseaux of Princess Philippa," 170–171.

26 Rosalind Kay Marshall, *Scottish Queens, 1034–1714* (East Linton: Tuckwell, 2003), 60.

27 Ana Rodrigues Oliveira, "Philippa of Lancaster: The Memory of a Model Queen," in *Queenship in the Mediterranean: Negotiating the Role of the Queen in the Medieval and Early Modern Eras*, ed. Elena Woodacre (New York: Palgrave Macmillan, 2013), 125–144, at 139.

28 András Kubinyi, "The Court of Queen Mary of Hungary and Politics between 1521 and 1526," in *Mary of Hungary: The Queen and Her Court 1521–1531*, ed. Orsolya Rethélyi et al. (Budapest: Budapest History Museum, 2005), 13–25, at 19.

29 Retha M. Warnicke, *The Marrying of Anne of Cleves: Royal Protocol in Early Modern England* (Cambridge: Cambridge University Press, 2000), 116–117.

30 However, the purchases made for the new queen's trousseaux show a few items of martial importance. There were two cargo ships filled with gunpowder, two warships, 26 pieces of brass ordinance for the creation of a battery, and 30 small pieces of brass for the field. On this, see James Balfour, *The Historical Works of Sir James Balfour* (Edinburgh: W. Atchinson, 1824–1825), 266–267; Marshall, *Scottish Queens, 1034–1714*, 106–107.

31 Amalie Fößel, "The Queen's Wealth in the Middle Ages," *Majestas* 13 (2005): 23–45, at 37–44.

32 David Abulafia, *Frederick II: A Medieval Emperor* (London: Pimlico, 1988), 106; Ernst Kantorowicz, *Frederick the Second 1194–1250* (New York: Frederick Ungar, 1957), 32–35.

33 Seymour Philipps, *Edward II* (New Haven, CT: Yale University Press, 2010), 500–501; Karl Petit, "Le Mariage de Philippa de Hainaut, reine d'Angleterre (1328)," *Le Moyen Age* 87 (1981): 373–385, at 377–378.

34 James Tait, ed. *Chronica Johannis de Reading et Anonymi Cantuariensis 1346–1367* (Manchester, 1914), 88–89, cited in Rosemary Horrox, ed., "44. Indecent Clothing as a Cause of the 1348–1349 Epidemic," in *The Black Death* (Manchester: Manchester University Press, 1995), 131.

35 Ana Maria Rodrigues, "For the Honor of Her Lineage and Body: The Dowers and Dowries of Some Late Medieval Queens of Portugal," *E-Journal of Portuguese History* 5, no. 1 (June 2007): 1–14, at 4.

36 Tiago Viúla de Faria, "Tracing the 'chemyn de Portyngale': English Service and Servicemen in Fourteenth-Century Portugal," *Journal of Medieval History* 37 (2011): 257–268, at 265–268.

37 Zoltán J. Kosztolnyik, *Five Eleventh Century Hungarian Kings: Their Policies and Their Relations with Rome*, East European Monographs 79 (New York: Columbia University Press, 1981), 76–81.

38 Talia Zajac, "Remembrance and Erasure of Objects Belonging to Rus' Princesses in Medieval Western Sources: The Cases of Anastasia Iaroslavna's 'Saber of Charlemagne' and Anna Iaroslavna's Red Gem," in *Moving Women Moving Objects (400–1500)*, ed. Tracy Chapman Hamilton and Mariah Proctor-Tiffany (Leiden: Brill, 2019), 33–58, at 42–46; Mielke, *The Archaeology and Material Culture of Queenship in Medieval Hungary*, 51–52.

39 Długosz, *The Annals of Jan Długosz*, 32.

40 Ibid., 33–34.

41 George Conklin, "Ingeborg of Denmark, Queen of France," in *Queens and Queenship in Medieval Europe*, ed. Anne Duggan (Woodbridge: Boydell, 2002), 39–52, at 40–41.

42 Ruth Mazo Karras, *Unmarriages: Women, Men, and Sexual Unions in the Middle Ages* (Philadelphia: University of Pennsylvania Press, 2012), 60; Jane Sayers, *Innocent III: Leader of Europe, 1198–1216* (London: Longman, 1994), 116.

43 Conklin, "Ingeborg of Denmark, Queen of France," 41–52.

44 Monica Azzolini, *The Duke and the Stars: Astrology and Politics in Renaissance Milan* (Cambridge: Harvard University Press, 2013), 162–163; William Coxe, *History of the House of Austria, from the Foundation of the Monarchy by Rhodolf of Habsburgh to the Death of Leopold II: 1218 to 1792*, vol. 1 (London: Henry G. Bohn, 1847), 294, 306.
45 Gerhard Bernecke, *Maximilian I (1459–1519): An Analytical Biography* (London: Routledge and Kegan Paul, 1982), 95. For further details, see pages 94–103.
46 Bernard Hamilton, "Eleanor of Castile and the Crusading movement," *Mediterranean Historical Review* 10, no. 1–2 (1995): 92–103, at 94–95.
47 Ane Bysted et al. *Jerusalem in the North: Denmark and the Baltic Crusades, 1100–1522* (Turnhout: Brepols, 2012), 19.
48 While she was most likely killed, it was nonetheless rumoured that she was captured and put into a harem where she gave birth to Zengi, the founder of the Zengid dynasty. Otto of Freising, *The Two Cities: A Chronicle of Universal History to the Year 1146 A.D., by Otto, Bishop of Freising*, trans. Charles Christopher Mierow (New York: Columbia University Press, 2002), 411; Steven Runciman, *A History of the Crusades*, vol. 2, *The Kingdom of Jerusalem and the Frankish East 1100–1187* (Cambridge: Cambridge University Press, 1957), 27–29.
49 Peter W. Edbury, *The Conquest of Jerusalem and the Third Crusade: Sources in Translation* (Aldershot: Ashgate, 1998), chap. 183, 142–143.
50 Howell, *Eleanor of Provence: Queenship in Thirteenth-Century England*, 59–60; Hamilton, "Eleanor of Castile and the Crusading Movement," 95.
51 This experience had a profound effect on her and she later commissioned a translation of Vegetius' *De re militari* from Latin into Old French. Hamilton, "Eleanor of Castile and the Crusading Movement," 101–103.
52 Conor Kostick, "Eleanor of Aquitaine and the Women of the Second Crusade," in *Medieval Italy, Medieval and Early Modern Women: Essays in Honour of Christine Meek*, ed. id. (Portland: Four Courts Press, 2010), 195–205, at 202–205.
53 Manasses of Hierges was from the Ardennes. Alan V. Murray, "Women in the Royal Succession of the Latin Kingdom of Jerusalem (1099–1291)," *Mächtige Frauen? Königinnen und Fürstinnen im europäischen Mittelalter (11–14. Jahrhundert)*, ed. Claudia Zey, Vorträge und Forschungen 81(Ostfildern: Jan Thorbecke Verlag, 2015), 131–162, at 142–143.
54 In particular, they include Teresa, Countess of Portugal (d. 1130), Urraca of Castile and Leon (r. 1109–26), Douce, Countess of Provence (d. 1127), and Emma, daughter of Roger I of Sicily (d. 1120). Myra Miranda Bom, *Women in the Military Orders of the Crusades* (New York: Palgrave Macmillan, 2012), 49.
55 Jaroslav Fulda, "Melisende of Jerusalem: Queen and Patron of Art and Architecture in the Crusader Kingdom," in *Reassessing the Roles of Women as "Makers" of Medieval Art and Architecture*, ed. Therese Martin (Leiden: Brill, 2012), 429–478, at 459–460; Helen Gaudette, "The Spending Power of a Crusader Queen," in *Women and Wealth in Late Medieval Europe*, ed. Therese Martin (New York: Palgrave Macmillan, 2010), 135–148, at 138–139.
56 Bom, *Women in the Military Orders of the Crusades*, 68–70.
57 Nicolaus von Jeroscin, *The Chronicle of Prussia by Nicolaus von Jeroschin: A History of the Teutonic Knights in Prussia, 1190–1331*, trans. Mary Fischer (Farnham: Ashgate, 2011), 47.
58 Aleksander Pluskowski, *The Archaeology of the Prussian Crusade: Holy War and Colonisation* (London: Routledge, 2013), 20, 337–338.
59 Janus Møller Jensen, *Denmark and the Crusades: 1400–1650* (Leiden: Brill, 2007), 37.
60 Bysted et al., *Jerusalem in the North: Denmark and the Baltic Crusades, 1100–1522*, 338–339.
61 Rodrigues Oliveira, "Philippa of Lancaster: The Memory of a Model Queen," 132–133; Jennifer Goodman, *Chivalry and Exploration: 1298–1630* (Woodbridge: Boydell Press, 1998), 138–144.
62 At present-day, Ceuta is held by the Spanish government. Bailey W. Diffie and George D. Winius, *Foundations of the Portuguese Empire, 1415–1580* (Minneapolis, MN: University of Minnesota Press, 1977), 54–56.

63 Werner Schulz, *Andreaskreuz und Christusorden: Isabella von Portugal und der burgundisch Kreuzzug* (Freiburg: Universitätsverlag, 1976), 101–114.
64 Charity Cannon-Willard, "Isabel of Portugal and the Fifteenth-Century Burgundian Crusade," in *Journey towards God: Pilgrimage and Crusade*, ed. Barbara N. Sargent-Baur (Kalamazoo, MI: Medieval Institute Publications, 1992), 205–214, at 207–212.
65 Of course, the first charter evidence is only from 1193, decades after it was originally founded. Zsolt Hunyadi, *The Hospitallers in the Medieval Kingdom of Hungary c. 1150–1387* (Budapest: CEU Medievalia, 2010), 24.
66 Miha Kosi, "The Age of the Crusades in the South-East of the Empire (between the Alps and the Adriatic)," in *The Crusades and the Military Orders: Expanding the Frontiers of Medieval Latin Christianity*, ed. Zsolt Hunyadi and József Laszlovszky (Budapest: CEU Department of Medieval Studies, 2001), 123–165, at 137; Mielke, *The Archaeology and Material Culture of Queenship in Medieval Hungary*, 84–87.
67 Her burial plaque is also indicative of her earlier struggles and eventual success towards the legitimization she spent her whole reign aiming for. Kathleen S. Schowalter, "The Ingeborg Psalter: Queenship, Legitimacy, and the Appropriation of Byzantine Art in the West," in *Capetian Woman*, ed. Kathleen Nolan (New York: Palgrave Macmillan, 2003), 99–135, at 105; Kathleen Nolan, *Queens in Stone and Silver: The Creation of a Visual Imagery of Queenship in Capetian France* (New York: Palgrave Macmillan, 2009), 116–119.
68 Bom, *Women in the Military Orders of the Crusades*, 82–87.
69 For example, Margaret of Provence, wife of Louis VII of France, gave birth to the duke of Alençon at a Templar fortress of Castle Pilgrim, and, despite prohibitions, the Master of the Temple even served as godfather. On this, see Helen Nicholson, *The Knights Templar: A New History* (Stroud: Sutton, 2001), 158.
70 Henric Bagerius and Christine Ekholst, "The Unruly Queen: Blanche of Namur and Dysfunctional Rulership in Medieval Sweden," in *Queenship, Gendered, and Reputation in the Medieval and Early Modern West, 1060–1600*, ed. Lisa Benz St. John and Zita Rohr (New York: Palgrave Macmillan, 2016), 99–118, at 104.

References

Abulafia, David. *Frederick II: A Medieval Emperor.* London: Pimlico, 1988.

Azzolini, Monica. *The Duke and the Stars: Astrology and Politics in Renaissance Milan.* Cambridge: Harvard University Press, 2013.

Bagerius, Henric, and Christine Ekholst. "The Unruly Queen: Blanche of Namur and Dysfunctional Rulership in Medieval Sweden." In *Queenship, Gendered, and Reputation in the Medieval and Early Modern West, 1060–1600*, edited by Lisa Benz St. John and Zita Rohr, 99–118. New York: Palgrave Macmillan, 2016.

Baildon, W. Paley. "The Trousseaux of Princess Philippa, Wife of Eric, King of Denmark, Norway, and Sweden." *Archaeologia* 67 (1916): 163–188.

Bak, János M. "Roles and Functions of Queens in Árpádian and Angevin Hungary (1000–1386 A.D.)." In: *Medieval Queenship*, edited by John Carmi Parsons, 13–24. New York: St. Martin's Press, 1993.

Balfour, James. *The Historical Works of Sir James Balfour.* Edinburgh: W. Atchinson, 1824–1825.

Bernecke, Gerhard. *Maximilian I (1459–1519): An Analytical Biography.* London: Routledge & Kegan Paul, 1982.

Bijvoet Williamson, Maya. *The Memoirs of Helene Kottanner (1439–1440).* Cambridge: D. S. Brewer, 1998.

Bom, Myra Miranda. *Women in the Military Orders of the Crusades.* New York: Palgrave Macmillan, 2012.

Bysted, Ane, Carsten Selch Jensen, Kurt Villads Jensen, and John H. Lind. *Jerusalem in the North: Denmark and the Baltic Crusades, 1100–1522*. Turnhout: Brepols, 2012.

Cannon-Willard, Charity. "Isabel of Portugal and the Fifteenth-Century Burgundian Crusade." In *Journey towards God: Pilgrimage and Crusade*, edited by Barbara N. Sargent-Baur, 205–214. Kalamazoo: Medieval Institute Publications, 1992.

Christ, Georg. "Diasporas and Diasporic Communities in the Eastern Mediterranean: An Analytical Framework." In *Union in Separation: Diasporas and Diasporic Groups in the Wider Mediterranean (1100–1800)*, edited by Georg Christ et al., 19–40. Rome: Viella, 2015.

Conklin, George. "Ingeborg of Denmark, Queen of France." In *Queens and Queenship in Medieval Europe*, edited by Anne Duggan, 39–52. Woodbridge: Boydell, 2002.

Coxe, William. *History of the House of Austria, from the Foundation of the Monarchy by Rhodolph of Hapsburgh to the Death of Leopold II: 1218 to 1792*. Vol. 1. London: Henry G. Bohn, 1847.

Diffie, Bailey W., and George D. Winius. *Foundations of the Portuguese Empire, 1415–1580*. Minneapolis: University of Minnesota Press, 1977.

Długosz, Jan. *The Annals of Jan Długosz: An English Abridgement*, edited and translated by Maurice Michael. Chichester: I M Publications, 1997.

Earenfight, Theresa. *Queenship in Medieval Europe*. New York: Palgrave Macmillan, 2013.

Edbury, Peter W. *The Conquest of Jerusalem and the Third Crusade: Sources in Translation*. Aldershot: Ashgate, 1998. Reprint Routledge 2017.

Fößel, Amalie. "The Queen's Wealth in the Middle Ages." *Majestas* 13 (2005): 23–45.

Fulda, Jaroslav. "Melisende of Jerusalem: Queen and Patron of Art and Architecture in the Crusader Kingdom." In *Reassessing the Roles of Women as "Makers" of Medieval Art and Architecture*, edited by Therese Martin, 429–478. Leiden: Brill, 2012.

Gaudette, Helen. "The Spending Power of a Crusader Queen." In *Women and Wealth in Late Medieval Europe*, edited by Therese Martin, 135–148. New York: Palgrave Macmillan, 2010.

Goodman, Jennifer. *Chivalry and Exploration: 1298–1630*. Woodbridge: Boydell Press, 1998.

Hamilton, Bernard. "Eleanor of Castile and the Crusading Movement." *Mediterranean Historical Review* 10, no. 1–2 (1995): 92–103.

Horrox, Rosemary, ed. *The Black Death*. Manchester: Manchester University Press, 1995.

Howell, Margaret. *Eleanor of Provence: Queenship in Thirteenth-Century England*. Oxford: Blackwell, 1998.

Hunyadi, Zsolt. *The Hospitallers in the Medieval Kingdom of Hungary c. 1150–1387*. Budapest: CEU Medievalia, 2010.

Jensen, Janus Møller. *Denmark and the Crusades: 1400–1650*. Leiden: Brill, 2007.

Kantorowicz, Ernst. *Frederick the Second 1194–1250*. New York: Frederick Ungar, 1957.

Kosi, Miha. "The Age of the Crusades in the South-East of the Empire (between the Alps and the Adriatic)." In *The Crusades and the Military Orders: Expanding the Frontiers of Medieval Latin Christianity*, edited by Zsolt Hunyadi and József Laszlovszky, 123–165. Budapest: CEU Department of Medieval Studies, 2001.

Kostick, Conor. "Eleanor of Aquitaine and the Women of the Second Crusade." In *Medieval Italy, Medieval and Early Modern Women: Essays in Honour of Christine Meek*, edited by id., 195–205. Portland: Four Courts Press, 2010.

Kosztolnyik, Zoltán J. *Five Eleventh Century Hungarian Kings: Their Policies and Their Relations with Rome*. East European Monographs 79. New York: Columbia University Press, 1981.

Kosztolnyik, Zoltán J. *From Coloman the Learned to Béla III (1095–1196)*. Eastern European Monographs 220. New York: Columbia University Press, 1987.

Kubinyi, András. "The Court of Queen Mary of Hungary and Politics between 1521 and 1526." In *Mary of Hungary: The Queen and Her Court 1521–1531*, edited by Orsolya Rethélyi, Beatrix Romhányi, Enikő Spekner, and András Végh, 13–25. Budapest: Budapest History Museum, 2005.

Marshall, Rosalind Kay. *Scottish Queens, 1034–1714*. East Linton: Tuckwell, 2003.

Mazo Karras, Ruth. *Unmarriages: Women, Men, and Sexual Unions in the Middle Ages*. Philadelphia: University of Pennsylvania Press, 2012.

Mielke, Christopher. *The Archaeology and Material Culture of Queenship in Medieval Hungary, 1000–1395*. Cham: Palgrave Macmillan, 2021.

Murray, Alan V. "Women in the Royal Succession of the Latin Kingdom of Jerusalem (1099–1291)." In *Mächtige Frauen? Königinnen und Fürstinnen im europäischen Mittelalter (11.–14. Jahrhundert)*, edited by Claudia Zey, 131–162. Vorträge und Forschungen 81. Ostfildern: Jan Thorbecke Verlag, 2015.

Nicholson, Helen. *The Knights Templar: A New History*. Stroud: Sutton, 2001.

Nolan, Kathleen. *Queens in Stone and Silver: The Creation of a Visual Imagery of Queenship in Capetian France*. New York: Palgrave Macmillan, 2009.

Otto, Bishop of Freising. *The Two Cities: A Chronicle of Universal History to the Year 1146 A.D., by Otto, Bishop of Freising*, edited by Austin P. Evans and Charles Knapp. Translated by Charles Christopher Mierow. New York: Columbia University Press, 2002.

Parsons, John Carmi. *The Court and Household of Eleanor of Castile in 1290*. Toronto: Pontifical Institute of Mediaeval Studies, 1977.

Petit, Karl. "Le Mariage de Philippa de Hainaut, reine d'Angleterre (1328)." *Le Moyen Age*, 87 (1981): 373–385.

Philipps, Seymour. *Edward II*. New Haven: Yale University Press, 2010.

Pluskowski, Aleksander. *The Archaeology of the Prussian Crusade: Holy War and Colonisation*. London: Routledge, 2013.

Prestwich, John Oswald. "The Military Household of the Norman Kings." *The English Historical Review*, 96, no. 378 (January 1981): 1–35.

Renoux, Annie. "Elite Women, Palaces, and Castles in Northern France (ca. 850–1100)." In *Reassessing the Roles of Women as "Makers" of Medieval Art and Architecture*, edited by Therese Martin, 739–782. Vol. 2. Leiden: Brill, 2012.

Rodrigues, Ana Maria. "For the Honor of Her Lineage and Body: The Dowers and Dowries of Some Late Medieval Queens of Portugal." *E-Journal of Portuguese History* 5, no.1 (June 2007): 1–14.

Rodrigues Oliveira, Ana. "Philippa of Lancaster: The Memory of a Model Queen." In *Queenship in the Mediterranean: Negotiating the Role of the Queen in the Medieval and Early Modern Eras*, edited by Elena Woodacre, 125–144. New York: Palgrave Macmillan, 2013.

Runciman, Steven. *A History of the Crusades*. Vol. 2, The Kingdom of Jerusalem and the Frankish East 1100–1187. Cambridge: Cambridge University Press, 1957.

Sayers, Jane. *Innocent III: Leader of Europe, 1198–1216*. London: Longman, 1994.

Schowalter, Kathleen S. "The Ingeborg Psalter: Queenship, Legitimacy, and the Appropriation of Byzantine Art in the West." In *Capetian Woman*, edited by Kathleen Nolan, 99–135. New York: Palgrave Macmillan, 2003.

Schulz, Werner. *Andreaskreuz und Christusorden: Isabella von Portugal und der burgundisch Kreuzzug*. Freiburg: Universitätsverlag, 1976.

Simon of Keza, Jenő Szűcs, and László Veszprémy, eds. *Gesta Hungarorum: The Deeds of the Hungarians*. Budapest: Central European University Press, 1999.

Slater, Colleen. "'So Hard Was It to Release Princes Whom Fortuna Had Put in Her Chains': Queens and Female Rulers as Hostage- and Captive-Takers and Holders." *Medieval Feminist Forum: A Journal of Gender and Sexuality* 45 (2009): 12–40.

Sniezynska-Stolot, Eva. "Queen Elizabeth as Patron or Architecture." *Acta Historiae Artium* 20 (1974): 13–36.

Tait, James, ed. *Chronica Johannis de Reading et Anonymi Cantuariensis 1346–1367*. Manchester: Manchester University Press, 1914.

Thuróczy, János. *Chronicle of the Hungarians*. Bloomington: Indiana University, 1991.

Tuzson, John. *István II (1116–1131): A Chapter in Medieval Hungarian History*. Boulder: East Europe Monographs, 2002.

Vajay, Szabolcs de. "*Dominae reginae milites*: Árpád-házi Jolánta magyarjainak meghonosodása Valencia visszavétele idején." [*Dominae reginae milites*: The Settlement of the Hungarians of Yolanda of the Árpád House and the Re-Conquest of Valencia] In *Királylányok messzi földről: Magyarország és Katalónia a középkorban [Princesses from Afar: Hungary and Catalonia in the Middle Ages]*, edited by Ramon Sarobe and Csaba Toth, 243–257. Budapest: Hungarian National Museum, 2009.

Van Houts, Elisabeth. "Queens in the Anglo-Norman/Angevin Realm 1066–1216." In *Mächtige Frauen? Königinnen und Fürstinnen im europäischen Mittelalter (11.–14. Jahrhundert)*, edited by Claudia Zey, 199–224. Vorträge und Forschungen 81. Ostfildern: Jan Thorbecke Verlag, 2015.

Viúla de Faria, Tiago. "Tracing the 'chemyn de Portyngale': English Service and Servicemen in Fourteenth-Century Portugal." *Journal of Medieval History* 37 (2011): 257–268.

von Jeroscin, Nicolaus, and Mary Fischer, trans. *The Chronicle of Prussia by Nicolaus von Jeroschin: A History of the Teutonic Knights in Prussia, 1190–1331*. Farnham: Ashgate, 2011.

Warnicke, Retha M. *The Marrying of Anne of Cleves: Royal Protocol in Early Modern England*. Cambridge: Cambridge University Press, 2000.

Zajac, Talia. "Remembrance and Erasure of Objects Belonging to Rus' Princesses in Medieval Western Sources: The Cases of Anastasia Iaroslavna's 'Saber of Charlemagne' and Anna Iaroslavna's Red Gem." In *Moving Women Moving Objects (400–1500)*, edited by Tracy Chapman Hamilton and Mariah Proctor-Tiffany, 33–58. Leiden: Brill, 2019.

Zsoldos, Attila. *The Árpáds and Their Wives: Queenship in Early Medieval Hungary 1000–1301*. Rome: Viella, 2019.

13

ENCOUNTERING THE HEATHEN ON THE BALTIC FRONTIER

The Order of the Sword Brethren and the Teutonic Order in Thirteenth-Century Livonia

Verena Schenk zu Schweinsberg[1]

The Christianization and conquest of the Baltic in the twelfth and thirteenth centuries was a complex process, in which many different indigenous and foreign, religious and secular groups were involved. One of them was the Order of the Sword Brethren (*Fratres Militiae Christi de Livonia*, Ger. *Schwertbrüderorden*), which was founded during the Livonian Crusade in 1202 and was later incorporated into the Teutonic Order. It can be considered a military diaspora according to this volume's definition because of the shared ethnical, religious, regional, and social background of its members, their distinct collective identity and the Order's status as one of the main territorial as well as military powers in Livonia. The Order's knight brothers were given a great amount of independence in the territories they conquered and ruled. During their presence in the Baltic, they became entangled in multiple ways, in cooperation and conflict, with the indigenous population and other foreign groups such as missionaries, crusaders, and bishops.

The chapter will examine to what extent the situation of permanent conflict, especially the fight against the heathen,[2] was constitutive for the knight brothers' identity in addition to the essential elements as a military diasporic group—shared ethnical background and service in a foreign land. It will also ask if, and which, further factors contributed to the collective identity of this military, but also religious, community, garrisoned in a foreign, forbidding land. How were heathen and Christian groups perceived by this order and how was their perception influenced by the knight brothers' diasporic experience? What forms of interaction and encounter were possible and how were these encounters evaluated? The analysis will focus on the "Livonian Rhymed Chronicle" (Ger. *Livländische Reimchronik*),[3] an extensive contemporary, vernacular source whose author was most likely an associate of the Teutonic Order. It provides valuable insight into the knight brothers' perceptions of themselves as well as of their enemies and allies. This chapter shall concentrate on two main areas: First, religion

DOI: 10.4324/9781003245568-14

and Christianization, including the estimation and transfer of religious ideas or practices; second, military conflict, including the description and assessment of heathen allies and enemies.

Christianization, Expansion, Conquest—The Order of the Sword Brethren and the Teutonic Order in Livonia

The geographical region that is called "Livonia" in the medieval sources roughly corresponds to the borders of modern day Estonia and Latvia. The medieval name derives from the settlement areas of the Livs at the mouth of the river Düna, but as a result of the thirteenth-century conquest, the term was used to describe a wider area, including several regions and ethnic groups like the Lettgallians, Estonians or Semgallians.[4] These groups were not only divided into many independent chiefdoms, which belonged to two different linguistic groups—Baltic and Finno-Ugric—but were also partially in conflict with each other.[5] Apart from archaeological evidence, all information on the social, religious, and political organization of the indigenous population, as well as on its perception of Christianization and conquest, derive from Christian sources, for there are no medieval indigenous accounts extant today.[6]

Foreign expansion in the fields of economy, religion, and rulership started with early missionary attempts in the region by Swedish missionaries and crusaders from ca. 1160 onwards but were promoted, especially with the arrival of German traders and missionaries along the river Düna around the year 1180. In 1186/1188, the missionary Meinhard was consecrated bishop of the see of Üxküll, which was subordinate to the archbishopric of Hamburg-Bremen. However, from the outset, missionaries and traders were confronted with apostatizing Livs and attacks from other tribes, so that armed forces were necessary to protect the Christian presence from early on.[7] In the 1190s, after Popes Celestine III and Innocent III granted full remission of sins for taking the cross to protect the Livonian mission against heathens and apostates, successive bishops brought an increasing number of crusaders with them to Livonia.[8]

Bishop Albert of Buxhövden (1199–1229), in particular, furthered the missionary, military, and institutional expansion in the region. In this context, he was also involved in the foundation of the city of Riga in 1201 and the Order of the Sword Brethren in 1202.[9] The military knightly order was established according to the model of the Knights Templar by the Cistercian monk Theoderic of Treiden, confirmed by Pope Innocent III in 1204, and put under the authority of the bishop of Riga.[10] Although the numbers of knight brethren remained much smaller than those of the bishop's seasonal crusaders and mercenaries, they established themselves successfully as a powerful, mounted, and heavily armed "elite unit" which commanded crusaders and indigenous forces in battle.[11]

The members were organized in convents, divided into knight brothers, priest brothers, and lay brothers, mostly originating from the lower and ministerial nobility of various German regions.[12] The higher offices were almost all held by German aristocrats who only remained in Livonia for a short period of time. However, there is also evidence for lay brothers of indigenous Baltic origin—an interesting exception from the definition of a military diaspora as a community with a common ethnic background.[13] Unlike the crusaders, who in most cases only stayed in Livonia for one season, the Sword Brethren were permanently garrisoned in their castles. This was one of the important elements constituting them as a military diaspora in contrast to the other Christian foreign military forces present in Livonia.

Due to an agreement reached with Albert of Buxhövden, awarding the Sword Brethren a third of conquered territory and booty, the Brethren were able to establish themselves as the main power in Livonia besides the local bishoprics and the Kingdom of Denmark.[14] The Brethren were not mainly involved in or commissioned with the conversion or baptism of the population, which was the responsibility of the bishops. However, to enforce their territorial interests, the Order also supported forced baptisms of the population as part of its subjection.[15] In addition, several priest brethren involved in the Christianization are mentioned in Henry of Livonia's *Chronicon Livoniae*.[16] The spheres of conquest and mission were deeply entangled and a clear separation of the "peacefully baptizing bishop" and the conquering and exploiting "Sword Brethren" is not possible.[17]

For the thirteenth century, it has to be pointed out that the conquest of Livonia and the adjoining territories was in no way a straightforward expansion but was accompanied by military setbacks, apostasy of neophytes, and changing coalitions on both sides. Many indigenous groups allied themselves in changing coalitions with the bishops and the Order and were thereby able to remain relatively independent. In other cases, subjugated and baptized parts of the population could be obliged to military service but remained personally free. The different forms of control and dependency could range from military alliances and formal submission to Christian law to direct foreign rule or enslavement.[18]

The long struggle against the Semgallians and Lithuanians that lasted for more than a century, in particular, had fatal consequences for the Sword Brethren. In the battle of Saule in 1236, they lost more than a third of their knights and had to be incorporated into the Teutonic Order in the following year to survive. A union with the Teutonic Order had been sought for several years, but the actual loss of independence after the battle did not proceed without internal and episcopal opposition.[19] It took the Brethren more than 20 years to reestablish their territorial base, just to be defeated again in 1260 in Curonia. The Kurs and Semgallians would be subjected by the end of the century, but Lithuania, where large parts of the Semgallian population fled to, remained independent.[20]

Furthermore, internal rivalry and conflicts with the Rigan bishop and Denmark challenged the Order's position in thirteenth-century Livonia. The Danish conquest and permanent presence in Northern Estonia, established from 1219 onwards, was contested by the Order but finally accepted in the Treaty of Stensby in 1238.[21] The Order's conflicts with the Rigan bishopric (archbishopric from 1255 onwards) over territories, seigniorial rights and rule were of more permanent nature and increased after the incorporation of the Sword Brethren into the Teutonic Order. The Brethren's relationship with the bishops was characterized on the one hand by their formal subordination to the bishopric and partial cooperation in military or church organization, and on the other by their efforts to expand influence and invert this balance of power. Many of the complaints from both sides concerned the violation of rights, such as the obstruction of missionary activity in the Order's territories, but also violent abuses against members of the other party, theft or damage of property.[22]

As the Sword Brethren lost their organizational independence and a lot of their members by becoming a branch of the Teutonic Order, both military orders in Livonia will now be considered as one military diaspora because of their consistent collective identity, reflected in their self-perception, world view, aims, and enemies.[23] Moreover, their shared origin and social background—mostly ministeriality or low nobility from Low German regions—made them a homogenous foreign military presence in Livonia.[24]

The Livonian Rhymed Chronicle

The most comprehensive account of the knight brothers' activities in thirteenth-century Livonia, as well as on their view of themselves and others, is the aforementioned "Livonian Rhymed Chronicle". It is the oldest vernacular (Middle High German) chronicle from Livonia and encompasses the period from the first trading contacts between Germans and Livs to the subjection of the Semgallians in 1290. The author is unknown, but it is thought that he was a knight brother of the Teutonic Order, who partly reported first-hand experiences and partly relied on unknown written and oral sources.[25] The function and intended audience of the chronicle are still under debate. It seems highly convincing though that it was written for members of the Order of the Sword Brethren and maybe other visiting crusaders to commemorate its past deeds. Thereby it could also strengthen the members' enthusiasm for battle and self-perception as a fighting community.[26]

The chronicle clearly presents the history of the Sword Brethren: The plot and descriptions are focused on the deeds, successes, and setbacks of the knights and on the continuous fights, which are described vividly and in great detail. Mission, Christianization, and the development of the Livonian church are only treated marginally and primarily in the first part of the chronicle that describes the early mission up to the foundation of the Order of the Sword Brethren.[27] Furthermore, little attention is paid to agents outside of the Order, such as the bishops, cities, or merchants, and conflicts between the different parties are marginalized. Although the author clearly sides with the Order and shows a slightly anticlerical position, he does not criticize the Livonian bishops directly. By masking the conflicts with them, he rather conveys an impression of joint action and common purpose in past times that contrasts the tense situation of his time (late thirteenth or early fourteenth century).[28]

In addition, by leaving out topics, events, and conflicts that are irrelevant or would shed an unfavourable light on the Order, the Chronicle reduces complexity and evokes a coherent, stable worldview based on clear constellations of friend and foe. This view is only given from the Order's perspective and, according to Volker Honemann and Hartmut Kugler, it reinforces a sense of shared identity.[29] Recently it has been argued that the collective identity of the Order is portrayed in the Chronicle through the narration of continuous battles against the heathen, which characterize the Order's members as a community of fighters on the frontier. From this point of view, religious or cultural motives for the Order's presence in Livonia are of subordinate importance.[30] Due to the narration of constant fighting against the heathens in Livonia, the incorporation of the Sword Brethren into the Teutonic Order is presented by the Chronicle as a smooth transition—which it was in no way according to parallel sources.[31]

In the following two sections, these considerations concerning the knight brothers' self-perception as a foreign fighting community in Livonia—a military diaspora—shall be reviewed on the basis of two thematic aspects in the "Livonian Rhymed Chronicle": Religion and military conflict were chosen as two areas, where the encounter of the knight brothers with other agents in Livonia could provoke statements on themselves and the other.

Encountering the Heathen: Religion and Christianization

Although the chronicler did not focus much on the Christianization of Livonia, we find a lot of statements about "heathen" religious ideas and practices in the text. Of course, no coherent account of the indigenous religion is given, for the chronicle's focus is on the history of conquest, but nevertheless, the occasional hints that he gives present an ambivalent picture of the brothers' perception of heathenism.

There are descriptions corresponding to the traditional biblical and patristic interpretation of heathen deities as demons that have to be combated.[32] They are called *apgôte*—idols—in several cases, a notion which appears to be associated or identical with the devil and carries a negative connotation.[33] Nevertheless, the existence and agency of these deities are not denied, as the following example shows: "I am convinced that the devil himself was leading them, for no army ever moved so arrogantly into foreign lands as did this one. [...] They crossed the Osterhap at Swurben, for Perkune, their idol, had made the sea freeze harder than it ever had before".[34] We have to distinguish between the devil and the god *Perkune*, whose existence is not contested. Both appear to be allies of the heathen Lithuanians and a real threat to the Christian troops, especially because *Perkune* is attributed the power to freeze the strait *Osterhap*. Each side seems to be supported by their own supernatural forces in battle. However, this should not be interpreted as tolerance for heathen religious beliefs but as a functional use of these beliefs in the narrative: By attributing power to the enemy's deity, the author could let his account appear even more dramatic—the knights' battles appear to be a part of a fateful divine conflict. Elements of heathen religion could therefore be used to serve the author's intentions: The fight on the Livonian frontier could appear even more urgent or necessary to the audience and reinforce a sense of community.

However, the author is inconsistent in his descriptions and judgements: In the majority of cases, heathen deities are not at all demonized but neutrally described as *gote*, and the author does not comment on the worship of them in a negative way.[35] Moreover, practices such as divination and making thank offerings after the battle, not only seem to be tolerated (in case of the Order's allies), but are also depicted as sensible and effective, as can be seen from the example of the allied Kurs: "They gladly joined the expedition, because they expected to succeed. Their oracle sticks had fallen propitiously, and their birds had sung favorably, and from all this, they had concluded that everything would go well for them".[36] For a better understanding, the context of this ritual has to be considered: It was performed after the Kurs joined the knight brothers so that the term "they" could also include the Christian troops. In this case, it is implied that the Christian troops would not only accept the practice but also believe in its efficacy. This interpretation is also supported by the fact that the forecast came true and the allies won the battle.[37] This acceptance and adoption of a heathen ritual is not even commented on by the author, which indicates that for his audience, it had to appear as a normal incident that did not need to be denigrated or explained.[38]

This remarkable process of transferring and integrating heathen religious beliefs and practices becomes even more apparent in the context of thank offerings. Occurrences like the following are described several times after successful expeditions:

> [...] they all immediately praised God in heaven for having mercifully defended Christendom with this expedition. With the counsel of his brothers, the master gave a part of the booty to our Lord, because He had given them a safe journey. He had earned His share, and they gave Him weapons and horses.[39]

The offering of horses and weapons, a typical heathen ritual several times depicted in the chronicle, is performed by the Livonian master in front of the knight brothers and citizens of Riga.[40] It is combined with a traditional Christian practice, a thanksgiving prayer—seemingly without contradiction for the author and audience, for no explanation or comment is given for the incident. In addition, the ritual is described exactly in the same way as the heathen thank offerings in the chronicle—except for the substitution of "gods" with "Lord" or "God".[41]

Even though the bishop's accusations against the Order actually included the performance of heathen rituals, such as divination or cremation of the dead, it cannot be said with certainty if such allegations were based on actual occurrences.[42] Syncretisms and mutual transfer of practices like these are quite probable in rather recently and superficially Christianized territories.[43] However, irrespective of the validity, it should rather be asked why the author would mention these adopted practices at all in his account and not simply conceal or omit them.

As already mentioned, the topic of Christianization seems to be of little relevance to the author and his audience, consisting of members of the Order. Moreover, for the Order's purpose and identity displayed in the chronicle—a fighting community on the frontier—Christian mission and religious zeal played only a marginal part.[44] Therefore, it was probably irrelevant to them if their heathen or formally Christianized allies cast lots before battle as long as they remained loyal. In a chronicle that presents a history of conquest and was part of the historiography of a military order, the intermixture of Christian and heathen practices or adoption of their rituals does not reveal the failure of missionary efforts and therefore does not diminish the "success story" presented. As Michael Neecke also points out, this "acculturation" is not presented as new, unknown, or deprecated in the chronicle: The brothers' self-perception as well as one of its major elements—the regional conflict against the heathen—could not be threatened by it, but remained stable and continuous.[45] In addition, the episodes could maybe even confirm the audience's feeling of belonging to a successful fighting community, with the rituals being obviously effective and thus helping the brothers and their allies in battle.[46]

Encountering the Heathen: Military Conflict

Military activity, the Order's expeditions into heathen territory and battles are the predominant topics of the "Livonian Rhymed Chronicle", and they are also an essential, if not the most important, constituent of the knight brothers' concept of collective identity. Thus, military conflict is considered as a second field of encounter between the brothers and the indigenous heathen population. As already mentioned, heathen or neophyte warriors could fight on the Order's and bishop's side as well as on the side of the indigenous tribes. Changing loyalties of individual fighters or whole tribes, even in battle, was not uncommon according to the chronicle and even a Christian renegade is mentioned.[47] Heathen or neophyte allies could be highly praised as well as depicted as disloyal or unsteady. However, the characterization is not based upon religion but rather on performance in battle and general reliability.[48] A good example of changing characterizations are the relatively independent Kurs, who are depicted as brave and skilled heroes in battle until they desert the knight brothers against the Lithuanians at the Battle of Durben (1260), whereupon the Christian forces are defeated.[49] Here, the author emphasizes that the Order had relied on the Kurs and had not known about their treason.[50] Thus, it becomes evident that the negative depiction of allies as disloyal was also used to explain some of the Order's defeats in battle.

Even more revealing for the Order's perception of itself and others is the portrayal of enemies in the chronicle. The heathen combatants are firmly incorporated in the continuous constellation of conflicts that structure the plot and that are essential to present this account of Livonian history as the historiography of a military order. Therefore, it could be expected that the author made use of negative stereotypes based on the enemy's heathenism to confirm or reinforce the Order's purpose. However, instead of elaborate crusading rhetoric only general, unchristian character traits like the *bôsheit*[51] (wickedness) of the heathens or *hôchvart*[52] (arrogance) can be found.

Apart from these general stereotypes, one finds very few religious or ideologically motivated characterizations in the chronicle. On the contrary, many descriptions express social or military equality, appreciation, or even admiration. The Christian forces are not depicted as generally superior because of their religion and their defeats are described in many cases without underplaying the enemy's military performance.[53] Both parties also resemble each other in tactics, aims, and warfare, such as plundering, devastating the enemy's territory, and robbing women and children. There is obviously no need to justify these actions in the chronicle.[54] Frequently, Christian and heathen armies are described using the same external attributes like golden helmets or glittering armour that let them appear similarly impressive and maybe also frightening.[55] In addition to this, equality in social rank is ascribed to heathen rulers or nobles in the context of military or political negotiations. By the use of titles, they are identified as members of the nobility: The title *kunic* (king) is used for heathen rulers such as the Lithuanian Mindaugas and the Semgallian Vesthard and they are treated with respect according to their noble rank, even when imprisoned.[56] Furthermore, positive character traits and knightly virtues are attributed to Christians and heathens alike, such as *stolz* (pride, splendidness, goodliness) and the striving for *êre* (fame/glory).[57] These virtues had, without a doubt, a certain relevance and positive connotation for the author and his audience since the members of both orders and the visiting crusaders originated for the most part from a noble background—the social group that generated and upheld this ideal in knightly culture.[58]

Christians and heathens are not only described using identical attributes but were even explicitly placed side by side by the chronicler to emphasize their equality in battle: "There was a wild hacking and hewing on both sides, Christian and heathen, and the blood of men from both armies spilled onto the ice. It was a fierce battle in which many brave and excellent men on both sides were piteously struck down".[59] Both sides are similarly depicted as bravely fighting and giving their lives on the battlefield. This does not lessen at all the brutality and aggressiveness of the conflict, but the enmity rather appears to be a natural and accepted basis for the Order's purpose and actions in Livonia than to be founded on feelings of religious alterity.

The middle-high German term for a formidable warrior, epitomizing knightly virtues is *helt* (hero).[60] It is used very often in the "Livonian Rhymed Chronicle" to distinguish members or masters of the Order for their achievements in the conquest of Livonia. However, in several cases, the term is also used to refer to individual heathen warriors or groups—a revealing fact for the Order's perception of their enemies. An example is the Samogitian named Aleman: "At this time there was a man in Samogitia named Aleman, a bold hero, best of all the Samogitians".[61] Aleman can be described as a hero because of his skills and abilities, even though he is said to hate the Christians and provoked an attack on the Christianized Kurs.[62]

Lastly, Christians and heathens could also appear side by side as heroes in battle, evoking the impression of a fated community of warriors.[63] The religious discrepancy as a cause of the conflict seems to be nearly irrelevant. The interaction of heroism on both sides and their community in death becomes particularly apparent from the example of a defeat against the Lithuanians in 1278: "Both sides began to fight. They hacked deep wounds and blood flowed across the snow. Many dauntless warriors [literally: heroes, V. S.], daring and excellent men from both sides, Christian and heathen, fell in grim death. The snow turned red with blood".[64]

Encountering the heathen in battle therefore could mean encountering an equal enemy, who could be esteemed or even admired for his fighting abilities.[65] This can again be explained with the irrelevance the topic of Christianization obviously had for the author. He was probably more interested in commemorating the conquering expeditions of the Order for an audience that was most likely familiar with and interested in battles anyway. For them, the heathen enemies were above all military adversaries who could be brave, cowardly, and heroic like all others.

However, this respect for the enemies of the Order, and even the adoption of some of their religious practices, should not to be misunderstood as religious or cultural tolerance. The heathens are treated without question as enemies to be fought with great brutality throughout the chronicle. In addition, the brothers and also the chronicler "needed" them as constant legitimation for the Brethren's existence and purpose since a major part of their self-image was based on the continuing regional fight against the heathen. Yet this enmity appeared so self-evident for the audience that it did not need an elaborate religious or ideological justification. Religious difference as an original cause for the conflict fades into the background over the course of the narrative.

Conclusion

In thirteenth-century Livonia, the Order of the Sword Brethren and later on the Teutonic Order can be considered as military diasporas especially with regard to their members' background and their consistent collective identity. In the "Livonian Rhymed Chronicle", this concept of identity is primarily shaped by the notion of the continuous fight against the heathen to secure and enlarge the Sword Brethren's territory in Livonia—a purpose, so fundamental and self-evident to its members that an actual religious motivation apart from stereotypical phrases is hard to trace in the text.

According to the chronicle, Christian encounters with the heathen could result in ambivalent depictions and estimations of religious ideas or practices: Gods are described as demons but also as powerful entities. Religious practices of heathen allies, like forecasts and thank offerings, could be tolerated or even adopted. Seemingly this way of presenting aspects of heathen religion to the audience was possible because missionary zeal and the promotion of Christianization did not play an important part in the self-perception of the Order—and did not pose any threat to its successes that should be commemorated in the chronicle. Powerful antagonistic deities could even underline the urgency of the Order's endeavour, which the chronicler wanted to support. Therefore, even if the author's descriptions were not based on actual incidents, they had to be intelligible to an audience most likely from within the Order and its direct environment. If the chronicle wanted to commemorate the Order's past, to entertain or to motivate its members and fighters, the audience had to approve and share its evaluations as well as to identify with the narrative.

Furthermore, the knight brothers' self-perception as a fighting community allows the author to portray heathen fighters in battle as coequal or sometimes even superior enemies and to appreciate their social rank and military abilities. Also, the Christians' indigenous allies are not judged on the basis of their religion but rather on their reliability and performance. The heathen combatants did not need to be condemned because of their religious difference and the original cause of the conflict seems to have been of little importance. On the contrary, both sides resemble each other and both are described using identical attributes, titles, and knightly virtues. Heathen warriors could be depicted as skilled fighters and could even be honoured as heroes who fought as bravely as the Christian knights and met the same fate in battle.

Notes

1 This chapter is based on the results of my master's thesis entitled "Christians and Heathens in Thirteenth-Century Livonian Chronicles", which was submitted at Heidelberg University in August 2013. I would like to thank Dr. Benjamin Müsegades and Katharina Reif for the critical revision of the manuscript and Prof. Dr. Julia Burkhardt for helpful references and advice.

2 The term "heathen" has a pejorative connotation in its historic and contemporary use and can only be defined in relation to Christianity, not as a meaningful concept in itself. Despite this difficulty, it should be used here, for this paper focuses on the medieval, Christian perspective on the "other"—a point of view implemented by the term "heathen", which is also used in the sources. In addition, an adequate substitutional term is still lacking since "pagan" or "non-Christian" similarly define the subject *ex negativo* from a Christian perspective.

3 Leo Meyer, ed., *Livländische Reimchronik* (Paderborn: Ferdinand Schöningh, 1876, repr. Hildesheim: Georg Olms Verlagsbuchhandlung, 1963). All quotations are cited from the—unfortunately not always accurate—English prose translation Jerry C. Smith and William L. Urban, trans., *The Livonian Rhymed Chronicle*, rev. ed. (Chicago: Lithuanian Research and Studies Center, 2001). Due to the focus and scope of this paper, Henry of Livonia's *Chronicon Livoniae*, the second comprehensive chronicle which was analyzed in my master's thesis, shall be considered only marginally here.

4 Markus Wüst, "Westliche Einflüsse auf den Verlauf der Christianisierung in Livland," in *Christianisierung Europas: Entstehung, Entwicklung und Konsolidierung im archäologischen Befund / Christianisation of Europe: Archaeological Evidence for Its Creation, Development and Consolidation*, ed. Orsolya Heinrich-Tamáska, Niklot Krohn, and Sebastian Ristow (Regensburg: Schnell+Steiner, 2012), 483–496, at 483–485; Ēvalds Mugurēvičs, "Die 'ältere Livländische Reimchronik' über die ethnische hyphenation: Situation im Baltischen Raum," in *Deutschsprachige Literatur des Mittelalters im östlichen Europa: Forschungsstand und Forschungsperspektiven*, ed. Ralf G. Päsler and Dietrich Schmidtke, Beiträge zur älteren Literaturgeschichte (Heidelberg: Winter, 2006), 267–273, at 270.

5 Alan V. Murray, "Henry the Interpreter: Language, Orality and Communication in the Thirteenth-Century Livonian Mission," in *Crusading and Chronicle Writing on the Medieval Baltic Frontier: A Companion to the Chronicle of Henry of Livonia*, ed. Marek Tamm, Linda Kaljundi, and Carsten Selch Jensen (Farnham: Ashgate, 2011), 107–134, at 209–211; Eva Eihmane, "The Baltic Crusades: A Clash of Two Identities," in *The Clash of Cultures on the Medieval Baltic Frontier*, ed. Alan V. Murray (Farnham: Ashgate, 2009), 37–51, at 46–47. See also Eric Christiansen, *The Northern Crusades: The Baltic and the Catholic Frontier, 1100–1525* (Minneapolis, MN: University of Minnesota Press, 1980), 34–39.

6 Tiina Kala, "The Incorporation of the Northern Baltic Lands into the Western Christian World," in *Crusade and Conversion on the Baltic Frontier, 1150–1500*, ed. Alan V. Murray (Aldershot: Ashgate, 2001), 3–20, at 4; Wüst, "Westliche Einflüsse," 485. For the archaeological findings concerning the social and political organization of the indigenous population, see Andris Šnē, "The Emergence of Livonia: The Transformations of Social and Political Structures in the Territory of Latvia during the Twelfth and Thirteenth Centuries," in *The Clash of Cultures*, 53–71.

7 Tiina Kala, "Verkündigung und Kreuzpredigt in und für Livland im 13. Jahrhundert," in *Leonid Arbusow (1882–1951) und die Erforschung des mittelalterlichen Livland*, ed. Ilgvars Misans and Klaus Neitmann, Quellen und Studien zur baltischen Geschichte 24 (Cologne: Böhlau, 2014), 187–208, esp. 189–194; Id., "The Incorporation," 7–8; Manfred Hellmann, "Die Anfänge christlicher Mission in den baltischen Ländern," in *Studien über die Anfänge der Mission in Livland*, ed. id., Vorträge und Forschungen 37 (Sigmaringen: Thorbecke, 1989), 7–38, esp. 12–30. For a critical discussion of the supposed "peacefulness" of the early mission see Carsten Selch Jensen, "The Early Stage of Christianisation in Livonia in Modern Historical Writings and Contemporary Chronicles," in *Medieval History Writing and Crusading Ideology*, ed. Tuomas M. S. Lehtonen and Kurt Villads Jensen, Studia Fennica, Historica 9 (Helsinki: Finnish Literature Society, 2005), 207–215 and Raoul Zühlke, "Bischof Meinhard von Üxküll: Ein friedlicher Missionar? Ansätze zu einer Neubewertung; Ein quellenkundlicher Werkstattbericht," *Hansische Geschichtsblätter* 127 (2009): 100–121, at 110–120.

8 Iben Fonnesberg-Schmidt, *The Popes and the Baltic Crusades: 1147–1254*, The Northern World 26 (Leiden: Brill, 2007), here esp. chapters 1 and 2. See also Marek Tamm, "How to Justify a Crusade? The Conquest of Livonia and New Crusade Rhetoric in the Early Thirteenth Century," *Journal of Medieval History* 39 (2013): 431–455, esp. 435–437; Kala, "Verkündigung," 189–191 and 198–199; Jensen, "The Early Stage," 211–213.

9 Jensen, "The Early Stage," 212–213; William Urban, *The Baltic Crusade*, rev. ed. (Chicago: Lithuanian Research and Studies Center 1994), 50–54; Christiansen, *The Northern Crusades*, 93–95.

10 Alan V. Murray, "The Sword Brothers at War: Observations on the Military Activity of the Knighthood of Christ in the Conquest of Livonia and Estonia (1203–1227)," *Ordines Militares* 18 (2013): 27–37, at 28–29. The most comprehensive study on the Order remains Friedrich Benninghoven, *Der Orden der Schwertbrüder: Fratres Milicie Christi de Livonia*, Ostmitteleuropa in Vergangenheit und Gegenwart 9 (Cologne: Böhlau, 1965).

11 Murray, "The Sword Brothers," 29–30; Christiansen, *The Northern Crusades*, 77 and 95. The estimated numbers of knight brothers range from ca. 120 (Benninghoven) up to 180 (Militzer). For detailed information and further literature on armament and fighting techniques of the different groups involved, see Murray, "The Sword Brothers," 30–34.

12 Only for a small percentage of the brothers and only for those holding higher offices, name and origin could be identified, see Klaus Militzer, "The Recruitment of Brethren for the Teutonic Order in Livonia, 1237–1562," in *The Military Orders. Volume I: Fighting for the Faith and Caring for the Sick*, ed. Malcolm Barber (Aldershot: Variorum, 1994), 270–277, at 270–273. See also Lutz Fenske and Klaus Militzer, ed., *Ritterbrüder im livländischen Zweig des Deutschen Ordens*, Quellen und Studien zur baltischen Geschichte 12 (Cologne: Böhlau, 1993) for a detailed analysis and catalogue of the knight brethren in Livonia.

13 Militzer, "The Recruitment," 270–273; Urban, *The Baltic Crusade*, 66–68 and 81–82; Kaspars Kļaviņš, "The Significance of the Local Baltic Peoples in the Defence of Livonia (Late Thirteenth–Sixteenth Centuries)," in *The Clash of Cultures*, 321–340, at 332–333.

14 Murray, "The Sword Brothers," 30; Benninghoven, *Der Orden*, 77–80.

15 Sven Ekdahl, "Die Rolle der Ritterorden bei der Christianisierung der Liven und Letten," in *Gli inizi del cristianesimo in Livonia-Lettonia: Atti del colloquio internazionale di storia ecclesiastica in occasione dell'VIII centenario della Chiesa in Livonia (1186–1986), Roma, 24–25 Giugno 1986*, Pontifico Comitato di scienze storiche: Atti e documenti 1 (Vatican City: Libreria Ed. Vaticana, 1989), 203–243, at 233; Marie-Luise Favreau-Lille, "Mission to the Heathen in Prussia and Livonia: The Attitudes of the Religious Military Orders towards Christianization," in *Christianizing Peoples and Converting Individuals*, ed. Guyda Armstrong and Ian N. Wood, International Medieval Research 7 (Turnhout: Brepols, 2000), 147–154, at 150. See also Bernd Ulrich Hucker, "Zur Frömmigkeit von Livlandpilgern und Ordensrittern," in *Die Spiritualität der Ritterorden im Mittelalter*, ed. Zenon Hubert Nowak, Ordines militares 7 (Torún: Uniwersytet Mikolaja Kopernika, 1993), 111–130, at 121–122 for baptism of the heathen population as a condition for peace.

16 Leonid Arbusow and Albert Bauer, eds., *Heinrichs Livländische Chronik*, MGH Scriptores rerum Germanicarum in usum scholarum separatim editi 31 (Hannover: Hahnsche Buchhandlung, 1955), i.e. chap. 18.7.120 and 19.4.127 (Otto) or chap. 24.6.176 (Hartwig).

17 See Favreau-Lille, "Mission," 148–150 and 152 for a critical discussion on the reality of this separation of spheres.

18 Kļaviņš, "The Significance," esp. 322–324, 328–330, and 333, Murray, "The Sword Brothers," 33–34. On enslavement, see Sven Ekdahl, "The Treatment of Prisoners of War during the Fighting between the Teutonic Order and Lithuania," in *The Military Orders* 263–269, at 265.

19 Benninghoven, *Der Orden*, 308–312 and 354–356; Klaus Militzer, *Die Geschichte des Deutschen Ordens*, 2nd ed. Kohlhammer Urban Taschenbücher 713 (Stuttgart: Kohlhammer, 2012), 115–117; Volker Hentrich, "Die Darstellung des Schwertbrüderordens in der Livländischen Reimchronik (Ordenschronik, Missionsgeschichte oder nur 'Kriegstagebuch'?)," in *Perzeption und Rezeption: Wahrnehmung und Deutung im Mittelalter und in der Moderne*, ed. Joachim Laczny, Nova mediaevalia 12 (Göttingen: V&R Unipress, 2014), 107–154, at 139–140 and 143–144.

20 Militzer, *Die Geschichte*, 118–120; Christiansen, *The Northern Crusades*, 98–100.

21 On the Danish conquest and mission in Estonia and the conflicts with the Sword Brethren, see Ane L. Bysted, Carsten S. Jensen, Kurt V. Jensen, and John H. Lind, *Jerusalem in the North: Denmark and the Baltic Crusade, 1100–1552*, Outremer 1 (Turnhout: Brepols, 2012), esp. 195–225 and 269–271; Urban, *The Baltic Crusade*, 117–125, 129–132, and 188–189.

22 Eihmane, "The Baltic Crusades," 40–42; Benninghoven, *Der Orden*, 113–116 and 198–204 on the papal legates' efforts to settle the conflicts. See also Militzer, *Die*

Geschichte, 123–125; Matthias Thumser, "Das Baltikum im Mittelalter: Strukturen einer europäischen Geschichtsregion," *Jahrbuch des baltischen Deutschtums* 58 (2011): 17–30, at 23–27 and Anti Selart, "The Use and Uselessness of the Chronicle of Henry of Livonia in the Middle Ages," in *Crusading*, 345–361, at 347–348 for valuable summaries on the conflicts and further literature.

23 For a detailed discussion on the Teutonic Order as a military diaspora, see the contribution of Mark Whelan in this volume.

24 See Militzer, "The Recruitment," 272 and 275–276; Fenske and Militzer, ed., *Ritterbrüder.*

25 The most valid arguments are his local knowledge for the Order's radius of action, precise descriptions of military tactics, weapons and the Order's customs as well as the chronicle's structuring according to the periods of office of the Order's masters: Lutz Mackensen, "Zur livländischen Reimchronik," in *Zur deutschen Literatur Altlivlands: Untersuchungen von Lutz Mackensen*, Ostdeutsche Beiträge aus dem Göttinger Arbeitskreis or Arbeits- kreis 18 (Würzburg: Holzner, 1961), 21–58; Hartmut Kugler, "Die 'Livländische Reimchronik' des 13. Jahrhunderts," in *Latvijas Zinātņu Akadēmijas vēstis* A 9, no. 554 (1993): 22–30, at 22–23; Alan V. Murray, "The Structure, Genre and Intended Audience of the Livonian Rhymed Chronicle," in *Crusade*, 235–251, at 238–241; Mugurēvičs, "Die ältere 'Livländische Reimchronik'," 268–269; Udo Arnold, "Livländische Reimchronik," in *Die deutsche Literatur des Mittelalters: Verfasserlexikon*, vol. 5 (Berlin: De Gruyter, 1985), col. 855–862, at col. 855 and 861; Norbert Angermann, "Die mittelalterliche Chronistik," in *Geschichte der deutschbaltischen Geschichtsschreibung*, ed. Georg von Rauch, Ostmitteleuropa in Vergangenheit und Gegenwart 20 (Cologne: Böhlau, 1986), 3–20, at 10. A summary on the discussion about authorship is provided by Edith Feistner, Michael Neecke, and Gisela Vollmann-Profe, *Krieg im Visier: Bibelepik und Chronistik im Deutschen Orden als Modell korporativer Identitätsbildung* (Tübingen: Niemeyer, 2007), 89–94.

26 For a critical summary of the debate and state of the art see Feistner, Neecke, and Vollmann-Profe, *Krieg*, 99–104; Michael Neecke, *Literarische Strategien narrativer Identitätsbildung: Eine Untersuchung der frühen Chroniken des Deutschen Ordens*, Regensburger Beiträge zur deutschen Sprach- und Literaturwissenschaft/B. Untersuchungen 94 (Frankfurt am Main: Lang, 2008), 22–24 and Volker Honemann, "Zu Selbstverständnis und Identitätsvorstellungen in der livländischen Geschichtsschreibung des Mittelalters," in *Geschichtsschreibung im mittelalterlichen Livland*, ed. Matthias Thumser, Schriften der Baltischen Historischen Kommission 18 (Berlin: Lit, 2011), 255–297, at 270–271.

27 Honemann, "Zu Selbstverständnis," 266–267; Arnold, "Livländische Reimchronik," col. 856–857; Murray, "The Structure," 237–238.

28 Neecke, *Literarische Strategien*, 56; Hentrich, "Die Darstellung," 145–146 (with a list of events omitted by the author) and 148–149; Mackensen, "Zur livländischen Reimchronik," 21–25 and 28; Angermann, "Die mittelalterliche Chronistik," 10.

29 Hartmut Kugler, "Über die 'Livländische Reimchronik': Text, Gedächtnis und Topographie," *Jahrbuch der Brüder-Grimm-Gesellschaft* 2 (1992): 85–104, at 91–92; Kugler, "Die 'Livländische Reimchronik'," 23; Honemann, "Zu Selbstverständnis," esp. 263 and 267–268.

30 Honemann, "Zu Selbstverständnis," 263–267; Neecke, *Literarische Strategien*, 54–57; Edith Feistner, Michael Neecke, and Gisela Vollmann-Profe, "Ausbildung korporativer Identität im Deutschen Orden: Zum Verhältnis von Bibelepik und Ordenschronistik, Werkstattbericht," in *Deutschsprachige Literatur*, 57–74, at 66. Feistner, Neecke and Vollmann-Profe reduce the concept of identity presented in the chronicle to a "military *esprit de corps*" ("militärischer Corpsgeist"). See also iid., *Krieg*, 26–40 for general remarks on the construction of identity within the Teutonic Order.

31 See note 19 above; Neecke, *Literarische Strategien*, 55–56.

32 For the Baltic deities, this view was also confirmed by Pope Innocent III, his successors, and contemporary theologians: Rasa Mažeika, "Granting Power to Enemy Gods in the Chronicles of the Baltic Crusades," in *Medieval Frontiers: Concepts and*

Practices, ed. David Abulafia und Nora Berend (Aldershot: Ashgate, 2002), 153–171, at 160–161; Marek Tamm, "A New World in Old Words: The Eastern Baltic Region and the Cultural Geography of Medieval Europe," in *The Clash of Cultures*, 11–35, at 28–29.

33 For example Meyer, *Livländische Reimchronik*, vv. 1276–1277, p. 30 or v. 339, p. 8.

34 noch wên ich, / daz sie der tûvel vûrte: / kein her sich nie gerûrte / sô vrevelîchen in vremde lant, / sô von den selben wart bekannt. [...] zû Swurben vûren sie ubir sê, / daz ist genant daz Ôsterhap, / als ez Perkune ir apgot gap / daz nimmer sô hart gevrôs. Meyer, *Livländische Reimchronik*, vv. 1426–1437, p. 33; Smith and Urban, *The Livonian Rhymed Chronicle*, 18. See also Mažeika, "Granting Power," 162–163 and Rasma Lazda-Cazers, "Landscape as Other in the Livländische Reimchronik," *Amsterdamer Beiträge zur älteren Germanistik* 65 (2009): 183–209, at 190 on this episode.

35 See for example the descriptions of the enemies' thank offerings: die Sameiten teilten dô / pferde und wâren vollen vrô / und saiten iren goten danc, / daz an deme strîte in gelanc.—"The Samogitians divided the horses of the vanquished and happily gave thanks to their gods that they had won". Meyer, *Livländische Reimchronik*, vv. 4873–4876, p. 112; Smith and Urban, *The Livonian Rhymed Chronicle*, 53. See also ibid., 31, note 137 for further examples of divination practices.

36 sie waren algemeine vrô / und ir mût der stûnt alsô, / daz ez in solde wol ergân. / in viel vil dicke wol ir spân, / ir vogel in vil wol sanc: / sô prûveten sie, daz in gelanc. Meyer, *Livländische Reimchronik*, vv. 7229–7234, p. 166; Smith and Urban, *The Livonian Rhymed Chronicle*, 74. Compare the similar passage in vv. 2484–2487, Meyer, *Livländische Reimchronik*, 57: ir hertze stûnt nâch strîte gar, / in was der spân gevallen vol. / des wâren sie alle sturmes vol.—"Since the fall of the stick had been propitious for them, they were full of fight and eager for war". Smith and Urban, *The Livonian Rhymed Chronicle*, 31. See also Mažeika, "Granting Power," 164–165.

37 Meyer, *Livländische Reimchronik*, vv. 7245–7280; Mažeika, "Granting Power," 164–165; Neecke, *Literarische Strategien*, 58.

38 Another example of a divination ritual can be found for the Order's enemies, the Samogitians: ir blûtekirl der warf zû hant / sîn lôz nâch ir alden site: / zû hant er blûtete allez mite / ein quek, als er wol wiste. [... The blûtekirl said, V. S.:] ir sult geloben daz dritte teil / den goten, sô geschiet ûch heil. / werden ûch die gote gût, / sô werdet ir vil wol behût. / die gote die sint wol wert, / daz man brunjen und pfert / und ouch rische man dâ mite / burne nâch unser site.—"The priest of sacrifice threw his lots according to the ancient custom and also expertly sacrificed a living animal [...The priest said, V. S.:] If you promise a third of the spoils to the gods, you will meet with success, for if the gods are kindly disposed toward you, you will be well-protected. The gods well deserve that one should burn armor and horses as well as brave men for them in accordance with our customs". Meyer, *Livländische Reimchronik*, vv. 4680–4700, p. 108; Smith and Urban, *The Livonian Rhymed Chronicle*, 52. Even though a human sacrifice is demanded, the ritual is not commented on—on the contrary, the heathen *blûtekirl* is depicted as expert and the practice as ancient custom. Also in this case the forecast comes true later on, the Christians are defeated in battle and the demanded thank offering marks the logical conclusion of the episode.

39 sie lobeten algemeine sân / got von himelrîche, / daz er genêdeclîche / in der selben herevart / die cristenheit hatte bewart. / von der brûdere râte / der meister gab vil drâte / des roubes unserm hêrren teil, / wen er hatte in gegeben heil. sînes teiles was er wert: / man gab im wâpen unde pfert. Meyer, *Livländische Reimchronik*, vv. 3394–3404, p. 78; Smith and Urban, *The Livonian Rhymed Chronicle*, 40.

Further examples of Christian thank offerings can be found in vv. 2669–2675, Meyer, *Livländische Reimchronik*, 62; vv. 11775–11781, p. 269 and vv. 11990–11995, p. 274.

40 Although sporadic evidence for Christian thank offerings is known and their inner logic was appreciated from the Old Testament tradition of sacrifice, it was at least highly unusual and was considered typically heathen: Kugler, "Die 'Livländische Reimchronik'," 24; Robert Bartlett, "Reflections on Paganism and Christianity

in Medieval Europe," *Proceedings of the British Academy* 101 (1999): 55–76, at 61–67, esp. 66–67.

41 Kugler, "Die 'Livländische Reimchronik'," 24. Compare, for example, the heathen thank offerings in vv. 4873–4876 (Samogitian episode cited above), Meyer, *Livländische Reimchronik*, 112 with ibid., vv. 6085–6089, p. 140 (Lithuanians).

42 See Mažeika, "Granting Power," 165–166 for further references and a discussion on the credibility of these accusations. We find a similar debate for the knights Templar in the Levant, cf. for example, Malcolm Barber, "Was the Holy Land Betrayed in 1291?," *Reading Medieval Studies* 34 (2008): 35–52, esp. 36–38 and 45–46.

43 Tiina Kala, "Rural Society and Religious Innovation: Acceptance and Rejection of Catholicism among the Native Inhabitants of Medieval Livonia," in *The Clash of Cultures*, 169–190, esp. 170–171; Wüst, "Westliche Einflüsse," 292–293.

44 See p. 322 above; Kugler, "Die 'Livländische Reimchronik'," 25.

45 Neecke, *Literarische* Strategien, 58–59.

46 Mažeika calls this process of transfer "going native" and interprets it as a pragmatic consequence of the fighters' suspicion that the enemy's gods could be powerful and their rituals effective as well. Mažeika, "Granting Power," 157–158.

47 The archer Bertholt crosses over to the Semgallians: dâ was ein schalc, der hieȥ Bertholt, / dem wâren die Semegallen holt, / wen er was ein schutze; / er wart in sint vil nutze. / deme lieȥen sie daȥ leben, / ob er sich wolde zû in geben. / er tet daȥ und was vrô.—"There was a knave there named Berthold, and the Semgallians spared him because he was an archer. He later proved to be of great use to them. They offered him his life if he would join them, and this he gladly did". Meyer, *Livländische Reimchronik*, vv. 8631–8637, p. 198; Smith and Urban, *The Livonian Rhymed Chronicle*, 87.

48 Juhan Kreem, "Der Deutsche Orden in Livland: Die Heiden, Landvolk und Undeutsche in der livländischen Heeresverfassung," in *L'Ordine Teutonico tra Mediterraneo e Baltico: Incontri e scontri tra religioni, popoli e culture / Der Deutsche Orden zwischen Mittelmeerraum und Baltikum: Begegnungen und Konfrontationen zwischen Religionen, Völkern und Kulturen*, ed. Hubert Houben and Kristjan Toomaspoeg, Acta Theutonica 5 (Galatina: Congedo, 2008), 237–251, at 241; Kļaviņš, "The Significance," 328–330.

49 die Kûren wolden des nicht lân, / sie enwerten kint und wîb / ir herren und irs selbes lîb, / dar zû burge und lant. / den brûderen quâmen sie zû hant—"The Kurs resolved to defend their women and children, themselves and their lords, their castles and land, and so they came to the brothers with a determined force"; die Kûren dâ mit heldes hant / werten wol ir selbes lant.—"The Kurs bravely [literally: with a hero's hand, V. S.] defended their own country […]". Meyer, *Livländische Reimchronik*, vv. 2478–2482 and 2541–2542, p. 57 and 59; Smith and Urban, *The Livonian Rhymed Chronicle*, 31.

50 die Kûren hatten vor gedâcht / ein ding, daȥ wart vollenbrâcht / zû den selben zîten: / sie enwolden nicht da strîten […] dô daȥ die Eisten sâhen, / sie begunden gâhen / vaste mit in von dannen—"[…] it happened that the Kurs refused to fight. They had decided this beforehand, […] And so they withdrew, and when the Estonians saw this, they fled as well"; dâ wurden in der nôt gelân / die brûdere und die Semen gût. / sie enhatten alle keinen mût, / daȥ iemant solde vlîhen dan.—"The brothers and the good Samites were abandoned in time of need. Not a one them had imagined that their allies might flee, […]". Meyer, *Livländische Reimchronik*, vv. 5601–5617 and 5636–5639, p. 129; Smith and Urban, *The Livonian Rhymed Chronicle*, 59 and 60.

51 For example, Meyer, *Livländische Reimchronik*, vv. 3214–3215, p. 74 and vv. 11807–11808, p. 270 (*bôse heidenschaft* as a threat to Christianity). See also Neecke, *Literarische Strategien*, 57.

52 For example Meyer, *Livländische Reimchronik*, vv. 4154–4155, p. 96 (the Livonian master wants to put an end to the Samogitian arrogance). See also Mary Fischer, *"Di Himels Rote": The Idea of Christian Chivalry in the Chronicles of the Teutonic Order*, Göppinger Arbeiten zur Germanistik 525 (Göppingen: Kümmerle, 1991), 182–183 and Lazda-Cazers, "Landscape," 191 for further examples.

53 For example in a battle against the Lithuanians: sie enmûsten von der walstat / wîchen, wen sie wâren mat / von den heiden wurden al. / dâ nam die cristenheit den val.—"[...] the brothers and the crusaders had no choice but to retreat from the battlefield. The heathens had overwhelmed them at all points. Christianity suffered a defeat". Meyer, *Livländische Reimchronik*, vv. 6063–6066, p. 139; Smith and Urban, *The Livonian Rhymed Chronicle*, 63. See also Lazda-Cazers, "Landscape," 189; Kugler, "Die 'Livländische Reimchronik'," 24.

54 Kugler, "Die 'Livländische Reimchronik'," 24.

55 Compare for example the description of the Lithuanian fighters' armour with that of the Christians: ouch was man wurden gewar / mancher brunjen wunneclîch. / ir helme wâren von golde rîch, / eʒ lûchte alsam ein spîgelglas. / waʒ gesmîdes an in was, / daʒ schien alleʒ silbervar.—"They reported seeing many wonderful breastplates, helmets of rich gold which glistened like mirrors, and weapons that shone like silver"; man sach helme und schilde / glîʒen ûf dem gevilde, / die brunjen blenken sam ein glas,—"One could see helmets and shields glistering on the field and countless breastplates shining like glass". Meyer, *Livländische Reimchronik*, vv. 5016–5021, p. 115 and vv. 3281–3284, p. 76; Smith and Urban, *The Livonian Rhymed Chronicle*, 55 and 39. See also Fischer, *"Di Himels Rote,"* 179 for further examples.

56 For Mindaugas see especially the episode Meyer, *Livländische Reimchronik*, vv. 3451–3576, pp. 80–82; for Vesthard ibid., v. 1700, p. 39. See also Rasa Mažeika, "When Crusader and Pagan Agree: Conversion as a Point of Honour in the Baptism of King Mindaugas of Lithuania (c. 1240–1263)," in *Crusade*, 197–214. For the treatment and ransom of noble prisoners, see the imprisonment of the Lithuanian Lengewin: Meyer, *Livländische Reimchronik*, vv. 2961–2961, 3018–3054, and 3069–3074, pp. 68, 70–71.

57 For *stolz* compare for example Meyer, *Livländische Reimchronik*, vv. 1420–1421, p. 33 and vv. 2498–2501, p. 58. For *êre* as an incentive to fight compare ibid., vv. 608–613, p. 15 and vv. 906–910, p. 21. See also Lazda-Cazers, "Landscape," 191 and Fischer, *"Di Himels Rote,"* 184 on secular motives for crusading.

58 Raisa Mažeika, "An Amicable Enmity: Some Peculiarities in Teutonic-Balt Relations in the Chronicles of the Baltic Crusades," in *The Germans and the East*, ed. Charles W. Ingrao and Franz A. J. Szabo, Central European Studies (West Lafayette: Purdue University Press, 2008), 49–58, at 49–50. For an overview on knightly virtues as an ideal see Werner Paravicini, *Die ritterlich-höfische Kultur des Mittelalters*, Enzyklopädie deutscher Geschichte 32 (München: Oldenbourg, 1999), esp. 7–8; Jörg Arentzen and Uwe Ruberg, eds., *Die Ritteridee in der deutschen Literatur des Mittelalters: Eine kommentierte Anthologie*, 2nd ed. (Darmstadt: WBG, 2011), 16–18.

59 ein ungevûgeʒ howen / von den heren beiden, / von cristen und von heiden. / der strît was starc und grôʒ, / daʒ blût ûf dem îse vlôʒ / von ir beider sîten. / dô gienc eʒ an ein strîten, / dô sach man manchen rischen man / ellenthaften howen an. Meyer, *Livländische Reimchronik*, vv. 7896–7904, p. 181; Smith and Urban, *The Livonian Rhymed Chronicle*, 80.

60 If the term *helt* was considered archaic and referred to the older German heroic epic at the time, the *Livonian Rhymed Chronicle* was composed, is being discussed by Murray, "The Structure," 246–247; Kugler, "Die 'Livländische Reimchronik'," 23 and Feistner, Neecke, and Vollmann-Profe, *Krieg*, 96–99.

61 Zû Sameiten was ein man / bie der zît hieʒ Aleman; / der was ein vil vromer helt / von Sameiten ûʒ erwelt. Meyer, *Livländische Reimchronik*, vv. 4085–4088, p. 94; Smith and Urban, *The Livonian Rhymed Chronicle*, 47.

62 A further example is the Semgallian Schabe, who advises his apostatizing tribe to let the Christian advocates leave the land unharmed: Meyer, *Livländische Reimchronik*, vv. 5246–5253, pp. 120–121. Examples of heathen troops described as heroes can be found ibid., vv. 8991–8993, p. 206 or vv. 9195–9199, pp. 210–211.

63 Neecke, *Literarische Strategien*, 60.

64 dô gienc eꝫ an ein strîten / von ir beider sîten. / die wunden hieb man dâ sô grôꝫ, / daꝫ ir blût durch den snê vlôꝫ. / des sach man von in beiden / von cristen und von heiden / manchen unverzageten helt, / beide rasch und ûꝫ erwelt, / sturtzen in den grimmen tôt. / der snê was dâ von blûte rôt. Meyer, *Livländische Reimchronik*, vv. 8393–8402, p. 192; Smith and Urban, *The Livonian Rhymed Chronicle*, 85. See also the following passage that emphasises the shared fate of both parties in the cold of winter: der winter der was alsô kalt, / daꝫ es manich mensche entkalt / von cristen und von heiden. / dâ ervrôs von in beiden / manich unverzageter helt, / kûne und dâ bie ûꝫ erwelt.—"[…] the winter was so very cold that many men, both Christian and heathen, suffered greatly. Many undaunted warriors, daring and excellent men on both sides, froze to death". Meyer, *Livländische Reimchronik*, vv. 8489–8494, pp. 194–195; Smith and Urban, *The Livonian Rhymed Chronicle*, 86. See also Fischer, *"Di Himels Rote,"* 182–184.

65 See Honemann, "Zu Selbstverständnis," 264; Feistner, Neecke, and Vollmann-Profe, *Krieg*, 87; Neecke, *Literarische Strategien*, 59 for different interpretations of this fact. According to Feistner, Neecke, and Vollmann-Profe (*Krieg*, 87) these descriptions exceed the well-known topos of the appreciation of the enemy.

References

Printed Sources

Arbusow, Leonid, and Albert Bauer, eds. *Heinrichs Livländische Chronik*. MGH Scriptores rerum Germanicarum in usum scholarum separatim editi 31. Hannover: Hahnsche Buchhandlung, 1955.

Meyer, Leo, ed. *Livländische Reimchronik*. Paderborn: Ferdinand Schöningh, 1876. Reprinted Hildesheim: Georg Olms Verlagsbuchhandlung, 1963.

Smith, Jerry C., and William L. Urban, trans. *The Livonian Rhymed Chronicle*. Rev. ed. Chicago: Lithuanian Research and Studies Center, 2001.

Studies

Angermann, Norbert. "Die mittelalterliche Chronistik." In *Geschichte der deutschbaltischen Geschichtsschreibung*, edited by Georg von Rauch, 3–20. Ostmitteleuropa in Vergangenheit und Gegenwart 20. Cologne: Böhlau, 1986.

Arentzen, Jörg, and Uwe Ruberg, eds. *Die Ritteridee in der deutschen Literatur des Mittelalters: Eine kommentierte Anthologie*. 2nd ed. Darmstadt: WBG, 2011.

Arnold, Udo. "Livländische Reimchronik." In *Die deutsche Literatur des Mittelalters: Verfasserlexikon*, col. 855–862. Vol. 5. Berlin: De Gruyter, 1985.

Barber, Malcolm. "Was the Holy Land Betrayed in 1291?" *Reading Medieval Studies* 34 (2008): 35–52.

Bartlett, Robert. "Reflections on Paganism and Christianity in Medieval Europe." *Proceedings of the British Academy* 101 (1999): 55–76.

Benninghoven, Friedrich. *Der Orden der Schwertbrüder: Fratres Milicie Christi de Livonia*. Ostmitteleuropa in Vergangenheit und Gegenwart 9. Cologne: Böhlau, 1965.

Bysted, Ane L., Carsten S. Jensen, Kurt V. Jensen, and John H. Lind. *Jerusalem in the North: Denmark and the Baltic Crusade, 1100–1552*. Outremer 1. Turnhout: Brepols, 2012.

Christiansen, Eric. *The Northern Crusades: The Baltic and the Catholic Frontier, 1100–1525*. Minneapolis: University of Minnesota Press, 1980.

Eihmane, Eva. "The Baltic Crusades: A Clash of Two Identities." In *The Clash of Cultures on the Medieval Baltic Frontier*, edited by Alan V. Murray, 37–51. Farnham: Ashgate, 2009.

Ekdahl, Sven. "Die Rolle der Ritterorden bei der Christianisierung der Liven und Letten." In *Gli inizi del cristianesimo in Livonia-Lettonia: Atti del colloquio internazionale di storia ecclesiastica in occasione dell'VIII centenario della Chiesa in Livonia (1186–1986), Roma, 24–25 Giugno 1986*, 203–243. Pontifico Comitato di scienze storiche: Atti e documenti 1. Vatican City: Libreria Ed. Vaticana, 1989.

Ekdahl, Sven. "The Treatment of Prisoners of War during the Fighting between the Teutonic Order and Lithuania." In *The Military Orders. Volume I: Fighting for the Faith and Caring for the Sick*, edited by Malcolm Barber, 263–269. Aldershot: Variorum, 1994.

Favreau-Lille, Marie-Luise. "Mission to the Heathen in Prussia and Livonia: The Attitudes of the Religious Military Orders towards Christianization." In *Christianizing Peoples and Converting Individuals*, edited by Guyda Armstrong and Ian N. Wood, 147–154. International Medieval Research 7. Turnhout: Brepols, 2000.

Feistner, Edith, Michael Neecke, and Gisela Vollmann-Profe. "Ausbildung korporativer Identität im Deutschen Orden: Zum Verhältnis von Bibelepik und Ordenschronistik, Werkstattbericht." In *Deutschsprachige Literatur des Mittelalters im östlichen Europa: Forschungsstand und Forschungsperspektiven*, edited by Ralf G. Päsler and Dietrich Schmidtke, 57–74. Beiträge zur älteren Literaturgeschichte. Heidelberg: Winter, 2006.

Feistner, Edith, Michael Neecke, and Gisela Vollmann-Profe. *Krieg im Visier: Bibelepik und Chronistik im Deutschen Orden als Modell korporativer Identitätsbildung.* Tübingen: Niemeyer, 2007.

Fenske, Lutz, and Klaus Militzer, eds. *Ritterbrüder im livländischen Zweig des Deutschen Ordens.* Quellen und Studien zur baltischen Geschichte 12. Cologne: Böhlau, 1993.

Fischer, Mary. *"Di Himels Rote": The Idea of Christian Chivalry in the Chronicles of the Teutonic Order.* Göppinger Arbeiten zur Germanistik 525. Göppingen: Kümmerle, 1991.

Fonnesberg-Schmidt, Iben. *The Popes and the Baltic Crusades: 1147–1254.* The Northern World 26. Leiden: Brill, 2007.

Hellmann, Manfred. "Die Anfänge christlicher Mission in den baltischen Ländern." In *Studien über die Anfänge der Mission in Livland*, edited by id., 7–38. Vorträge und Forschungen 37. Sigmaringen: Thorbecke, 1989.

Hentrich, Volker. "Die Darstellung des Schwertbrüderordens in der Livländischen Reimchronik (Ordenschronik, Missionsgeschichte oder nur 'Kriegstagebuch'?)." In *Perzeption und Rezeption: Wahrnehmung und Deutung im Mittelalter und in der Moderne*, edited by Joachim Laczny, 107–154. Nova mediaevalia 12. Göttingen: V&R Unipress, 2014.

Honemann, Volker. "Zu Selbstverständnis und Identitätsvorstellungen in der livländischen Geschichtsschreibung des Mittelalters." In *Geschichtsschreibung im mittelalterlichen Livland*, edited by Matthias Thumser, 255–297. Schriften der Baltischen Historischen Kommission 18. Berlin: Lit, 2011.

Hucker, Bernd Ulrich. "Zur Frömmigkeit von Livlandpilgern und Ordensrittern." In *Die Spiritualität der Ritterorden im Mittelalter*, edited by Zenon Hubert Nowak, 111–130. Ordines militares 7. Torún: Uniwersytet Mikolaja Kopernika, 1993.

Kala, Tiina. "The Incorporation of the Northern Baltic Lands into the Western Christian World." In *Crusade and Conversion on the Baltic Frontier, 1150–1500*, edited by Alan V. Murray, 3–20. Aldershot: Ashgate, 2001.

Kala, Tiina. "Rural Society and Religious Innovation: Acceptance and Rejection of Catholicism among the Native Inhabitants of Medieval Livonia." In *The Clash of Cultures on the Medieval Baltic Frontier*, edited by Alan V. Murray, 169–190. Farnham: Ashgate, 2009.

Kala, Tiina. "Verkündigung und Kreuzpredigt in und für Livland im 13. Jahrhundert." In *Leonid Arbusow (1882–1951) und die Erforschung des mittelalterlichen Livland*, edited by Ilgvars Misans and Klaus Neitmann, 187–208. Quellen und Studien zur baltischen Geschichte 24. Cologne: Böhlau, 2014.

Kļaviņš, Kaspars. "The Significance of the Local Baltic Peoples in the Defence of Livonia (Late Thirteenth–Sixteenth Centuries)." In *The Clash of Cultures on the Medieval Baltic Frontier*, edited by Alan V. Murray, 321–340. Farnham: Ashgate, 2009.

Kreem, Juhan. "Der Deutsche Orden in Livland: Die Heiden, Landvolk und Undeutsche in der livländischen Heeresverfassung." In *L'Ordine Teutonico tra Mediterraneo e Baltico: Incontri e scontri tra religioni, popoli e culture/Der Deutsche Orden zwischen Mittelmeerraum und Baltikum: Begegnungen und Konfrontationen zwischen Religionen, Völkern und Kulturen*, edited by Hubert Houben and Kristjan Toomaspoeg, 237–251. Acta Theutonica 5. Galatina: Congedo, 2008.

Kugler, Hartmut. "Über die 'Livländische Reimchronik': Text, Gedächtnis und Topographie." *Jahrbuch der Brüder-Grimm-Gesellschaft* 2 (1992): 85–104.

Kugler, Hartmut. "Die 'Livländische Reimchronik' des 13. Jahrhunderts." *Latvijas Zinātņu Akadēmijas vēstis* A 9, no. 554 (1993): 22–30.

Lazda-Cazers, Rasma. "Landscape as Other in the Livländische Reimchronik." *Amsterdamer Beiträge zur älteren Germanistik* 65 (2009): 183–209.

Mackensen, Lutz. "Zur livländischen Reimchronik." In *Zur deutschen Literatur Altlivlands: Untersuchungen von Lutz Mackensen*, 21–58. Ostdeutsche Beiträge aus dem Göttinger Arbeitskreis 18. Würzburg: Holzner, 1961.

Mažeika, Rasa. "When Crusader and Pagan Agree: Conversion as a Point of Honour in the Baptism of King Mindaugas of Lithuania (c. 1240–1263)." In *Crusade and Conversion on the Baltic Frontier, 1150–1500*, edited by Alan V. Murray, 197–214. Central European Studies. Aldershot: Ashgate, 2001.

Mažeika, Rasa. "Granting Power to Enemy Gods in the Chronicles of the Baltic Crusades." In *Medieval Frontiers: Concepts and Practices*, edited by David Abulafia and Nora Berend, 153–171. Aldershot: Ashgate, 2002.

Mažeika, Rasa. "An Amicable Enmity: Some Peculiarities in Teutonic-Balt Relations in the Chronicles of the Baltic Crusades." In *The Germans and the East*, edited by Charles W. Ingrao and Franz A. J. Szabo, 49–58. Central European Studies. West Lafayette: Purdue University Press, 2008.

Militzer, Klaus. "The Recruitment of Brethren for the Teutonic Order in Livonia, 1237–1562." In *The Military Orders. Volume I, Fighting for the Faith and Caring for the Sick*, edited by Malcolm Barber, 270–277. Aldershot: Variorum, 1994.

Militzer, Klaus. *Die Geschichte des Deutschen Ordens*. 2nd ed. Kohlhammer Urban Taschenbücher 713. Stuttgart: Kohlhammer, 2012.

Mugurēvičs, Ēvalds. "Die 'ältere Livländische Reimchronik' über die ethnische Situation im Baltischen Raum." In *Deutschsprachige Literatur des Mittelalters im östlichen Europa: Forschungsstand und Forschungsperspektiven*, edited by Ralf G. Päsler and Dietrich Schmidtke, 267–273. Beiträge zur älteren Literaturgeschichte. Heidelberg: Winter, 2006.

Murray, Alan V. "The Structure, Genre and Intended Audience of the Livonian Rhymed Chronicle." In *Crusade and Conversion on the Baltic Frontier, 1150–1500*, edited by id., 235–251. Aldershot: Ashgate, 2001.

Murray, Alan V. "Henry the Interpreter: Language, Orality and Communication in the Thirteenth-Century Livonian Mission." In *Crusading and Chronicle Writing on the Medieval Baltic Frontier: A Companion to the Chronicle of Henry of Livonia*, edited by Marek Tamm, Linda Kaljundi, and Carsten Selch Jensen, 107–134. Farnham: Ashgate, 2011.

Murray, Alan V. "The Sword Brothers at War: Observations on the Military Activity of the Knighthood of Christ in the Conquest of Livonia and Estonia (1203–1227)." *Ordines Militares* 18 (2013): 27–37.

Neecke, Michael. *Literarische Strategien narrativer Identitätsbildung: Eine Untersuchung der frühen Chroniken des Deutschen Ordens*. Regensburger Beiträge zur deutschen Sprach- und Literaturwissenschaft/B. Untersuchungen 94. Frankfurt am Main: Lang, 2008.

Paravicini, Werner. *Die ritterlich-höfische Kultur des Mittelalters*. Enzyklopädie deutscher Geschichte 32. München: Oldenbourg, 1999.

Selart, Anti. "The Use and Uselessness of the Chronicle of Henry of Livonia in the Middle Ages." In *Crusading and Chronicle Writing on the Medieval Baltic Frontier: A Companion to the Chronicle of Henry of Livonia*, edited by Marek Tamm, Linda Kaljundi, and Carsten Selch Jensen, 345–361. Farnham: Ashgate, 2011.

Selch Jensen, Carsten. "The Early Stage of Christianisation in Livonia in Modern Historical Writings and Contemporary Chronicles." In *Medieval History Writing and Crusading Ideology*, edited by Tuomas M. S. Lehtonen and Kurt Villads Jensen, 207–215. Studia Fennica, Historica 9. Helsinki: Finnish Literature Society, 2005.

Šnē, Andris. "The Emergence of Livonia: The Transformations of Social and Political Structures in the Territory of Latvia during the Twelfth and Thirteenth Centuries." In *The Clash of Cultures on the Medieval Baltic Frontier*, edited by Alan V. Murray, 53–71. Farnham: Ashgate, 2009.

Tamm, Marek. "A New World in Old Words: The Eastern Baltic Region and the Cultural Geography of Medieval Europe." In *The Clash of Cultures on the Medieval Baltic Frontier*, edited by Alan V. Murray, 11–35. Farnham: Ashgate, 2009.

Tamm, Marek. "How to Justify a Crusade? The Conquest of Livonia and New Crusade Rhetoric in the Early Thirteenth Century." *Journal of Medieval History* 39 (2013): 431–455.

Thumser, Matthias. "Das Baltikum im Mittelalter: Strukturen einer europäischen Geschichtsregion." *Jahrbuch des baltischen Deutschtums* 58 (2011): 17–30.

Urban, William. *The Baltic Crusade*. Rev. ed. Chicago: Lithuanian Research and Studies Center 1994.

Wüst, Markus. "Westliche Einflüsse auf den Verlauf der Christianisierung in Livland." In *Christianisierung Europas: Entstehung, Entwicklung und Konsolidierung im archäologischen Befund/Christianisation of Europe: Archaeological Evidence for Its Creation, Development and Consolidation*, edited by Orsolya Heinrich-Tamáska, Niklot Krohn, and Sebastian Ristow, 483–496. Regensburg: Schnell+Steiner, 2012.

Zühlke, Raoul. "Bischof Meinhard von Üxküll: Ein friedlicher Missionar? Ansätze zu einer Neubewertung; Ein quellenkundlicher Werkstattbericht." *Hansische Geschichtsblätter* 127 (2009): 100–121.

14

A MILITARY DIASPORA IN MEDIEVAL CHRISTENDOM

The Teutonic Order

Mark Whelan[1]

The political and diplomatic horizons of the Teutonic Order stretched well beyond their fortresses in Prussia and estates in Germany—in fact, beyond Latin Christendom itself.[2] Writing in his headquarters at Marienburg in 1407, Grand Master Konrad von Jungingen addressed a letter to the king of Ethiopia, whom he also held to be Prester John, the mythical Christian king in the Orient, in which he described the purpose of his Order.[3] The Order, he stated, was dedicated to serving in Christ's militia and battling his dreadful and nocturnal enemies, and he went on to ask for the king's support in uniting the Church and ridding the Holy Land of its occupiers.[4] Leaving aside the grand master's zeal and his range of diplomatic contacts (real or imagined), Jungingen's letter draws attention to the martial nature of the Order, which by his time had fought non-Christian and Christian alike in the Holy Land, Iberia, Armenia, Hungary, Gotland, and, most notably, Prussia.[5] The Order's conquest of the latter proved enduring, involving the settlement of largely German-speaking colonists in newly founded cities, clearing forests and extending arable agriculture, constructing castles, canals, and roads, and creating historical and devotional literature that could legitimate their rule.[6] The purpose of this chapter, in line with the volume's definition of a military diaspora, is to assess how the diasporic character of the Teutonic Order shaped the nature of its rule in Prussia. After outlining its most salient features, this piece will examine the Order's character as a military diaspora by focusing on its ability to implant the technologies and techniques of its native German-speaking lands into Prussia. It will then assess the enduring links that the Order's members maintained with their homelands, especially its most important recruiting grounds in Franconia and Swabia, and move on to scrutinize external perceptions of the Order. Such an analysis underlines how the diasporic qualities of the Teutonic Order enabled their leaders to transplant more effectively the technology, manpower, and political structures of

DOI: 10.4324/9781003245568-15

their homelands into Prussia and reinforces the utility of diasporic studies when studying pre-modern military collectives and movements.

As the introduction to this volume explains, the term military diaspora can be contentious, and it is worth clarifying why the Teutonic Knights warrant inclusion in this collection. Fundamentally, the brethren of the Order constituted an elite sharing a common confession, common tongue, common geographical origins, and supposedly a common degree of nobility. The Teutonic Order of the House of St Mary in Jerusalem (to use a version of its longer name) traced its foundations to Acre in the 1190s and was a military-religious order with a monastic rule initially modelled on those of the Templars and Hospitallers.[7] The Order retained its presence in the Holy Land until Acre fell in 1291 and eventually transferred its headquarters to Marienburg in Prussia, which became its primary theatre of conquest.[8] From the outset, the Order's membership was almost exclusively German-speaking. It may be that historians have exaggerated, if only slightly, the Germanic nature of the organization in its first decades, for in the thirteenth century, members are known to have originated from Italy, Poland, Sweden, Livonia, Prussia, and Sweden, among other locations, but the "Germanness" of the Order soon became a principal plank of the leadership's rhetoric.[9] Writing in the 1250s to King Alfonso X of Castile, the author (probably the Marshal of the Order, then in the Holy Land) explained how the Order's name (*domus Theutonicorum*) stemmed from their foundation by a German prince (*per principis Theutoniae*).[10] When expelling a foreigner in 1450, the grand master stated that "the Order is a German Order, which up until now no non-German, but only Germans, healthy and trained people ... are customarily received".[11] In 1494, the Order addressed a letter to the imperial diet stating that it was "founded on the German language and thus not mixed and invaded by foreign tongues".[12] In 1509, the grand master went one step further and appeared at the imperial diet himself (then at Worms), and proclaimed the close connection between the Empire and the Order and how the latter had been founded as a house for the knights of the German nation.[13] The grand masters tailored their rhetoric, of course, with a clear audience in mind and emphasizing their Germanic nature aimed perhaps to encourage both the material and moral support they relied upon in the Holy Roman Empire for the exercise of their rule in Prussia. The fact remains, however, that by the fourteenth century, the Order was almost entirely German-speaking in membership and language, and this exclusive linguistic and ethnic nature would continue into the early modern period and beyond.[14]

As well as being almost exclusively German, the Order increasingly restricted its recruitment throughout the fourteenth and fifteenth centuries to the "well born" (*wolgeboren*), the nobility, and these recruits were drawn more and more exclusively from southern German families in Franconia and Swabia.[15] That the Order's monastic rule imposed celibacy meant the organization was in constant need of replenishment, and this fashioned a strong attachment to the lands and families—mostly of lower noble stock—from which it drew recruits. Probably numbering around 700 by 1400, the brethren of the Teutonic Order came to rule a complex society formed of German-speaking colonists and a largely subjugated native Prussian population, and they vigorously held on to positions of authority and attempted to monopolize political and economic power.[16] Living standards in the Teutonic Order have been debated, and they no doubt fluctuated as the fortunes of the Order waxed and waned, but the financial records in the fifteenth century often suggest wealth and well-being.[17] While not exactly luxurious, the living arrangements appear comfortable, at least for the leaders of the Order. The brethren at Marienberg enjoyed a central heating system unrivalled until the modern era, and their monastic rule was modified to allow them to hunt, a traditional pursuit of the medieval aristocracy.[18] Heated bathing facilities for all brethren were present in even the smallest of houses, and when these were not up to scratch, brothers were known to complain.[19] A shared sense of religious purpose, as well as shared language, birth, geographical origins, and engrained privilege, endowed the Teutonic Order with a clear sense of corporate identity and means—in light of the volume's focus and definition—that the term military diaspora can be applied to the Teutonic Order in Prussia.[20]

It is important to note, however, that similarity in language, religion, and status, did not necessarily mean that brethren lived in harmony. Vicious disputes could and did arise within the Order, often along geographic and linguistic lines.[21] In 1440, for example, a group of Rhinelanders, annoyed at the tight hold of southern Germans on the important posts of the Order, broke into Marienburg and posted an inflammatory notice on their leader's door, forbidding him from offering high office to Bavarians, Swabians, and Franconians.[22] The grand master at the time, Paul von Rusdorf, a Rhinelander himself, was so concerned that he apparently fled from his fortress by hiding in a sleigh and riding over the winter ice to safety in Gdańsk. Disagreements ranging from arguments among the ruling clique over the direction of policy to the often petty tensions aroused with brothers living in close quarters also left their mark on the Order's history.[23] On a micro level, disputes within individual Order houses could turn nasty. In the foundation at Schwetz in 1428, for example, one brother—a certain *Sporneke*l—returned from visiting a nearby town one evening and, after resting a while, began throwing little sticks (*kleynen spenchen*) into the refectory at his brethren. This seemed to have annoyed his colleague, *Willemerten*, who later that night stabbed him, "giving him a slash or two" (*gab im noch eynen slag oder czwene*) and leaving him "very bloodied" (*sere bluttende*).[24] On a grander level, the belligerence and incompetence of Grand Master Rusdorf pushed the masters of the German and Livonian branches of the Order into open revolt in 1438, a situation that was only reconciled after his resignation in the early 1440s.[25] Nevertheless, divisions within the Order should not be exaggerated: The leadership was ruthless in excluding the native populations from high office, and the Order's unwillingness to share power with German settlers, often concentrated in the wealthy urban settlements of Prussia, provoked a brutal war in the mid-fifteenth century—the so-called Thirteen Years' War—resulting in substantial territorial losses to the Kingdom of Poland and the permanent weakening of the Order's state.[26] For all intents and purposes, and despite the diversity within its ranks, the Teutonic Order was a self-perpetuating military and religious clique eager to hold onto its powers and privileges.

The source base for the study of the Teutonic Order as a military diaspora is rich indeed, particularly in the fifteenth century. The detail on offer in the array of account books, financial documentation, and memoranda generated by the Order's leadership and subjects can be overwhelming.[27] One can discover the cost of the onions stable boys fed their horses, the cost and quantity of the wax used to finish the strings on crossbows, and even the very names of the hunting dogs used by the officers—Reece (*Rÿs*) the sighthound (*wind*) seems to have been in particular demand in the early 1430s.[28] This source base has enabled historians to delve deeply into the Order's military organization, its relationships (friendly or not) with both neighbouring and distant powers, the management of its economic resources, and its ideology and self-perception.[29] In this context, it is important to remember Jürgen Sarnowsky's observation that the source materials produced by the Teutonic Order maintain a relevance for the history of medieval Europe in general.[30] This is not just because these materials enable valuable comparisons with concurrent political structures and developments in Christendom (and not to mention the case studies embodied in other chapters in this volume), but also on account of the fact that the political and economic influence of the grand master could stretch widely indeed. The Order's leadership corresponded with emperors, kings, and princes across Europe, Asia, and Africa, their campaigns against their enemies in Prussia and Lithuania attracted figures of European standing, and the Order's lands contained valuable natural resources traded across Christendom.[31] Prussia and its forested hinterlands produced wares enjoying high contemporary demand, including tar, pitch, furs, and wood. The timbers exported from the region were needed for shipbuilding, in construction, and for equipping armies—in 1427, for example, an English cardinal specifically ordered bow staves from the Order to furnish his crusading force destined for Bohemia.[32] Baltic amber was desired across Europe for its use in jewellery, especially rosary and paternoster beads, and Baltic wax—exported principally through cities under the aegis of the Order, such as Riga, Gdańsk, and Tallinn—was renowned for its quality and used in ecclesiastical foundations across western Europe for divine service.[33] The animals raised by the Order were similarly renowned, and between 1449 and 1453, the grand master dispatched no less than 316 falcons and three dogs as gifts to various nobles and prelates in the Holy Roman Empire.[34] This source base means that the diasporic nature of the Teutonic Knights can be examined from several angles, and the remainder of this article will examine the Order's nature as a military diaspora more closely, exploring the ability of the Knights to transplant the technologies of their native German lands into Prussia, the maintenance of close links with their homelands, and external perceptions of the Order in the fifteenth and sixteenth centuries.

* * *

From the first campaigns in their region in the 1220s and throughout the 1300s, the Teutonic Knights fought their enemies with several distinct advantages that stemmed from their character as a military diaspora. These included access to technologies, specialists, and natural resources to which their enemies did not have access to, at least for a while. As native enemies learnt to adapt, and as the Knights came increasingly to clash with Christian monarchies operating on a similar technological level, many of their advantages receded.[35] Sven Ekdahl has highlighted the importance of crossbows and horses to the Order's war-making, and exploring the Order's attitude towards the latter underlines its character as a military diaspora and the advantages it drew from its German connections.[36]

Horses were not unknown to the Baltic region in the thirteenth century, but the equines the Order brought to this region were heavier and stronger than the breeds native to Prussia, and these gave it a decisive advantage in warfare throughout the thirteenth and fourteenth centuries.[37] The horses' bigger size allowed the Teutonic Knights to wear heavy armour into battle and to armour the horse too, protecting both from missile shots and increasing the shock power of their charge. These animals could be especially tough and resilient, and in a skirmish in the fifteenth century, one war horse apparently suffered over 40 wounds from arrow shots before perishing.[38] Initially, these horses were brought by members of the Order from their native counties, particularly from in and around Saxony, but by the fourteenth century, they were sourced increasingly from Franconia, Swabia, and the Rhineland, as well as other parts of the Empire, many probably from the Order's own estates.[39] It would appear that the Order eventually established stud farms in Prussia—in 1376, the Lithuanians reportedly laid waste to a stud farm (*equirream*) on the Sambian Peninsula—giving them the ability to raise war horses much closer to where they were needed for warfare.[40]

The importance which these great warhorses played in the Order's mindset is reflected in the design and management of its property, its economic policies, and even its literature and art. Their convents had to accommodate enough stabling and be sited adjacent to suitable meadowland and pasture, and rather than collect taxes in cash, many of the Order's estates collected their dues in oats in order to accumulate sufficient stocks of fodder.[41] Payments for equine doctors (*pferdeartcz*) also appear in the Order's financial records.[42] These powerful steeds were jealously guarded, and they usually had their sperm cords strangled or crushed by specialists before entering military service, so if they were captured by the Order's enemies, they at least could not be used as studs.[43] To maintain the Order's technological advantage in cavalry, the grand master expressly banned the export of horses to Lithuania and Poland on several occasions, and even when relations with the Lithuanian grand duke warmed, the grand master proved reluctant to give him fertile stallions with which more could be bred.[44] In 1429, for example, he gifted the Lithuanian ruler a valuable heavy horse, but he ensured he was gelded. The Order's leadership was reluctant, too, to allow native Prussians access to strong horses, despite the fact that they could help improve agricultural productivity on the Teutonic Knights' own estates. Writing in the 1420s, an officer reported to the grand master that the Prussians in his commandery lacked horses strong enough to drag loads of timber and therefore were not suitable to perform the labour services he required.[45] The names of some of the leadership's beloved horses survive and include colourful appellations such as "Doctor", "Mouse", and "Jester", and some convent walls were decorated with paintings of equines.[46] It is not surprising, therefore, that the first manual on equine care in the German language was produced in Prussia and dedicated to the grand master of the Teutonic Order in 1408—the author, after all, was assured of an interested audience.[47]

As well as heavier horses, the Order's conquest of the Baltic also introduced crossbows to the region.[48] The crossbow was more powerful than the bows used by natives and was widely deployed in armies raised by the Teutonic Knights, where often, every second and third man would be equipped with one.[49] As using the crossbow required less training than either the sword or the traditional bow, it was widely used not just by the brethren of the Order but also its servants and townsmen, and along with the spear, played a vital role in equipping the rank-and-file in the Knights' armies.[50] When the Order ordered contingents for military service from Prussian towns—such as from Gdańsk c. 1400—they provided troops of crossbowmen.[51] In their negotiations with mercenary companies, moreover, the contracts of service often stipulated that the mercenaries should arrive armed with crossbows and spears.[52] Crossbows and crossbow bolts were stockpiled by the Order in significant quantities: Inventories record that the Order's fortress at Königsberg held 1,736 crossbows alone in 1392 and that Elbląg held no less than 156,000 bolts in 1396.[53] The majority of these were probably produced in Prussia, as around 1400, there were perhaps 16 crossbow manufactories (*snichzhus*) in Order territory, and the crossbow would continue to play an important role in the Knights' warfare even after the introduction of artillery and handguns.[54]

The Order's exploitation of heavy war horses of German provenance and crossbowmen points more broadly to the importance of its links with its native homelands, and in line with its character as a military diaspora, the Teutonic Knights also brought to Prussia new building technologies and forms of architecture. The Order made effective use of the landscape from the outset of its campaigns, even fortifying large trees with stockades and platforms at strategic sites on rivers and hills.[55] The ingenuity of the Knights in exploiting the local conditions is reflected hundreds of years later in a charter of 1429, authored by Sigismund of Luxemburg, then the Holy Roman Emperor elect, in which he traced the beginnings of the Order to an Oak tree in the vicinity of Toruń (*die von Anfang von einer Eychen zu alden Thorun*), one of its first fortified sites in Prussia.[56] Fortified trees, however, did not serve the longer-term purposes of the Order, and its leaders soon recruited experts from Germany and Silesia to build castles and fortify bases with strong towers and walls.[57] Quality building stone was expensive to transport to the Baltic, so the Knights built in brick instead, introducing thereby a novel architectural feature into the region as the manufacture of bricks and mortar had been hitherto unknown in the region.[58] Even today, the (often much restored) castles of the Order still impress, and items of correspondence circulated between the highest officers attest to the significant resources required to build such large structures. In a letter probably written in 1423, the treasurer (*Treßler*) of the Order reported that over 150,000 bricks (*wol anderhalbhunderttußunt zcigels*) alone had been used in a nearby building project and that a further 100,000 were awaiting transport in Toruń.[59]

The Order eventually became renowned for its talent in building in brick: In the later 1420s, for example, a diplomatic brief produced in Hungary by a Teutonic Knight reported to the grand master that the queen of Hungary was "greatly desirous of a brick-maker", and that one could not well refuse her (*unser frawe konigynne eynes czigelstreichers so groslich begerende, das her ir nicht wol stedt czuversagen*), with the implication being that she wanted one from the Order.[60] This is hardly surprising, given that the Order's fortresses benefitted from the latest that contemporary technology had to offer. The aforementioned heating systems installed in fortresses across Prussia in the fourteenth century used furnaces in the cellars to spread warm air throughout the convent with a system of vaults and ceramic piping under the floor and in the walls. The fourteenth-century system in the fortress of Marienburg was still in partial working order before its nineteenth-century restoration, and experiments in 1824 succeeded in raising the temperature of some rooms to a respectable 22.5 degrees Celsius.[61] Deep wells were dug in the castle and convent courtyards to provide clean drinking water, with those in Mewe and Stuhm (modern-day Gniew and Sztum) over 30 metres deep.[62] Numerous sites were outfitted too with a so-called "Danzker", a high protruding brick corridor built on strong arches. As well as serving defensive purposes, there was often a latrine tower at its farthest end, where the brethren could relieve themselves shielded from the elements.[63] The fortresses built by the Order were understandably strong. Marienburg, the greatest fortress of the Order, survived two significant Polish sieges supported with artillery in the fifteenth century (1410 and 1454) and was only taken in 1457 through treachery when the mercenaries employed to defend the castle turned it over to the king of Poland in return for a significant sum of money.[64]

The Order's frequent reliance on links with its native lands for specialist workmen and engineers to accomplish its building projects is made clear by the predicament of the master of Livonia in 1419.[65] In February of that year, he wrote to the grand master in Prussia asking for Master Johannes, the latter's head builder (*murmeister*).[66] The master of Livonia stated that he had specifically arranged for someone from Germany (*us Dutsche land*) to come and lay the foundations for one of his planned buildings, but that this man had fallen ill (*krang geworden*) and so was no longer available. The master went on to explain that he had the materials stockpiled for the construction, including "stone and lime and other things", but that now he had no one in his land (*hir im lande niemande gehaben*) capable of laying the complicated foundations.[67] The import of specialists *us Dutsche land* was not just restricted to builders. Doctors, for example, were in high demand and could not be easily sourced from within Prussia. The Order's commitment to running hospitals aside, doctors were needed both to tend to the health of the senior officers and to accompany forces on the campaign.[68] One of the first identifiable doctors in the sources, the personal doctor to Grand Master Ludolf König (r. 1342–1345), came from the diocese of Cologne.[69] At the Council of Constance (1414–1418), an officer of the Teutonic Order tried—and failed—to recruit a doctor to serve the grand master in Prussia. The evidence suggests, however, that the grand master's attempts to recruit medical expertise from Germany generally met with a little more success. A knight-brother of the Order in the 1440s based in Danzig was described as *eyn ryman und doctor* ("a Rhinelander and a doctor"), and a series of high-profile doctors in the Order in the fifteenth and sixteenth century came from its more traditional recruiting grounds, in the vicinity of Eichstadt, Ulm, and Nuremberg.[70]

The Order was, of course, not entirely reliant on German-speaking lands to provide its manpower and made effective use of local populations when and where it could. This included the use of local scouts and the arming of select groups of (converted) Prussians to aid in the Order's raids and campaigns.[71] The Teutonic Knights also proved more than willing to adapt their warfare and equipment to suit their new surroundings, abandoning many of the conventions practised in their native German lands in the process. Thus the long lance was dropped in favour of a shorter spear that could be wielded more effectively in forest settings, and some of the Order's shields and helmets drew upon Prussian and Lithuanian rather than western-European design.[72] The Order used native horses as pack animals, and over time its heavier breeds interbred with Prussian stock.[73] Rather than fight in the summer and rest in the winter, the Knights learned to campaign in the winter.[74] In fact, frozen rivers and lakes often proved easier to traverse than boggy swamps and fetid wilderness in the height of summer, and the Order, in general, was forced to adapt many of its military practices to suit the landscape and harsher climate.

While the Order certainly adapted its behaviour to meet the demands of a new region, it is also clear that where possible it adapted the landscape to serve its own ends. Whereas the men and buildings of the Knights have long been the subject of historical enquiry, scholars such as Aleks Pluskowski are increasingly demonstrating how the Order's conquest of Prussia had wide-reaching consequences for the wider Baltic environment.[75] The Teutonic Knights—and the German settlers that followed them—introduced agricultural and industrial methods from their native regions into Prussia, permanently changing the ecosystem and affecting native flora and fauna. This included the carving of large canals into the landscape, such as those between Königsberg and Memel or Kaunas and Gdańsk, and the damming and manipulation of waterways to make rivers navigable and to create defensive moats and waterways for mills.[76] Under the aegis of the Order, German settlers introduced swathes of Prussia to the heavy iron plough and the three-field system of crop rotation, with the extension of arable agriculture contributing to sometimes significant deforestation.[77] The fact that the Order was not just able to feed itself but also to export grain to western Europe demonstrates the rise in agricultural productivity that came in the wake of the conquest.[78] The intensive management of the landscape was made possible by the Order's dense administrative structure in Prussia, a structure which has been explored in depth by Sarnowsky in the fifteenth century and included the close management of forests, harbours, and incoming and outgoing trade as well as their landed estates.[79] The landscape was subject to intensive spiritual management too. Sites dedicated to the native Prussian Goddess Kurko, for example, were re-dedicated to the Virgin Mary, and recent work by Gregory Leighton has shed new light on how the Teutonic Order conceptualized their conquest and created in their literature and landscape an image of a deeply sacralized Prussia.[80]

In summary, the technologies and resources the Order used to establish its rule in Prussia draw attention to the importance of its diasporic links with its native German lands. The Order's own estates throughout the Holy Roman Empire, in particular, merit mention in this respect. Its landholdings in the Empire were organized into twelve bailiwicks, four of which were governed by the grand master himself and from which he drew important incomes and resources.[81] The bailiwick of Koblenz, for example, provided the grand masters with valuable Rhenish wine, an important resource that supplemented both the leading officers' diet and was also used to entertain foreign visitors to Prussia in the style to which they were accustomed.[82] These bailiwicks were all in place by the end of the thirteenth century, and similarly to its lands in Prussia, its close administration was designed to ensure that the Order squeezed as much as possible out of its surfeit of scattered properties, rents, and incomes. Wine aside, its lands in Germany were vital to sustaining the Order's rule in Prussia, both in terms of recruiting the specialists it needed to offer the logistical and technical support called for to manage its conquests and also the very men constantly needed to fill the ranks of an organization that was ostensibly celibate and could not reproduce itself—the subject to which we now turn.

⋆ ⋆ ⋆

In theory, the brethren of the Teutonic Order were all celibate and consequently had no sons to provide to the Order. The Order therefore constantly needed to replenish its ranks with fresh recruits from German lands, and its bailiwicks proved active recruitment centres. It is difficult to generalize about membership and recruitment for a complex organization across several hundred years, but some common themes do emerge. Fundamentally, geographic location and family tradition were the most significant factors determining who became a knight-brother. In terms of geographic location, the bailiwicks were the first and most important stage in recruitment, as postulants were generally vetted by either the German master (the Deutschmeister—the leading officer in charge of the Order within the Holy Roman Empire, but ultimately subordinate to the grand master in Prussia) or one of his provincial commanders. It was generally they who ensured that the candidate had noble ancestry and that they had suitable armour and weaponry and enough money for the journey to Prussia.[83] Family connections were always important avenues for recruitment. With no sons to recruit, there are plenty of cases where members of the Order recruited their nephews instead—in 1432, for example, one knight in the bailiwicks succeeded in recruiting two nephews at the same time.[84] Brothers and cousins from the same families could join the Order simultaneously, and the same families provided recruits again and again. A brief look at the list of grand masters, for example, sees several family names reappear at intervals, including Heinrich and Gottfried Hohenlohe (relatives), Siegfried and Konrad Feuchtwangen (relatives), Konrad and Ulrich Jungingen (brothers), and Konrad and Ludwig Erlichshausen (uncle and nephew).[85] Similar trends can be observed in the other top offices and below. Five families from the vicinity of Wurzburg provided fifteen senior officers across the period 1250–1450.[86] Repeated recruitment from the same families often bred entitlement, and some family names could develop bad reputations. In the later fifteenth century, the grand master apparently kept a black list of such troublesome aristocratic families.[87]

Even after joining the Order and dedicating their lives to service in the name of Christ, family attachments to Germany remained. In 1424, the commander of the house at Fellin asked the grand master's permission to allow Heinrich von Augshem to return to Germany in order to settle the debts attached to his recently-deceased father's estate.[88] In support, the commander added that he was the only son in the family and also that his mother and friends had been writing to Heinrich about this. Despite having a new home in the Order (and not to mention the rules forbidding brethren from fraternizing with laity), many clearly retained links with their families in Germany. One gets the impression that attachment to their native German lands could sometimes go a little too far. In 1420 one brother, a certain Friedrich von Limburg, wrote repeatedly to the master of Livonia about his desire to return to Germany, despite having already been told he could not go.[89] His perseverance was so great, however, that the master eventually wrote to the grand master asking to allow him some "holiday" (*orlop*).

While Prussia may have formed one of the frontiers of the German-speaking world in the late medieval period, the Order did not remain isolated from the Holy Roman Empire, and official travel went both ways. Allowances for holiday aside, brethren could be dispatched westwards from Prussia for any manner of reasons: To deliver messages and take part in negotiations, to assume office in one of the bailiwicks, and to gather resources and men to send back east. The aforementioned falcons and dogs dispatched as gifts to nobles and prelates in the Holy Roman Empire so liberally by the grand master presumably needed accompanying too.[90] Prussia lacked a university, so men were also occasionally sent abroad to study for degrees.[91] The reasons for sending men west could be very specific: Instructions drawn up in Elbląg in 1443 for an emissary heading west ordered them to visit an officer in the New Mark in order to consult "an old book written by a priest from Berlin" (*eyne aldt buch das etwan von eynen prebyste czu Borlyn*), just in case it could help the Order in upcoming political negotiations.[92] That some brethren were sent away simply to get rid of them also features in contemporary texts.[93] In a letter written c. 1430, for example, the commander at Memel asked the grand master simply to send his useless artilleryman away (*her off zcu schicken*) and to replace him with someone useful (*thuchtigen*) instead.[94] Messengers were frequently dispatched back and forth to Rome, where the grand master maintained a permanent representative—the procurator—to represent the Order's interests at the Curia.[95] The Knights also stationed officers in Flanders, who were in charge of selling the rich natural resources exported from Prussia—especially grain, furs, and wax—on to the international markets in Bruges for the profit of the Order.[96]

Perhaps the most important traffic moving to and from Prussia were the crusaders, who came in significant numbers from across Christendom to join the campaigns (*reysen*) of the Teutonic Knights.[97] The international flavour of these campaigns—attracting high-profile participants from across Europe, including reigning monarchs, such as King Louis of Hungary, who visited in 1344–1345—has drawn significant attention from scholars, and the campaigns have been compared to modern tourist excursions.[98] But the important contributions to the *reysen* made by English, French, and Scottish knights (among others) should not obscure the fact that the most consistent and significant contributions came from the princes and knights of the Holy Roman Empire. Henry, Earl of Derby, and later Henry IV of England (r. 1399–1413), took around 100 men (13 of whom were knights) to Prussia in the 1390s, but this was dwarfed five times over by the margrave of Meissen's entourage in 1391, which numbered 500 knights alone.[99] It would be otiose to detail every major contingent led to Prussia by German nobility in the period c. 1250–c. 1410, but the evidence assembled by Paravicini points to the continual and constant support the Teutonic Order drew from their native lands.[100] Contingents came from across the Holy Roman Empire, but the areas of Thuringia, Saxony, Bavaria, and Austria, as well as the middle parts of the Rhineland, appear to have been the main contributors. Many German nobles were repeated visitors too—Duke William of Jülich (c. 1325–1393), for example, fought in Prussia no less than seven times.[101] The aforementioned charter of Sigismund of Luxemburg, written in 1429, granted the Teutonic Order a portion of territory known as the New Mark, part of the electorate of Brandenburg (*land der Newen Mark zu Brandenburg*), which connected the Teutonic Order with the Holy Roman Empire. The charter aptly described the importance of this slither of territory, stressing that secure possession of the New Mark would mean the Order had a gate and safe road (*ein pforten und offen strassen hat*), through which could come princes, lords, knights, and servants from German and other lands (*us Deutschen und andern Landen*) to help the Teutonic Knights when they were attacked.[102] The close links between the Teutonic Knights in the Baltic with their native lands were given human and material form on such roads, which for over a century saw German nobility and knights travel with their retinues to risk their lives in extending and securing the conquests of the Order.

As the fifteenth century progressed, however, the Teutonic Order suffered a series of setbacks, including a heavy defeat at the hands of a joint Polish-Lithuanian army at the Battle of Tannenberg (or Grunwald) (1410) and the brutal Thirteen Years' War (1454–1466). The latter significantly damaged the prestige of the Knights and resulted in the secession of the wealthier western portion of Prussia, with its important cities of Gdańsk and Toruń, to the Kingdom of Poland and the sale of territories such as the New Mark to neighbouring rulers.[103] During this time, moreover, military support from the Holy Roman Empire began to dry up, forcing the Order to rely on expensive mercenaries it could now ill afford. This trend was apparent to the grand masters at the time, with Grand Master Rusdorf lamenting to the master of Livonia in 1422 how "our neighbours are all ready for war" (*unser nokebur vaste zu krige stellet*), and that it was clear to all that no visitors (*geste*) were at hand to help.[104] The reasons behind the apparent reluctance of the German nobility to support the Teutonic Order have been much debated. Certainly, the conversion of the duke of Lithuania to Catholic Christianity in 1387 shed doubt on the Order's claims to be waging spiritually meritorious warfare in that region, for they were now fighting fellow Christians.[105] But Paravicini has drawn attention to events of perhaps more immediate significance to contemporary German nobility: In particular, the eruption of Bohemia into rebellion in 1419 and the resulting Hussite crusades, which both diverted support from the Order.[106] To take just three examples, the dukes of Bavaria, the bishops of Meissen, and the dukes of Austria—all of whom belonged to families with a history of crusading in Prussia—directed their martial energies to fighting the Hussites, and it is perhaps against this background that we should understand the weakening links—in military terms, at least—between the Teutonic Order and its German homelands in the fifteenth and sixteenth centuries.[107]

Nevertheless, regard for the Teutonic Order was not just limited to the nobility and knights of Germany, even in the mid and later fifteenth century. Across the social order of the Holy Roman Empire, individuals and institutions developed attachments to the Order. At some point around 1450 in Hamburg, the Brotherhood of St Martin, a fraternity composed of merchants with trading interests in the Baltic, paid for a new window to be installed in the cloisters (*cruceganghe*) of the house of St Mary Magdalene, "upon which stood the shield of the Order of Prussia" (*dar steit inne de schilt des orden van prusen*).[108] The "Teutonic" in the Teutonic Order was no empty moniker but accurately reflected the institution's deep and reciprocal connections with its German-speaking homelands upon which it depended for survival.

* * *

Despite a diminished territorial base and military capability after the travails of the early and mid-fifteenth century, the Teutonic Knights remained on the political agenda of Christendom's elites.[109] Some contemporaries still saw in the institution a discreet and effective fighting machine that could be moved profitably to a new theatre where they could—once again—conquer and hold territories in the name of Christ. Some contemporaries suggested that the Order could retain its base in Prussia and tackle enemy threats through long-range campaigns. The crusade propagandist Philippe de Mézières (c. 1327–1405), for example, optimistically suggested in one of his crusade plans that "the lords of Prussia, with the king of Lithuania, should proceed through the kingdom of Russia [...] and go towards Constantinople with all of their force".[110] This, of course, never came to pass, although matters to contemporaries were not so clear cut: Adam of Usk, a Welsh chronicler, writing perhaps in the 1420s, believed that the grand master had in fact invaded the "kingdom of the Turks" in 1410 and that he had battled the Turkish king himself before putting him and "five hundred thousand others to flight".[111] Fantastical as Usk's account was, his belief that the Order was still a military power capable of battling enemies of the Latin Church was shared by others—not least Sigismund of Luxemburg, the Holy Roman Emperor elect and King of Hungary. In the later 1420s, he convinced the grand master to establish a base in southern Hungary where the Knights could help Sigismund defend his kingdom against the Ottoman Turks. For reasons that continue to be unclear, but probably because of shortages in manpower and funding, the detachment appeared incapable of defending the fortresses allotted to them, and by 1433 Sigismund had relieved the Knights of their command.[112]

As the fifteenth century progressed, however, some proposed that the Order needed to be transplanted en masse to somewhere else in Christendom.[113] At the Council of Basle in 1437, Sigismund of Luxemburg suggested that the Order should be united with the Hospitallers and moved to Hungary in order to fight "against the Turks".[114] In the 1450s, the Polish court implausibly suggested Tenedos, the island guarding the western entrance to the Dardanelles, as a new base for the Order, although this was perhaps a Polish ploy to place the troublesome knights as far away from their own borders as possible.[115] By the early 1500s, Podolia (in Ukraine), Crimea, Cyprus, and Dalmatia, had all been raised as possibilities for a new base, but possibilities they remained.[116] As well as the logistical complexities involved in such a relocation, the leaders of the Teutonic Order proved reluctant to relinquish their territorial base in Prussia—a fact that must have been clear to contemporaries. Emperor Frederick III, for example, unable or unwilling the lure the Order to his homelands of Austria to help fight the Turks, founded in the late 1460s the Order of St George in Carinthia on the model of the Teutonic Knights and charged it with defending important fortresses against the Ottomans. A papal grant of 1487 allowed members of the Teutonic Order to transfer to Frederick's foundation, but whether any chose to do so remains unclear.[117] In 1525 the Grand Master, Albrecht von Hohenzollern, took the remarkable step of paying homage to Sigismund I, King of Poland, marking a turning point in the history of Prussia and of the Order [118] As the German master put it in a contemporary letter, Albrecht wished "to bring our Order's land [in] Prussia into secularity" (*unsers Ordens land Preußen in weltlichkeit zu bringen*), dispose of the Order's monastic habit, and "commit the land [of] Prussia to secular rule" (*land Preußen zu weltlicher regierung zu vererben*).[119] Albrecht did just this, establishing Lutheranism as the official religion, secularizing all of the Order's possessions (that is, taking them into his personal ownership), and ruling the now-former monastic state of Prussia as a duke, with the newly formed duchy now held in fief from the Polish Crown.[120] In regard to the Order's remaining holdings in Catholic lands, the Holy Roman Emperor Charles V installed a new grand master—effectively the German master—who now oversaw a reduced territorial base restricted largely to estates in Germany and Italy, which were tied much more closely to the policies of the imperial court. Emperors would often appoint members of their own family as grand masters, and the Order was used to provide troops and leadership for imperial military campaigns throughout the early modern period, whether against the Turks, the French, or even fellow German-speakers, such as the Prussians.[121]

A similar fate befell the Hospitallers, who when relocating to Malta in 1531, also tied themselves much more closely to the Holy Roman Emperor, although the tribute demanded of them in return for the rocky island remained nominal, at a mere one falcon per year.[122] But whereas the Hospitallers were stripped of any meaningful military capability after Napoleon's invasion of Malta in 1798, the Teutonic Order continued to provide contingents and officers for Austrian armies throughout the nineteenth and twentieth centuries, retaining influence in Austro-Hungarian military structures right until the empire's collapse after the First World War. Archduke Eugen of Austria, for example, grand master between 1887 and 1923, commanded Austro-Hungarian forces in the Balkans and Italy in the First World War, and in 1916 was even raised to field marshal.[123] The Order's close identification with the Habsburgs left it exposed after the dynasty lost control of its central-European possessions, but the organization survived by reforming itself in 1929 and focusing solely on charitable endeavours. Stripped of any military role, its history as a military diaspora correspondingly came to an end.

* * *

Exploring the character of the Teutonic Order through the lens of a military diaspora brings to the fore numerous themes of relevance to this volume. First and foremost, the Teutonic Knights underline the ability of military diasporas to mobilize and project resources, technologies, and manpower, across significant distances. This in turn highlights the strong attachments that military diasporas could fashion and maintain with their native homelands; in the case of the Teutonic Knights, this attachment was made peculiarly strong by their members' vow of celibacy, barring intermarriage with local families and meaning their ranks required constant replenishment in manpower from their centres of recruitment in Germany. Commonalities in tongue, birth, confession, and outlook conditioned a strong corporate identity among the Order's brethren and reinforced their members' attachment to their native lands. Although the military support offered by the nobility and knights of the Holy Roman Empire may have lessened as the fifteenth century progressed, diplomatic and spiritual connections remained firm. The last Grand Master of the Teutonic Order in Prussia, Albrecht of Hohenzollern (1490–1568), was precisely the last grand master because of his links to his German homelands. In 1523 he met the radical German theologian Martin Luther at Wittenberg, who appears to have convinced him of the merits of Protestantism.[124] By 1525, Albrecht had begun the process of converting the Order's lands in Prussia into a secular state and taken the title of duke of Prussia, becoming the first European ruler to establish the Lutheran faith as a state religion. Prussia's close links with Germany would endure, and their rulers would continue to interact with the Holy Roman Empire and its leaders, although now in a different fashion.

Notes

1 I would like to thank the Prussian Cultural Foundation (Preußischer Kulturbesitz) for their award of a post-doctoral research grant to undertake archival research in the Geheimes Staatsarchiv, Berlin, upon which portions of this work are based. I completed this chapter while working as a Research Associate on the Leverhulme-funded project, "Bees in the Medieval World: Economic, Environmental and Cultural Perspectives", at King's College, London (Leverhulme Trust RPG-2018-080), led by Alexandra Sapoznik, to whom I am grateful for allowing me to use in this chapter materials gathered in the Staatsarchiv Hamburg while pursuing research for the project.

2 The literature on the Teutonic Order is vast, and the following references cannot be exhaustive. For important sources and recent historiographical developments, the following are starting points: Jürgen Sarnowsky, "Die Quellen zur Geschichte des Deutschen Ordens in Preußen," in *Edition Deutschsprachiger Quellen aus dem Ostseeraum (14.–16. Jahundert)*, ed. Matthias Thumser, Janusz Tandecki, and Dieter Heckmann (Toruń: Uniwersytetu Mikołaja Kopernika, 2001), 171–200; Beata Możejko, "Introduction," in *New Studies in Medieval and Renaissance Poland and Prussia: The Impact of Gdańsk*, ed. ead. (Abingdon: Routledge, 2017), 1–17. The following abbreviations are used: OBA = Berlin, Geheimes Staatsarchiv Preußischer Kulurbesitz, XX, Ordensbriefarchiv. *LEC* = *Liv-, Est- und Curländisches Urkundenbuch nebst Regesten*, 17 vols., ed. Friedrich Georg von Bunge et al. (Aalen: Scientia-Verlag, 1967–1981). For reasons of space, this article largely restricts itself to examining the Teutonic Order in Prussia, omitting detailed discussion of the Order's relationships with its branches in the Holy Roman Empire and Livonia. On the Livonian background, see Johannes Götz, "Verbunden mit der Marienburg: Livländischer und preußischer Deutschordenszweig bis zum Ausbruch des Zungenstreits 1438," in *Livland—eine Region am Ende der Welt? Forschungen zum Verhältnis zwischen Zentrum und Peripherie im späten Mittelalter*, ed. Anti Selart and Matthias Thumser (Cologne: Böhlau, 2017), 371–414, as well as the contribution of Verena Schenk zu Schweinsberg in this volume.

3 Kurt Forstreuter, "Der Deutsche Orden und Südosteuropa," *Kyrios: Vierteljahresschrift für Kirchen- und Geistesgeschichte Osteuropas* 1 (1936): 245–272, at 272 (nr. 5). On the Prester John Legend, see Matteo Salvadore, *The African Prester John and the Birth of Ethiopian-European Relations, 1402–1555* (London: Routledge, 2017). I am grateful to Adam Simmons for this reference.

4 Forstreuter, "Der Deutsche Orden," 272. For background, Kurt Forstreuter, *Der Deutsche Orden am Mittelmeer* (Bonn: Wissenschaftliches Archiv Bad Godesberg, 1967), 68–69.

5 On the Order's activities in Hungary, see the contribution of László Veszprémy in this volume.
6 The environmental and ecological impact of the Teutonic Order's conquest of the Baltic is surveyed in Aleksander Pluskowski, ed., *Ecologies of Crusading, Colonization, and Religious Conversion in the Medieval Baltic* (Turnhout: Brepols, 2019).
7 For background and detailed references, Gregory Leighton, "Did the Teutonic Order Create a Sacred Landscape in Thirteenth-Century Prussia?," *Journal of Medieval History* 4 (2018): 457–483, at 462–466.
8 Ibid., 462–463. On these conquests, see Eric Christiansen, *The Northern Crusades* (London: Penguin, 1997), chap. 3, 6.
9 Klaus Militzer, "Die Aufnahme von Ritterbrüdern in den Deutschen Orden: Ausbildungsstand und Aufnahmevoraussetzungen," in *Das Kriegswesen der Ritterorden im Mittelalter*, ed. Zenon Hubert Nowak (Toruń: Universitas Nicolai Copernici, 1991), 7–17, at 9–10.
10 For the letter and debates surrounding its authorship, J. M. Rodríguez García and A. Echevarría Arsuaga, "Alfonso X, la Orden Teutónica y Tierra Santa: Una nueva fuente para su estudio," in *Las Órdenes Militares en la Península Ibérica*, 2 vols. (Cuena: Ediciones de la Universidad de Castilla-La Mancha, 2000), 1:489–510 (letter on 1:507–509). For a helpful glossary of the major offices of the Order, see Michael Burleigh, *Prussian Society and the German Order: An Aristocratic Corporation in Crisis, c. 1410–1466* (Cambridge: Cambridge University Press, 1984), 187–189.
11 Quoted in Johannes Voigt, *Geschichte des Deutschen Ritterordens in seinen zwölf Balleien in Deutschland*, 2 vols. (Berlin: Georg Reimer, 1857–1859), 1:274. Translated and discussed in Burleigh, *Prussian Society*, 37–40.
12 Quoted in David Nicholas, *The Northern Lands: Germanic Europe, c. 1270–c. 1500* (Oxford: Wiley-Blackwell, 2009), 203–204.
13 Stephan Flemmig, "Der Anteil sächsischer Berater an der Außenpolitik von Hochmeister Friedrich (1498–1510)," in *Akteure mittelalterlicher Außenpolitik: Das Beispiel Ostmitteleuropas*, ed. id. and Norbert Kersken (Marburg: Herder-Institut, 2017), 145–168, at 154–155.
14 On the early modern period: Bernhard Demel, "Welfare and Warfare in the Teutonic Order: A Survey," in *The Military Orders*, vol. 2: *Welfare and Warfare*, ed. Helen Nicholson (Aldershot: Ashgate, 1998), 61–73, at 65, 72.
15 Militzer, "Die Aufnahme," 8–9; Burleigh, *Prussian Society*, 39.
16 Robert Frost, *The Oxford History of Poland-Lithuania*, vol. 1: The Making of the Polish-Lithuanian Union, 1385–1569 (Oxford: Oxford University Press, 2015), 209. On the status of Prussians and other native peoples, Burleigh, *Prussian Society*, 10–36; for comments on Livonia, Kaspars Kļaviņš, "Reorganizing the Livonian Landscape: Some Issues and Research Perspectives," in *Ecologies of Crusading, Colonization, and Religious Conversion in the Medieval Baltic*, ed. Aleksander Pluskowski (Turnhout: Brepols, 2019), 197–208, at 198–204.
17 Burleigh, *Prussian Society*, 62–64.
18 Christian Probst, *Der Deutsche Orden und sein Medizinalwesen in Preussen: Hospital, Firmarie und Arzt bis 1525* (Bonn: Wissenschaftliches Archiv Bad Godesberg, 1969), 136–139; Aleksander Pluskowski, *The Archaeology of the Prussian Crusade: Holy War and Colonisation* (London: Routledge, 2012), 310.
19 Probst, *Der Deutsche Orden*, 132.
20 For further discussion of corporate identity in the Teutonic Order, see Edith Feistner, Michael Neecke, and Gisela Vollmann-Profe, "Ausbildung korporativer Identität im Deutschen Orden: Zum Verhältnis zwischen Bibelepik und Ordenschronistik, Werkstattbericht," in *Deutschsprachige Literatur des Mittelalters im östlichen Europa: Forschungsstand und Forschungsperspektiven*, ed. Ralf G. Päsler and Dietrich Schmidtke (Heidelberg: Universitätsverlag Winter, 2006), 57–74; Marcus Wüst, *Studien zum Selbstverständnis des Deutschen Ordens im Mittelalter* (Weimar: Verlag und Datenbank für Geisteswissenschaften, 2013), esp. chap. 9.

21 German linguistic divisions are explained in Nicholas, *The Northern Lands,* 191–198.
22 Anonymous, "Die Danziger Chronik vom Bunde," in *Scriptores Rerum Prussicarum: Die Geschichtsquellen der preussischen Vorzeit bis zum Untergange der Ordensherrschaft*, ed. Theodor Hirsch, Max Töppen and Ernst Strehlke (Leipzig: Hirschel, 1870), 405–489, at 4:413–415.
23 Intra-communal tensions are explored with wit in Burleigh, *Prussian Society*, 111–133.
24 At least, that was one side of the story; for the differing versions of the incident, see the original report: OBA, 4984; Burleigh, *Prussian Society*, 118.
25 William Urban, *Tannenberg and After* (Chicago: Lithuanian Research and Studies Center, 1999), 360–365.
26 For exceptions as regards Prussians, see Christiansen, *The Northern Crusades*, 212; Klaus Neitmann, "Die Preußischen Stände und die Außenpolitik des Deutschen Ordens vom I. Thorner Frieden bis zum Abfall des Preußischen Bundes (1411–1454): Formen und Wege ständischer Einflußnahme," in *Ordensherrschaft, Stände und Stadtpolitik: Zur Entwicklung des Preußenlandes im 14. und 15. Jahrhundert*, ed. Udo Arnold (Lüneburg: Nordostdeutsches Kulturwerk, 1985), 27–71, at 33–36. On the course and consequences of the Thirteen Years' War, see note 103 below.
27 Burleigh, *Prussian Society*, 51–52.
28 For onions, Jürgen Sarnowsky, *Die Wirtschaftsführung des Deutschen Ordens in Preußen (1382–1454)* (Cologne: Böhlau, 1993), 814 (nr. 33); on the preparation of crossbows, *Nowa Księga Rachunkowa Starego Miasta Elbląga, 1404–1414*, ed. Markian Pelech, 2 vols. (Toruń: Państowe wydawnictwo naukowe, 1987–9), 2:79 (nr. 1432). Reece appears in an item of correspondence between the Vogt of Dirschau and the grand master in April 1432: OBA, 6069.
29 For representative examples: Sven Ekdahl, *Die Schlacht bei Tannenberg 1410: Quellenkritische Untersuchungen* (Berlin: Duncker & Humblot, 1982); Sebastian Kubon, *Die Außenpolitik des Deutschen Ordens unter Hochmeister Konrad von Jungingen (1393–1407)* (Göttingen: V&R Unipress, 2016); Sarnowsky, *Wirtschaftsführung*; Burleigh, *Prussian Society*. For research into the literature and devotional works produced by the Order, see the contributions in Ralf G. Päsler and Dietrich Schmidtke, eds., *Deutschsprachige Literatur des Mittelalters im östlichen Europa: Forschungsstand und Forschungsperspektiven* (Heidelberg: Universitätsverlag Winter, 2006).
30 Sarnowsky, "Die Quellen," 171–172.
31 On correspondence with emperors and sultans, Forstreuter, *Der Deutsche Orden*, 68–69. On the Order's European visitors, Alan V. Murray, "The Saracens of the Baltic: Pagan and Christian Lithuanians in the Perception of English and French Crusaders to Late Medieval Prussia," *Journal of Baltic Studies* 41 (2010): 413–430, at 413–415.
32 Mark Whelan, "Between Papacy and Empire: Cardinal Henry Beaufort, the House of Lancaster, and the Hussite Crusades," *English Historical Review* 133 (2018): 1–31, at 27–28.
33 Werner Böhnke, "Der Binnenhandel des Deutschen Ordens in Preußen und seine Beziehung zum Außenhandel um 1400," *Hansische Geschichtsblätter* 80 (1962): 26–95, at 67–70; Alexandra Sapoznik, "Bees in the Medieval Economy: Religious Observance and the Production, Trade, and Consumption of Wax in England, c. 1300–1555," *Economic History Review* 72 (2019): 1152–1174, at 1153–1154; On the prestigious Baltic wax exported from the Order's lands, see Mark Whelan, "'On Behalf of the City': Wax and Urban Diplomacy in the Late Medieval Baltic and North Sea," *Urban History*, forthcoming.
34 Sarnowsky, *Wirtschaftsführung*, 839–840. On the giving of falcons, see Werner Paravicini, *Adlig leben im 14. Jahrhundert: Weshalb sie fuhren; Die Preußenreisen des europäischen Adels. Teil 3* (Göttingen: V&R Unipress, 2020), 572–596. The grand master also gave bear cubs and elk as gifts: see Chris Given-Wilson, *Henry IV* (New Haven, CT: Yale University Press, 2016), 69.
35 Sven Ekdahl, "Horses and Crossbows: Two Important Warfare Advantages of the Teutonic Order in Prussia," in *The Military Orders*, vol. 2: Welfare and Warfare, ed. Helen Nicholson (Aldershot: Ashgate, 1998), 119–151, at 120–121.

36 Ibid., 120–130 (for horses), 130–151 (for crossbows).
37 Ibid., 126.
38 Discussed in Sven Ekdahl, "Das Pferd und seine Rolle im Kriegswesen des Deutschen Ordens," in *Das Kriegswesen der Ritterorden im Mittelalter*, ed. Zenon Hubert Nowak (Toruń: Universitas Nicolai Copernici, 1991), 29–48, at 39.
39 In more detail, Edkahl, "Horses and Crossbows," 126–127, and more generally, id., "Das Pferd und seine Rolle im Kriegswesen des Deutschen Ordens," 29–47.
40 Ekdahl, "Horses and Crossbows," 126. For recent archaeological work on excavated horse bones in Prussia, Aleksander Pluskowski, "The Ecology of Crusading: Investigating the Environmental Impact of Holy War and Colonisation at the Frontiers of Medieval Europe," *Medieval Archaeology* 55 (2011): 192–225, at 208–209.
41 Ekdahl, "Horses and Crossbows," 127.
42 Sarnowsky, *Wirtschaftsführung*, 817.
43 Processes described in Ekdahl, "Horses and Crossbows," 128–129.
44 Ekdahl, "Das Pferd," 38.
45 Burleigh, *Prussian Society*, 35–36.
46 Hartmut Boockmann, "Pferde auf der Marienburg," in Vera lex historiae*: Studien zu mittelalterlichen Quellen, Festschrift für Dietrich Kurze zu seinem 65. Geburtstag am 1. Januar 1993*, ed. Stuart Jenks, Marie-Luise Laudage and Jürgen Sarnowsky (Cologne: Böhlau 1993), 117–126, at 121; Ekdahl, "Das Pferd," 33.
47 See Ottomar Bederke, ed., *Liber de Cura Equorum: Bearbeitungen von Albertus Magnus und Jordanus aus dem Deutschen Ritterorden (1408)* (Hannover: Tierärztliche Hochschule, 1962).
48 Ekdahl, "Horses and Crossbows," 137–150.
49 Ibid., 149.
50 Andrzej Nowakowski, "Some Remarks about Weapons Stored in the Arsenals of the Teutonic Order's Castles in Prussia by the End of the 14th and Early 15th Centuries," in *Das Kriegswesen der Ritterorden im Mittelalter*, ed. Zenon Hubert Nowak (Toruń: Universitas Nicolai Copernici, 1991), 75–88, at 83–84.
51 Andrzej Nowakowski, *Arms and Armour in the Medieval Teutonic Order's State in Prussia*, trans. Maria Abramowicz (Łódź: Oficyna Naukowa, 1994), 101.
52 Ekdahl, "Horses and Crossbows," 149–150.
53 Nowakowski, "Some Remarks," 84.
54 Nowakowski, *Arms and Armour*, 99; for the employment of servants involved in crossbow manufacture (*snitzknechte* and *snitzjungene*), see Sarnowsky, *Wirtschaftsführung*, 799.
55 Seweryn Szczepański, "*Arbor custodie que vulgariter dicitur Wartboumi*: The Function and Existence of the so-Called 'Watchtower Trees' in Pomesania and Żuławy Wiślany in the 13th–14th Centuries," *Zapiski Historyczne* 76 (2011): 5–18; see also Leighton, "Did the Teutonic Order," 476.
56 This charter of Sigismund's is quoted and discussed in Jürgen Sarnowsky, "The Military Orders and Crusading in the Fifteenth Century: Perception and Influence," in *Reconfiguring the Fifteenth-Century Crusade*, ed. Norman Housley (London: Palgrave Macmillan, 2017), 123–160, at 127.
57 Pluskowski, *The Archaeology of the Prussian Crusade*, 152.
58 Ekdahl, "Horses and Crossbows," 119.
59 OBA, 4124.
60 OBA, 5251.
61 Probst, *Der Deutsche Orden*, 137.
62 Ibid., 133.
63 Ibid., 134–139.
64 Sarnowsky, "The Military Orders," 128.
65 Pluskowski, *The Archaeology of the Prussian Crusade*, 152.
66 OBA, 2922. This document is published, though with the wrong date, in *LEC*, 5:457 (nr. 2300).

67 OBA, 2922.
68 Probst, *Der Deutsche Orden*, 164–165.
69 Ibid., 160–161.
70 Ibid., 165, 169–175.
71 William Urban, "The Frontier Thesis and the Baltic Crusade," in *Crusade and Conversion on the Baltic Frontier, 1150–1500*, ed. Alan V. Murray (Aldershot: Ashgate, 2001), 45–71, at 67.
72 Nowakowski, "Some Remarks," 78, 82–83.
73 Ekdahl, "Horses and Crossbows," 124.
74 Nowakowski, "Some Remarks," 67–68.
75 For much of what follows, see Pluskowski, *The Archaeology of the Prussian Crusade* (esp. chap. 4).
76 Bernhart Jähnig, "Zur Wirtschaftsführung des Deutschen Ordens in Preußen vornehmlich vom 13. bis zum frühen 15. Jahrhundert," in *Zur Wirtschaftsentwicklung des Deutschen Ordens im Mittelalter*, ed. Udo Arnold (Marburg: Elwert, 1989), 113–147, at 121; Pluskowski, *The Archaeology of the Prussian Crusade*, 326.
77 Pluskowski, *The Archaeology of the Prussian Crusade*, 315.
78 Ibid., "The Ecology of Crusading," 207.
79 See Sarnowsky, *Die Wirtschaftsführung*, which also includes a rich appendix of source materials.
80 Ibid., 71; Leighton, "Did the Teutonic Order."
81 The Teutonic Order had estates and properties throughout Christendom, including Iberia, Italy, and Greece. On these, see Forstreuter, *Der Deutsche Orden am Mittelmeer.*
82 Burleigh, *Prussian Society*, 52. On the uses of wine in the Order, see Udo Arnold, "Weinbau und Weinhandel des Deutschen Ordens im Mittelalter," in *Zur Wirtschaftsentwicklung des Deutschen Ordens im Mittelalter*, ed. Udo Arnold (Marburg: Elwert, 1989), 71–102.
83 Burleigh, *Prussian Society*, 39.
84 Ibid., 40.
85 Valuable biographies of each grand master can be found in *Die Hochmeister des Deutschen Ordens*, ed. Udo Arnold (Marburg: Elwert, 1998).
86 Christiansen, *The Northern Crusades*, 85.
87 Burleigh, *Prussian Society*, 40.
88 *LEC*, 7:132 (nr. 184).
89 *LEC*, 5:633–634 (nr. 2470).
90 Sarnowsky, *Wirtschaftsführung*, 840.
91 See, for example, the cases in *LEC*, 7:562 (nr. 800), and Probst, *Der Deutsche Orden*, 162–163, 172.
92 OBA, 8266. Discussed in Burleigh, *Prussian Society*, 50.
93 Militzer, "Die Aufnahme," 12.
94 OBA, 28544.
95 Kurt Forstreuter et al., eds., *Die Berichte der Generalprokuratoren des Deutschen Ordens an der Kurie*, 4 vols. (Göttingen: Vandenhoeck & Ruprecht, 1960–1976).
96 The records of these officers, called "Lieger" in German, have been published: see volumes three and four in the series *Schuldbücher und Rechnungen der Großschäffer und Lieger des Deutschen Ordens in Preußen*, vol. 3 = ed. Christina and Jürgen Sarnowsky (Böhlau: Cologne, 2008); vol. 4 = ed. Cordula A. Franzke (Berlin: Duncker & Humblot, 2018).
97 Fundamental to the study of these campaigns is Werner Paravicini, *Die Preußenreisen des europäischen Adels*, 3 vols. (Sigmaringen: Jan Thorbecke, 1989–1995; Göttingen: V&R Unipress, 2020).
98 Urban, "The Frontier Thesis," 67.
99 Christiansen, *The Northern Crusades*, 156–157; On Henry's crusades, see most recently Given-Wilson, *Henry IV*, 61–76.

100 Cf. the list of crusaders heading to Prussia of princely rank: Paravicini, *Die Preußenreisen*, 1:147–150.
101 Paravicini, *Die Preußenreisen*, 1:175.
102 Sarnowsky, "The Military Orders," 127.
103 Frost, *The Oxford History of Poland-Lithuania*, 209–215. The impact of the Thirteen Years' War is explored in detail in Joachim Laczny, *Schuldenverwaltung und Tilgung der Forderungen der Söldner des Deutschen Ordens in Preußen nach dem Zweiter Thorner Frieden. Ordensfoliant 259 und 261, Zusatzmaterial* (Göttingen: V&R Unipress, 2019), 93–105.
104 *LEC*, 5:824 (nr. 2603).
105 For an overview, Sobiesław Szybowski, "Two Local Empires in Confrontation: Poland and Lithuania Competing with the Teutonic State to Dominate the Southeastern Region of the Baltic Sea (the End of the Fourteenth to 1409)," in *Studies in Medieval and Renaissance Poland and Prussia: The Impact of Gdańsk*, ed. Beata Możejko (Abingdon: Routledge, 2017), 60–73.
106 See Werner Paravicini, "Von der Preußenfahrt zum Hussitenkreuzzug," in *Beiträge zur Militärgeschichte des Preußenlandes von der Ordenszeit bis zum Zeitalter der Weltkriege*, ed. Bernhart Jähnig (Marburg: Elwert, 2010), 121–159.
107 On the international reverberations of the Hussite wars and for further references, see Whelan, "Cardinal Henry Beaufort."
108 Hamburg, Hamburg Staatsarchiv, 612–2/3, 7, p. 8 (from the Schafferbuch (c. 1400–1550) of the Schonenfahrer).
109 For an overview, Sarnowsky, "The Military Orders," 123–160.
110 Giedrė Mickūnaitė, *Making a Great Ruler: Grand Duke Vytautas of Lithuania* (Budapest: Central European University Press, 2006), 39, 89 (fn. 127).
111 Adam of Usk, *The Chronicle of Adam Usk*, ed. and trans. Chris Given-Wilson (Oxford: Clarendon, 1997), 216–219.
112 Carl August Lückerath, *Paul von Rusdorf: Hochmeister des Deutschen Ordens, 1422–1441* (Bonn: Wissenschaftliches Archiv Bad Godesberg, 1969), 81–85; Mark Whelan, "Sigismund of Luxemburg and the Imperial Response to the Ottoman Turkish Threat" (PhD diss., Royal Holloway, University of London, 2014), 105–111, 178–183.
113 Malgorzata Dąbrowska, "From Poland to Tenedos: The Project of Using the Teutonic Order in the Fight against the Turks after the Fall of Constantinople," in *Byzanz und Ostmitteleuropa, 950–1453: Beiträge zu einer table-ronde des XIX International Congress of Byzantine Studies, Copenhagen, 1996*, ed. Günter Prinzing and Maciej Soloman (Wiesbaden: Harrassowitz, 1999), 165–176; see also Veszprémy's chapter in this volume.
114 Mark Whelan, "Dances, Dragons, and a Pagan Queen: Sigismund of Luxemburg and the Publicizing of the Ottoman Turkish Threat," in *The Crusade in the Fifteenth Century: Converging and Competing Cultures* (London: Routledge, 2016), 49–63, at 54–55.
115 Dąbrowska, "From Poland to Tenedos," 171–172. See also, Sarnowsky, "The Military Orders," 126–127.
116 Forstreuter, *Der Deutsche Orden*, 223.
117 Norman Housley, *Crusading and the Ottoman Turkish Threat, 1453–1505* (Oxford: Oxford University Press, 2014), 104.
118 For background and consequences, Frost, *The Oxford History of Poland-Lithuania*, 393.
119 Marian Biskup and Irena Janosz-Biskupowa, eds., *Protokolle der Kapitel und Gespräche des Deutschen Ordens im Reich* (Marburg: Elwert, 1991), 186–187.
120 On the secularization process, see Bernhart Jähnig, "Albrecht von Brandenburg-Ansbach und die Säkularisation des Deutschen Ordens in Preußen," in *Vorträge und Forschungen zur Geschichte des Preußenlandes und des Deutschen Ordens im Mittelalter*, ed. Hans-Jürgen Kämpfert and Barbara Kämpfert (Münster: Nicolaus-Copernicus, 2011), 90–99.

121 See, for example, Heinz Noflatscher, *Glaube, Reich und Dynastie: Maximilian der Deutschmeister (1558–1618)* (Marburg: Elwert, 1987).
122 See the comments regarding the Knights of St. John in the volume's introduction.
123 Demel, "Welfare and Warfare," 72.
124 Frost, *The Oxford History of Poland-Lithuania*, 393.

References

MS Sources

Berlin, Geheimes Staatsarchiv Preußischer Kulturbesitz, XX, Ordensbriefarchiv [OBA], 2922, 4124, 5251, 4984, 6069, 8266, 28544.

Hamburg, Hamburg Staatsarchiv, 612-2/3, 7.

Printed Sources

Adam of Usk. *The Chronicle of Adam Usk*, edited and translated by Chris Given-Wilson. Oxford: Clarendon, 1997.

Anonymous. "Die Danziger Chronik vom Bunde." In *Scriptores Rerum Prussicarum: Die Geschichtsquellen der preussischen Vorzeit bis zum Untergange der Ordensherrschaft*, edited by Theodor Hirsch, Max Töppen, and Ernst Strehlke, 405–489. Vol. 4. Leipzig: Hirzel, 1870.

Bederke, Ottomar, ed. *Liber de Cura Equorum: Bearbeitungen von Albertus Magnus und Jordanus aus dem Deutschen Ritterorden (1408)*. Hannover: Tierärztliche Hochschule, 1962.

Biskup, Marian, and Irena Janosz-Biskupowa, eds. *Protokolle der Kapitel und Gespräche des Deutschen Ordens im Reich*. Marburg: Elwert, 1991.

Bunge, Friedrich von, et al. *Liv-, Est- und Curländisches Urkundenbuch nebst Regesten*. 17 vols. Aalen: Scientia-Verlag, 1967–1981.

Forstreuter, Kurt, et al., eds. *Die Berichte der Generalprokuratoren des Deutschen Ordens an der Kurie*. 4 vols. Göttingen: Vandenhoeck & Ruprecht, 1960–1976.

Pelech, Markian, ed. *Nowa Księga Rachunkowa Starego Miasta Elbląga, 1404–1414*. 2 vols. Toruń: Państowe wydawnictwo naukowe, 1987–1989.

Sarnowsky, Jürgen, Cordelia Heß, Joachim Laczny, Cordula A. Franzke, and Christina Linke, eds. *Schuldbücher und Rechnungen der Großschäffer und Lieger des Deutschen Ordens in Preußen*. 4 vols. Cologne: Böhlau, 2008–2013; Berlin: Duncker & Humblot, 2018.

Studies

Arnold, Udo. "Weinbau und Weinhandel des Deutschen Ordens im Mittelalter." In *Zur Wirtschaftsentwicklung des Deutschen Ordens im Mittelalter*, edited by id., 71–102. Marburg: Elwert, 1989.

Arnold, Udo, ed. *Die Hochmeister des Deutschen Ordens*. Marburg: Elwert, 1998.

Böhnke, Werner. "Der Binnenhandel des Deutschen Ordens in Preußen und seine Beziehung zum Außenhandel um 1400." *Hansische Geschichtsblätter* 80 (1962): 26–95.

Boockmann, Hartmut. "Pferde auf der Marienburg." In *Vera lex historiae: Studien zu mittelalterlichen Quellen, Festschrift für Dietrich Kurze zu seinem 65. Geburtstag am 1. Januar 1993*, edited by Stuart Jenks, Marie-Luise Laudage, and Jürgen Sarnowsky, 117–126. Cologne: Böhlau, 1993.

Burleigh, Michael. *Prussian Society and the German Order: An Aristocratic Corporation in Crisis, c. 1410–1466*. Cambridge: Cambridge University Press, 1984.

Christiansen, Eric. *The Northern Crusades*. London: Penguin, 1997.

Dąbrowska, Malgorzata. "From Poland to Tenedos: The Project of Using the Teutonic Order in the Fight against the Turks after the Fall of Constantinople." In *Byzanz und Ostmitteleuropa, 950–1453: Beiträge zu einer table-ronde des XIX International Congress of Byzantine Studies, Copenhagen, 1996*, edited by Günter Prinzing and Maciej Soloman, 165–176. Wiesbaden: Harrassowitz, 1999.

Demel, Bernhard. "Welfare and Warfare in the Teutonic Order: A Survey." In *The Military Orders, Volume II Welfare and Warfare*, edited by Helen Nicholson, 61–73. Aldershot: Ashgate, 1998.

Ekdahl, Sven. *Die Schlacht bei Tannenberg 1410: Quellenkritische Untersuchungen*. Berlin: Duncker & Humblot, 1982.

Ekdahl, Sven. "Das Pferd und seine Rolle im Kriegswesen des Deutschen Ordens." In *Das Kriegswesen der Ritterorden im Mittelalter*, edited by Zenon Hubert Nowak, 29–48. Toruń: Universitas Nicolai Copernici, 1991.

Ekdahl, Sven. "Horses and Crossbows: Two Important Warfare Advantages of the Teutonic Order in Prussia." In *The Military Orders, Volume II Welfare and Warfare*, edited by Helen Nicholson, 119–151. Aldershot: Ashgate, 1998.

Feistner, Edith, Michael Neecke, and Gisela Vollmann-Profe. "Ausbildung korporativer Identität im Deutschen Orden: Zum Verhältnis zwischen Bibelepik und Ordenschronistik, Werkstattbericht." In *Deutschsprachige Literatur des Mittelalters im östlichen Europa: Forschungsstand und Forschungsperspektiven*, edited by Ralf G. Päsler and Dietrich Schmidtke, 57–74. Heidelberg: Universitätsverlag Winter, 2006.

Flemmig, Stephan. "Der Anteil sächsischer Berater an der Außenpolitik von Hochmeister Friedrich (1498–1510)." In *Akteure mittelalterlicher Außenpolitik: Das Beispiel Ostmitteleuropas*, edited by Stephan Flemmig and Norbert Kersken, 145–168. Marburg: Herder-Institut, 2017.

Forstreuter, Kurt. "Der Deutsche Orden und Südosteuropa." *Kyrios: Vierteljahresschrift für Kirchen- und Geistesgeschichte Osteuropas* 1 (1936): 245–272.

Forstreuter, Kurt. *Der Deutsche Orden am Mittelmeer*. Bonn: Wissenschaftliches Archiv Bad Godesberg, 1967.

Frost, Robert. *The Oxford History of Poland-Lithuania: Volume 1:* The Making of the Polish-Lithuanian Union, *1385–1569*. Oxford: Oxford University Press, 2015.

Given-Wilson, Chris. *Henry IV*. New Haven: Yale University Press, 2016.

Götz, Johannes. "Verbunden mit der Marienburg: Livländischer und preußischer Deutschordenszweig bis zum Ausbruch des Zungenstreits 1438." In *Livland—eine Region am Ende der Welt? Forschungen zum Verhältnis zwischen Zentrum und Peripherie im späten Mittelalter*, edited by Anti Selart and Matthias Thumser, 371–414. Cologne: Böhlau, 2017.

Housley, Norman. *Crusading and the Ottoman Turkish Threat, 1453–1505*. Oxford: Oxford University Press, 2014.

Jähnig, Bernhart. "Zur Wirtschaftsführung des Deutschen Ordens in Preußen vornehmlich vom 13. bis zum frühen 15. Jahrhundert." In *Zur Wirtschaftsentwicklung des Deutschen Ordens im Mittelalter*, edited by Udo Arnold, 113–147. Marburg: Elwert, 1989.

Jähnig, Bernhart. "Albrecht von Brandenburg-Ansbach und die Säkularisation des Deutschen Ordens in Preußen." In *Vorträge und Forschungen zur Geschichte des Preußenlandes und des Deutschen Ordens im Mittelalter*, edited by Hans-Jürgen Kämpfert and Barbara Kämpfert, 90–99. Münster: Nicolaus-Copernicus, 2011.

Kļaviņš, Kaspars. "Reorganizing the Livonian Landscape: Some Issues and Research Perspectives." In *Ecologies of Crusading, Colonization, and Religious Conversion in the Medieval Baltic*, edited by Aleksander Pluskowski, 197–208. Turnhout: Brepols, 2019.

Kubon, Sebastien. *Die Außenpolitik des Deutschen Ordens unter Hochmeister Konrad von Jungingen (1393–1407)*. Göttingen: V&R Unipress, 2016.

Laczny, Joachim. *Schuldenverwaltung und Tilgung der Forderungen der Söldner des Deutschen Ordens in Preußen nach dem Zweiter Thorner Frieden: Ordensfoliant 259 und 261, Zusatzmaterial*. Göttingen: V&R Unipress, 2019.

Leighton, Gregory. "Did the Teutonic Order Create a Sacred Landscape in Thirteenth-Century Prussia?" *Journal of Medieval History* 4 (2018): 457–483.

Lückerath, Carl August. *Paul von Rusdorf: Hochmeister des Deutschen Ordens, 1422–1441*. Bonn: Wissenschaftliches Archiv Bad Godesberg, 1969.

Mickūnaitė, Giedrė. *Making a Great Ruler: Grand Duke Vytautas of Lithuania*. Budapest: Central European University Press, 2006.

Militzer, Klaus. "Die Aufnahme von Ritterbrüdern in den Deutschen Orden: Ausbildungsstand und Aufnahmevoraussetzungen." In *Das Kriegswesen der Ritterorden im Mittelalter*, edited by Zenon Hubert Nowak, 7–17. Toruń: Universitas Nicolai Copernici, 1991.

Możejko, Beata. "Introduction." In *New Studies in Medieval and Renaissance Poland and Prussia: The Impact of Gdańsk*, edited by ead., 1–17. Abingdon: Routledge, 2017.

Murray, Alan V. "The Saracens of the Baltic: Pagan and Christian Lithuanians in the Perception of English and French Crusaders to Late Medieval Prussia." *Journal of Baltic Studies* 41 (2010): 413–430.

Neitmann, Klaus. "Die Preußischen Stände und die Außenpolitik des Deutschen Ordens vom I. Thorner Frieden bis zum Abfall des Preußischen Bundes (1411–1454): Formen und Wege ständischer Einflußnahme." In *Ordensherrschaft, Stände und Stadtpolitik: Zur Entwicklung des Preußenlandes im 14. und 15. Jahrhundert*, edited by Udo Arnold, 27–71. Lüneburg: Nordostdeutsches Kulturwerk, 1985.

Nicholas, David. *The Northern Lands: Germanic Europe, c. 1270–c. 1500*. Oxford: Wiley-Blackwell, 2009.

Noflatscher, Heinz. *Glaube, Reich und Dynastie: Maximilian der Deutschmeister (1558–1618)*. Marburg: Elwert, 1987.

Nowakowski, Andrzej. "Some Remarks about Weapons Stored in the Arsenals of the Teutonic Order's Castles in Prussia by the End of the 14th and Early 15th Centuries." In *Das Kriegswesen der Ritterorden im Mittelalter*, edited by Zenon Hubert Nowak, 75–88. Toruń: Universitas Nicolai Copernici, 1991.

Nowakowski, Andrzej. *Arms and Armour in the Medieval Teutonic Order's State in Prussia*, translated by Maria Abramowicz. Łódź: Oficyna Naukowa, 1994.

Paravicini, Werner. "Von der Preußenfahrt zum Hussitenkreuzzug." In *Beiträge zur Militärgeschichte des Preußenlandes von der Ordenszeit bis zum Zeitalter der Weltkriege*, edited by Bernhart Jähnig, 121–159. Marburg: Elwert, 2010.

Paravicini, Werner. *Die Preußenreisen des europäischen Adels*. 3 vols. Sigmaringen: Jan Thorbecke, 1989–95; Göttingen: V&R Unipress, 2020.

Päsler, Ralf G., and Dietrich Schmidtke, eds. *Deutschsprachige Literatur des Mittelalters im östlichen Europa: Forschungsstand und Forschungsperspektiven*. Heidelberg: Universitätsverlag Winter, 2006.

Pluskowski, Aleksander. "The Ecology of Crusading: Investigating the Environmental Impact of Holy War and Colonisation at the Frontiers of Medieval Europe." *Medieval Archaeology* 55 (2011): 192–225.

Pluskowski, Aleksander. *The Archaeology of the Prussian Crusade: Holy War and Colonisation*. London: Routledge, 2012.

Pluskowski, Aleksander, ed. *Ecologies of Crusading, Colonization, and Religious Conversion in the Medieval Baltic*. Turnhout: Brepols, 2019.

Probst, Christian. *Der Deutsche Orden und sein Medizinalwesen in Preussen: Hospital, Firmarie und Arzt bis 1525*. Bonn: Wissenschaftliches Archiv Bad Godesberg, 1969.

Rodríguez García, J. M., and A. Echevarría Arsuaga. "Alfonso X, la Orden Teutónica y Tierra Santa: Una nueva fuente para su estudio." In *Las Órdenes Militares en la Península Ibérica*, edited by Ricardo Izquierdo Benito and Ruiz Gómez Francisco, 489–510. Vol. 1. Cuena: Ediciones de la Universidad de Castilla-La Mancha, 2000.

Salvadore, Matteo. *The African Prester John and the Birth of Ethiopian-European Relations, 1402–1555*. London: Routledge, 2017.

Sapoznik, Alexandra. "Bees in the Medieval Economy: Religious Observance and the Production, Trade, and Consumption of Wax in England, c. 1300–1555." *Economic History Review* 72 (2019): 1152–1174.

Sarnowsky, Jürgen. *Die Wirtschaftsführung des Deutschen Ordens in Preußen (1382–1454)*. Cologne: Böhlau, 1993.

Sarnowsky, Jürgen. "Die Quellen zur Geschichte des Deutschen Ordens in Preußen." In *Edition Deutschsprachiger Quellen aus dem Ostseeraum (14.–16. Jahrhundert)*, edited by Matthias Thumser, Janusz Tandecki, and Dieter Heckmann, 171–200. Toruń: Uniwersytetu Mikołaja Kopernika, 2001.

Sarnowsky, Jürgen. "The Military Orders and Crusading in the Fifteenth Century: Perception and Influence." In *Reconfiguring the Fifteenth-Century Crusade*, edited by Norman Housley, 123–160. London: Palgrave Macmillan, 2017.

Szczepański, Seweryn. "*Arbor custodie que vulgariter dicitur Wartboumi*: The Function and Existence of the so-Called 'Watchtower Trees' in Pomesania and Żuławy Wiślany in the 13th–14th Centuries." *Zapiski Historyczne* 76 (2011): 5–18.

Szybowski, Sobiesław. "Two Local Empires in Confrontation: Poland and Lithuania Competing with the Teutonic State to Dominate the Southeastern Region of the Baltic Sea (the End of the Fourteenth to 1409)." In *Studies in Medieval and Renaissance Poland and Prussia: The Impact of Gdańsk*, edited by Beata Możejko, 60–73. Abingdon: Routledge, 2017.

Urban, William. *Tannenberg and After*. Chicago: Lithuanian Research and Studies Center, 1999.

Urban, William. "The Frontier Thesis and the Baltic Crusade." In *Crusade and Conversion on the Baltic Frontier, 1150–1500*, edited by Alan V. Murray, 45–71. Aldershot: Ashgate, 2001.

Voigt, Johannes. *Geschichte des Deutschen Ritterordens in seinen zwölf Balleien in Deutschland*. 2 vols. Berlin: Georg Reimer, 1857–1859.

Whelan, Mark. "Sigismund of Luxemburg and the Imperial Response to the Ottoman Turkish Threat." PhD diss., Royal Holloway, University of London, 2014.

Whelan, Mark. "Dances, Dragons, and a Pagan Queen: Sigismund of Luxemburg and the Publicizing of the Ottoman Turkish Threat." In *The Crusade in the Fifteenth Century: Converging and Competing Cultures*, edited by Norman Housley, 49–63. London: Routledge, 2016.

Whelan, Mark. "Between Papacy and Empire: Cardinal Henry Beaufort, the House of Lancaster, and the Hussite Crusades." *English Historical Review* 133 (2018): 1–31.

Whelan, Mark. "'On Behalf of the City': Wax and Urban Diplomacy in the Late Medieval Baltic and North Sea." *Urban History*, forthcoming.

Wüst, Marcus. *Studien zum Selbstverständnis des Deutschen Ordens im Mittelalter*. Weimar: Verlag und Datenbank für Geisteswissenschaften, 2013.

15

THE COLD WINTER CAMPAIGN OF 1511

Swiss Military Autonomy and Heteronomy during the Transalpine Campaigns

Anna Katharina Weltert and Georg Christ[1]

Introduction

"A free view to the Mediterranean!" is the ironic response given by Swiss people when asked how they would like to improve their landlocked country. In the late Middle Ages, Swiss soldiers were not so far from realizing this vision. The Italian Wars, in which the Old Swiss Confederation was actively involved between 1499 and 1515, opened the prospect of enlarging Swiss territory south of the Alps and to position the confederation as a notable player among the great European powers such as the Holy Roman Empire, Spain, France, or the Holy See. Yet the tensions between élites and commoners, city and country, mountains and plains, stifled the Swiss expansionist drive. The resulting stalemate was further exacerbated by the Reformation and fostered an outcome of what one might call unfinished state formation. The Battle of Marignano in 1515 is often considered to be the turning point that stopped Swiss expansionism. From then onwards, and for almost three centuries, Swiss soldiers only fought each other in Switzerland or served in foreign armies within a diplomatic framework tightly controlled by the cantonal élites.

The fault lines between both Swiss cantons and Swiss troops serving great powers abroad, however, were older. Four years before Marignano, another unsuccessful expedition into the Duchy of Milan had taken place. The Swiss crossed the Alps in November 1511 in order to avenge the imprisoning and killing of two heralds by the French. This campaign, the so called Cold Winter Campaign, turned into a disaster. Whereas the march to Milan was harsh but disciplined, on the return, the battlefield-hardened, hungry, cold, and penniless warriors plundered and burned indiscriminately. Instead of wealth and honour they brought home dishonour and shame.

DOI: 10.4324/9781003245568-16

Whereas this often-overlooked episode of the transalpine campaigns is undoubtedly viewed as a black spot in the allegedly glorious history of Swiss soldiery, it provides us with valuable information about power, authority, and the monopoly over the use of military force beyond the confederation's borders during this founding period of Switzerland.[2] It offers a case study through which we can explore the fault lines jeopardizing Swiss joint political and military action in the years before Marignano. Which factors negatively affected Swiss engagement in the Cold Winter Campaign? To what extent was the Swiss Diet (*Tagsatzung*) in control against both the agendas of individual cantons but also the military decision makers in the field? To which extent did the pope as the main Swiss ally hold a monopoly over recruitment of Swiss mercenaries? Who made decisions in the field between cantonal troop assemblies and cantonal or federal field commanders including the fateful decision to chance a spontaneous siege of Milan? How strong were other foreign influences, especially of France and the Empire, both in the field and back home?

This chapter argues that although different authorities (not only the Swiss cantons and Swiss Diet but also France and the Papacy) tried to control the Swiss military potential and harness it for their political interests, the Swiss military, as a community abroad, a temporary Swiss diasporic group, could exploit a power vacuum, which offered a certain degree of agency to the soldiers in the field. This could enhance but also jeopardize their military success and incur the wrath of foreign powers, which was not in the interest of at least some of the Swiss polities and their élites. It was this risk and lack of control that motivated the Diet's repeated attempts to regulate Swiss service in foreign armies. This eventually gave rise to a system of increasingly more regulated Foreign Service (*Fremde Dienste* that is military service abroad) and renunciation of any independent Swiss expansionist military policy. The Cold Winter Campaign thus is an important catalyzer that established the fault lines, which were then fully opened at Marignano and shaped a long-term divided post-Reformation confederation.

The cantonal contingents of the Cold Winter Campaign can hardly be considered a military diaspora or at least only a temporary one. Rather, these Swiss soldiers were transhumant; their warring followed a pattern of seasonal campaigning beyond the Alps. Their wanderings were akin and perhaps related to those of Swiss transhumant dairy farmers.[3] Nevertheless, the (deliberately, as we shall argue) unclear status of the Swiss in the Cold Winter Campaign makes them an interesting object of analysis within a volume on military diasporas. It was unclear whether these troops were loosely coordinated cantonal troops, a confederate army, or a mercenary contingent of the papal army. It is not clear whether they avenged an insult, negotiated a treaty with the French, expanded Swiss (or, rather, the central Swiss cantons') territory or fought the pope's war. This undetermined status allowed both the pope and, to a lesser extent, the Swiss Confederation to leave the troops to their own resources, quite literally, i.e. not paying or supplying them properly. This situation was further exacerbated by the grim winter weather, which complicated logistics and communications across the Alps.

To better understand the role and status of Swiss troops abroad, we propose the following analytical categories: We distinguish between cantonal contingents in Foreign Service within the framework of a Swiss treaty with a foreign power (type I), Swiss corps as a fixed (more or less permanent, "diasporic") part of foreign armed forces (type II), life (body) guards (type III), free companies of mercenaries (type IV), free individual mercenaries (type V). These are ideal types, of course, and we shall see how the status of the Swiss forces, in reality, was fluid in the case of the Cold Winter Campaign. The *Pensionsherren* or Swiss military entrepreneurs were mediators operating across these types; they could be instrumental in the brokering of military alliances (type I) but also recruit soldiers for a corps (increasingly, this would become the standard e.g. recruitment for the various Swiss infantry regiments in France, type II and III) but also for free companies (type IV). In terms of payments, we have to distinguish between *Pensionen/Jahrgelder,* i.e. regular payments to influential personalities or to a state/ canton, and the monthly pay of the soldiers.[4] We also have to consider the payments necessary to outfit and maintain troops, whereby it seems to have been a matter of disagreement to which extent the soldiers or contingents had to cover these expenses themselves.

But first, we will provide some background to the Cold Winter Campaign, especially the end of the French-Swiss alliance and the papal-Swiss entente that came to replace it in 1510. We will also look at the command structures. The next section will discuss the actual Cold Winter Campaign and the related preceding campaign within the context of the new alliance with the pope, the *Chiasserzug* of 1510. Then, we will analyze problems of communication, especially with the pope on the one hand and the Swiss Diet on the other, and how this shaped the agency of the soldiers in the field and their decision-making. Finally, in conclusion, we will argue that the experiment of a stand-alone Swiss confederate force operating independently failed despite showing some undoubted promise. Henceforth, Swiss military personnel abroad increasingly operated within a diplomatic framework tightly regulating service in foreign militaries. In this context, we will also respond to the questions proposed in the introduction of this volume regarding military groups' diasporic nature, military edge, integration into host armies, role in state formation, and diasporic solidarities.

Historical Background: The Old Swiss Confederation as a Global Player in the Shaping of Modern Europe?

The Old Swiss Confederacy and Its Foreign Relations

The Old Swiss Confederation was a loose network of independent small states. They joined in different treaties for mutual peace-keeping and military support in times of internal or external conflicts and support in upholding legitimate authority and jurisdiction. At the time of the Italian Wars, the Confederation consisted of twelve cantons, including the city of Zurich, the city and republic of Bern, the city of Lucerne, the land cantons of Uri, Schwyz and Unterwalden, the city of Zug, the canton of Glarus, and the cities of Fribourg, Solothurn, Schaffhausen, and Basel (Appenzell only joined in 1513). All of them enjoyed imperial immediacy, i.e. relative independence within the framework of the Holy Roman Empire.[5] Yet, for some (Basel, Schaffhausen, and the Argovian polities bordering the Rhine) affiliation with the empire was more important than others. Although belonging to the *Eidgenossenschaft* was of special importance, it did not mean that the individual entities did not pursue their own interests and foreign policy.

Bern, and to an extent Fribourg, for instance, rather looked westwards than northwards. The cities were eager to maintain good relations with the Duke of Savoy, which in turn provoked the hostility of the French king. Since the French were fighting for hegemony in Northern Italy, Bern also tended to side with the Duke of Milan in defence of his territory. Bern nevertheless was in this particular period keen to preserve good relations with France—maybe because Savoy also maintained peace with France and because of French payments to Bernese patricians.[6]

South looking Uri, on the other hand, was mainly interested in protecting and controlling the trade routes (chiefly the Gotthard Pass) over the Alps towards Italy in order to ensure access to the Northern Italian (cattle) markets. It therefore aspired to enlarge its territory southwards at the expense of the Duchy of Milan. Thus, Uri, supported by Schwyz, Unterwalden, and, to an extent, Lucerne, Zurich, and Solothurn, tended to favour alliances with the French king against the Duke of Milan and emperor Maximilian I.[7] In 1511, however, after France had occupied Milan, Uri and its supporters were ready to change sides if it furthered their interest in southward expansion. Matthäus Schiner, bishop of Sion from 1499, papal legate, and cardinal of Pudenziana from 1511,[8] represented the interests of the Holy See in the Confederation. He supported the ambitions of his native Valais and also the Three Leagues (the Grisons), both allied to the Confederation, which similarly tended to look southwards.[9]

The need to access international markets was increased by an agricultural shift that had taken place in the late Middle Ages and that greatly influenced the further development of the economy of central Switzerland and of the mountainous areas of Fribourg and the Bernese Oberland. Farmers gave up their self-sufficiency and increasingly produced for the interregional market by turning from mixed agriculture to export-oriented livestock breeding and cheese production while importing grain in return.[10] As a result of this, the yearly livestock fairs in Lugano became one of the most important markets for farmers, to which they regularly journeyed across the Alps.[11] The precarious food supply in combination with a population growing since the fifteenth century raised the spectre of malnutrition and famine. The limited resources raised unemployment especially among the poor day-labourers. They now heavily relied on the fortunes of war: Military service abroad, therefore, was not only a military and political issue but also an economic one. Soldiers became one of the Confederation's most important export items and mercenary service became a crucial way to alleviate the pressures on resources exerted by a growing population.[12]

Mercenaries: The Confederation's Most Important Export Item

These political, economic, and military factors meant the Swiss Confederation was strongly involved in international politics of the great powers, which had a strong interest in the Confederation because of the Swiss military potential, i.e. mercenaries.

Indeed, by the early 1500s, thousands of Swiss men fought as mercenaries for foreign powers. Since the spectacular victories over the Burgundian Duke Charles the Bold (Grandson and Murten 1476, Nancy 1477), the Swiss infantry had become famous for their ability to fend off cavalry attacks, their swift marches, and their courage. Consequently, mercenaries from the Old Swiss Confederation were fighting for foreign rulers on the main European battlefields. The reasons for Swiss men to enter into Foreign Service were manifold. In addition to the aforementioned lack of opportunities at home, adventure, the chance to win a share of the spoils of war, glory, and honour were further motives. The military success of the mercenaries, however, also whetted the Swiss cantons' appetite for their own expansionist projects.

Charles VIII of France had enlisted 8,000 Swiss mercenaries for his conquest of Naples in 1494. Five years later, he signed a treaty with the Swiss Confederation that granted him the exclusive right to enlist Swiss mercenaries for the next ten years. Yet this did not mean a full-fledged French monopoly over Swiss mercenary potential, as the Pontifical Swiss Guard, the personal guard of the pope, was founded precisely in this period. The particular arrangement between France and the Swiss confederates also questions the use of the term "mercenary" to describe these troops. They were, of course, paid, but—contrary to mercenaries in a more narrow sense—they did not enter directly into a contractual relationship with the French sovereign but indirectly or, rather, collectively through treaties between cantonal élites and the kingdom of France. This arrangement clashed and coexisted with the "free" mercenaries who entered into foreign military service individually, but less so with the different guards, who were, at least officially, only used for close protection of the respective rulers.

Charles' successor Louis XII, however, was not in favour of continuing the alliance with the Swiss. Therefore, and due to monetary constraints, he did not renew the treaty after its expiration in 1509. Bishop Matthäus Schiner exploited the resulting anger of the Swiss and induced the Confederation to instead sign an agreement with Pope Julius II, which granted His Holiness a monopoly on recruiting Swiss mercenaries.[13] The Pope's aim was to re-establish a powerful Papal State, which would dominate a more unified and independent Italy. As a first step, Julius II allied with Louis XII and Emperor Maximilian I in a campaign against Venice, moving against the expanding republic in 1508.[14] After

Venice had been soundly defeated at the battle of Agnadello in 1509, the Pope then turned against his erstwhile ally, the king of France, and sought an alliance with Venice, Spain, and England in order to evict the French from northern Italy. This alliance was formalized in 1511 with the foundation of the "Holy League". Julius also needed soldiers and turned to the permanent Pontifical Swiss Guard, which he had formed in 1506,[15] and now hoped to expand.[16] In March 1510, Schiner succeeded in convincing the Swiss Diet to sign a five-year treaty, which ensured the Pope a contingent of 6,000 Swiss soldiers.[17] Probably, the pope was following the French model of a permanent or semi-permanent corps of Swiss troops within his armed forces.[18] Furthermore, the treaty obliged the Confederation to remain neutral towards possible opponents of the Pope.[19]

The Pike Square: Key to Military Success

Why were foreign powers, e.g. the Holy See and France, so eager to have Swiss troops in their armies and willing to spend a fortune on their pay? The reason is that the Swiss enjoyed a particularly strong military reputation at the time. Fostered by the famously hardy mountainous environment from which they (or rather: some of them) hailed, they were considered to be particularly tough and courageous warriors. The role of infantry had become more important within the context of the so-called infantry revolution, which saw foot soldiers increasingly challenge the dominance of heavy cavalry on the battlefield since the fourteenth century.[20] The Swiss had perfected precisely such infantry skills, armament, and tactics. In their various encounters, mostly with the Habsburgs and then the Burgundians during the fourteenth and fifteenth centuries, the Swiss infantry honed the pike square or *Gewalthaufen* tactic. A tightly packed, square-shaped infantry unit was armed with swords, daggers, and, importantly, pikes and halberds. These latter were versatile weapons combining elements of a pike with a hook and an axe blade, which were ideal for engaging mounted horseman but exposed their handlers considerably. This was aggravated by the fact that these foot soldiers wore only light and partial armour. Several (three to five) outer ranks, therefore, did not carry halberds but increasingly long pikes, which reached up to five metres in order to fend off attacking cavalry.[21] The *Gewalthaufen* tactic thus required the unit to act in unison and with a high level of discipline. Soldiers abandoning their ranks meant the loss of the order of battle and, possibly, the battle itself. If the soldiers stuck together, however, they were difficult to overcome, as enemy cavalry found it hard to breach their hedgehog-like formation. Thus, Swiss infantrymen found a way to effectively compete with the previously dominant military force; the feudal, armoured cavalry. When the square managed to breach the attacked formation, the warriors from within the square resorted to their halberds, swords, and daggers for close combat (*mêlée*).[22] The relative lightness of armament allowed the soldiers to rapidly march long distances with relative ease. This was not only crucial when it came to chasing the enemy and plundering at the end of a battle—which was one of the main interests of mercenaries—but also to exploit the weakness of an enemy formation not yet pitched for battle.[23]

The Swiss soldiers passed their test 1476–1477 when they defeated the feudal army of the powerful Burgundian duke, Charles the Bold, three times in succession.[24] Their tactics proved highly effective against the Burgundian cavalry, archers, and artillery. Even after the Swiss infantry experienced a crushing defeat at the hands of the Spanish military reformer Gonzalo de Cordoba in the Battle of Cerignola 1503, who used massed harquebus fire against them, the Swiss updated their *Gewalthaufen*-tactics only moderately by lengthening the pikes and by including handguns.[25]

Europe's great powers were henceforth willing to pay high pensions to the authorities and/or private stakeholders (*Pensionsherren*) in order to secure Swiss soldiers for their armies. Also, the mercenaries themselves were paid well: One source states that they were paid 4.5 instead of the usual four guilders a month, which corresponded to the salary of a master carpenter.[26] The Burgundian Wars not only made the Swiss soldiers famous and highly demanded mercenaries, but they also had a huge impact on their self-confidence and self-perception. The victories, it was generally believed, were a sign from God: The simple and modest mountain people were chosen as His instrument to punish an idle and sinful nobility.[27]

Yet this newly acquired reputation for military prowess created conflicts of interest and widespread unease. Serving abroad deteriorated the men physically and morally; it also negatively impacted their communities upon their return—if they returned. Military service abroad fostered the emergence of the above-mentioned *Pensionsherren*. This military entrepreneurial élite aligned its interests with foreign powers, thus jeopardizing the cantons' freedom of action. These feelings came to a climax shortly after the Burgundian Wars and were discussed by the Swiss Diet in the wake of the trial of the *Pensionsherr* and prominent mayor of Zurich, Hans Waldmann, in 1489.[28] Also, the first Battle of Novara in 1500, that pitched French against Milanese forces both relying on Swiss infantry, highlighted the problem of unregulated Foreign Service. The dilemma, in this case, was resolved by the infamous betrayal of Novara, whereby the Swiss mercenaries abandoned the duke of Milan, citing a passage in their contract that they could not be ordered to fight compatriots. The Cold Winter Campaign thus needs to be seen within the wider context of inner-Swiss conflicts of visions over how to control and organize Foreign Service. Some (such as the Zurich reformer Zwingli a few years later) wanted to abolish Foreign Service altogether. The Cold Winter Campaign was perhaps the last attempt at a stand-alone Swiss expeditionary force with some elements of (or hopes to be) a contingent (type I) and contrasted with a more laissez-faire attitude favoured by others (types II–V). Eventually, a Swiss compromise emerged whereby the cantons controlled a tightly regulated system of military service abroad, Foreign Service, that was aligned with cantonal and Swiss federal interests (types II (while retaining some elements such as cantonal standards from type I) and III). Free companies and individual soldiering (types IV and V) by contrast declined and the development of a more robust Swiss military organization stalled until well into the nineteenth century.[29]

The Cold Winter Campaign

The newly acquired honour and respect for the Swiss soldier on an international level was challenged in September 1510, when three Swiss heralds (*Standesläufer*), one each from Schwyz, Fribourg, and Bern, were taken hostage by the French governor of Lugano.[30] Two of them were killed; one narrowly escaped. The Cold Winter Campaign was a failed attempt to avenge this insult.

The incident occurred during the preceding Chiasso Campaign (*Chiasser-Zug*), which, although not officially part of the papal campaign against France, was the effect of the rapprochement between Venice and Pope Julius II and seemed directed against Milanese, i.e. currently French, possessions in Lombardy. A contingent of about eight thousand soldiers from the Swiss confederation formally entered (individually but within the framework of the papal-Swiss treaty) into papal mercenary service, i.e. they were directly paid by the Pope through Matthäus Schiner but marched as a Swiss contingent, or rather, a coalition of cantonal units (*Haufen* or *Fähnlein*) under cantonal standards. These Swiss were thus a de facto military contingent in Foreign Service (type I, see above) but *de jure* insisted on being a Papal corps (II), an extended life guard (type III), with a more robust but, as the Swiss believed, purely defensive mandate to protect the person, palace and state of the pope.

The hostilities with the Milanese erupted when French-Milanese troops prevented the Swiss contingent from entering Milanese territory at Ponte Tresa on 9 September 1510. The Swiss demanded free passage in order to reach Papal territories, but the French denied this request doubting the defensive mandate of the Swiss and suspecting that they would be used in an aggressive role (type I or, perhaps, II). This was later confirmed by Pope Julius himself, who admitted that he had planned to use the Swiss troops against the duke of Ferrara, his unruly vassal.[31] The Swiss, in any case, showed enough aggressive spirit to force their entry and march onto Varese.[32]

We should see the treatment of the captured messengers in this context. The French needed evidence of the suspected papal plots, and thus the interception of the messengers was probably not a mistaken move by the French governor of Lugano (as was claimed later) but actually done on the behest of the French commanders. The different fates of the three messengers might indicate that the French found what they were looking for but probably not uniformly in all the three messenger boxes.[33] Indeed, according to Zurich reformer Heinrich Bullinger (although hardly a disinterested observer), the Schwyzer messenger carried correspondence between the pope and Matthäus Schiner.[34] That might explain why he was stabbed. The messenger from Fribourg (probably carrying incriminating material as well) was drowned in the lake of Lugano.[35] The messenger from Bern, however, survived. Did he escape relatively unscathed because the missives in his box were not as anti-French, mirroring a Bernese unease about breaking with France? Or was his release intended to fan tensions between the pro- and anti-French parties in the cantons? Whatever the answer, it was indeed a fragile coalition that supported the treaty with the pope, and the French party in the Swiss cantons remained strong. When the French presented their written proof of the papal plans and applied pressure, seconded also by the emperor, the Swiss Diet caved in and recalled the troops only two weeks after the start of the campaign.[36]

When the messenger from Bern returned to his homeland after six months of tribulation, he reported the sorry tale of murder, torture, and humiliation. His uniform—in the colours of the canton and therefore officially representing it—had been destroyed and his tin letter box, showing the coat of arms of the canton of Bern, had been ridiculed and broken by French soldiers.[37] On 9 September 1511, almost a year after the incident, the Swiss Diet met in Lucerne to discuss the matter.[38] Not only were diplomatic relations with the powerful Kingdom of France at stake, but the Diet also had to placate the people of Schwyz, who called for immediate revenge. France had committed a serious violation of diplomatic immunity as heralds/official messengers were protected under imperial/international law.[39] A stark condemnation of the crime and the demand for compensation was inevitable, lest the Confederacy lose face. Yet, until recently, France had been a strong ally and a crucial source of revenue, and it was hoped that good relations could be restored in the future.[40] It seemed thus unwise to provoke this powerful neighbour. The unwillingness to move against France might have been bolstered by French gold, which certainly found its way not only officially but also unofficially to members of the Swiss political élite.[41] Therefore, to solve the problem diplomatically seemed more attractive despite the outcry by the people and their call for revenge.

First, the Diet tried to act diplomatically and sent a letter to Schwyz, asking the people to keep calm until matters become clearer. A second letter was written to Gaston de Foix, the French governor of Milan, requesting compensation for the slain messengers but at the same time reassuring him that their intention was to keep the peace.[42] Two months passed until de Foix's answer reached the Diet. It stated succinctly that the King of France likewise did not wish to engage in warfare against the Confederation but was eager to renew their friendship. Another letter by the Holy Roman Emperor urged the Swiss not to provoke his French ally on which's side he was prepared to fight should war break out.[43] This thinly veiled imperial threat, combined with a lack of contrition on the part of the French, and the ignored request for compensation, was not to be misinterpreted: In the eyes of the Swiss, the French did not take the matter as seriously as they ought to. The Diet called for another meeting without delay and met in Lucerne on 4 November 1511. After hearing the different opinions regarding the appropriate reaction, the representatives decided to attend the highest authority of the canton of Schwyz, the people's assembly (*Landsgemeinde*), at the market place of Schwyz on 9 November. The people were to be informed about the content of both letters and officially asked to be patient until the end of the month. This would give the Diet time to organize a coordinated attack against the French if a violent response was unavoidable.[44]

The Schwyzer, however, refused to wait. In fact, on 28 October the canton had gathered its troops and started to train them.[45] The Schwyzer answer to the Diet left no doubt: They would not delay marching towards Bellinzona, and if the other cantons remembered their oaths, they would not hesitate to also send their troops. Matthäus Schiner, by then promoted to cardinal and orchestrating a Papal-Spanish-Venetian alliance to expel the French from Italy, provided papal endorsement and blessing for the campaign. To fight under the protection of the head of the Church was a strong incentive for the soldiers. It gave their campaign a greater legitimization than merely revenge; it turned it into a holy war. As such, the Swiss soldiers were promised that all sins committed during the campaign would be forgiven because they were fighting for the Holy See and defending the Christian faith. Also, the French presence in Northern Italy was seen as a threat to the Confederation, endangering the independence of the Valais as well as the Bernese Vaud and Savoy. Furthermore, the Duchy of Milan, especially Lugano, was of vital economic importance.[46] Although the pro-French forces in the Confederation were still strong, this combination of incentives was difficult to resist.[47] Shortly afterwards Schiner travelled to Venice to raise funds in support of the campaign.[48] The other cantons had little choice but to join, although some of them did so rather half-heartedly.[49]

This half-heartedness might explain the poor coordination of the cantonal contingents, some of which departed much later than others. By the time the Diet in Zurich officially approved the campaign on 17 November, the contingent from Fribourg (with some field artillery and, probably, horsemen)[50] had already crossed the Alps, joined on 14 November by 1,500 men from Schwyz.[51] They marched to the river Tresa and reconstructed and crossed the bridge that had been destroyed by the French.[52] It was only when they reached Varese on 30 November that they were joined by the Bernese contingent. Heavy rain had destroyed much of the roads and rumours had it that the governor of Milan had gathered his troops against the Swiss in Gallarate.[53] One after another, the remaining contingents now finally arrived; between 30 November and 4 December the contingents from Lucerne, Uri, Unterwalden, Glarus, Schaffhausen and some other allies (*Zugewandte Orte*) including a second contingent from Fribourg.[54] Meanwhile, ambassadors were sent to Venice to coordinate the attack. It was planned that the Venetian army should be ready in the area of Verona, while the papal troops should attack the French from Bologna and Parma. The Swiss, for their part, were to march from Bergamo to Brescia and then to Peschiera, at the southern point of Lake Garda, where they would rendezvous with the Venetian troops.[55]

The Swiss field army now consisted of more than 5,000 men, who were still waiting for the last troops to join them. Since food had become scarce by this point, they moved further south to Gallarate, where they were briefly encircled by the French troops of Gaston de Foix but soon relieved by the contingents from Basel (probably including also some cavalry),[56] Zurich, and the subject towns of Bremgarten, Baden and Farnburg, which finally had caught up with them on 6 to 8 December.[57] The Zurichers had also brought some cannon, which were decisive in the break-through, and probably also cavalry.[58] After evading the French encirclement, the Swiss then moved onto Legnano, less than 20 miles to the northwest of Milan. This is where they were finally joined by another 4,000 soldiers from Bern and 600 from Solothurn, with more artillery.[59] There was thus some artillery assembled and most likely some cavalry (although its strength remains unclear), mainly for communications, intelligence gathering, foraging, protection of troop gatherings, flank attacks, pursuit, and support of the vanguard as it had been customary since the Burgundian wars.[60]

At this point, a papal herald arrived in the name of Cardinal Schiner and brought several letters signed by the Pope himself, in which he gave the mercenaries his blessing, granted them absolution of sins, and promised generous pay.[61] The army now consisted of 10,000–16,000 men.[62] The Swiss received no more communications from their allies, and since they were now at full strength and very close to Milan, the commanders decided to march onto the city rather than head eastward to join the Venetian army, as had previously been agreed.[63] The fact that they had gathered a substantial amount of ordnance (for a Swiss force) might have influenced this decision, although, of course, the light field cannons, while useful on the battlefield, were not sufficient for such a siege.[64] Under the (nominal) command of Jakob Stapfer from Zurich,[65] the army moved south without major French interference. The Swiss hope that the people of Milan would rise up against the French in their support did not come true.[66]

Knowing that the Swiss army had neither stamina, ordnance nor logistical capacity for a long winter siege, the French cut off their supplies. Due to the lack of food and the unusually cold temperatures, the Swiss withdrew to Monza, a few miles northeast of Milan.[67] There they awaited the Venetian and Papal reinforcements rather than moving towards them, while Venetian troops moved from the Friuli towards Vicenza, thus closing in on the agreed meeting point. The Swiss, however, seemed to be unaware of this and thought that the Venetians had abandoned them.[68] Meanwhile, Gaston de Foix sent an envoy to the Swiss troops to negotiate peace exploiting the rift among pro- and anti-French Swiss leaders.[69] The French offer consisted of one month's pay for the soldiers totalling 25,000 guilders, plus 8,000 guilders compensation for the canton of Schwyz.[70] Even after De Foix significantly improved the offer, the Swiss declined. They wanted much more: Three month's pay for the soldiers and the territories of Lugano and Locarno.[71] The French could not agree to this and the negotiations broke down.

As the hope for reinforcements shrank, growing hunger, the bitter cold and failed negotiations fuelled discontent in the Swiss forces that eventually turned into blind fury. The commanders were not able to control their troops any longer and therefore decided to return home as soon as possible. There were rumours afterward that French money had found its way to some agitators within the Swiss ranks and that these people, in combination with weak and unexperienced commanders, triggered the break-down of the negotiations and the hasty retreat.[72] The Venetian diarist Marino Sanuto even reports that four thousand Swiss took French service.[73] The (rest of the) Swiss withdrew but not without leaving destruction in their wake. Around 20 December most of the units had left Monza.[74] The soldiers hoped to be back home for Christmas and therefore hurried northwards without maintaining marching order or discipline. According to Wernher Schodoler, a military commander and eyewitness, they burnt every house they passed and if they needed shelter for the night, the houses were set on fire the next morning: "[the land] was burning behind, in front and on both sides, for a good mile far and wide" around the withdrawing troops.[75] Schodoler claimed that 8,000 houses were destroyed in total. He describes how, at times, the smoke was so thick and dark that the sun could not break through.[76] No civilian—be it a man, woman or child—was safe. Many families who survived the raids but lost their homes and provisions faced a cruel death by cold, sickness, and hunger.[77] Churches and monasteries were also destroyed, which was strictly against the rules of war,[78] Schodoler stresses that the mercenaries did not stop burning and plundering even after they reached Swiss soil. He notes that "when the confederates came home, they took also from the poor what was theirs and burnt countless houses, which belonged to the confederates".[79] Valerius Anshelm tells us about the legal consequences of this shocking behaviour. He states that an investigation was held "in order to punish the disobedient and severe desecraters of countless churches and raiders of monasteries. But despite the large numbers involved, only a few were found and even less were punished. The reason for this is easy to work out: gentle, fearsome lords—[stood against] hard, outraged soldiers".[80]

Between Autonomy and Heteronomy: Who Controlled and Maintained Military Power in the Swiss Confederacy?

The Cold Winter Campaign of 1511 was not a success, neither from a military, economic nor political point of view. Therefore it may be considered a negligible part of Swiss military history. Yet the anatomy of the failure reveals insights into the tensions jeopardizing joint Swiss military action. The struggle for control over this campaign and military action abroad more generally reveals a political constellation that stifled effective military power projection abroad and a further integration of the Old Swiss Confederacy into an early modern state. The failure of the Cold Winter Campaign showed a fundamental inability to mount robust military action through an integrated Swiss army composed of the mix of specialists (including artillery, cavalry, pioneers, and logisticians) necessary for independent military action including siege warfare. There was no robust federal command structure and communications and logistics were deficient. This insight might have been shared at the time and catalyzed a consensus among and between cantonal élites that independent military force projection abroad was too costly and risky to pursue in the future. Yet that might be too rash a conclusion as the campaign could also be seen as a remarkable success, considering how reluctantly and half-heartedly the Diet and many of the cantons supported it.

The Diet

The Diet was the political and symbolic centre of the Confederation in the early sixteenth century.[81] One or, more generally, 2 representatives of the 12 and then 13 cantons plus representatives of the *Zugewandten Orte* (allies) like the Grisons, the Valais, Biel, the abbot of St. Gall, Mulhouse and Rottweil met several times a year in order to administer the common subject territories and discuss internal and external political issues. The decisions, theoretically, did not come to effect immediately but had to be "carried home" (taken *ad referendum*)[82] and be approved by the cantonal authorities.[83]

The Diet, therefore, was above all a meeting place and communication platform for the élites.[84] If an envoy from abroad wished to meet the most important Swiss politicians in one place, he would attend the Diet.[85] Thus, if the Diet decided unanimously, the decisions were almost guaranteed to be carried out by the cantons. However, although reaching a consensus on condemning the attack on the messengers was one thing (the cantons agreed that this offense needed to be dealt with), agreeing how to put this into practice was quite another. The Diet was divided about the choice between diplomacy and war. Even when it reluctantly settled for the latter, it had no war chest or military means of its own. It depended entirely on the cantons even in matters of the deployment of a federal army—its actual strength, its training, its composition (force mix, especially the provision of artillery), and logistics were determined by the cantons. The cantons hardly coordinated their military efforts and although most of the cantons and their troops were willing to avenge an offense, not all were interested in the connected expansion into Lombardy.

The Diet, as a kind of supreme war council of the confederation at home, included many militarily experienced representatives. They knew each other, had socialized, and perhaps even fought together. These personal entanglements may have mitigated communication and consensus-making problems, but they also potentially reinforced oligarchic structures. The uniformity of military and political power, if formalized by a written warrant (as happened in the case here), had the advantage that political decisions could be immediately put into action on the battlefield.[86] Yet even this accumulation of power among members of the cantonal élites could not overcome the deep rift between pro- and anti-French parties dividing the cantons and their élites. It did not fully iron out the problems of communications between the homeland and field army either. The Diet lost all contact with its armed forces once the soldiers had crossed the Alps. The communication between the Diet and the military contingents during the Italian wars was more complicated than in previous conflicts. Whereas the battlefields in previous campaigns could be reached within one day's march, now the messengers had to cross the Alps in order to reach the front lines.[87] During the Cold Winter Campaign, communications were effectively cut by heavy snowfalls. Therefore, it became almost impossible for the Diet to receive timely intelligence once the troops had crossed the Gotthard Pass and into Italy.[88] The Diet's lack of awareness of the conditions on the ground can be illustrated by two decisions regarding the campaign, which came too late to influence events. First, the commanders were given full power to conclude a peace treaty by the Diet, but the respective instruction reached them only after negotiations with the French governor in Milan had broken down. Second, the Diet ordered 4,000 reinforcements to join the main army across the Alps,[89] but this decision was made after the Swiss had already retreated from Milan and started their disastrous way back home.[90]

Although the symbolic supreme authority of the Diet remained unquestioned, the Cold Winter Campaign shows how its actual power over both the declaration and the conduct of war was limited. Vengeance for the murdered messengers triggered the campaign and explains why the troops from Fribourg and Schwyz left first without waiting for the Diet's enforcement. But what about Bern? Was their messenger not also kidnapped? Despite this, Bern sent its declaration of war to the governor of Milan unwillingly two weeks later and phrased in such a polite way that it provoked the Schwyzers' indignation.[91] Berne's long hesitation suggests that political and economic interests divided the members of the city's governing élite.[92] As mentioned above, Bern was mainly interested in conducting trade with its western neighbours and saw little benefit in a war in Italy. For central Switzerland, however, it was crucial to control the passes over the Alps in order to secure access to the Lombard markets. Thus, for them, avenging an offense quite naturally blended with and mutated into a war of expansion. Hence the Swiss states or cantons found it difficult to enjoin a common grand strategy and did not act in unison. There were considerable tensions, latent conflicts between rural communities/regions and cities both within and between cantons, but also between common people and an élite that was said to take French money and to defend French interests stifling joint federal action.[93] The (although unsuccessful) Bernese motion of January 1512 that no canton should start a war without the consent of the majority highlights the resulting stalemate and scaling down of Swiss military ambitions.[94]

Military Command Structures

The Swiss federal force thus remained very much a composite, ad hoc coalition of cantonal armies. Command structures were complicated by the many cantonal commands. The bigger contingents, e.g. the one of Berne, were led by a cantonal captain, who, however, would consult with the captains (*Hauptleute*, *venner*) of the main cantonal units. These leaders, usually members of the cantonal élites, formed the cantonal war councils. There was also a strong bottom-up element of decision-making. Even if the commanders of the many cantonal and allied contingents agreed on a joint course of action, they could be challenged by their soldiers. Thus the commanders of the various cantonal contingents were not only in consultation with one another as well as with the federal and cantonal authorities but also with their contingents' war councils and soldiers' assemblies. The federal commanders appointed by the Diet, therefore, remained relatively weak. Important decisions had to be taken by the federal war council reuniting all cantonal commanders, but ultimately these decisions needed to be approved by the contingents' war councils. Even so, they could be challenged by the bottom-up decision-making bodies of the cantonal soldiers' assemblies (the *Kriegsgemeinden*).[95]

These assemblies could effectively argue that once the soldiers had departed, the canton itself was on the move and that the cantonal soldiers' assembly in the field therefore not only substituted but fully embodied the cantonal citizens' assembly and thus became the supreme authority of the state.[96] Thus, bottom-up pressure could force the commanders to change military operations in the field. How dangerous and unpredictable impecunious and enraged people from the countryside, organized in bottom-up assemblies, could be, is illustrated by the Hog-Banner Campaign (*Saubannerzug*) of 1477, when angry soldiers from central Switzerland marched towards Geneva in order to claim what they asserted to be their fair share of the booty from the Burgundian Wars. The authorities could only stop them by agreeing to their demands and requesting immediate pay from the Burgundians.[97] In the Cold Winter Campaign, the Swiss command structure and decision-making mechanism proved its limits. Even the field commanders, in contrast to the Diet in direct contact with the troops, could not stop the excesses of the retreat. This indicates the limits of their command authority. Once unleashed, the battle-hardened soldiers were uncontrollable and acted contrary to the intentions of the federal and even cantonal war councils and disobeyed their commanders in the field.[98] To effectively *lead* the soldiers in war, the Swiss federal army, for the better or the worse, seems to have lacked adequate institutionalization and apparatus of oppression.[99]

Logistics

In addition to the organizational problems discussed above, the Cold Winter campaign was also beset by logistical difficulties, which were a great challenge for any army throughout the Medieval and Early Modern Period.[100] As long as the supply of food, fodder, horses, and weapons was insufficient, the soldiers were not only willing but in fact forced to partly organize logistics independently and to rely on foraging and plunder, which impacted operative decisions not only in a military organization shaped so strongly by bottom-up decision-making.[101] Living off the land, in any case, meant that troops had to spread out widely in order to find enough resources, which produced its very own communications and thus command and control challenges.

Documents from the Bernese contingent illustrate the considerable costs of the campaign. These included the shipping of troops over Lake Lucerne (*Vierwaldstättersee*), the clearance of paths over the Alpine passes by local inhabitants, the transportation of 74 bags of flour and 66 bags of oats from Uri to Bellinzona, and also various other activities such as the provision of cannonballs, delivering of messages, or the transport of a captain's halberd.[102] Logistical challenges were amplified by the late season and also because the forces included artillery and, probably, some cavalry, which were more complex to provision. The cantons, spearheaded by the Schwyzer, probably somewhat underestimated the problems of such a "joint" campaign[103] while also trusting Schiner's promises and thus assuming that joint action with Venetian and papal troops would also entail logistical, cavalry, and artillery support.[104]

Once the troops had left Swiss territory, their resupply from the homeland became virtually impossible due to the wintery conditions, while Venetian support did not materialize. As a result of these conditions, more than a hundred men died in the cold and the snow of the Gotthard Pass.[105] As the chronicler Anshelm recounts, the troops from Bern in particular suffered on their return. The hesitation of Bern to join the expedition had not gone unnoticed and when the Bernese troops reached the Gotthard Pass, other Swiss soldiers insulted them and prevented them from baking their bread or using pack horses.[106]

Finances

The financing of the Swiss war effort was overshadowed by the fact that members of the Swiss military élite took French pay and thus allied with French interests. Under the alliance with France and the respective pension system in place until 1509, the military leaders of most cantons had profited from military service for France. Pensions and salaries plus unofficial payments flowed directly in to the pockets of the élites, while only a small part made its way to the simple foot soldier.[107]

The new ally, the pope, however, was different. Not only was his reputation for punctual payment of wages already badly damaged by the preceding 1510 *Chiasserzug*, but the modus operandi of these payments was also different. The pope transferred yearly payments to the cantons, which then had to recruit the required mercenaries. He did not pay pensions to the entrepreneurs organizing military contingents to serve abroad (of which some seem to have continued taking French pay) and effectively cut out these intermediaries by stipulating that all payments to the soldiers would be handled by his plenipotentiary, cardinal Schiner, or the latter's representatives.[108] Some of the traditional leaders thus, perhaps, felt cut out and did not want to lead troops—at least not against the French. This might explain Anselm's remark about unexperienced commanders.[109]

For the soldiers, however, these payment modalities were more attractive, and, in any case, not much income could be expected during the winter. The salary of a military campaign was more than welcome—not to mention the expected booty. When the mercenaries realized that the promised salary would not arrive and not much booty was forthcoming, they either changed camp and joined the French or returned home seizing as much plunder as possible to make up for their hardship and lost pay. The cantons apparently did not provide sufficiently for the troops, probably because they counted on the promised papal support. Not only did this support not materialize, but there was also no leadership or guidance from the pope in the field, which strengthened the lure of French pay and, thus, the French party's position. This may have triggered the unorderly retreat, as shown above. After the promises of papal support, the troops and their leaders, perhaps, also lacked clarity about their status and purpose: Were they restoring the honour of the confederation, furthering particular central-Swiss interests in southwards expansion, or fighting the pope's war? This lack of clarity also allowed the pope to renege on promises to pay for the campaign arguing that this was the Swiss' private affair that had nothing to do with him as they had returned before even joining the papal troops, thus effectively abandoning the papal campaign and cause.[110]

Conclusion: The Influence of Military Diasporas on State Formation

The Cold Winter Campaign is an example of how political and military authorities can easily lose control over the use of force. It shows how a lack of clear objectives, unclear political-legal framing and insufficient logistical and financial support jeopardize the success of military action. It further illustrates how violence was seen as a state prerogative but that the notion of state and thus the question, who exactly should control violence, remained layered and contested between bottom-up and top-down concepts and aspirations of rule.[111] Swiss authorities allowed military service for foreign rulers and maybe even welcomed it. Pensions were crucial sources of revenue for the cantons (in some cases more than half of the canton's budget) and their élites, i.e. the military entrepreneurs and *Pensionsherren*.[112] Mercenary service was an important outlet for poor, discontented, and rebellious young men. There was, however, also resistance against Foreign Service fuelled by failures and excesses such as the Cold Winter Campaign.[113]

In the case of the Cold Winter Campaign, the ambiguity in the command structure was perhaps also intentional in order to mitigate the risks inherent to military action. Janice Thomson points out two advantages of delegating violence to non-state actors. The first is "plausible deniability" (if a campaign was successful, the state could claim a share of the profit; if it failed, it could not be held responsible).[114] This logic applies to both the Confederation, which sent the troops and the pope, who convinced them to fight for his interests. As the episode demonstrates, neither pope nor confederation took responsibility for the failed expedition, while the middleman Cardinal Schiner was accused of turning a small spark into a conflagration for no discernible benefit.[115] The second advantage is the motivation of the mercenaries, which is disproportional to state control. The less the state controls the soldiers, the more they are willing to take risks.[116] Viktor Hofer concludes that "with elemental, ebullient force through the blindfold recklessness of the mercenaries many victories have been won, which did not leave space for rational considerations".[117] If successful, the victory could then be adopted and owned by the authorities or, as Walter Schaufelberger puts it, "the councils under the pressure of the circumstances were often not disinclined to make a tumultuous irregular campaign their own".[118]

Abroad, soldiers enjoyed a certain degree of autonomy reinforced by the above-mentioned strongly embedded political structures of bottom-up communal decision-making. Along with the motivation to fight at their own risk and for their own good, the individual or, rather, the *Gemeinde*'s collective responsibility increased the political and military awareness beyond the immediate necessities of the battlefield. The Cold Winter Campaign had been continuously evaluated not only by the authorities but also by the soldiers themselves, who wondered whether they should stay with the pope or side with France. Thus acquired war experience and the knowledge of other countries including their rulers and armies then fed into the political decision-making of the cantonal councils and the Diet.[119]

Can we thus speak of a Swiss military diaspora? In the case of the Cold Winter Campaign, we would probably not. These were ad hoc formations (type I) which, contrary to the guards (type III), did not develop any permanent structures abroad beyond the immediate purpose of conducting a military campaign. The campaign was seasonal and the cantonal contingents remained strictly regional in composition, mingled only slightly with other cantonal troops and probably (and contrary to the later, e.g. eighteenth-century Swiss infantry regiments in France) did not include non-Swiss mercenaries. Nevertheless, soldiers from different cantons did meet (if only to trade in insults) and thus became gradually more familiar with one another and some might have bonded across cantonal affiliation in their decision to abandon camp and join the French. They also shared, to an extent, (infantry) tactics and thus gradually formed a more united Swiss military identity and reputation, grounded in their specific, telluric, i.e. local military tradition born out of the need to defend relatively small polities against heavy cavalry. Thus their specific infantry tactics (*Gewalthaufen*) and weapons (halberd and pike) became the Swiss mercenaries' hallmark and unique selling proposition, their perceived military edge. The Swiss soldiers managed to turn their original weakness (little to no cavalry) into a strength, which made them highly sought-after mercenaries. This in turn gave the Swiss commanders abroad the chance to learn about military command, strategy, and tactics from the best European armies and to apply this knowledge to their cantonal armies.[120] These, contrary to what Taylor argued, did adapt to the latest developments in warfare, and they did understand the importance of force-mix and especially artillery on the battlefield.[121] Especially the city-cantons added fire weapons to their arsenals, invested in artillery and even maintained cavalry units. The Cold Winter

Campaign is of particular interest because it shows, perhaps for the last time, that the Swiss tried to act as an (almost) independent military player, defending their honour while pursuing southwards expansion with their own combined armed force.[122]

The pope, however, refused to recognize the Swiss as a European sovereign power and treated them as subordinates of the emperor.[123] And when the Swiss took the field with their own combined force including artillery and cavalry, he was perhaps not entirely unhappy to see them run into a strategic, icy void. Arguably the French agreed with their papal enemy as far as Swiss status was concerned. The Swiss for them should, however, be subservient to France as the great sovereign power, and sovereign themselves only vis à vis their former overlord and French nemesis: the emperor.[124] Henceforth the Swiss seemed to have abstained from independent action and the outfitting of a force capable of combined arms warfare. In that respect, it may have been the Cold Winter Campaign and not the often-quoted defeat of Marignano (1515), which was the turning point. It was perhaps not the French guns at Marignano that taught the Swiss the importance of combined arms warfare and thus stopped Swiss expansionism. Some Swiss, on the one hand, understood the importance of combined arms warfare well before Marignano and realized the limits of Swiss federal military power projection abroad. They knew that ordnance was key to take and defend fortresses and thus to wage war independently. It was monetary considerations (and extreme venality, perhaps) that made them accept a role as junior partner of the big European powers. Swiss mercenaries in the field, on the other hand, remained even after Marignano firm in their conviction that swift infantry could pull off a stand-alone victory on the open battlefield. Had it not worked before so brilliantly, e.g. at Novara in 1509 and 1513? And so they tried it again and failed, e.g. in Biccoca 1525.

In any case, operating a federal combined armed force split between different cantons was costly and difficult. Also, and crucially, the contrast between high-tech artillery-strong city cantons with some cavalry, and infantry-strong countryside cantons, mapped perfectly onto the landscape of rifts within the confederation. The central Swiss cantons which benefitted the most from a southwards expansion needed the heavy weaponry and expertise of the little-trusted, allegedly pro-French and/or pro-imperial city states in order to realize their objectives. The Cold Winter Campaign's lack of success did nothing to overcome but a lot to exacerbate these tensions and weaken the willingness to invest in joint, independent Swiss power projection.

The alternative was more lucrative, especially for the poorer country- and mountain-side cantons. They embraced the infantry role that the great powers had carved out for them. Thus, one could take their pay canton by canton within a loose federal framework that did not upset the (especially after the reformation) increasingly fragile inner-Swiss balance of power. The cantons could gain money and perhaps even influence without much expense. Of course, this arrangement still came at a price; a price that already was felt a year later during the Pavia campaign (*Pavierzug*) of 1512. The Swiss, this time, did not march under their cantonal banners (as in the Cold Winter Campaign) but used the contingent standards (*Fähnlein*,[125] type I) as in the *Chiasserzug* of 1510. In 1512 the army was again only a lightly armed infantry force akin to the one mustered in 1510, although much bigger in size. Despite the increased size of the force, it seemed to have brought only little ordnance (two Bernese cannons).[126] For the sieges of Lugano and Locarno later in the campaign, the Swiss seemed to have used borrowed and captured ordnance in small numbers but to little avail, not least because they lacked the necessary specialist personnel.[127] Thus the Swiss were severely handicapped in pursuing their own interests, which meant that they did not reap the full reward of their military exploits. The fact that the conquered cities of Cremona, Pavia, etc. paid homage to the four main allies only—the papacy, the emperor, Venice, and Spain—but not the Swiss indicates how the Swiss were (not incorrectly) seen as a type I contingent in papal service and subjects of the emperor- junior partners at best.[128] The Diet protested but could hardly change this state of affairs.[129]

Hence, the Swiss became very much a mercenary non-state; contributing to the state formation of others while buttressing their own oligarchic cantonal micro-state structures. The modus operandi of the *Pavierzug* continued, although, by 1515, the Swiss turned back to France as their strategic military ally (or, one might say, overlord). This alliance (alongside other alliances of smaller size) provided a symbiotic outlet for the opportunities and risks posed by the great Swiss military potential and aggressive energy. The Diet regulated and, hence, tried to control the mercenaries. This was not new; attempts to do this date back to the fourteenth century (e.g. the *Sempacherbrief* of 1393), but the system became more rigid increasingly pushing a type II service with some elements (cantonal standards and recruitment regions) of type I, thus side-lining type IV and V free services.[130] This arrangement led Switzerland onto a specific path of fiscal-military state formation, one might call it a rental-mercenary, cantonal micro-state formation. The Swiss used their military resources in order to tap into the fiscal income of other (fiscal military) states but not by taking money through the brute force of the soldiery, as was once advocated by Machiavelli,[131] but rather by selling soldiers (and their brute force)—yet another variation of Thomson's "plausible deniability". This system sponsored an oligarchic élite and fuelled a system of rule combining military coercion (against uprisings or other cantons) and cash-lined consensus (with the soldiers). This came, however, at the expense of political and military dependence. The system worked reasonably well until it collapsed in the wake of the French revolution. It left Switzerland exposed and ill-prepared to counter the French invasion of 1798.

Of course, there was a whole range of other negative consequences of foreign military service. Death, injury, and trauma made the reintegration of the mercenaries hard, if not impossible. The soldiers remained a threat to society and a challenge to the authorities, as illustrated by the Cold Winter Campaign. Failed campaigns could damage reputations and cause high costs. Thousands of men lost their lives or remained maimed or traumatized. However, as long as money and booty lured, neither the simple soldier nor the élite had an interest in stopping Foreign Service. It would take another three and a half centuries until Switzerland effectively prohibited foreign military service in 1859 without, however, fully stopping it.

Notes

1 Many thanks to Adrian Wettstein for reading a draft and providing detailed and most helpful feedback. All remaining errors are ours.

2 The only treatments of the episode in any depth are by the venerable Ildefons Fuchs, *Die mailändischen Feldzüge der Schweizer*, vol. 2 (St. Gallen: Huber, 1812) and the more recent but still rather dated Siegfried Frey, "Der Chiasserzug 1510 und der kalte Winterfeldzug 1511," in *Schweizer Kriegsgeschichte*, ed. Ulrich Wille et al. (Bern: Oberstkriegskommissariat, 1915), 1:314–320; see also Robert Durrer, *Die Schweizergarde in Rom und Die Schweizer in Päpstlichen Diensten* (Luzern: Räber, 1927).

3 While the *Chiasserzug* started in September 1510, the Cold Winter Campaign commenced even later in the season, November 1511, after moving cattle to the winter pasture, harvest and vintage, not so, however, the *Pavierzug* of late spring and summer 1512.

4 Valentin Groebner, "Pensionen," in *Historisches Lexikon der Schweiz (HLS)*, s.v., Schweizerische Akademie der Geistes- und Sozialwissenschaften, 2011, accessed 11 May 2020, https://hls-dhs-dss.ch/de/articles/010241/2011-11-03/; Philippe Rogger, *Geld, Krieg und Macht: Pensionsherren, Söldner und eidgenössische Politik in den Mailänderkriegen 1494–1516* (Baden: Hier und Jetzt, 2015), 14.

5 A modern take of Swiss history is available in the *Historisches Lexikon der Schweiz*, online on www.hls.ch; see also: Volker Reinhardt, *Die Geschichte der Schweiz: Von den Anfängen bis heute* (München: C. H. Beck, 2011), 31–33; Thomas Maissen, *Geschichte der Schweiz*, 3rd ed. (Baden: Hier und Jetzt, 2011), 24–25. Dieter Fahrni, *Schweizer Geschichte: Ein historischer Abriss von den Anfängen bis zur Gegenwart* (Zürich: Pro Helvetia, 2002), 18–20.

6 Fuchs, *Die mailändischen Feldzüge*, 262.

7 The Habsburg emperor was nominally the suzerain but had been defeated by the Swiss (as a Habsburg lord rather than the emperor) in a hard-fought war in 1499. The Swiss and the empire coexisted henceforth along the terms laid out in the peace of Basel of 1499, mediated by the Duke of Milan.

8 See Bernhard Truffer, "Schiner, Matthäus," in *Historisches Lexikon der Schweiz (HLS)*, s.v., Schweizerische Akademie der Geistes- und Sozialwissenschaften, 2012, accessed 23 March 2022, https://hls-dhs-dss.ch/de/articles/012963/2012-11-20.

9 See Walter Schaufelberger, "Spätmittelalter," in *Handbuch der Schweizer Geschichte* (Zürich: Berichthaus, 1972), 2:336–338.

10 Bernhard Stettler, *Die Eidgenossenschaft im 15. Jahrhundert: Die Suche nach einem gemeinsamen Nenner* (Zürich: Markus-Widmer-Dean, 2004). This transformation is ultimately rooted in the climate change of the late Middle Ages and the demographic changes in the wake of the fourteenth-century famines and subsequent outbreaks of the plague and other epidemics that left alpine grain production more vulnerable, thus driving up production costs while interregional grain prices declined, Bruce Campbell, *The Great Transition: Climate, Disease and Society in the Late-Medieval World* (Cambridge: Cambridge University Press, 2016), 333; Wilhelm Abel, *Agrarkrisen und Agrarkonjunktur eine Geschichte der Land- und Ernährungswirtschaft Mitteleuropas seit dem hohen Mittelalter*, 3rd ed. (Hamburg: Paul Parey, 1978), 55–57, 68.

11 André Holenstein, *Mitten in Europa: Verflechtung und Abgrenzung in der Schweizer Geschichte* (Baden: Hier und Jetzt, 2014), 79–81.

12 Stettler, *Die Eidgenossenschaft im 15. Jahrhundert*, 85.

13 Julius II created the Swiss Guard in 1506. See Urban Fink, Hervé de Weck, and Christian Schweizer, *Hirtenstab und Hellebarde: Die Päpstliche Schweizergarde in Rom* (Zürich: Theologischer Verlag, 2006). See also Schaufelberger, "Spätmittelalter," 351–352, and for a more recent overview Walter Schaufelberger, *Spurensuche: Siebzehn Aufsätze zur Militärgeschichte der Schweiz*, ed. Claudia Miller (Lenzburg: Merker im Effingerhof, 2008).

14 See Nicolas Morard, "Auf der Höhe der Macht: 1394–1536," in *Geschichte der Schweiz und der Schweizer*, ed. Ulrich Im Hof et al. (Basel: Helbling & Liechtenhahn, 1986), 211–352, at 338.

15 Philippe Henry, "Service étranger," in *Historisches Lexikon der Schweiz (HLS)*, s.v., Schweizerische Akademie der Geistes- und Sozialwissenschaften, 2017, accessed 11 May 2020, https://hls-dhs-dss.ch/de/articles/008608/2017-12-08/.

16 Probably modelled after the French king's Swiss life guards founded in 1497, Fuchs, *Die mailändischen Feldzüge*, 34, n. 69; Henry, "Service étranger."

17 Which, much reduced in numbers, still exists: *The Pontifical Swiss Guard*, accessed 27 October 2021, https://www.guardiasvizzera.ch/paepstliche-schweizergarde/it/chi-siamo/.

18 The Swiss Corps within the French armed forces, founded in 1480, Henry, "Service étranger."

19 See Bruno Koch, "Heilige Liga," in *Historisches Lexikon der Schweiz (HLS)*, s.v., Schweizerische Akademie der Geistes- und Sozialwissenschaften, 2011, accessed 23 March 2021, https://hls-dhs-dss.ch/de/articles/017171/2011-03-09/.

20 *Encyclopaedia Britannica*, "The Infantry Revolution, c. 1200–1500," accessed 27 October 2021, https://www.britannica.com/technology/military-technology/The-infantry-revolution-c-1200-1500. Conventionally this has been explained by technological innovation of, say, the halberd, pike, crossbow, longbow or archebuse, see Clifford J. Rogers, "The Military Revolution in History and Historiography," in *The Military Revolution Debate*, ed. id. (London: Routledge, 1995), although Rogers later somewhat modified his stance. As pointed out above, it also had to do with drill, and that, in return, might suggest that social developments, not least those triggered by climate change and later the plague, preceded and fostered this change (see John Stone, "Technology, Society, and the Infantry Revolution of the Fourteenth Century," *The Journal of Military History* 68, no. 2 (2004): 361–380, accessed 27 October 2021, https://www.proquest.com/docview/195630871?pq-origsite=primo&accountid=12253).

21 See Walter Schaufelberger, *Der Alte Schweizer und sein Krieg: Studien zur Kriegführung vornehmlich im 15. Jahrhundert*, 3rd ed. (Frauenfeld: Huber, 1987), 17. For a more recent overview see Schaufelberger: *Spurensuche*.

22 Schaufelberger, *Der Alte Schweizer und sein Krieg*, 17.

23 See Ibid., 22.

24 See Gottlieb Friedrich Ochsenbein, *Die Urkunden der Belagerung und Schlacht von Murten* (Freiburg: E. Bielmann, 1876), https://archive.org/details/dieurkundender-b03ochsgoog, accessed 23 March 2022. See also Benjamin Geiger, *Militärgeschichte zum Anfassen*, vol. 4, *Die Burgunderkriege* (Au: Militärische Führungssch., 1994).

25 Fuchs, *Die mailändischen Feldzüge*, 192, n. 163; Frederick Lewis Taylor, *The Art of War in Italy, 1494–1529* (Cambridge: Cambridge University Press, 1921), 38.

26 See Stettler, *Die Eidgenossenschaft im 15. Jahrhundert*, 262.

27 See Holenstein, *Mitten in Europa*, 172–173.

28 *Encyclopaedia Britannica Online*, Academic ed., s.v. "Hans Waldmann," accessed 27 April 2020, https://www.britannica.com/biography/Hans-Waldmann.

29 Hermann Romer, "Militärunternehmer," in *Historisches Lexikon der Schweiz (HLS)*, s.v., Schweizerische Akademie der Geistes- und Sozialwissenschaften, 2009, accessed 23 March 2022, https://hls-dhs-dss.ch/de/articles/024643/2009-11-10/. They, however, never fully disappeared, not even in the eighteenth century, see for instance

Ulrich Bräker, *The Life Story and Real Adventures of the Poor Man of Toggenburg*, trans. Derek Bowman (Edinburgh: University Press, 1970).

30 Since 1500 Northern Italy was under the rule of the French king Louis XII. At the same time, Bellinzona, wanting to escape the grip of the French, joined the Confederation as a protectorate of Uri, Schwyz, and Unterwalden. Bullinger maintains (and other sources seem to suggest) that the heralds were taken later, in November, and that they did not travel to but from Switzerland with letters from Schiner to the pope, Fuchs, *Die mailändischen Feldzüge*, 253–254.

31 Ibid., 205, 217, 239.

32 Ibid., 177.

33 Ibid., 189, 250, 254–255.

34 Ibid., 254, n. 309b.

35 See ibid., 253–254.

36 Ibid., 178–194.

37 Ibid., 225–226, 254; Valerius Anshelm, *Die Berner Chronik*, ed. Historischer Verein des Kantons Bern (Bern: K. J. Wyss, 1888), 3:256.

38 Ibid.

39 The *Standesläufer* had the status of a *nuntius* within the framework of the *ius legationis,* i.e. Roman diplomatic law, and thus enjoyed a sort of diplomatic immunity, Garrett Mattingly, *Renaissance Diplomacy* (Baltimore, MD: Penguin Books, 1955), 26, 28; Montell Ogdon, "The Growth of Purpose in the Law of Diplomatic Immunity," *The American Journal of International Law* 31, no. 3 (1937): 449–465, here 454, also Hans Strahm, *Die Standesläufer* (Bern: PTT, 1942), n.v.

40 During the sixteenth century, foreign pensions made up from 15.2% (Zurich) to 80% (Appenzell) of the state revenue, the largest share coming from France. See Holenstein, *Mitten in Europa*, 146.

41 A state of affairs that endured with variations from the mid-fifteenth to the late eighteenth centuries.

42 See Anshelm, *Berner Chronik*, 257.

43 See ibid., 257–258.

44 See ibid., 257.

45 See Frey, "Der Chiasserzug 1510 und der kalte Winterfeldzug 1511," 316.

46 Lombardy was an important market for livestock; and the region could become, as Schiner argued, very valuable for winter pastures and food supplies in case of bad harvests and famines on the northern side of the Alps. See Morard, "Auf der Höhe der Macht," 337.

47 See ibid.

48 See Daniel Reichel, "Matthieu Schiner (vers 1465–1522): Cardinal et homme de guerre," in *Marignano 1515–2015: Von der Schlacht zur Neutralität*, ed. Roland Haudenschild (Lenzburg: Merker im Effingerhof, 2014), 65–78, here 72.

49 Fuchs, *Die mailändischen Feldzüge*, 241 argues that the papal refusal to carry the full costs of the *Chiasserzug* had reinforced anti-papal feelings.

50 Durrer, *Die Schweizergarde*, 86; Hervé de Weck, "Cavalerie," in *Dictionnaire historique de la Suisse (DHS)*, s.v. Académie suisse des sciences humaines et sociales, 2007, accessed 23 March 2022, https://hls-dhs-dss.ch/fr/articles/008579/2007-03-01/. The size of the Freiburger cavalry in the early fifteenth century would have been customarily 100 horses but as this was not the full *Auszug*, we can assume it was much less.

51 See Anshelm, *Berner Chronik*, 258; Frey, "Der Chiasserzug 1510 und der kalte Winterfeldzug 1511," 316.

52 Frey, "Der Chiasserzug 1510 und der kalte Winterfeldzug 1511," 316–317; Fuchs, *Die mailändischen Feldzüge*, 283.

53 See Frey, "Der Chiasserzug 1510 und der kalte Winterfeldzug 1511," 317.

54 Ibid.

55 Ibid.; the Venetian sources do not mention Peschiera but insinuate rather a meeting in Vincenza, Marino Sanuto, *I diarii*, vol. 13 (01.10.1511–28.02.1512) (Venezia: Visentini, 1879), 340, 342.

56 Rudolf Wackernagel, *Geschichte der Stadt Basel* (Basel: Helbing und Lichtenhahn, 1924), 1:19: "Am 1. Dezember [1510] trafen die Basler in Bellinzona ein, nach mühevoller Überwindung des Passes, dessen winterlicher Rauheit ihre Pferde nicht gewachsen waren". It might only refer to horses for the transport of provisions and booty or for the artillery train although we do not read of Basler artillery in the sources.

57 Fuchs, *Die mailändischen Feldzüge*, 279.

58 There is no clear evidence for cavalry in the sources. For some indirect, weak evidence that the contingent of Baden, maybe others, might have contained horse, see Werner Schodoler and Pascal Ladner, *Die eidgenössische Chronik des Wernher Schodolerum 1510 bis 1535*, ed. Walther Benz – Kommentar zur Faksimile-Ausgabe der dreibändigen Handschrift MS 62 in der Leopold-Sophien-Bibliothek Überlingen, MS 2 im Stadtarchiv Bremgarten, MS Bibl. Zurl. Fol. 18 in der Aargauischen Kantonsbibliothek Aarau (Luzern: Faksimile-Verlag, 1983), 276: "unnd ordnet man am abint dru tusent man der ringisten uß allem zug, unnd gab mann in en das vennly von Baden zß, dann man gewusse kundtschafft hatt, das die vyend zil Granpusch lagenn". The separate mention of Baden might insinuate that this contingent was of a different nature as it would not have made sense to select the fastest marchers from all contingents and then top them with Badeners regardless of how well they marched. Was their entire Fähnlein cavalry? It was common that besides burger-aristocrats cities would provide cavalry, see Emanuel v. Rodt, *Bernisches Kriegswesen* Band I (Bern: E. A. Jenni, 1831), 36–40, including see note 65 on p. 39 on the composition of the Bernese cavalry ("Rossbanner") although arguing (p. 40) that the Swiss cavalry essentially disappeared in the Italian Wars. Again, archival research would have to be undertaken to shed light on this. Then, Schodoler, *Chronik*, 276: "Da meinten dieselben, die harnach zugen, es werend ouch vyend, do die vyend all so uff beyden siten still hiellten, unnd wellten mit inen schlaehen. Si vernamen aber balld, das dasselb eydgnossen waren". The shadowing force was clearly described as mounted: "die vyend, (…) hiellten ouch uff beyden sidten uff iren hengsten in guotter ordnung," which might suggest that so was the Swiss force they were confusing them with.

59 Fuchs, *Die mailändischen Feldzüge*, 289–291 and Frey, "Der Chiasserzug 1510 und der kalte Winterfeldzug 1511," 318. The fact that the provisions included considerable amounts of oats does not necessarily suggest there was cavalry as the artillery, train and officers all used horses, Arnold Esch, "Mit Schweizer Söldnern auf dem Marsch nach Italien: Das Erlebnis der Mailänderkriege 1510–1515 nach bernischen Akten," *Quellen und Forschungen aus italienischen Archiven und Bibliotheken* 70 (1990): 348–440, here 426. It would be useful to check the Bernese accounts for this campaign, StA Bern, UP e.g. 16 Nr. 31. Nobles and urban patricians, from whom horsemen would be recruited, often were members of both the noble society of the Distelzwang but also one of the quarter-guilds, Eduard von Wattenwyl von Diesbach, "Die Gesellschaft zum Distelzwang," *Berner Taschenbuch* 14 (1865): 174–200, here 177. For the seventeenth century we know for sure that the society was providing troopers, ibid., 191, see also notes above and below.

60 Max E. Ammann, *Der Eidgenoss* (Luzern: Reich Verlag, 1975), 18–19; for the Swiss cavalry's role and importance in the late fifteenth century, cf. the following statement from the Twingherrenstreit (1470) with thanks to RA Dr. Jürg Gassmann for bringing it (and the excellent v. Rodt) to my attention: "Dass auf Tagen von den Eidgenossen kein Berner geschätzt werde, denn die Edlen; weil jene, wie sie heiter bekennen, im Zürcher – Krieg und wider den Kayser, und wider die Oestrychischen, nicht hätten bestehen mögen, wenn die reisigen Edelsleute von Bern nit gesin wären; diese hätten ihnen die Speis erhalten, alle Ding erkundiget," von Rodt, *Bernisches Kriegswesen*,

39–40, based on Thüring Frickart/Fricker's report in Gottlieb Studer (ed.), *Quellen zur Schweizergeschichte* vol. 1 (Basel: Schneider, 1877), 137–138: "(...) reisigen volks und houptlüten werend sy notwendig gsin, indem habind ir sy erhalten; rümendt, wie sy inen die spyß erhalten, den fyenden verhalten, alle ding erkundiget, — grosse ding, die sy inen zügebendt".

61 Fuchs, *Die mailändischen Feldzüge*, 281 and Frey, "Der Chiasserzug 1510 und der kalte Winterfeldzug 1511," 317.

62 Fuchs, *Die mailändischen Feldzüge*, 292; for the higher estimate: Francesco Guicciardini, *History of Italy*, ed. Austin Parke Goddard (London: John Towers, 1753), 5:338; Sanuto, *Diarii*, 13:307 with an even higher estimate: 25,000.

63 Frey, "Der Chiasserzug 1510 und der kalte Winterfeldzug 1511," 318; Fuchs, *Die mailändischen Feldzüge*, 284.

64 Cf. the heavy siege artillery employed by emperor Maximilian against Padua in 1509, Leopold von Ranke, *Geschichten der romanischen und germanischen Völker von 1494 bis 1535* (Leipzig: G. Reimer, 1824), 319–320.

65 See Martin Lassner, "Stapfer, Jakob," in *Historisches Lexikon der Schweiz (HLS)*, s.v., Schweizerische Akademie der Geistes- und Sozialwissenschaften, 2012, accessed 23 March 2021, https://hls-dhs-dss.ch/de/articles/018197/2012-02-17/.

66 See Frey, "Der Chiasserzug 1510 und der kalte Winterfeldzug 1511," 319; Fuchs, *Die mailändischen Feldzüge*, 293.

67 Monza being, by papal edict, home of the Iron Crown and symbol of imperial power in Italy.

68 Sanuto, *Diarii*, 13:342–343.

69 See Fuchs, *Die mailändischen Feldzüge*, 295.

70 Ibid.

71 See Frey, "Der Chiasserzug 1510 und der kalte Winterfeldzug 1511," 319.

72 Anshelm, *Berner Chronik*, 260: "Denen kriegslüten, so lieber dem küng, dan den iren gedient hätten, glowten die frommen aber kriegs unerfarnen hoptluet"; Fuchs, *Die mailändischen Feldzüge*, 298, 300, the same had already been suspected regarding the *Chiasserzug*, 239; cf. also the tenacious rumours and contradictory news regarding whether the Swiss struck a deal with the French, Sanuto, *Diarii*, 13:346–383; cf. also a similar French attempt in the subsequent *Pavierzug*, Durrer, *Schweizergarde*, 132–133.

73 Sanuto, *Diarii*, 13:383.

74 See Frey, "Der Chiasserzug 1510 und der kalte Winterfeldzug 1511," 320; Fuchs, *Die mailändischen Feldzüge*, 296.

75 Translation by A. Weltert. In the original it says: "so bran es hinden, vor unnd nebenthalb einer guotten mil wegs wit und breyt". Schodoler, *Die eidgenössische Chronik*, 276: "Unnd fiengen die eydgnossen allso an unnd verbranndten alle dörfer unnd flecken unnd darzuo die húpschen lusthúser unnd sicz, so die grossen herren hatten, die si dann am strich mochten erlangen. Allso wo si am abint hinkamen, so brandt man dann mornndes, wann man uffbrach [...]. Unnd was das nit ein wunder, dann zuo vilen malen eins tags me dann zwey oder drú tusend húser leyder verbrent wurden". Cf. also Sanuto, *Diarii*, 13:362, 367.

76 Schodoler, *Die eidgenössische Chronik*, 276.

77 Anshelm, *Berner Chronik*, 261.

78 Bernhard Stettler, "Sempacherbrief," in *Historisches Lexikon der Schweiz (HLS)*, s.v., Schweizerische Akademie der Geistes- und Sozialwissenschaften, 2011, accessed 23 March 2022, https://hls-dhs-dss.ch/de/articles/009804/2011-11-22/; Anshelm, *Berner Chronik*, 261; Fuchs, *Die mailändischen Feldzüge*, 295–297; Frey, "Der Chiasserzug 1510 und der kalte Winterfeldzug 1511," 320.

79 Translation by the authors. In the original it says: "alls die eydgnossen heimb kamen, und namen ouch armen lútten das ir unnd verbranten etliche húser, den eydgnossen zuogehörig". Schodoler, *Chronik*, 276.

80 Translation by A. Weltert. In the original it says: "Item und ersuochung in allen orten, die unghorsamen und sweren kilchenroeuber und klosterbrecher zestrafen. und wie wol iro viel waren, so wurden doch wenig funden und noch weniger gestraft; worum, ist liecht zu ermessen: lind, forchtsam herren—hart fraefel knecht". Anshelm, *Berner Chronik*, 263–264.
81 See Andreas Würgler, *Die Tagsatzung der Eidgenossen: Politik, Kommunikation und Symbolik einer repräsentativen Institution im europäischen Kontext 1470–1798*, Frühneuzeit-Forschungen 19 (Epfendorf: Bibliotheca academica, 2013), 20.
82 The usual mode for alliances, cf. the Hanseatic league, Rolf Hammel-Kiesow, *Die Hanse* (München: Beck, 2004), 68–69.
83 See Andreas Würgler, "Krieg und Frieden organisieren: Eidgenossen und Gesandte europäischer Mächte an der Tagsatzung," in *Berner Zeitschrift für Geschichte* 74, no. 2 (2012): 87–105, here 94.
84 Andreas Würgler, "Tagsatzung," in *Historisches Lexikon der Schweiz (HLS)*, s.v., Schweizerische Akademie der Geistes- und Sozialwissenschaften, 2014, accessed 23 March 2022, https://hls-dhs-dss.ch/de/articles/010076/2014-09-25/.
85 See Holenstein, *Mitten in Europa*, 140.
86 See Viktor Hofer and Georges Rapp, *Der Schweizerische Generalstab* (Basel: Helbing & Lichtenhahn, 1983), 1:26.
87 See Klara Hübner, "Ein falscher Sieg und falsche Boten: Nachrichtenübermittlung und -verbreitung zur Zeit von Marignano," in *Marignano 1515–2015: Von der Schlacht zur Neutralität*, ed. Roland Haudenschild (Lenzburg: Merker im Effingerhof, 2014), 163–174, here 164.
88 This already had been a problem in the preceding *Chiasserzug*, although it took place much earlier in September. Weather, of course, was not the only problem—the other was French-Milanese control over the *passages obligés* Lugano and Locarno enabling them to cut communications by imprisoning messengers. This did not only happen, famously, during the *Chiasserzug* (see above) but also a year later during the *Pavierzug*, Fuchs, *Die mailändischen Feldzüge*, 377.
89 Ibid., 288; Sanuto, *Diarii*, 346 even speaks of 17,000. Contingents from Lucerne and Bern were supposed to leave on 23 and 27 December, Frey, "Der Chiasserzug 1510 und der kalte Winterfeldzug 1511," 320.
90 See Frey, "Der Chiasserzug 1510 und der kalte Winterfeldzug 1511," 320.
91 Fuchs, *Die mailändischen Feldzüge*, 274, n. 352.
92 Richard Feller, *Geschichte Berns* (Bern: Feuz, 1946), 1:518.
93 These tensions, despite the victories in the following *Pavierzug* and of Novara in 1512–1513 became stronger and errupted in the so-called *Zwiebelnkrieg* (Onion War) of 1513 and the dividing of the Swiss forces ahead of Marignano 1515 that significantly contributed to the defeat.
94 Feller, *Geschichte Berns*, 520; Durrer, *Schweizergarde*, 110.
95 Walter Schaufelberger and Jürg Stüssi-Lauterburg, *Marignano: Strukturelle Grenzen eidgenössischer Militärmacht zwischen Mittelalter und Neuzeit* (Frauenfeld: Ed. ASMZ im Huber Verlag, 1993), 60–70; in this case, however, some cantons and allies did not send the *Auszug* with the cantonal banner but only a contingent with a cantonal standard thus contributing to the war effort without declaring war themselves, Fuchs, *Die mailändischen Feldzüge*, 279; see also Durrer, *Schweizergarde*, 159, 161.
96 Schaufelberger and Stüssi-Lauterburg, *Marignano*, 60–70.
97 Thomas Schibler, "Saubannerzug," in *Historisches Lexikon der Schweiz (HLS)*, s.v., Schweizerische Akademie der Geistes- und Sozialwissenschaften, 2011, accessed 23 March 2022, https://hls-dhs-dss.ch/de/articles/008887/2011-02-16/; Schaufelberger, *Der Alte Schweizer und sein Krieg*, 162.
98 See Schaufelberger, *Der Alte Schweizer und sein Krieg*, 153 and Hofer, *Der Schweizerische Generalstab*, 28.
99 See Hofer, *Der Schweizerische Generalstab*, 26.

100 John H. Pryor, ed., *Logistics of Warfare in the Age of the Crusades* (Aldershot: Ashgate, 2006); Martin L. van Creveld, *Supplying War: Logistics from Wallenstein to Patton* (Cambridge: Cambridge University Press, 1977).
101 See Hofer, *Der Schweizerische Generalstab*, 28.
102 See Esch, "Mit Schweizer Söldnern," 426; the Bernese were famous for their large *tross* (bagage train) designed to bring home rich booty, Carl von Elgger, *Kriegswesen und Kriegskunst der Schweizerischen Eidgenossen im XIV., XV., und XVI. Jahrhundert* (Luzern: Militärisches Verlags-Bureau, 1873).
103 The reports of the Fribourg contingent imply that bringing along field ordnance was an exciting novelty for them; also for the Solothurner it was the first time that they brought cannon with them, according to Fuchs, *Die mailändischen Feldzüge*, 280 seqq.
104 An important part of the negotiations with Venice: Sanuto, *Diarii*, 13:301: "vituarie, artellarie al bisogno et 600 homeni d'arme [*gens d'arme, i.e. heavy cavalry*]".
105 See Fuchs, *Die mailändischen Feldzüge*, 295.
106 See Anshelm, *Berner Chronik*, 263.
107 Rogger, *Geld, Krieg und Macht*.
108 Fuchs, *Die mailändischen Feldzüge*, 158–159, 165, 175–177.
109 Anshelm, *Berner Chronik*, 260, cf. note 75.
110 Ibid., 326.
111 Cf. Janice Thomson, *Mercenaries, Pirates, and Sovereigns: State-Building and Extraterritorial Violence in Early Modern Europe* (New Jersey: Princeton University Press, 1994), 15.
112 Groebner, "Pensionen".
113 For instance Huldrich Zwingli and other reformers came out strongly against Foreign Service.
114 Thomson, *Mercenaries, Pirates, and Sovereigns*, 21.
115 See Fuchs, *Die mailändischen Feldzüge*, 300. These accusations, however. did not affect his ability to send Swiss troops into another Milanese war only a few months later—which led to a great victory and the establishment of a Swiss protectorate over the Duchy of Milan from 1512 to 1515. See Reichel, "Matthieu Schiner," 73, and Hans Stadler, "Ennetbirgische Feldzüge," in *Historisches Lexikon der Schweiz (HLS)*, s.v., Schweizerische Akademie der Geistes- und Sozialwissenschaften, 2016, accessed 23 March 2022, https://hls-dhs-dss.ch/de/articles/024649/2016-04-21/.
116 See Thomson, *Mercenaries, Pirates, and Sovereigns*, 43.
117 Translation by the authors. In the original it states: "Mit elementarer, überschäumender Kraft, durch das unüberlegte Draufgängertum der Knechte, wurden oft Siege errungen, die rationalen Überlegungen keinen Raum liessen". Hofer, *Der Schweizerische Generalstab*, 28, following Christian Padrutt, *Staat und Krieg im alten Bünden* (Chur: Verlag Bündner Monatsblatt, 1991) and Schaufelberger, *Der Alte Schweizer und sein Krieg*, 157.
118 Translation by the authors. In the original it says: "Wie wir wissen, zeigten sich die Räte unter dem Druck der Umstände oft nicht abgeneigt, einen tumultösen Freischarenzug zur eigenen Sache zu machen". See Schaufelberger, *Der Alte Schweizer und sein Krieg*, 157.
119 The so called *Ämterbefragungen* by the canton of Berne is proof of such inquiries among the people. See Esch, "Mit Schweizer Söldnern," 431–433.
120 See Hofer, *Der Schweizerische Generalstab*, 29.
121 Taylor, *The Art of War*, 38, 65 and passim.
122 Including some artillery and possibly also cavalry.
123 Durrer, *Schweizergarde*, 72, also 165; also Fuchs, *Die mailändischen Feldzüge*, 366, n. 183.
124 Cf. the French encouraging the Swiss to claim sovereignty from the empire in 1648 while maintaining strong control over the Swiss military potential and, to an extent, foreign policy.

125 The *Fähnlein* is a complicated term with multiple meanings: it is a standard clearly distinct from the *banner* or *panner* (the cantonal official standard) but can also stand for the unit fighting under said standard, normally a company (80–150 men) but sometimes a bigger cantonal contingent in a federal army or a type I contingent.
126 Esch, "Mit Schweizer Söldnern," 397.
127 In the end, they obtained these two places through negotiations, Fuchs, *Die mailändischen Feldzüge*, 394, 396, cf. 322, 366; the Venetians seemed to have provided the bulk of the artillery during the campaign: 358.
128 Although Schiner tried his best to placate the Swiss, Durrer, *Schweizergarde*, 162 and they were otherwise honoured (banners, indulgences etc., Anshelm, *Berner Chronik*, 326.
129 Fuchs, *Die mailändischen Feldzüge*, 366.
130 Such attempts can be found already earlier in the *Pfaffenbrief* (1370), cf. for the later period, the *Stanserverkommnis* (1481).
131 *Discorsi* II, chap. 10.

References

Abel, Wilhelm. *Agrarkrisen und Agrarkonjunktur: Eine Geschichte der Land- und Ernährungswirtschaft Mitteleuropas seit dem hohen Mittelalter.* 3rd ed. Hamburg: Paul Parey, 1978.

Ammann, Max E. *Der Eidgenoss.* Luzern: Reich Verlag, 1975.

Anshelm, Valerius: *Die Berner Chronik*, edited by Historischer Verein des Kantons Bern. Vol. 3. Bern: K. J. Wyss, 1888.

Bräker, Ulrich. *The Life Story and Real Adventures of the Poor Man of Toggenburg.* Translated by Derek Bowman. Edinburgh: University Press, 1970.

"Bundesgesetz betreffend die Werbung und den Eintritt in den fremden Kriegsdienst vom 30. Heumonat 1859." In *Amtliche Sammlung der Bundesgesetze und Verordnungen der schweizerischen Eidgenossenschaft.* Vol. 6. Bern: Bundesversammlung, 1860.

Campbell, Bruce. *The Great Transition: Climate, Disease and Society in the Late-Medieval World.* Cambridge: Cambridge University Press, 2016.

Durrer, Robert. *Die Schweizergarde in Rom und Die Schweizer in Päpstlichen Diensten.* Luzern: Räber, 1927.

Encyclopaedia Britannica Online, "The Infantry Revolution, c. 1200–1500." N. d., s. v. Accessed 27 October 2021. https://www.britannica.com/technology/military-technology/The-infantry-revolution-c-1200-1500.

Encyclopaedia Britannica Online, n. d., s.v. "Hans Waldmann." Accessed 27 April 2020. https://www.britannica.com/biography/Hans-Waldmann.

Esch, Arnold. "Mit Schweizer Söldnern auf dem Marsch nach Italien: Das Erlebnis der Mailänderkriege 1510–1515 nach bernischen Akten." *Quellen und Forschungen aus italienischen Archiven und Bibliotheken* 70 (1990): 348–440. Reprinted in *Alltag der Entscheidung: Beiträge zur Geschichte der Schweiz an der Wende vom Mittelalter zur Neuzeit*, 249–328. Bern: Haupt, 1998.

Fahrni, Dieter. *Schweizer Geschichte: Ein historischer Abriss von den Anfängen bis zur Gegenwart.* Zürich: Pro Helvetia, 2002.

Feller, Richard. *Geschichte Berns.* Vol. 1. Bern: Feuz, 1946.

Fink, Urban, Hervé de Weck, and Christian Schweizer. *Hirtenstab und Hellebarde: Die Päpstliche Schweizergarde in Rom.* Zürich: Theologischer Verlag, 2006.

Frey, Siegfried. "Der Chiasserzug 1510 und der kalte Winterfeldzug 1511." In *Schweizer Kriegsgeschichte*, edited by Ulrich Wille et al., 314–320. Vol 1. Bern: Oberstkriegskommissariat, 1915.

Fuchs, Ildephons. *Die mailändischen Feldzüge der Schweizer.* Vol. 2. St. Gallen: Huber, 1812.
Geiger, Benjamin. *Militärgeschichte zum Anfassen.* Vol. 4, *Die Burgunderkriege.* Au (ZH): Militärische Führungsschule, 1994.
Groebner, Valentin. "Pensionen." In *Historisches Lexikon der Schweiz (HLS)*, s.v. Schweizerische Akademie der Geistes- und Sozialwissenschaften, 2011. Accessed 11 May 2020. https://hls-dhs-dss.ch/de/articles/010241/2011-11-03/.
Hammel-Kiesow, Rolf. *Die Hanse.* München: Beck, 2004.
Henry, Philippe. "Service étranger." In *Historisches Lexikon der Schweiz (HLS)*, s.v. Schweizerische Akademie der Geistes- und Sozialwissenschaften, 2017. Accessed 11 May 2020. https://hls-dhs-dss.ch/de/articles/008608/2017-12-08/.
Hofer, Viktor and Georges Rapp. *Der Schweizerische Generalstab.* Vol. 1. Basel: Helbing & Lichtenhahn, 1983.
Holenstein, André. *Mitten in Europa: Verflechtung und Abgrenzung in der Schweizer Geschichte.* Baden: Hier und Jetzt, 2014.
Hübner, Klara. "Ein falscher Sieg und falsche Boten: Nachrichtenübermittlung und -verbreitung zur Zeit von Marignano." In *Marignano 1515–2015: Von der Schlacht zur Neutralität*, edited by Roland Haudenschild, 163–174. Lenzburg: Merker im Effingerhof, 2014.
Koch, Bruno. "Heilige Liga." In *Historisches Lexikon der Schweiz (HLS)*, s.v. Schweizerische Akademie der Geistes- und Sozialwissenschaften, 2011. Accessed 23 March 2022. https://hls-dhs-dss.ch/de/articles/017171/2011-03-09/.
Lassner, Martin. "Stapfer, Jakob." In *Historisches Lexikon der Schweiz (HLS)*, s.v. Schweizerische Akademie der Geistes- und Sozialwissenschaften, 2012. Accessed 23 March 2022. https://hls-dhs-dss.ch/de/articles/018197/2012-02-17/.
Maissen, Thomas. *Geschichte der Schweiz.* 3rd ed. Baden: Hier und Jetzt, 2011.
Mattingly, Garrett. *Renaissance Diplomacy.* Baltimore, MD: Penguin Books, 1955.
Morard, Nicolas. "Auf der Höhe der Macht: 1394–1536." In *Geschichte der Schweiz und der Schweizer*, edited by Ulrich Im Hof et. al., 211–352. Basel: Helbling & Liechtenhahn, 1986.
Ochsenbein, Gottlieb Friedrich. *Die Urkunden der Belagerung und Schlacht von Murten.* Freiburg: E. Bielmann, 1876. Accessed 23 March 2022. https://archive.org/details/dieurkundenderb03ochsgoog.
Ogdon, Montell. "The Growth of Purpose in the Law of Diplomatic Immunity." *The American Journal of International Law* 31, no. 3 (1937): 449–465.
Padrutt, Christian. *Staat und Krieg im alten Bünden.* Chur: Verlag Bündner Monatsblatt, 1991.
Pryor, John H., ed. *Logistics of Warfare in the Age of the Crusades.* Aldershot: Ashgate, 2006.
Reichel, Daniel. "Matthieu Schiner (vers 1465–1522): Cardinal et homme de guerre." In *Marignano 1515–2015: Von der Schlacht zur Neutralität*, edited by Roland Haudenschild, 65–78. Lenzburg: Merker im Effingerhof, 2014.
Reinhardt, Volker. *Die Geschichte der Schweiz: Von den Anfängen bis heute.* München: C. H. Beck, 2011.
Rogers, Clifford J. "The Military Revolution in History and Historiography." In *The Military Revolution Debate*, edited by id., 2–10. London: Routledge, 1995.
Rogger, Philippe. *Geld, Krieg und Macht: Pensionsherren, Söldner und eidgenössische Politik in den Mailänderkriegen 1494–1516.* Baden: Hier und Jetzt, 2015.
Romer, Hermann. "Militärunternehmer." In *Historisches Lexikon der Schweiz (HLS)*, s.v. Schweizerische Akademie der Geistes- und Sozialwissenschaften, 2009. Accessed 23 March 2022. https://hls-dhs-dss.ch/de/articles/024643/2009-11-10/.

Schaufelberger, Walter. "Spätmittelalter." In *Handbuch der Schweizer Geschichte*, 239–388. Vol. 2. Zürich: Berichthaus, 1972.

Schaufelberger, Walter. *Der Alte Schweizer und sein Krieg: Studien zur Kriegführung vornehmlich im 15. Jahrhundert*. 3rd ed. Frauenfeld: Huber, 1987.

Schaufelberger, Walter. *Marignano: Strukturelle Grenzen eidgenössischer Militärmacht zwischen Mittelalter und Neuzeit*. Frauenfeld: Ed. ASMZ im Huber Verlag, 1993.

Schaufelberger, Walter. *Spurensuche: Siebzehn Aufsätze zur Militärgeschichte der Schweiz*, edited by Claudia Miller. Lenzburg: Merker im Effingerhof, 2008.

Schibler, Thomas. "Saubannerzug." In *Historisches Lexikon der Schweiz (HLS)*, s.v. Schweizerische Akademie der Geistes- und Sozialwissenschaften, 2011. Accessed 23 March 2022. https://hls-dhs-dss.ch/de/articles/008887/2011-02-16/.

Schodoler, Wernher. *Die eidgenössische Chronik des Wernher Schodoler um 1510 bis 1535: Kommentar zur Faksimile-Ausgabe der dreibändigen Handschrift MS 62 in der Leopold-Sophien-Bibliothek Ueberdingen, MS 2 im Stadtarchiv Bremgarten*, edited by Walther Benz. Luzern: Faksimile-Verlag, 1983.

Stadler, Hans. "Ennetbirgische Feldzüge." In *Historisches Lexikon der Schweiz (HLS)*, s.v. Schweizerische Akademie der Geistes- und Sozialwissenschaften, 2016. Accessed 23 March 2022. https://hls-dhs-dss.ch/de/articles/024649/2016-04-21/.

Stettler, Bernhard. "Sempacherbrief." In *Historisches Lexikon der Schweiz (HLS)*, s.v. Schweizerische Akademie der Geistes- und Sozialwissenschaften, 2011. Accessed 23 March 2022. https://hls-dhs-dss.ch/de/articles/009804/2011-11-22/.

Stettler, Bernhard. *Die Eidgenossenschaft im 15. Jahrhundert: Die Suche nach einem gemeinsamen Nenner*. Zürich: Markus-Widmer-Dean, 2004.

Stone, John. "Technology, Society, and the Infantry Revolution of the Fourteenth Century." *The Journal of Military History* 68, no. 2 (2004): 361–380. Accessed 27 October 2021. https://www.proquest.com/docview/195630871?pq-origsite=primo&accountid=12253.

Strahm, Hans. *Die Standesläufer*. Bern: PTT, 1942.

Studer, Gottlieb, ed. *Quellen zur Schweizergeschichte*. Vol. 1. Basel: Schneider, 1877.

Taylor, Frederick Lewis. *The Art of War in Italy, 1494–1529*. Cambridge: Cambridge University Press, 1921.

The Pontifical Swiss Guard, n. d. Accessed 27 October 2021. https://www.guardiasvizzera.ch/paepstliche-schweizergarde/it/chi-siamo/.

Thomson, Janice. *Mercenaries, Pirates, and Sovereigns: State-Building and Extraterritorial Violence in Early Modern Europe*. Princeton, N.J.: Princeton University Press, 1994.

Truffer, Bernhard. "Schiner, Matthäus." In *Historisches Lexikon der Schweiz (HLS)*, s.v. Schweizerische Akademie der Geistes- und Sozialwissenschaften, 2012. Accessed 23 March 2022. https://hls-dhs-dss.ch/de/articles/012963/2012-11-20.

van Creveld, Martin L. *Supplying War: Logistics from Wallenstein to Patton*. Cambridge: Cambridge University Press, 1977.

von Elgger, Carl. *Kriegswesen und Kriegskunst der Schweizerischen Eidgenossen im XIV., XV., und XVI. Jahrhundert*. Luzern: Militärisches Verlags-Bureau, 1873.

von Ranke, Leopold. *Geschichten der romanischen und germanischen Völker von 1494 bis 1535*. Leipzig: G. Reimer, 1824.

von Rodt, Emanuel. *Bernisches Kriegswesen*. Vol. 1. Bern: E. A. Jenni, 1831.

Wattenwyl von Diesbach, Eduard von. "Die Gesellschaft zum Distelzwang." *Berner Taschenbuch* 14 (1865): 174–200.

Weck, Hervé de. "Cavalerie." In *Dictionnaire historique de la Suisse (DHS)*, s.v. Académie suisse des sciences humaines et sociales, 2007. Accessed 23 March 2022. https://hls-dhs-dss.ch/fr/articles/008579/2007-03-01/.

Würgler, Andreas. "Krieg und Frieden organisieren: Eidgenossen und Gesandte europäischer Mächte an der Tagsatzung." In *Berner Zeitschrift für Geschichte* 74, no. 2 (2012): 87–105.

Würgler, Andreas. *Die Tagsatzung der Eidgenossen: Politik, Kommunikation und Symbolik einer repräsentativen Institution im europäischen Kontext 1470–1798*. Frühneuzeit-Forschungen 19. Epfendorf: Bibliotheca Academica, 2013.

Würgler, Andreas. "Tagsatzung." In *Historisches Lexikon der Schweiz (HLS)*, s.v. Schweizerische Akademie der Geistes- und Sozialwissenschaften, 2014. Accessed 23 March 2022. https://hls-dhs-dss.ch/de/articles/010076/2014-09-25/.

INDEX

Persons

Places

Topics

For Product Safety Concerns and Information please contact our EU
representative GPSR@taylorandfrancis.com
Taylor & Francis Verlag GmbH, Kaufingerstraße 24, 80331 München, Germany

www.ingramcontent.com/pod-product-compliance
Ingram Content Group UK Ltd.
Pitfield, Milton Keynes, MK11 3LW, UK
UKHW020317270926
475702UK00050B/944
9781032157566